1006621906

DEFENDING COPERNICUS AND GALILEO

BOSTON STUDIES IN THE PHILOSOPHY OF SCIENCE

Editors

ROBERT S. COHEN, *Boston University*
JÜRGEN RENN, *Max Planck Institute for the History of Science*
KOSTAS GAVROGLU, *University of Athens*

Editorial Advisory Board

THOMAS F. GLICK, *Boston University*
ADOLF GRÜNBAUM, *University of Pittsburgh*
SYLVAN S. SCHWEBER, *Brandeis University*
JOHN J. STACHEL, *Boston University*
MARX W. WARTOFSKY†, *(Editor 1960–1997)*

VOLUME 280

DEFENDING COPERNICUS AND GALILEO

Critical Reasoning in the Two Affairs

Maurice A. Finocchiaro
University of Nevada, Las Vegas

Maurice A. Finocchiaro
University of Nevada, Las Vegas
Las Vegas, NV
USA
maurice.finocchiaro@unlv.edu

ISBN 978-90-481-3200-3   e-ISBN 978-90-481-3201-0
DOI 10.1007/978-90-481-3201-0
Springer Dordrecht Heidelberg London New York

Library of Congress Control Number: 2009931215

© Springer Science+Business Media B.V. 2010
No part of this work may be reproduced, stored in a retrieval system, or transmitted in any form or by any means, electronic, mechanical, photocopying, microfilming, recording or otherwise, without written permission from the Publisher, with the exception of any material supplied specifically for the purpose of being entered and executed on a computer system, for exclusive use by the purchaser of the work.

Printed on acid-free paper

Springer is part of Springer Science+Business Media (www.springer.com)

Contents

Preface and Acknowledgments ix

Introduction: A Galilean Approach to the Galileo Affair xiii

Part I Defending Copernicus 1

1 The Geostatic World View 3

    1.1 Terminology 3
    1.2 Cosmology 4
    1.3 Physics 10
    1.4 Astronomy 12

2 The Copernican Controversy 21

    2.1 Copernicus's Innovation 21
    2.2 The Anti-Copernican Arguments 24
    2.3 Responses to Copernicanism 34

3 Galileo's Stances Toward Copernican Astronomy 37

    3.1 Historical Testing of Methodological Models 37
    3.2 Conceptual Clarifications 38
    3.3 Historiographical Considerations 43
    3.4 Periodization 45
    3.5 Indirect Pursuit (Before 1609) 46
    3.6 Full-Fledged Pursuit (1609–1616) 51
    3.7 The Post-1616 Period 61
    3.8 Conclusion 63

4 Galilean Critiques of the Biblical Objection 65

    4.1 Preliminary Considerations 65
    4.2 Copernicanism and Scripture 69

	4.3	Ingoli	72
	4.4	Foscarini	76
	4.5	Galileo	79
	4.6	Campanella	89
	4.7	Conclusion	93
5	**Galileo on the Mathematical Physics of Terrestrial Extrusion**		**97**
	5.1	Introduction	97
	5.2	The Extruding Power of Whirling	99
	5.3	Restatement of the Anti-Copernican Argument	101
	5.4	Tangential Extrusion Versus Secant Fall	102
	5.5	Linear Versus Angular Speed	103
	5.6	Physical Processes Versus Mathematical Entities	105
	5.7	Escape Extrusion Versus Orbital Extrusion	106
	5.8	Exsecants Versus Tangents, or Achilles and the Tortoise	108
	5.9	Distance Fallen, Distance To Be Fallen, and Speeds of Fall	112
	5.10	Exsecants Versus Exsecants	113
	5.11	A Definition of Physical–Mathematical Reasoning	114
	5.12	Galileo's Reflections on Physical–Mathematical Reasoning	115
6	**Galilean Rationality in the Copernican Revolution**		**121**
	6.1	The Copernican Revolution and the Role of Criticism	121
	6.2	Copernicus and Explanatory Coherence	123
	6.3	Physics and Reasoning	124
	6.4	The Telescope and the Role of Judgment	129
	6.5	Critical Reasoning	132
Part II	**Defending Galileo**		**135**
7	**The Trial of Galileo, 1613–1633**		**137**
	7.1	The Earlier Proceedings and the Condemnation of Copernicanism	138
	7.2	The Later Proceedings and the Condemnation of Galileo	143
8	**The Galileo Affair, 1633–1992**		**155**
	8.1	The Condemnation of Galileo (1633)	156
	8.2	Diffusion of the News (1633–1651)	161
	8.3	Emblematic Reactions (1633–1642)	163
	8.4	Polarizations (1633–1661)	166
	8.5	Compromises (1654–1704)	171
	8.6	Myth-making or Enlightenment? (1709–1777)	175
	8.7	Incompetence or Enlightenment? (1740–1758)	179

8.8	New Criticism (1770–1797)	185
8.9	Napoleonic Wars and Trials (1810–1821)	188
8.10	The Settele Affair (1820)	190
8.11	The Torture Question and the Demythologizing Approach (1835–1867)	195
8.12	The Documentation of Impropriety (1867–1879)	198
8.13	Theological Developments (1893–1912)	203
8.14	Tricentennial Rehabilitation (1941–1947)	207
8.15	Secular Indictments (1947–1959)	211
8.16	The Paschini Affair (1941–1979)	216
8.17	John Paul II's Rehabilitation (1979–1992)	220

9 Galileo Right for the Wrong Reasons? 229

9.1	The Problem	229
9.2	The *Dialogue* and Its Critics	235
9.3	The *Letter to Christina* and Its Critics	243
9.4	Conclusion	248

10 Galileo as a Bad Theologian? 251

10.1	Relative Calm (1633–1784)	252
10.2	Mallet du Pan's Thesis (1784)	254
10.3	Text of an Apocryphal Letter (1785)	256
10.4	Gaetani's Forgery	258
10.5	Diffusion and Development of a Myth (1790–1908)	260
10.6	Metamorphosis of the Myth (1909–1959)	273
10.7	Demise of the Myth (1979–1992)	274
10.8	Conclusion	275

11 Galileo as a Bad Epistemologist? 277

11.1	Introduction: Duhem on Saving the Phenomena	277
11.2	Unificationism	280
11.3	The Condemnation of Galileo	281
11.4	Metaphysics	282
11.5	Biblical Authority	284
11.6	Certainty	286
11.7	Proof Strategies	287
11.8	Conclusion	289

12 Galileo as a Symbol of Science Versus Religion? 291

12.1	The "Interaction" Between "Science" and "Religion"	291
12.2	Conflictual Accounts	293
12.3	John Paul II's Harmony Thesis	296
12.4	Morpurgo-Tagliabue's Version of Harmony	298

	12.5	Feyerabend's Version of Conflict	300
	12.6	Heresy or Disobedience?	301
	12.7	Science Versus Religion in the Subsequent Affair?	305
	12.8	Conclusion	313

Selected Bibliography 315

Index 339

Preface and Acknowledgments

The literature on the Galileo affair is so extensive that it may be useful to preface this book with a sketch of the author's own view of its scope, orientation, theme, argument, thesis, and limitations. It will also be useful to clarify how this book differs from the author's earlier books. The sketch and the clarification will hopefully help the reader understand what is distinctive about this book vis-à-vis the many others on the subject.

First, there is the topical and chronological scope of this book. The subject matter is the *Galileo affair in its twofold connotation*: the original affair climaxing with the trial and condemnation of Galileo by the Inquisition in 1633, and the subsequent controversy following his condemnation and acquiring a life of its own and continuing to our own day.

Next, the book stresses a particular, although crucial, conceptual orientation, namely the notion of *critical reasoning*. Here, critical reasoning means reasoning aimed at the interpretation, evaluation, or self-reflective presentation of arguments; and an argument is a piece reasoning that justifies a conclusion by supporting it with reasons or defending it from objections. In this sense, the book provides a case study of the role of critical reasoning in the Copernican Revolution and the Galileo affair.

The book's title attempts to capture a simple but elegant theme meant to avoid both oversimplifications and inflated complications: that an essential aspect of the Copernican Revolution was the need to defend the geokinetic hypothesis from many apparently conclusive objections; that an essential aspect of Galileo's work consisted of undertaking and carrying out such a defense; that an essential aspect of the trial of Galileo was the Church's attempt to stop his defense of Copernicus; that an essential aspect of the subsequent Galileo affair has been the emergence of a host of objections to his defense of Copernicus; and that equally essential is the fact that Galileo has been or can be defended from such objections. Finally, the title is also meant to suggest the project of comparing and contrasting Galileo's defense of Copernicus and the defense of Galileo, with an eye toward understanding both affairs better, as well as toward providing the groundwork and the framework for a resolution of the second affair or ongoing controversy.

Less obvious than this subject matter, conceptual orientation, and simplifying theme, just described, the reader can find the thread of the following main argument.

The Copernican Revolution required that the geokinetic hypothesis be not only supported with new arguments and evidence, but also defended from many powerful old and new objections. This defense in turn required not only the destructive refutation but also the appreciative understanding of those objections in all their strength. One of Galileo's major accomplishments was not only to provide new evidence and arguments supporting the earth's motion, but also to show how those objections could be refuted, *and* to elaborate their power before they were answered. In this sense, Galileo's defense of Copernicus was *reasoned, critical, open-minded,* and *fair-minded*. Now, an essential thread of the subsequent Galileo affair has been the emergence of many anti-Galilean criticisms, from the point of view of astronomy, physics, theology, hermeneutics, logic, epistemology, methodology, law, morality, and social responsibility. It is important to understand both that such criticisms arise naturally and legitimately and that Galileo can be defended from them. Accordingly, this book advances the following particular and yet *overarching thesis*: that today in the context of the Galileo affair and the controversies over the relationship between science and religion and between institutional authority and individual freedom, *the proper defense of Galileo should have the reasoned, critical, open-minded, and fair-minded character which his own defense of Copernicus had*. And this is a thesis that has both interpretive and evaluative dimensions.

With regard to the book's limitations, it should be noted that the treatment is not meant to be complete or exhaustive, but is rather illustrative and selective. It is obvious that one could select other points of view (besides critical reasoning) from which to study the same material, and that even from the point of view of critical reasoning, such an account could cover much more material and include many more studies than are included in this book.

Furthermore, this book does not pretend to be a comprehensive, definitive, or final synthesis, if for no other reason than that such a synthesis would have to wait for the assimilation and digestion of the *overarching thesis* advanced here. I have indeed begun working on such a synthesis, whose primary aim would be the systematic exposition and polished presentation of an account based on a body of established historical and philosophical theses. However, the present book operates more in the context of the discovery, formulation, and justification of the *overarching thesis* mentioned above. To paraphrase Kant's eloquent and colorful words at the beginning of the *Prolegomena to Any Future Metaphysics* (Academy edition, vol. 4, p. 255), here the aim is "not ... the systematic exposition of a ready-made science, but ... the discovery of the science itself ... [the former] must wait till those who endeavor to draw from the fountain of reason itself have completed their work; it will then be [its] turn to inform the world of what has been done."

However, on a personal, subjective, and autobiographical note, this book *is* a kind of synthesis of my previous work on Galileo. Let me explain. On the one hand, there is a partial overlap in subject matter with *The Galileo Affair: A Documentary History* (1989), which covers the original trial in the years 1613–1633, and with *Retrying Galileo, 1633–1992* (2005), which is an introductory survey of the subsequent controversy. And there is a partial overlap in conceptual orientation with *Galileo and the Art of Reasoning* (1980) and *Galileo on the World Systems* (1997), both of which focus on critical reasoning. On the other hand, one novelty in the present book is that it applies the conceptual orientation of the latter two books (critical reasoning) to the subject matter of the former two (the Galileo affairs). Another novelty is this: *The Galileo Affair* and *Retrying Galileo* are primarily presentations of documents, sources, and facts that largely stay away from advancing controversial interpretations and evaluations; but this book is primarily concerned with advancing interpretations and evaluations, which can be expected to engender controversy.

This feature is worth stressing and elaborating. The primary aim of *The Galileo Affair* was to make available in one volume the essential documents pertaining to Galileo's trial (1613–1633), so as to enable readers to arrive at their own interpretations and evaluations, as well as to allow the author to do so at a later stage of inquiry. The primary aim of *Retrying Galileo* was analogous, allowances being made for the longer duration and greater complexity of the subsequent affair (1633–1992). That is, its key aim was to establish the essential sources, facts, and issues (including, as space allowed, the presentation of a few of the most essential documents); and such an introductory survey was designed to avoid, as much as possible, any deeper or more controversial interpretations or evaluations, which were left to readers or postponed to a later stage of inquiry. Those more advanced stages of inquiry adumbrated and postponed in those two earlier works are now undertaken in the present book.

Finally, there is little likelihood that this book would be confused with my *Essential Galileo* (2008), which is an anthology of Galileo's most significant writings in physics, astronomy, methodology, and epistemology and of the most important documents relating to his Inquisition trial. But it is worth mentioning a similarity between these two works: the integration which *The Essential Galileo* aims to provide for Galileo's legacy, life, and works is analogous to the synthesis which the present work aims to develop for historical interpretations and philosophical evaluations of them.

In researching, writing, and publishing this book, I have benefited from the support and encouragement of many persons and institutions, and they deserve acknowledgment here. For financial support, I acknowledge the Guggenheim Foundation and the National Science Foundation; they funded a previous project which resulted in my book *Retrying Galileo, 1633–1992*, and during which I conceived and began exploring the present one. For comments on various parts of the manuscript at various stages of its development, I thank Albert DiCanzio, Hilary Gatti, Owen Gingerich, Michael Hoskin, Joao Medeiros, Ron Naylor, Michael Segre, Peter Slezak, and Albert van Helden. The University of Nevada, Las Vegas, its Philosophy Department, its chairman Todd Jones, the other departmental colleagues, and my former student Aaron Abbey have continued to provide institutional and moral support, even after I decided to retire from formal teaching to work full time on research, scholarship, and writing. I owe a special debt of gratitude to Jürgen Renn, who, as series co-editor of the Boston Studies in the Philosophy of Science, has been appreciative and gracious.

For the opportunity to present, disseminate, and discuss various versions of the key ideas of this book, and for assistance with the necessary travel expenses and practical arrangements, I am grateful to the following hosts at the following institutions. I list them in chronological order, spanning the first decade of the twenty-first century: José Montesinos at the Fundación Canaria Orotava de Historia de la Ciencia (Spain); Ted Porter and Margaret Jacob at UCLA; Carlos Alvarez and Carlos López Beltrán at the National Autonomous University of Mexico (UNAM); John Brooke at Oxford University; Piyo Rattansi and Hasok Chang at University College, London; Nicholas Jardine at Cambridge University; Paolo Galluzzi at the Istituto e Museo di Storia della Scienza, Florence; Giuliano Pancaldi at the University of Bologna; Fabio Bevilacqua at the University of Pavia; Bill Shea and Adelino Cattani at the University of Padua; Michael Segre at the University of Chieti; Rosario Moscheo at the University of Messina and the Accademia Peloritana dei Pericolanti; Hilary Gatti and Giorgio Stabile at the Department of Philosophy, University of Rome "La Sapienza"; Gennaro Auletta and Arcangelo Rossi at the Program in Science and Philosophy, Pontifical Gregorian University, Rome; Paula Findlen at Stanford University; Roger Hahn at the University of California, Berkeley; Jitse van der Meer at the Pascal Centre for Advanced Studies in Faith and Science, Redeemer College; Cynthia Pyle at the Renaissance Seminar, Columbia University; Monte Johnson at the University of California, San Diego; Paolo Palmieri at the University of Pittsburgh; Adelino Cattani and the International Society for the Study of Controversies at the University of Padua; Michele Camerota at the University of Cagliari; Andrea Frova at the Department of Physics, University of Rome "La Sapienza"; Arcangelo Rossi, Mauro Di Giandomenico, and Francesco De Ceglia at the University of Bari and the Accademia Pugliese delle Scienze; Francesco Coniglione at the University of Catania; Virginia Trimble and David DeVorkin at the 2009 convention of the American Association for the Advancement of Science; Steve Snyder and the Templeton Foundation at the Franklin Institute, Philadelphia; Neil Thomason at the University of Melbourne; and Peter Slezak at the University of New South Wales.

Acknowledgments are also due to the following publishers and copyright holders for various parts of this book that were published in earlier versions reflecting various occasions and contexts: the University of California Press for Chapters 1, 2, 7, and 8; Springer Science and Business Media for Chapters 3, 5, and 6; Brill Academic Publishers for Chapter 4; Catholic University of America Press for Chapter 9; Elsevier Science Ltd. for Chapter 10; the journal *Revue internationale de philosophie* for Chapter 11; and the History of Science Society for Chapter 12.

These acknowledgments to publishers suggest one final caveat, concerning the novelty of this book. Since, as just mentioned, the substance of each chapter has been previously published, clearly the book's novelty does not lie in the simple sum of those parts. Rather the novelty lies in the fact that when collected together and revised into their present form, these twelve essays constitute an argument for the *overarching thesis* stated above (and elaborated in the Introduction), and the novelty of that thesis is palpable.

Maurice A. Finocchiaro
June 2009

Introduction: A Galilean Approach to the Galileo Affair

The Copernican Controversy

In 1543, Nicolaus Copernicus published an epoch-making book, *On the Revolutions of the Heavenly Spheres*. In it he elaborated an idea that had been first advanced by the followers of Pythagoras and by Aristarchus in ancient Greece: that the earth moves by rotating on its own axis daily and by revolving around the sun yearly; this means that the earth is *not* standing still at the center of the universe, with all the heavenly bodies revolving around it; instead, the earth is a planet and the sun is the center of the planets' orbits. In its essentials this geokinetic and heliocentric idea turned out to be true, as we know today beyond any reasonable doubt, after five centuries of accumulating evidence. At the time, however, the knowledge situation was very different.

In fact, Copernicus's accomplishment was really to give a *new argument* in support of an *old idea* that had been considered and almost universally rejected for two millennia. He was able to demonstrate that the *known facts* about the motions of the heavenly bodies could be explained in *quantitative detail* if the sun rather than the earth is assumed to be at the center and the earth is taken to be the third planet circling the sun; and further that this geokinetic and heliocentric explanation was *more coherent* (and also simpler and more elegant) than the geostatic and geocentric account (Chapter 2). This detailed quantitative demonstration had never been carried out before, either by the ancient Pythagoreans or by the few heliocentrists who sporadically emerged in the course of the centuries. What was significant about the demonstration and gave it the character of a new argument, was precisely the details, for most astronomers since Ptolemy would have paid lip service to the general qualitative idea that the known facts could probably be explained heliocentrically and geokinetically.

Despite its novelty and significance, however, as a proof of the earth's motion Copernicus's argument was far from conclusive. First of all, his argument was hypothetical, that is, based on the claim that *if* the earth were in motion *then* the observed phenomena would result; but from this it does not follow with logical necessity that the earth is in motion, on pain of committing the logical fallacy of affirming the consequent. Moreover, and more importantly, there were many powerful arguments against terrestrial motion that had accumulated for 2,000 years (Chapter 2).

These counter-arguments and objections can be classified into various groups, depending on the branch of learning or type of principle from which they stemmed. In fact, the anti-Copernican arguments reflected the various traditional beliefs (Chapter 1) which contradicted or seemed to contradict the Copernican system. Thus, there were epistemological, philosophical, religious, theological, scriptural, physical, mechanical, astronomical, and empirical objections. In summary, the earth's motion seemed epistemologically absurd because it flatly contradicted direct sense experience, and thus it undermined the normal procedure in the search for truth. It seemed empirically untrue because it had astronomical consequences that were not observed to happen. It seemed physically impossible because it had consequences that contradicted the most incontrovertible mechanical phenomena, and because it directly violated many of the most basic principles of the available (Aristotelian) physics. And it seemed religiously heretical or suspect because it conflicted with the literal meaning of Scripture and with the biblical interpretations of the Church Fathers, and because it could be taken to undermine belief in the omnipotence of God.

Copernicus was acutely aware of these difficulties.[1] He realized that his novel argument did not conclusively prove the earth's motion and that there were many counter-arguments of apparently greater strength. I believe that this was an important reason why he delayed publication of his book until he was almost on his deathbed, although his motivation was complex and is not yet completely understood and continues to be the subject of serious research.[2]

However, Copernicus's argument was so important that it could not be ignored, and various attempts were made to come to terms with it, to assimilate it, to amplify it, or to defend it.[3] Some tried to exploit the mathematical advantages of the Copernican theory without committing themselves to its cosmological claims, by adopting an instrumentalist interpretation according to which the earth's motion was just a convenient instrument for making mathematical calculations and astronomical predictions, and not a description of physical reality. Tycho Brahe undertook an unprecedented effort to systematically collect new observational data, and he also devised another compromise, a theory that was partly heliocentric and partly geocentric, although fully geostatic: the planets revolved around the sun, but the sun moved daily and annually around the motionless and central earth. Giordano Bruno undertook a multi-faceted defense of Copernicanism

---

[1] For example, he discussed some of these objections in Book One of his great work; see Copernicus (1992, 7–50).
[2] See, for example, Goddu (2006).
[3] Cf. Barker and Goldstein (1998), Gatti (1997; 1999; 2002; 2008), Kozhamthadam (1994), Lattis (1994), Westman (1972; 1975b,c,d; 1980; 1987; 1990; forthcoming).

that addressed epistemological, metaphysical, theological, and empirical issues; Bruno's defense is in some ways similar but in important ways different from Galileo's; in any case, for complex reasons, it remained largely unknown, disregarded, or unappreciated. Johannes Kepler accepted Copernicanism for metaphysical reasons and then undertook a research program to prove its empirical adequacy, by meticulously analyzing Tycho's data; the result was an improvement of the Copernican theory such that the planets (including the earth) revolved around the sun in elliptical, rather than circular, orbits.

Galileo's Defense of Copernicus and the Original Affair

My focus here is on Galileo's response to the Copernican challenge, on how and why he went about "defending Copernicus" (Chapters 3–6). In his early career (1589–1609), Galileo's stance toward Copernicanism is best described as one of *indirect pursuit*, an attitude that is not only weaker than *acceptance* but also weaker than *direct pursuit* (Chapter 3). In fact, during this period his research focused on the general physics of motion rather than on astronomy and cosmology. He was critical of Aristotelian physics and favorably inclined toward Archimedean statics and mathematics. He followed an experimental approach, that is, active intervention into and exploratory manipulation of physical phenomena, combining empirical observation with quantitative mathematization and conceptual theorizing. It was then that he formulated, justified, and to some extent systematized various mechanical principles: an approximation to the law of inertia; the principle of the composition or superposition of motions; the law that in free fall the distance fallen increases as the square of the time elapsed; the (equivalent) law that the velocity acquired by a freely-falling body is directly proportional to the time elapsed; and the parabolic path of projectiles. However, he did not publish any of these results during that earlier period; and indeed he did not publish a systematic account of them until the *Two New Sciences* (Leiden, 1638).

A main reason for this delay was that in 1609 Galileo became actively involved in astronomy. He was already acquainted with Copernicus's theory of a moving earth and appreciative of the fact that Copernicus had advanced a novel and important argument. Galileo also had intuited that the geokinetic theory was more in accordance with his new physics than was the geostatic theory; and in particular he had been attracted to Copernicanism because he felt that the earth's motion could best explain why the tides occur.

In 1609 he perfected the telescope to such an extent as to make it an astronomically useful instrument that could not be easily duplicated by others for some time. By its means he made several startling discoveries which he immediately published in *The Sidereal Messenger* (Venice, 1610): that the moon's surface is rough, full of mountains and valleys; that innumerable other stars exist besides those visible with the naked eye; that the Milky Way and the nebulas are dense collections of large numbers of individual stars; and that the planet Jupiter has four moons revolving around it at different distances and with different periods.

Other discoveries quickly followed. At the end of 1610 and beginning of 1611, he started observing that the appearance of the planet Venus, in the course of its orbital revolution, changes regularly from a full disc, to half a disc, to crescent, and back to a half and a full disc, in a manner analogous to the phases of the moon. This discovery was announced in private correspondence.[4] And in the next few years he discovered that the surface of the sun is dotted with dark spots that are generated and dissipated in a very irregular fashion and have highly irregular sizes and shapes, like clouds on earth; but that while they last, these spots move regularly in such a way as to imply that the sun rotates on its axis with a period of about one month. These observations were described and discussed in *History and Demonstrations Concerning Sunspots* (Rome, 1613).[5]

The new telescopic evidence led Galileo to a re-assessment of the status of Copernicanism, by removing most of the empirical-astronomical objections against the earth's motion and adding new arguments in its favor. He now believed not only that the geokinetic theory had greater explanatory coherence than the geostatic theory (as Copernicus had shown); not only that it was physically and mechanically more adequate (as he himself had been discovering in the 20 years of his early research); but also that it was empirically and observationally more accurate in astronomy (as the telescope now revealed). However, he realized that this strengthening of Copernicanism was not equivalent to settling the issue because there was still some astronomical counter-evidence (mainly, the lack of annual stellar parallax); because the mechanical objections had not yet been explicitly refuted and the physics of a moving earth had not yet been articulated; and because the theological objections had not yet been answered. The case in favor of the earth's motion was still not conclusive.

That is, for the next several years (1609–1616), Galileo's attitude toward Copernicanism is best described as one of *direct pursuit* and *tentative acceptance*, by contrast with the *indirect pursuit* of the pre-telescopic situation, and with the *settled acceptance* toward which he was moving but never really reached (Chapter 3). Whereas before 1609 Galileo judged that the anti-Copernican arguments outweighed the pro-Copernican ones, afterwards he judged the reverse to be the case.

Thus, it was during this period that Galileo conceived a work on the system of the world in which all aspects of the question would be discussed. But this synthesis of astronomy, physics, methodology, and epistemology was not published until his *Dialogue on the Two Chief World Systems, Ptolemaic and Copernican* (Florence, 1632). It is there that we find not only a systematic reworking of his answers to the astronomical objections to Copernicanism and a criticism of the epistemological objections, but also the first explicit and articulated answers to the mechanical objections, such as the argument from the extruding power of whirling (Chapter 5) and the argument from vertical fall (Chapter 6).

---

[4] In December 1610 and January and February 1611; see Favaro (10: 483, 499–505; 11: 11–12, 46–50, 61–63). Cf. Palmieri (2001).
[5] For details, cf. Reeves and van Helden (forthcoming).

Besides realizing that the pro-Copernican arguments were still not absolutely conclusive, Galileo must have also perceived the potentially explosive character of the biblical and religious objections. In fact, for a number of years he did not get involved despite the fact that his *Sidereal Messenger* had been attacked by several authors on biblical grounds, among others. Eventually, however, he was dragged into the theological discussion.

This forced involvement into the theological controversy did lead to intellectually brilliant results, namely to the cogent refutation of the scriptural argument against Copernicanism. And in this context, Galileo's efforts were parallel to and complementary with those of some progressive Catholic theologians and philosophers, such as Paolo Antonio Foscarini and Tommaso Campanella (Chapter 4). However, despite winning the intellectual argument, Galileo lost the practical struggle.

In fact, in a series of incidents that make up the first phase of Galileo's trial, several tragic events happened (Chapter 7). In 1615, after some formal complaints were filed against Galileo, the Roman Inquisition launched an investigation. The proceedings lasted about a year, and the results were the following. In 1616, the Congregation of the Index (the Catholic Church's department of book censorship) issued a decree declaring that the doctrine of the earth's motion was physically false and contrary to Scripture; condemning and permanently banning Foscarini's *Letter on the Earth's Motion* (1615), which had argued that the earth's motion was probable and not contrary to Scripture; and temporarily prohibiting Copernicus's *Revolutions* until and unless it was revised. Although Galileo was not mentioned at all in the decree, in private he was given an oral warning by Cardinal Robert Bellarmine, on orders from the Inquisition, to refrain from holding or defending the geokinetic idea. Galileo agreed to comply.

As a result, Galileo's cognitive stance toward Copernicanism reverted back to a kind of *indirect pursuit* analogous to the one he had been practicing in his early career, before the telescopic discoveries. Of course, the situation was now much more complicated, and so after 1616 his indirect pursuit of the Copernican research program was simultaneously more substantial and more careful: more substantial because there was much more material to draw upon, and more careful because his own earlier internal cognitive scruples had been replaced by external ecclesiastical constraints. Nevertheless, his contributions to the controversy on comets, which resulted in his publication of *The Assayer* (Rome, 1623), can be interpreted in this light.[6] The same applies to the *Two New Sciences* (1638), which pursues the Copernican program even more indirectly, by avoiding astronomical topics altogether and limiting itself to mechanics.

Even the *Dialogue* (1632) fits this general pattern (indirect pursuit). For it can be interpreted as a *discussion* of the (geokinetic) opinion which he had been prohibited to *hold* or *defend*; a discussion that took the form of a critical examination of all the astronomical, physical, and philosophical (but *not* the theological) arguments

---

[6] Beltrán Marí (2006, 369–381, especially 373), Biagioli (1993, 267–311), Bucciantini (2003, 261–287), Camerota (2004, 363–398), Speller (2008, 111–123).

for and against the opinion. To be sure, the critical examination revealed that the geokinetic arguments were much stronger than the geostatic ones, and when the book's central thesis is so formulated, it is clear that Galileo holds and defends it, indeed that he demonstrates it successfully. However, this is a comparative, relative, and contextual thesis; the geokinetic opinion is discussed and evaluated vis-à-vis the geostatic opinion and vis-à-vis the available arguments. Galileo felt that this discussion and this comparative thesis did not amount to holding or defending the geokinetic opinion in an absolute, objectionable, or illegitimate sense; at most it amounted to defending it as probable. Or at least, this was his gamble in writing and publishing the *Dialogue*. Unfortunately, he lost this gamble, for complicated reasons that involve the later Inquisition proceedings of the trial (Chapter 7).

In 1633, the Inquisition concluded these proceedings by issuing a sentence with the following verdict and penalties. The verdict was that Galileo had been found guilty of "vehement suspicion of heresy," a crime intermediate between the more serious one of formal heresy and the less serious one of slight suspicion of heresy. Two main errors, and not just one, were imputed. The first involved holding an astronomical doctrine that was false and contrary to Scripture, namely the heliocentric and geokinetic thesis. The second alleged error was the methodological principle that it is permissible to defend as probable a doctrine contrary to Scripture. There were several penalties. First, Galileo had to immediately recite an "abjuration" of these beliefs. Second, the *Dialogue* was permanently banned. Third, he had to recite the seven penitential psalms once a week for three years. Finally, he was to be kept under imprisonment indefinitely, but this turned out to mean not detention in an actual jail, but rather house arrest.

The Subsequent Galileo Affair

Although the Inquisition's condemnation in 1633 ended the original Galileo affair, it gave rise to a new one that continues to our own day. The original affair is co-extensive with the trial of Galileo (1613–1633), that is, it is the aspect of the Copernican controversy consisting of his efforts to come to terms with Copernicanism, his attempt to defend Copernicus. On the other hand, the subsequent Galileo affair is largely co-extensive with the many post-trial and posthumous attempts by some to defend and re-affirm the Inquisition's condemnation, and by others to defend Galileo and criticize the Inquisition. The subsequent affair is much more complex than the original one because of the longer historical span, the broader interdisciplinary relevance, the greater international and multi-linguistic involvement, and the ongoing cultural import. To begin to make sense of it, it is useful to stress that the subsequent Galileo affair has three principal aspects: the historical aftermath; the reflective commentary; and the critical issues.

The *historical aftermath* of the original episode (Chapter 8) consists of facts and events directly stemming from the trial and condemnation of Galileo. Some of these involve actions taken by the Catholic Church, such as: the partial unbanning first of Galileo's *Dialogue* and later of Copernican books in general during the papacy of

Benedict XIV (1740–1748); the total repeal of the condemnation of the Copernican doctrine in the period 1820 to 1835; the implicit theological vindication of Galileo's hermeneutics by Pope Leo XIII's encyclical *Providentissimus Deus* (1893); the beginning of the rehabilitation of Galileo himself, occasioned by the commemoration in 1942 of the tricentennial of his death; and most recently the further rehabilitation of Galileo by Pope John Paul II (between 1979 and 1992). The historical aftermath also includes actions by various non-ecclesiastic actors, such as: René Descartes's decision (in 1633) to abort the publication of his own cosmological treatise *The World*; Gottfried Leibniz's indefatigable efforts (1679–1704) to convince the Church to withdraw its condemnation of Copernicanism and of Galileo; the Tuscan government's reburial of Galileo's body in a sumptuous mausoleum in the church of Santa Croce in Florence (1737); Napoleon's seizure of the Vatican file of the Galilean trial proceedings and his plan to publish its contents (between 1810 and 1814); the publication of those proceedings by lay scholars in France, Italy, and Germany between 1867 and 1878; and the attempts in the middle of the twentieth century by various secular-minded and left-leaning intellectuals (e.g., Bertolt Brecht, Arthur Koestler, and Paul Feyerabend) to blame Galileo for such things as the abuses of the industrial revolution, the social irresponsibility of scientists, the atomic bomb, and the rift between the two cultures.

The *reflective commentary* on the original trial consists of countless interpretations and evaluations advanced in the past four centuries by astronomers, physicists, theologians, churchmen, historians, philosophers, cultural critics, playwrights, novelists, and journalists (Chapter 8). These comments have appeared sometimes in specialized scholarly publications, sometimes in private correspondence or confidential ecclesiastical documents, and sometimes in classic texts. Among the latter are Descartes's *Discourse on Method*; John Milton's *Areopagitica;* Blaise Pascal's *Provincial Letters;* Leibniz's *New Essays on Human Understanding;* Voltaire's *Age of Louis XIV;* Denis Diderot and Jean D'Alembert's French *Encyclopedia;* Auguste Comte's *Positive Philosophy;* John Henry Newman's writings; Pope Leo XIII's *Providentissimus Deus;* Brecht's *Galileo;* and Koestler's *Sleepwalkers.* Here we have a historiographical or meta-historical labyrinth in which it is easy to get lost unless one uses some tentative guidelines. For example, it is useful to distinguish the following types of account: surface-structural versus deep-structural, circumstantial versus principled, one-dimensional versus multi-dimensional, pro-Galilean versus anti-Galilean, pro-clerical versus anti-clerical, and neutral versus evaluatively overcharged.[7]

The *critical issues* of the subsequent controversy in part reflect the original issues, which involved questions like the following: whether the earth is located at the center of the universe; whether the earth moves, around its own axis daily and around the sun annually; whether and how the earth's motion can be proved, experimentally or theoretically; whether the earth's motion contradicts Scripture; whether a contradiction between terrestrial motion and a literal interpretation of Scripture would constitute a valid reason against the earth's motion; whether Scripture must always be interpreted

---

[7] For more details on this taxonomy, see Finocchiaro (1999a).

literally; and, if not, when Scripture should be interpreted literally and when figuratively. However, the subsequent controversy has also acquired a life of its own, with debates over new issues such as (Chapter 8): whether Galileo's condemnation was right; why he was condemned; whether science and religion are incompatible; how science and religion do or should interact; whether individual freedom and institutional authority must always clash; whether cultural myths can ever be dispelled with documented facts; whether political expediency must prevail over scientific truth; and whether scientific research must bow to social responsibility.

Although distinct, these three principal aspects of the subsequent Galileo affair are obviously interrelated. For example, much of the reflective commentary consists of attempts to formulate or resolve one or more critical issues, and such formulations often represent important developments of the historical aftermath. I now want to focus on one set of interrelationships involving the theme of what I call "defending Galileo."

After the condemnation of Galileo in 1633, it was only natural to want to know why he was condemned, namely what were the reasons or causes for his condemnation. It was equally natural to ask oneself whether his condemnation was right or wrong, namely whether he had been unjustly persecuted or whether the Inquisition had acted justly in prosecuting him; and this question subdivides into several, depending on whether one is taking the point of view of science, philosophy, theology, law, morality, or practical utility. The first question pertains to the explanation (or interpretation) of the condemnation, the second pertains to its evaluation (or justification of Galileo or the Inquisition). These two interrelated issues have been persistent themes of the subsequent Galileo affair.

Defending Galileo: Criticisms and Replies

One initial response by critics of Galileo and pro-clerical thinkers was to hope or try to show that he had been scientifically wrong. For example, in 1642–1648, a controversy developed regarding the correctness of his science of motion; a controversy that has been called "the Galilean *affaire* of the laws of motion … a second 'trial'"[8] of Galileo. It started when Pierre Gassendi's *De motu impresso a motore translato* (1642) elaborated the connection between the new Galilean physics of motion and Copernican astronomy, strengthening each. The focal point was Marin Mersenne, whose correspondence and contacts facilitated and encouraged discussion. The anti-Galilean critics were: Jean-Baptiste Morin, Jesuit Pierre Le Cazre, Jesuit Honoré Fabri, Pierre Mousnier, Jesuit Niccolò Cabeo, and Giovanni Battista Baliani. The pro-Galilean exponents were primarily Gassendi himself, Jacques-Alexandre Le Tenneur,

---

[8] Galluzzi (2000, 539). Cf. Baliani (1638; 1646), Cabeo (1646), Caramuel Lobkowitz (1644), Galluzzi (1993a), Gassendi (1642; 1646; 1649), Huygens (1673), Le Cazre (1645a,b), Le Tenneur (1646; 1649), Mersenne (1647), Morin (1643), Mousnier (1646), Mousnier and Fabri (1648), Palmerino (1999), Torricelli (1644).

Evangelista Torricelli, and Christiaan Huygens. The result of this first "retrial" was a vindication of Galileo, who of course was dead by then, the controversy having ironically started the same year as his death (1642).

Similarly, in 1651, Jesuit Giovanni Battista Riccioli claimed that the Inquisition had been right and wise in condemning Galileo, both scientifically and theologically. Scientifically speaking, Riccioli argued that this was so chiefly because neither the Ptolemaic nor the Copernican, but rather the Tychonic, system was the correct one, and so Galileo was wrong in holding that the earth moves. Riccioli made a comprehensive examination of all the arguments to support his scientific choice. He even invented a new geostatic argument based on Galilean ideas, a Galilean argument against Galileo, so to speak. Riccioli accepted Galileo's law of acceleration of falling bodies, and he also took at face value the passage on semicircular fall in the *Dialogue*; there Galileo says that on a rotating earth a body falling freely from the top of a tower would follow a circular path in absolute space defined by the semicircle whose diameter is the line from the earth's center to the point of release. Riccioli's reasoning was that the earth cannot rotate because if it did, a body falling freely from the top of a tower would in reality be following the Galilean semicircular trajectory, and motion along this trajectory is uniform (as measured from center of the semicircle); but observation reveals that the motion of falling bodies is accelerated. In 1665, this argument engendered a controversy that involved Giovanni Alfonso Borelli and a mathematics professor at the University of Padua named Stefano degli Angeli; before subsiding four years later, this dispute had spawned at least nine books.[9] Once again, the objections of Galileo's scientific critics backfired against them, and they ended up being discredited, and he vindicated.

We may highlight as follows the rest of the history of the defense of Galileo regarding the earth's motion – the substantive astronomical thesis involved in his trial. In 1687, Isaac Newton completed the Copernican Revolution when he published his *Mathematical Principles of Natural Philosophy*. The Newtonian system of celestial mechanics has two important geokinetic consequences, among others. First, the relative motion between the earth and the sun corresponds to the actual motion of both bodies around their common center of mass; but the relative masses of the sun and the earth are such that the center of mass of this two-body system is a point inside the sun; so, although both bodies are moving around that point, the earth is circling the body of the sun. Second, the daily axial rotation of the earth has the centrifugal effect that terrestrial bodies weigh less at lower latitudes, and least at the equator, and that the whole earth is slightly bulged at the equator and slightly flattened at the poles; and these consequences were verified by observation; in other words, these observational facts can be explained in no other way than by terrestrial rotation.

These Newtonian proofs were still relatively indirect, and so the search for more direct evidence continued. In 1729, English astronomer James Bradley discovered

---

[9] Angeli (1667; 1668a,b; 1669), Borelli (1668a,b), Riccioli (1668; 1669), Zerilli (1668). For more details, see Galluzzi (1977), Koyré (1955).

the aberration of starlight, providing direct observational evidence that the earth has translational motion. In 1789–1792, an astronomer and priest of Bologna named Giambattista Guglielmini was the first to directly confirm terrestrial rotation by means of experiments detecting an eastward deviation of falling bodies; his work was soon confirmed and refined further by other experimenters and theoreticians.[10] In 1806, Giuseppe Calandrelli,[11] director of the astronomical observatory at the Roman College, claimed to have measured the annual parallax of star Alpha in the constellation Lyra; the variation was about 5 seconds of arc, yielding a distance of about 250 light days. This value is about 15 times smaller than the actual 11 light years, and so the discovery of annual stellar parallax is usually attributed to German astronomer and mathematician Friedrich W. Bessel (1784–1846), who observed it for the star 61 Cygni in 1838. Annual stellar parallax provides direct proof that the earth revolves annually in a closed orbit. In 1851, Léon Foucault in Paris invented the pendulum that bears his name and provided a spectacular demonstration of the earth's rotation; the experiment was ceremoniously repeated in many other places.[12] In 1903–1910, physicist Edwin Hall, from the Jefferson Laboratory at Harvard University, gave an updated sophisticated experimental confirmation of Guglielmini's eastward deviation of falling bodies and also of a predicted negligible southward deviation.[13] Finally, in 1910–1911, Jesuit J.G. Hagen, at the Vatican Observatory in Rome, invented a new instrument (called isotomeograph) to demonstrate and measure the earth's rotation in a new way.

By now, even the Jesuits were siding with Galileo on the astronomical issue. However, a long time before that, as it was becoming clearer that Galileo had been right in believing that the earth moves, another genre of anti-Galilean criticism and apologia of the Inquisition had been emerging. He started being charged with believing what turned out to be true for the wrong reasons, on the basis of flawed arguments, or with the support of inadequate evidence.

For example, even during Galileo's own lifetime, his geokinetic argument from tides had seemed not completely convincing. After Newton's correct explanation of the tides as caused by the gravitational attraction of the moon (and also of the sun), one could also claim that there was definitely an error in Galileo's theory that the tides were caused by the earth's motion. And so the anti-Galilean critics started to mention the tidal argument as one of Galileo's bad reasons for believing what turned out to be true. Today this criticism continues to be one of the most common charges against Galileo.[14]

In 1841, an anonymous article in the Munich journal *Historisch-politische Blätter für das katholische Deutschland* inaugurated this kind of apologia in an explicit

---

[10] Borgato (1996), Guglielmini (1789; 1792).

[11] Cf. Calandrelli (1806a,b), Brandmüller and Greipl (1992, 163, 168), Baldini (1996b, 50–51), Wallace (1999, 8).

[12] Müller (1911, 504), Hagen (1911, 42–43).

[13] Acloque (1982, 27–31), Borgato (1996, 257–57), Hagen (1911, 32–33), Hall (1903; 1904; 1910).

[14] See, for example, Shea (1972; 2005).

manner. It argued that the Inquisition rendered a service to science by condemning the Copernican theory when it had not yet been demonstrated to be true, and by condemning Galileo for supporting it with scientifically incorrect arguments. This anonymous article was originally attributed by some to a certain professor Clemens of the University of Bonn, but it was later shown by Karl von Gebler (1878) to be a translation of an Italian essay authored by Maurizio B. Olivieri, commissary of the Roman Inquisition and former general of the Dominicans; this essay circulated widely in manuscript form and was published only posthumously in 1872.[15] Olivieri claimed that the mechanical objections to the earth's motion depended crucially on the assumption that air has no weight; that therefore they could not be answered until the discovery that air has weight; that Galileo was not aware of this fact; and that of course the discovery was made after his death by Torricelli and Pascal.[16]

This type of criticism raises a crucial and valid point. That is, there is more to being right than that one's beliefs and conclusions happen to be true, i.e., correspond to reality. It is also important that one's own motivating reasons and supporting arguments be right. In other words, one's reasoning is at least as important as the substantive content of one's beliefs. However, most such anti-Galilean charges are misapplied and can be refuted (Chapter 9). Galileo's reasoning can be successfully defended; indeed it can be shown to be a model of critical thinking.[17]

In any case, other issues were bound to arise, and did arise, in the process of coming to terms with the condemnation of Galileo. As we have seen, he got into trouble with the Church, and was formally condemned in part, for holding the principle that Scripture is not a scientific authority, that Scripture is irrelevant to the assessment of provable or probable claims about nature. This methodological-theological principle is much more elusive and controversial than the astronomical-scientific claim that the earth moves, and so the corresponding issues are more complex (Chapter 8).

At first, some anti-Galilean critics mentioned this principle as one of Galileo's main errors. For example, in 1651 Riccioli, besides criticizing the geokinetic theory scientifically, elaborated explicitly and systematically a very conservative version of biblical fundamentalism, according to which the literal meaning of biblical statements must be held to be physically true and scientifically correct; thus, allegedly, the Inquisition had been theologically wise in upholding the fundamentalist view against Galileo.[18]

---

[15] Gebler (1879a, 244–246), Martin (1868, 383, 405), Olivieri (1840; 1841a,b).

[16] This position was historically untenable insofar as Galileo was clearly aware that air has weight; see Galileo's letters to Giovanni Battista Baliani of 1614 (Favaro 12: 33–36) and 1630 (Favaro 14: 158), and the discussion in *Two New Sciences* (Favaro 8: 123–124); cf. Roberts (1870, 103–104) and Favaro (1908). The position was also scientifically misconceived because most of the mechanical difficulties depended on such questions as conservation and composition of motion and the principle of inertia, and not on the weight of air; see Galileo's discussions in the *Dialogue*, in Finocchiaro (1980, 206–22; 1997a, 155–71, 212–20); cf. Govi (1872).

[17] See also Finocchiaro (1980; 1997a).

[18] Cf. Pesce (1987, 266–268).

Eventually, however, it turned out that Galileo was right regarding this principle as well. A crucial episode in this hermeneutical and theological vindication of Galileo came in 1893: in the encyclical *Providentissimus Deus*, Pope Leo XIII put forth a view of the relationship between biblical interpretation and scientific investigation that corresponds to the one advanced by Galileo in the *Letter to the Grand Duchess Christina*. Although Galileo was not even mentioned in the encyclical, the correspondence was easy to detect for anyone acquainted with both documents, and so the encyclical has been widely interpreted as an *implicit* vindication of Galileo's meta-hermeneutical principle.[19] About a century later, such a vindication was made explicit in Pope John Paul II's rehabilitation of Galileo in 1979–1992. Although this rehabilitation was incomplete, informal, and problematic in several ways (Section 8.17), on the hermeneutical issue John Paul was clear and emphatic. In his 1979 speech, he declared that "Galileo formulated important norms of an epistemological character, which are indispensable to reconcile Holy Scripture and science. In his letter to the grand-duchess mother of Tuscany, Christine of Lorraine, ... Galileo introduces the principle of an interpretation of the sacred books which goes beyond the literal meaning but is in conformity with the intention and the type of exposition characteristic of them."[20] And in his 1992 speech, the pope reiterated: "the new science, with its methods and the freedom of research that they implied, obliged theologians to examine their own criteria of scriptural interpretation. Most of them did not know how to do so. Paradoxically, Galileo, a sincere believer, showed himself to be more perceptive in this regard than the theologians who opposed him."[21]

However, once again, as it became increasingly clear that Galileo's meta-hermeneutical principle was correct, his critics started to emphasize the reasons and arguments he had given to justify it. They tried to find all sorts of incoherences and inconsistencies in it: that his essay contains not only assertions denying the scientific authority of Scripture, but also assertions affirming it;[22] that on the one hand he objects to the use of biblical passages against his own astronomical claims, but on the other hand he tries to interpret the passage in Joshua (10: 12–13) in geokinetic terms (i.e., he wants to have it both ways);[23] that he tries to illegitimately shift the burden of proof by a "sleight of hand ... it is no longer Galileo's task to prove the Copernican system, but the theologians' task to disprove it" (Koestler 1959, 436–437); and that he wants both to appeal to the theological tradition (for example, by frequent quotations from St. Augustine) and to overturn it by a radically new principle.

---

[19] Denzinger and Schoenmetzer (1967, no. 1948), Dubarle (1964, 25), Fantoli (2003b, 362), Finocchiaro (2005b, 263–266), Langford (1971, 66), Martini (1972, 444), Pesce (1987, 283–284), Poupard (1984, 13), Viganò (1969, 234). Cf. Section 8.13.

[20] John Paul II (1979, Section 8).

[21] John Paul II (1992, Section 5, paragraphs 4–5).

[22] E.g., McMullin (1998; 2005c), Carroll (1997; 1999; 2001).

[23] Cf. Pesce (2000, 48–50; 2005, 1–2, 226–229), McMullin (2005c, 101–102, 110–111), Lerner (1999, 81; 2005, 20), Biagioli (2003; 2006a, 219–259).

Again, it is important to know about the possibility of raising such objections and to understand them, but Galileo can be defended from this criticism of his reasoning, for the criticism is itself criticizable as invalid (Chapters 4, 9, and 10).

On the other hand, the greater complexity of the scriptural issue created new possibilities for anti-Galilean criticism. Independently of the truth or falsity of the principle denying the scientific authority of Scripture, and independently of the validity or invalidity of Galileo's reasoning to justify it, he is sometimes criticized for his theological intrusion, for his pastoral imprudence, and so on. The criticism of theological intrusion objects that Galileo was not a professional theologian, and so he had no right to interfere with, or get involved in, exegetical and hermeneutical discussions. One reply to this criticism is that Galileo did stay away from theological discussions until his scientific ideas were attacked on scriptural grounds; after that, he had every right to defend himself by refuting those attacks as fully as he did. The criticism of pastoral imprudence objects that it was irresponsible for Galileo to loudly proclaim to the popular masses the limitations of the literal or patristic interpretation of Scripture at a time when the Catholic Church was in a vital struggle with the Protestant Reformation, given that scriptural interpretation was a key aspect of that struggle.

Something even stranger occurred in the history of the hermeneutical aspect of the Galileo affair. At one point he was blamed for holding and doing the *opposite* of what he actually held and did; that is, it was alleged that he preached and practiced the principle that biblical passages be used to confirm astronomical theories. This criticism got started in 1784–1785 with an apologia of the Inquisition by Jacques Mallet du Pan in the *Mercure de France* and the printing in Girolamo Tiraboschi's *Storia della letteratura italiana* of an apocryphal letter attributed to Galileo but forged by Onorato Gaetani. The view proved to be long-lasting and widely accepted for more than a century; it became a slogan, namely that "Galileo was persecuted not at all insofar as he was a good astronomer, but insofar as he was a bad theologian" (Mallet du Pan 1784, 122). The myth seems to have acted as a catalyst insofar as its creation encouraged the proliferation of pro-clerical accounts and the articulation of pro-Galilean ones, thus making the discussion of Galileo's trial the cause célèbre it is today (Chapter 10).

Another strand of anti-Galilean criticism focused on his alleged legal or judicial culpability. It claimed that the trial did not really deal with the just discussed astronomical-geokinetic or hermeneutical-methodological issues. Galileo was condemned neither for being a good astronomer nor for being a bad theologian, but rather for something else – disobedience or insubordination. His crime was the violation of the ecclesiastical admonition which he received in February 1616. Admittedly, it is uncertain whether this admonition amounted simply to Cardinal Bellarmine's warning not to hold or defend the earth's motion, or to the Inquisition's special injunction not to discuss the topic in any way whatever. However, in either case Galileo's *Dialogue* violated the admonition. The violation of the special injunction is clear, direct, and incontrovertible. And a violation of Bellarmine's warning can be claimed to have occurred because the book does defend the earth's motion by criticizing all arguments against it and advocating some arguments in favor.

Whether valid or invalid, this criticism is relevant, important, and cannot be summarily dismissed. It can be dated as far back as Tiraboschi's apologia in 1793 (Section 8.8),[24] and it continues to be repeated, refined, and embellished (Section 12.6). In my opinion, however, such criticism is untenable. Two distinct points need to be made here.

First, there is the special injunction from which point of view it would seem that Galileo can have no defense. It turns out, however, that this time the relevant defense is contained in the documents and manuscripts that make up the special Vatican file of Galilean trial proceedings. This was discovered and established in the decade 1867–1878, when these proceedings were opened to scholars and published by them in their entirety. A consensus soon emerged that the special-injunction document has enough irregularities that this aspect of the proceedings must be regarded as embodying a legal or judicial impropriety. From this point of view, the legal criticism of Galileo again backfired against the critics. It emerged that Galileo had been the victim of an injustice in a way that had been previously unsuspected (Section 8.12). One could almost say that the trial documents suggest that he was framed.

There remains, of course, the criticism that Galileo violated Bellarmine's milder warning. A possible answer to this criticism relates to what I said earlier about Galileo's *indirect pursuit* of the Copernican research program after the prohibition of 1616. That is, the *Dialogue* discusses the earth's motion by examining all the arguments on both sides; the examination includes not only a presentation and an analysis of the arguments, but also their evaluation or assessment. Galileo was indeed taking the liberty of *evaluating* the arguments; he was hoping that if he carried out the evaluation fairly and validly, his having engaged in argument assessment would not be held against him. He was taking the gamble that a correct assessment of arguments would not be seen as an objectionable defense of Copernicanism. Although such a defense of Galileo has never, to my knowledge, been fully articulated, traces of it can be found in the historical record. In February 1633, soon after Galileo reached Rome to stand trial, Francesco Niccolini (the Tuscan ambassador) had a meeting with Cardinal Francesco Barberini (the Vatican secretary of state and a member-judge of the Inquisition tribunal) to discuss the forthcoming proceedings; to the cardinal's charge that Galileo's *Dialogue* amounted to "reporting much more validly what favors the side of the earth's motion than what can be adduced for the other side," the ambassador replied that "perhaps the nature of the situation indicated this, and therefore he was not to blame."[25] In 1635, Nicholas Claude Fabri de Peiresc (1580–1637) hinted at it when he interpreted the *Dialogue* as a "philosophical play," by which he meant a problem-oriented discussion of the arguments, evidence, and reasons on both sides (Section 8.3).[26] And in

---

[24] Tiraboschi (1782–1797, 10: 373–383), Finocchiaro (2005b, 164–174).
[25] Niccolini to Cioli, 27 February 1633 (ii), in Favaro 15: 55–56.
[26] *Scherzo problematico*, in Favaro (16: 170, 247). In a less defensive context, Campanella also used a similar notion, *comedia filosofica*, in Favaro 14: 366. Cf. Westman (1984, 334), Finocchiaro (2005b, 52–56).

1943, Pio Paschini explicitly formulated such a defense of Galileo by stating that "it was not his fault if the arguments for the heliocentric system turned out to be more convincing" (Section 8.14).[27]

Moreover, there is another issue to be raised in regard to Galileo's alleged disobedience of Bellarmine's warning. Was that warning legitimate? I know of no convincing argument justifying its legitimacy.[28] It may have been one of the many abuses of power in this story. If the warning was not legitimate, then Galileo disobeyed an illegal order. And even if the warning was proper from the point of view of canon law, we may ask whether it was also proper from the *moral* point of view. Again, at worst Galileo may have committed a legal "misdemeanor" while in pursuit of a morally desirable aim, or while exercising a basic human or civil right.

One might think that the implicit theological vindication of Galileo by an influential pope in 1893, coming soon after his judicial rehabilitation by the meticulous scholarship of the 1870s, on top of the older and more gradual scientific vindication provided by the proofs of the earth's motion climaxing with Foucault's pendulum (1851), that such developments would prevent or discourage further indictments or retrials of the victim. But to think so would be to underestimate the power of human ingenuity or the unique complexity of the Galileo affair. In fact, a novel apologia was soon devised by a great scholar who combined knowledge of physics, history, and philosophy – Pierre Duhem (1861–1916). In 1908 he advanced the new charge that Galileo was a bad epistemologist.

The criticism of Galileo as a bad epistemologist is often confused with, and is indeed related to, the criticism that he was a bad arguer or reasoner, that he did a poor job in proving the earth's motion, in defending Copernicanism. However, the two criticisms are distinct. The epistemological criticism of Galileo attributes to him wrong or untenable epistemological principles and practices, and then it connects such epistemological errors or naïveté with the tragedy of the trial. Epistemology may be defined as the study of the nature of knowledge in general, and scientific knowledge in particular, as well as of the principles and procedures that are useful in the acquisition of knowledge.

The epistemological doctrine which Duhem found especially objectionable is "realism": it states that science aims at the truth about the world, and scientific theories are descriptions of physical reality that are true, probably true, or potentially true. Duhem was an advocate of epistemological "instrumentalism," according to which scientific theories are merely instruments for making mathematical calculations and observational predictions, and not descriptions of reality, and so they are not the sort of things that can be true or false, but only more or less convenient. Duhem tried to blame Galileo's trial on epistemological realism, allegedly shared by Galileo and his Inquisition persecutors, and also on their failure to appreciate

---

[27] Paschini (1943, 97); cf. Finocchiaro (2005b, 280–284). In one of his unpublished writings dated 1980, Stillman Drake made a similar point; cf. DiCanzio (1996, 309).
[28] But see the important technicalities discussed in Mereu (1979, 435–437), Beretta (1998, 239–248).

instrumentalism, which in that historical context was being allegedly advocated by Cardinal Bellarmine and Pope Urban VIII. Duhem's own memorable words are "that logic was on the side of Osiander, Bellarmine, and Urban VIII, and not on the side of Kepler and Galileo; that the former had understood the exact import of the experimental method; and that, in this regard, the latter were mistaken."[29] To avoid being misled, here Duhem's word "logic" should be taken to mean primarily "epistemology," and not reasoning, as already clarified.

Duhem's epistemological criticism of Galileo is interesting, important, and influential. Nevertheless, it is untenable (Chapter 11), primarily because under the heading of Galilean realism Duhem subsumes too many other epistemological principles besides the ideal of truth and description of reality; but these other attributions are conceptually arbitrary and textually inaccurate.

Another example of this genre of criticism may be gleaned from the work of more recent scholars.[30] It claims that Galileo subscribed to the traditional Aristotelian ideal of science as strictly demonstrative, which he was never able to give up despite some flirtings with fallibilism or probabilism; that he believed he had provided a strict demonstration of the earth's motion (with arguments such as his explanation of tides); that much opposition to him was an attempt to make him aware of the nondemonstrative status of his arguments or the nonviability of the demonstrative ideal; and that the operative role of this problem is visible in such documents as Bellarmine's letter to Foscarini (1615), Galileo's "Discourse on the Tides" (1616), the *Dialogue* (1632), and the consultants' reports on this book produced during the 1633 proceedings. Such an account could be labeled the criticism of Galileo as a failed Aristotelian, or a failed demonstrativist. This criticism overlaps with both Duhem's epistemological criticism and the criticism of Galileo's reasoning mentioned earlier. For in part this criticism faults Galileo's epistemological doctrine of demonstration or his epistemological awareness of the nature of his own geokinetic arguments; and in part this criticism impugns the reasoning used by Galileo to arrive at or to justify his geokinetic beliefs. However, such criticism can be rebutted, and it emerges that rather than being a failed Aristotelian demonstrativist, Galileo is someone who was able to assimilate and transcend the Aristotelian ideal of science as demonstration (Chapters 3, 9, 11, and 12).[31]

A measure of Duhem's influence is the fact that it has spawned a genre of anti-Galilean criticism in which Galileo is charged with having held all kinds of implausible epistemological doctrines, and then a questionable connection with the trial is made. For example, a recent popular book revealingly entitled *Galileo's Mistake* portrays Galileo as a kind of positivist who held that only science provides the truth about reality, and that this mistake was the root cause of his condemnation (Rowland 2003, 6, 137).

---

[29] Duhem (1908, 136); cf. Duhem (1969, 113).
[30] E.g., Wallace (1981b; 1984a), McMullin (1967c; 1978; 1998; 2005c).
[31] See also Morpurgo-Tagliabue's (1981) important and pioneering contribution to this thesis; cf. Section 12.7.

Next, a crucial critical issue of the subsequent Galileo affair has been whether the original trial shows that science and religion are incompatible. In fact, the trial has been usually perceived as epitomizing the conflict between science and religion. In this case the result is not so much criticism of Galileo but criticism of the Church. That is, this received interpretation of Galileo's trial is really an anti-clerical but pro-Galilean position. Galileo is seen as a victim of religion's war on science, and such a picture makes Galileo into a secularist hero and icon. Such a portrayal of Galileo is intended to be a favorable one, at least by those who advance or advocate it. The first explicit advocate of this view was perhaps Jean D'Alembert, who put it forth in his introduction to the French *Encyclopedia* (Section 8.6).[32] It was later elaborated and popularized by authors such as Andrew D. White and John W. Draper.[33]

One criticism of this view takes the form of attempting to reverse it. It argues that the study of Galileo's trial demonstrates instead the *harmony* between science and religion, primarily because Galileo himself believed in such harmony and formulated some pretty good arguments for such harmony. This is the view most significantly advanced by Pope John Paul II and provided him with the stimulus for his partial rehabilitation in 1979–1992. However, this account goes back at least to 1942, when it was advanced by clergyman Agostino Gemelli in the context of what must be regarded as an earlier rehabilitation attempt by the Church (Section 8.14).[34] This interpretation is also meant to be pro-Galilean, in the sense that it sees Galileo as a hero and is intended to be favorable to him, but the interpretation is also pro-clerical, in the sense that the message or lesson learned is one favorable to the Church. However, for the anti-clerical conflictualists such a clerical appreciation of Galileo is distasteful and represents an anti-Galilean position.

In my opinion, both the conflictual and the harmonious interpretations are partly right and partly wrong. My first reason is that insufficient attention has been paid to the fact that there is a minimal but irreducible conflict in the history of the original trial: it is the conflict between those (like Galileo) who claimed that Copernicanism was *not* contrary to Scripture, and those (like the Index and the Inquisition) who declared that the earth's motion *was* contrary to the Bible. In other words, if in this controversy we take the Copernican theory of the earth's motion to represent science and Scripture to represent religion, then Galileo was the one claiming that there is no real conflict between the science and religion, whereas the Church was the one claiming that the apparent conflict was real. The irony of the situation is that it was the erstwhile loser or victim who held the view which the erstwhile winner of that battle would now like to advocate; and insofar as Galileo may be said to have become the historical and final winner of the war, then his view suggests an important harmonious element in the original affair.

---

[32] In Diderot and D'Alembert (1751–1780, 1: i–xlv), D'Alembert (1963). Cf. Finocchiaro (2005b, 120–125).
[33] White (1869; 1876; 1896; 1915; 1965), Draper (1875).
[34] Gemelli (1942); cf. Finocchiaro (2005b, 275–280).

Secondly, both the conflict and the harmony exist at the level of what I call the surface structure of the situation. But we should also ponder a deeper aspect. I am referring to the fact that Galileo was not the only one who held that there was no conflict. And the important thing is that many of those who agreed with him on this question of principle were themselves churchmen, two of the most important examples being Foscarini and Campanella. In other words, at the time of Galileo, there was a division within the Catholic Church between those who did and those who did not accept the scientific authority of Scripture. A similar split existed in scientific circles. A further division existed in regard to the other main issue of Galileo's trial, namely the physical proposition of the earth's motion. Thus, rather than having an ecclesiastic monolith on one side clashing with a scientific monolith on the other, the real conflict was between two attitudes that crisscrossed both.[35] I believe the most fruitful way of describing the two attitudes is to label them conservative or traditionalist on one side and progressive or innovative on the other. The real conflict was between these two groups. In this sense, Galileo's trial illustrated the clash between conservation and innovation and involves an episode where the conservatives happened to win one particular battle. This conflict is one that operates in such other domains of human culture as politics, art, economy, and technology. It cannot be eliminated on pain of stopping cultural development; it is a moving force of human history.[36] In short, in the original Galileo affair, the conflict between science and religion (or their harmony as well) is a less important feature than the conflict between conservation and innovation.

Thirdly, after Galileo's condemnation, the predominant view became the interpretation of the trial as epitomizing the conflict between science and religion. Even those who advocate the harmonious account of the trial do not deny that the key feature of the *subsequent* Galileo affair was indeed a conflict between science and religion. For example, Pope John Paul II believed that the lesson from Galileo's original trial was the harmony between science and religion, and he wanted to stress and elaborate this lesson in order to try to put an end to the subsequent, very real, but presumably unjustified science versus religion conflict. As regards this *subsequent controversy*, I claim that the science versus religion conflict is indeed an essential feature of it. One may wish to argue that there ought not to be a conflict between science and religion because, for example, this was not the key conflict in the trial of Galileo. Or one may wish to argue that the key conflict in the trial of Galileo was not the conflict between science and religion because the key conflict there was the one between conservation and innovation. But clearly one cannot argue that there was *not* a conflict between science and religion in the *subsequent*

---

[35] The non-monolithic character of the Catholic Church has been explicitly stressed in various ways by other authors, such as Segre (1991b, 30), Feldhay (1995), and Speller (2008); Feldhay emphasizes the disputes between Jesuits and Dominicans, in regard to which I would want to point out that these two orders were not themselves monolithic either.

[36] One author who has recognized the importance of the dialectic of conservation and innovation in the history of science is Kuhn (1977).

Galileo affair because there was *not* one in the *original* affair; nor that there *was* a conflict between science and religion in the *trial* of Galileo because there *was* one in the *subsequent* affair.

The Current Spectacle: Catholic Hero or Socialist Villain?

The controversy shows no signs of abating to date (A.D. 2009). This is obvious from the recent rehabilitation efforts by the Catholic Church and from new anti-Galilean critiques by left-leaning social critics. These two clusters of developments deserve some elaboration.

In 1942, the tricentennial of Galileo's death provided the occasion for a first partial and *informal rehabilitation*. This was done in the period 1941–1946 by several clergymen who held the top positions at the Pontifical Academy of Sciences, the Catholic University of Milan, the Pontifical Lateran University in Rome, and the Vatican Radio. They published accounts of Galileo as a Catholic hero who upheld the harmony between science and religion; who had the courage to advocate the truth in astronomy even against the Catholic authorities of his time; and who had the religious piety to retract his views outwardly when the 1633 trial proceedings made his obedience necessary (Section 8.14).

In 1979, Pope John Paul II began a further informal and partial rehabilitation of Galileo that was not concluded until 1992. In two speeches to the Pontifical Academy of Sciences, and in other statements and actions, the pope admitted that Galileo's trial was not merely an error but also an injustice; that Galileo was theologically right about scriptural interpretation, as against his ecclesiastical opponents; that even pastorally speaking, his desire to disseminate novelties was as reasonable as his opponents' inclination to resist them; and that he provides an instructive example of the harmony between science and religion (Section 8.17).

At about the same time that Galileo was being rehabilitated by various Catholic officials and institutions, he became the target of unprecedented criticism on the part of various representatives of *secular culture* (Section 8.15). It was almost as if a reversal of roles was occurring, with his erstwhile enemies turning into friends, and his former friends becoming enemies. Several other circumstances add interest and significance to such a development. These critics elaborated what may be called social and cultural criticism of Galileo; that is, they tried to blame Galileo by holding him personally or emblematically responsible for such things as the abuses of the industrial revolution, the social irresponsibility of scientists, the atomic bomb, and the rift between the two cultures. They were mostly writers with backgrounds and sympathies subsumable under the left wing of the political spectrum. The most outstanding and original examples of such criticism were central-European German-speaking personalities: Bertolt Brecht was a German playwright who authored a play entitled *Galileo* that was first written in 1938, then revised into a second version in 1947, then into a third version in 1955, and became a classic of twentieth-century theater; Arthur Koestler was a Hungarian-born writer, novelist, and intellectual who in 1958 published a book that became an international best-seller, entitled *The Sleepwalkers: A History of Man's Changing Vision of the Universe*; and Paul Feyerabend was an Austrian-born philosophy professor at the University of California, Berkeley, who advanced his version of social criticism in a book entitled *Against Method*, first published in 1975 and later revised in 1988 and in 1993.

These developments have not been properly assimilated yet. For example, the Catholic "rehabilitations" tend to be either *unfairly criticized* (even by Catholics) or *uncritically accepted* (even by non-Catholics). Moreover, the current pope, Benedict XVI, seems to have displayed an ambivalent attitude toward this issue; his ambivalence is emblematic and revealing, but continues to polarize and confuse. This will be elaborated presently. And the left-leaning social critiques tend to be summarily dismissed by practicing scientists, whose professional identity is thereby threatened, or dogmatically advocated by self-styled progressives, who seem not to have learned much from Galileo and to want to turn the clock back to pre-Galilean days.

For reasons that will emerge shortly, it is useful to examine Feyerabend's criticism more closely. He portrays Galileo's trial as involving a conflict between two philosophical attitudes toward, and historical traditions about, the role of experts. That is, Galileo allegedly advocated the uncritical acceptance by society of the views of experts, whereas the Church advocated the evaluation by society of the views of experts in the light of human and social values. Feyerabend extracts the latter principle from Cardinal Bellarmine's letter to Foscarini, asserting that "the Church would do well to revive the balance and graceful wisdom of Bellarmine, just as scientists constantly gain strength from the opinions of ... their own pushy patron saint Galileo" (Feyerabend 1985, 164). More generally, Feyerabend claims that "the Church at the time of Galileo not only kept closer to reason as defined then and, in part, even now; it also considered the ethical and social consequences of Galileo's views. Its indictment of Galileo was rational and only opportunism and a lack of perspective can demand a revision."[37]

I believe Feyerabend's criticism is untenable. In part, it is not really supported by the texts to which he refers. However, the principal difficulty is that he seems to perpetrate a fallacy of equivocation. For the principle in question could mean either that *social and political leaders* should evaluate the *use* of experts' views in light of human and social values, or that *scientists* should evaluate the *truth* of each other's views in light of human and social values.

Now under the first interpretation, Galileo did not reject the principle, but rather would have agreed with it. Moreover, when Feyerabend attributes this principle to Bellarmine, the documentation is unclear and unconvincing. In any case, in this regard, their difference was not one of principle but of application. For example, they would have disagreed on who the relevant experts were, in particular whether theologians should be counted as experts in physics and astronomy; another disagreement would have been whether the views of theological experts should be subject to the same requirement.

---

[37] Feyerabend (1988, 129; 1993, 125).

Introduction: A Galilean Approach to the Galileo Affair xxxiii

Under the second interpretation, the principle was indeed rejected and criticized by Galileo. However, it is in fact untenable. For this version of the principle cannot survive the objections (which we have inherited from Galileo) against teleological and anthropomorphic ways of thinking; such thinking reduces to arguing that something is true because it is useful, beneficial, or good, and false because it is useless, harmful, or bad.

However, whether untenable or not, Feyerabend's criticism is important because of the effects it has had. In fact, it has become involved in the very latest twist to the controversy that brings the story to our own day.

On the one hand, Feyerabend's apologia was politely rejected in 1989–1990 by Cardinal Joseph Ratzinger, who at the time was the chairman of the Congregation for the Doctrine of the Faith (the new name of the Inquisition), and who in 2005 became Pope Benedict XVI. In a scholarly essay, in the context of an analysis of the role of faith in the revolutionary geopolitical changes happening in 1989–1990, Cardinal Ratzinger quoted several anti-Galilean critiques, including Feyerabend's. However, Ratzinger went on to criticize such views as expressions of skepticism and philosophical insecurity, asserting that "it would be foolish to construct an impulsive apologetic on the basis of such views; faith does not grow out of resentment and skepticism with respect to rationality, but only out of a fundamental affirmation and a spacious reasonableness ... I mention all this only as a symptomatic case that permits us to see how deep the self-doubt of the modern age, of science and of technology goes today" (Ratzinger 1994, 98).

On the other hand, there seems to be a very widespread tendency that confuses or conflates Feyerabend's view with Ratzinger's. Some authors have claimed simply that Cardinal Ratzinger or Pope Benedict *accepts* Feyerabend's view.[38] Other authors have gone so far as to attribute this claim directly to Cardinal Ratzinger or Pope Benedict, without giving any indication that he was quoting Feyerabend.[39] There have been some attempts to clarify the situation,[40] but apparently to no avail.

In fact, in January 2008 such confusion triggered the following clash.[41] A few months earlier, Pope Benedict XVI had accepted an invitation by the rector of the University of Rome to deliver the keynote address at the formal ceremony inaugurating the new academic year. This plan, however, triggered protests by students and faculty, especially in the university's distinguished department of physics. They objected primarily on the grounds of the principle of separation of Church and State, but also in part because, as they stated, they felt offended and humiliated by

---

[38] Socci (1993, 62), Sinke Guimarães (2005, 6).
[39] Machamer (2005), Saka (2006).
[40] Accattoli (1990), Feyerabend (1993, 133–134 n. 20), Finocchiaro (2008a, 274 n. 19).
[41] See: Marcello Cini, "Lettera aperta al Rettore dell'Università La Sapienza di Roma (14 November 2007)," at www.sinistra-democratica.it, consulted on 4 February 2008; "Pope Quotes Feyerabend and Gets in Big Trouble at Leading Italian University," in *Leiter Reports: A Philosophical Blog*, at http://leiterreports.typepad.com/blog/2008/01/pope-quotes-f-1.html, consulted on 17 January 2008.

the pope's view of Galileo's condemnation, expressed some twenty years earlier when the pope was still a cardinal; that is, by his sharing Feyerabend's view. In the light of such opposition, and the potential for unrest and violence, the pope cancelled his speech.

This controversy is not helped, but rather exacerbated, by what seems to be a recurrent pattern of thinking or lecturing on the part of Benedict XVI, namely flirting with equivocation by means of quoting a controversial view. For example, an analogous issue arose as a result of a lecture he delivered at the University of Regensburg on 12 September 2006, in which he quoted a remark made by Byzantine emperor Manuel II Paleologus in 1391 regarding Islam and holy war.[42] Now, given the current geopolitical situation, Benedict did make a sustained effort to clear up the latter misunderstanding. But it appears that he has made no such effort regarding the approval of Galileo's condemnation.

On the other hand, this appearance may not correspond to reality. In fact, a few months later a story surfaced in the global news media that there was a plan to erect a statue to Galileo within the Vatican walls.[43] Without knowing more, one could speculate that such a statue is a gesture to suggest that Cardinal Ratzinger had really meant to criticize Feyerabend and that today's Church does not really approve the 1633 condemnation of Galileo. However, some time afterwards it emerged that the Vatican statue to Galileo had been proposed by a private firm, who wanted to pay for the cost but to remain anonymous. But the latest development in this episode is that the private donor has withdrawn the offer, perhaps afraid of the unwanted publicity and controversy which the idea was generating.

However, the story did not end there. On 21 December 2008, pope Benedict XVI delivered the weekly Sunday speech at noon from a window of the Vatican palace to the people assembled in St. Peter's Square. Besides the usual pieties, the pope exploited the time and place to mention Galileo and the International Year of Astronomy. The pretext was provided by the fact that the feast of Christmas was originally scheduled to come around the winter solstice, and by the fact that the obelisk at the center of St. Peter's Square casts its longest shadow on the winter solstice. Then the pope went on to "greet all those who will be taking part in various capacities in the initiatives for the World Year of

---

[42] "Faith, Reason, and the University: Memories and Reflections," Lecture delivered at the University of Regensburg, 12 September 2006. At http://www.vatican.va/holy_father/benedict_xvi/speeches/2006/september/documents/hf_ben_xvi_spe_20060912_university-regensburg_en.html.

[43] "Vatican to Erect Statue of Galileo," *Catholic News Service*, 5 March 2008, at www.cathnews.com/article.aspx?acid=6123, consulted on 9/2/08; Carol Glatz, "After Four Centuries, Galileo to Return to the Vatican," *Catholic News Service*, 7 March 2008, at http://www.catholicnews.com/data/stories/cns/0801299.htm, consulted on 3/18/08; Gabriel Kahn and Andrew Higgins, "Galileo Still Sends Church Spinning as Statue at the Vatican Is Considered," *Wall Street Journal*, 28 August 2008, pp. A1, A12; "Vaticano; Sfuma Progetto Statua di Galileo nei Giardini Vaticani," 28 August 2008, at http://notizie.alice.it/notizie/articolo/stampa.html?filter=foglia&nsid, consulted on 9/2/08.

Astronomy, 2009, established on the fourth centenary of Galileo's first observations by telescope."[44]

As usual, such a brief reference was widely reported, commented on, and amplified. The Associated Press compiled and circulated an article entitled "Good Heavens: Vatican Rehabilitating Galileo."[45] And Church critics sprinkled the blogosphere with their share of invective and abuse.

Then on 30 January 2009, the Pontifical Council for Culture announced several projects related to the International Year of Astronomy.[46] One project would be an exhibition at the Vatican Museums entitled "Astrum 2009: The Historical Legacy of Italian Astronomy from Galileo to Today," organized jointly with the Italian Institute of Astrophysics and the Vatican Observatory, and running from October 2009 to January 2010. Another project would be a conference entitled "1609–2009: From the Birth of Astrophysics to Evolutionary Cosmology," to be held in November 2009 at the Pontifical Lateran University in Rome. At the time of this writing, these events have not yet taken place, and so their success and impact cannot be judged. However, it is obvious that Church is attempting to exploit the International Year of Astronomy to project an image of herself as more friendly to and harmonious with Galileo and science.

Finally on 26–30 May 2009, there was a conference in Florence entitled "The Galileo Affair: A Historical, Philosophical and Theological Re-examination." It had been conceived and organized by the Florentine Jesuits, who are based at the Niels Stensen Institute in the Tuscan capital,[47] and who had worked on the project for about a year and one-half. Its institutional sponsors represented a who's who of Italian and Vatican academic, cultural, and political institutions, for example: the Lincean Academy, National Research Council, Arcetri Astrophysical Observatory, Pontifical Academy of Sciences, Pontifical Council for Culture, Vatican Observatory, as well as the President, Prime Minister, and Culture Ministry of the Italian

---

[44] The relevant passage reads: "The Christmas festivity is placed within and linked to the winter solstice when, in the northern hemisphere, the days begin once again to lengthen. In this regard, perhaps not everyone knows that in St. Peter's Square there is also a meridian; in fact the great obelisk casts its shadow in a line that runs along the paving stones toward the fountain beneath this window and in these days, the shadow is at its longest of the year. This reminds us of the role of astronomy in setting the times of prayer ... The fact that the winter solstice occurs exactly today, 21 December, and at this very time, offers me the opportunity to greet all those who will be taking part in various capacities in the initiatives for the World Year of Astronomy, 2009, established on the fourth centenary of Galileo's first observations by telescope ... If the heavens, according to the Psalmist's beautiful words, 'are telling the glory of God' (Ps 19[18]:1), the laws of nature which over the course of centuries many men and women of science have enabled us to understand better are a great incentive to contemplate the works of the Lord with gratitude." Cf. www.vatican.va/news_services/or/or_eng/text.html§1, consulted on 24 December 2008.

[45] By Nicole Winfield. See, for example, www.washingtonpost.com/wp-dyn/content/article/2008/12/23.

[46] Cf. "Vatican to Celebrate Galileo in Year of Astronomy," at http://www.indcatholicnews.com/news.php?viewStory=1049, consulted on 1 June 2009.

[47] Cf. the website of this Institute, http://www.stensen.it/, or of the conference, http://www.galileo2009.org, last consulted on 2 June 2009.

Republic. Moreover, besides an organizing committee consisting of members and staff of the Stensen Institute, there was a scientific committee consisting of several well-known Galileo scholars. And many public and private institutions provided funding and financial support.

This conference had a very ambitious agenda, as its mere title suggests. The various announcements available at the website of the Stensen Institute indicated that the underlying motivation of the organizers was not only the celebration of the International Year of Astronomy, but also the continuing and lingering dissatisfaction with some aspects of Pope John Paul II's attempt at rehabilitation in 1979–1992. The program included keynote addresses, presentations, and panel discussions by many distinguished scientists, historians, theologians, and philosophers. The variety of sponsoring organizations indicated that there was a general desire for fruitful dialogue not only across the divide of science and religion, but also across the separation of Church and State. The last point is especially significant, given that in Italy the Galileo affair has an additional significant complication which is absent or minor in other national contexts, namely the historical enmity between the Church and the political ideal of a unified Italian state.

We will have to wait for the published proceedings of this conference to see whether its stated ambition, initial promise, and obvious potential were realized, and whether there resulted some significant intellectual and scholarly substance, above and beyond the phenomenon defined in terms of rhetorical appearance, public relations, and tourist travel. Such events have a way of frustrating the aims and expectations of even the most astute planners and efficient organizers. In any case, the dialogue between science and religion, between Church and State, and even between scholarly disciplines (such as history and philosophy) is easier said than done. Too often, instead of a real dialogue, the result is primarily a proliferation of monologues. The preliminary results and effects of this conference are not promising. In fact, just two weeks before the conference, three of its leading participants[48] published in the popular press what may be regarded as expressions of their positions and previews of their contributions. And the impression they convey is precisely that little progress is being made, despite their exemplifying varying degrees of a valiant and eloquent struggle with the problem.

And so here we are in the first decade of the twenty-first century, still trying to come to terms with the trial of Galileo and the Galileo affair – with whether and why the 1633 condemnation was right or wrong. Must this controversy continue forever? Is there not a way of resolving it? I believe this controversy is likely to continue for the foreseeable future. Nevertheless, I also believe I have devised a framework that paves the way for coming to terms with it and eventually resolving it.

In my approach, one interprets the controversy in terms of arguments for and against the rightness of Galileo's condemnation; then one displays toward these arguments the same attitude which Galileo displayed toward the arguments for and against the earth's motion; and the key elements of this Galilean attitude (labeled

---

[48] Cabibbo (2009), Galluzzi (2009), and Bucciantini (2009).

reasoned, critical, open-minded, and fair-minded)[49] are to know and understand the arguments against one's own view and to appreciate their strength before refuting them. In short, my *overarching thesis*, interpretive as well as normative, is that today in the context of the Galileo affair and the controversies over science versus religion and over institutional authority versus individual freedom, *the proper defense of Galileo should have the reasoned, critical, open-minded, and fair-minded character which his own defense of Copernicus had*. Let me elaborate.

An Overarching Thesis

In this introduction I have been suggesting (and this book will argue) that an important aspect of the Copernican Revolution was the defense of the geokinetic hypothesis from a host of objections based on astronomical observation, Aristotelian physics, scriptural passages, and traditional epistemology (Chapters 1–2). A major contribution to this defense was provided by Galileo. He answered the observational astronomical objections once the telescope revealed new celestial phenomena and revolutionized astronomical observation by making it instrument-based (Chapter 3). He answered the scriptural objections by arguing that Scripture is not a scientific authority, and so scriptural passages should not be used to invalidate astronomical claims that are proved or provable (Chapter 4). He answered the mechanical objections by articulating a new mathematical physics centered on the principles of conservation and composition of motion and the application of mathematical abstractions to physical reality (Chapter 5). More generally, Galileo's key contribution to the Copernican Revolution was to elaborate a defense of Copernicanism that stressed reasoning and argumentation judiciously guided by the ideals of fallibility, open-mindedness and fair-mindedness (Chapter 6).

Despite Galileo's prudence and indirectness, and the support from many churchmen, his defense of Copernicanism was hindered by key officials and institutions of the Catholic Church. In fact, the trial of Galileo (Chapter 7) can be interpreted as a series of ecclesiastic attempts to stop him from defending Copernicus. In 1616, the Index decreed that the geokinetic doctrine was contrary to Scripture, and this decree amounted to a general prohibition on defending Copernicus from scriptural objections. At the same time, Cardinal Bellarmine officially warned Galileo to

---

[49] The meaning of these terms will be explicitly elaborated in the rest of this Introduction and in Section 6.5 below. Labels aside, the conceptual content of these terms will be shown to correspond to principles advocated and practiced by Galileo himself. Both the meaning and the content also correspond in large measure to notions I have used previously to analyze Galileo's work; see, for example, Finocchiaro (1980, 145–179; 1997, 309–356). It is also worth mentioning, especially with regard to *fair-mindedness*, that I have adapted these notions from the literature on argumentation, informal logic, and critical thinking; see, for example, Ennis (1996, 171), Fisher (1991), Fisher and Scriven (1997, 90–91, 137–143), Paul (1990, 110, 11, 198), and Scriven (1976, 166–167). Finally, the ideas, although not under these labels, can also be found explicitly discussed in Chapter 2 of John Stuart Mill's *On Liberty*; see for example, Mill (1997, 52–84), Finocchiaro (2007a).

cease defending the earth's motion, and this warning amounted to a personal prohibition on defending Copernicus from an astronomical, physical, or philosophical point of view. In 1633, the Inquisition condemned Galileo as a suspected heretic, and this sentence amounted to condemning him for defending Copernicus indirectly and probably in the *Dialogue* of 1632; for this book was primarily a critical discussion, examining the arguments on both sides, showing that the Copernican arguments were stronger than the geostatic ones, implying that the geokinetic hypothesis was probably true, and thus defending Copernicanism only indirectly and implicitly.

These condemnations, which represent the two principal phases of Galileo's trial (the original Galileo affair), in turn generated a much more protracted, complex, and controversial cause célèbre that continues to our own day. The subsequent Galileo affair is an intricate web of historical after-effects, reflective commentaries, and critical issues (Chapter 8). However, a key interpretive idea for making sense of it is to focus on the many criticisms of Galileo's defense of Copernicanism (i.e., the many apologias of his condemnation) that have been advanced by his critics and on the various replies and counter-arguments put forth by his defenders. Such an interpretation then readily enables one to adopt one's own evaluative position about such anti-Galilean criticisms. These criticisms can be systematized into the following sequence.

For a while, various questions were raised about the physical truth of the earth's motion; but gradually, the historical development of science established incontrovertibly that Galileo had been right on this issue. As this realization was emerging, questions began to be raised about whether his supporting reasons, arguments, and evidence had been correct; this is an instructive issue, but Galileo's reasoning can be defended from this criticism (Chapter 9). For some time, Galileo was also criticized for his hermeneutical principle that Scripture is not a scientific authority; but historical and cultural development also vindicated him in this regard; at least this is what happened from the point of view of what has become the official position of the modern Catholic Church (Chapters 8 and 10). However, before this theological vindication became clear, the occasion arose for the construction and diffusion of the myth that Galileo had been condemned for being a bad theologian, using Scripture to justify astronomical claims (i.e., the opposite of what he did); it took about a century before this myth was dispelled (Chapter 10). In any case, on the hermeneutical issue too, it is important to check the correctness of his argument justifying that Scripture is not scientific authority; although this Galilean reasoning has been the target of many objections, I believe it can be defended from them (Chapters 4, 9, and 10).

As it became increasingly clearer that Galileo could not be validly accused of being a bad scientist, a bad theologian, or a bad logician, he started being blamed for other reasons. Some authors began to stress the legal aspect of the trial, charging that he had been guilty of disobeying the Church's admonition regarding Copernicanism. However, the content of this admonition is ambiguous. If the admonition is interpreted to be a prohibition on discussion, the existence or occurrence of such a special injunction is undermined by the record of the trial proceedings, which was first published in 1867–1878 (Section 8.12); whereas if the admonition is taken to be a prohibition on defending Copernicanism, then the issue reduces to the questions whether such a prohibition was legitimate, and if it was whether Galileo's defense of Copernicanism was scientifically and logically fair and valid (Chapter 9).

Whether or not Galileo can be defended from such scientific, theological, logical, and legal criticisms, he can be and has been the target of epistemological criticism. In 1908 Duhem tried to blame him for his epistemological realism and argued that the condemnation would have been avoided if epistemological instrumentalism had prevailed. I believe Galileo can be defended from this charge that he was a bad epistemologist (Chapter 11).

Next, there is the issue of whether Galileo is to be credited or blamed for helping us understand that science and religion are in conflict or that they are in harmony. The resolution of this issue requires that we reflect properly on three things (Chapter 12): that the trial embodied a minimal but irreducible historical conflict between those who affirmed, and those who denied, that Copernicanism contradicted Scripture; that the trial epitomized more the conflict between conservation and innovation than the conflict between science and religion; and that because of how the trial was subsequently perceived, the conflict between science and religion is indeed an essential feature of the subsequent affair.

Finally, there is the current spectacle of the Galileo affair. One the one hand, we see the phenomenon of the rehabilitation movement within the Catholic Church. On the other hand, one can witness the rise of socially oriented critiques of Galileo by leftist sympathizers and self-styled progressives. And we also observe the conflict between these two points of view, as well as the irony of the switching of sides.

In short, the Copernican Revolution required that the geokinetic hypothesis be not only supported with new arguments and evidence, but also defended from many powerful old and new objections. This defense in turn required not only the destructive refutation but also the appreciative understanding of those objections in all their strength. One of Galileo's major accomplishments was not only to provide new evidence and arguments supporting the earth's motion, but also to show how those objections could be refuted, and to elaborate their power before they were answered. In this sense, Galileo's defense of Copernicus was *reasoned, critical, open-minded*, and *fair-minded*. Now, an essential thread of the subsequent Galileo affair has been the emergence of many anti-Galilean criticisms, from the point of view of astronomy, physics, theology, hermeneutics, logic, epistemology, methodology, law, morals, and social awareness. It is important to understand both that such criticisms arise naturally and legitimately and that Galileo has been, or can be, effectively defended from them. That is, the proper and effective way of defending him is by ensuring that we know and understand the anti-Galilean criticisms, and that we appreciate their strength before refuting them, thus modeling our own approach to the defense of Galileo on his approach to the defense of Copernicanism. Thus, as already mentioned, the *overarching thesis* formulated and justified in this book is then that today in the context of the Galileo affair and the controversies over the relationship between science and religion and between institutional authority and individual freedom *the proper defense of Galileo should have the reasoned, critical, open-minded, and fair-minded character which his own defense of Copernicus had.* This thesis may be regarded as a critical interpretation of the Galileo affair in the following sense.

First, the subject matter is the Galileo affair broadly understood: it includes not only the trial (1613–1633), with its old and deep roots in Galileo's earlier career (1589–1613), in Copernicus (1543), and earlier; but also the protracted and ongoing subsequent cause célèbre (from 1633 to 1992 and beyond). These are the two affairs: the original one rooted in the Copernican controversy and ending with his trial and condemnation; and the subsequent one following his condemnation and acquiring a life of its own and continuing to our own day.

Second, the interpretive aspect of this thesis stems from the fact that this book stresses a particular, although crucial, point of view, namely the element of reasoning, argumentation, and evidence: the reasons for accepting the ancient geostatic world view; the reasons for taking Copernicus seriously; the reasons for rejecting the physical truth of the geokinetic hypothesis; the reasons that led Galileo to pursue it, indirectly at first and directly later, so as to come to think that it was much more probable than the geostatic thesis; the reasons that led the Inquisition to prosecute and persecute Galileo; the reasons why his condemnation could be claimed and has been claimed to be right (first from the astronomical and hermeneutical points of view, then from the logical, legal, and epistemological points of view, and finally from the points of view of the cultural interaction between science and religion and between science and society); and the reasons why such anti-Galilean criticisms can be and have been criticized. In short, the principal interpretive thesis is that the defense of Copernicus was Galileo's chief offense in the trial and his key contribution to the Copernican Revolution, and that the defense of Galileo from various attempts to justify his condemnation is the most central issue in the subsequent affair.

Third, this book possesses a critical, evaluative, or normative dimension. This aspect can be seen by reflecting on the *overarching thesis* formulated above. For it amounts to saying that defending Copernicanism in the reasoned and critical way in which Galileo did is instructive and suggests the proper way in which Galileo himself can and should be defended from the many attempts to justify his condemnation. This is a lesson that results if, besides trying to understand and interpret what really happened in the Copernican Revolution and the Galileo affair, we also try to assess and evaluate what is right or wrong from various nuanced points of view. In short, the principal evaluative thesis is that just as Galileo's defense of Copernicus owed its success to its being reasoned, critical, open-minded, and fair-minded, so our defense of Galileo can succeed if it possesses these same qualities.

In the light of the three clarifications just made, my talk of "defending" Galileo should not be misunderstood. To defend Galileo does not mean to show that he was completely or perfectly right; it only means to show that he was essentially right, or more nearly correct than not. The defense of Galileo is not an attempt to show that criticisms of him are without foundation; on the contrary, the defense cannot even get started unless one first knows and understands that there are reasons for attributing to him various errors or improprieties; in such a context one tries to show that such anti-Galilean arguments are ultimately invalid, or at least weaker than the pro-Galilean ones. Defending Galileo is not meant to be a one-sided exercise pointing out only his merits and virtues; rather merits and virtues are meant to be inherently comparative properties whose positive aspects are seen only vis-à-vis the negative ones. Nor is the defense of Galileo an hagiographic exercise exaggerating the number or importance of his scientific, intellectual, and cultural achievements; in this regard, I want to reiterate that my *overarching thesis* has a historical or interpretive component, besides the philosophical or evaluative one, and that the main thrust of the interpretive component is the historical reality of the anti-Galilean criticisms.

In other words, the defense of Galileo is interpreted as, and is meant to be, an exercise in critical reasoning, just as his defense of Copernicus was. Critical reasoning, at the level of a nominal definition, is simply reasoning aimed at the interpretation, evaluation, or self-reflective formulation of arguments; and an argument is a piece of reasoning aiming to justify a conclusion by supporting it with reasons or defending it from objections.[50] And at a deeper level, critical reasoning, at least as practiced by Galileo and as aspired to in this book, is guided by a number of principles. The most relevant ones in this context are some recurrent themes which he both preached and practiced in his Copernican campaign.[51]

For example, Galileo considered *rational-mindedness* as essential; that is, the requirement that, in his own words, "one examine with the utmost severity what the followers of this doctrine know and can advance, and that nothing be granted them unless the strength of their arguments greatly exceeds that of the reasons for the opposite side."[52] He regarded *open-mindedness* as extremely significant and exemplified by the fact that, as he said, "the followers of the new system produce against themselves observations, experiments, and reasons much stronger than those produced by Aristotle, Ptolemy, and other opponents of the same conclusions."[53] And he took *fair-mindedness* to be equally important and formulated it as the principle that "when one presents arguments for the opposite side with the intention of confuting them, they must be explained in the fairest way and not be made out of straw to the disadvantage of the opponent."[54]

Thus, in my view, on the one hand the proposition that Galileo's defense of Copernicus was wrong (i.e., that Galileo's condemnation was right) is almost as false and untenable as the proposition that the earth stands still at the center of the universe. On the other hand, the arguments purporting to justify various Galilean errors or improprieties are in appearance almost as plausible as the anti-Copernican arguments seemed to be in the sixteenth century. But ultimately the anti-Galilean arguments can be shown to be in fact almost as weak and invalid as the anti-Copernican arguments

---

[50] For more details, see Finocchiaro (1980, 27–45, 311–431; 1997a, 309–335; 2005a, 292–326).

[51] For more details, see Finocchiaro (1980, especially 114–115; 1997a, especially 339–341), and Chapter 6.

[52] In Finocchiaro (1989, 85), or Galilei (2008, 165); cf. Favaro 5: 368–369.

[53] In Galilei (1997, 147–148); cf. Favaro 7: 153–154.

[54] In Finocchiaro (1989, 278), or Galilei (2008, 283); cf. Favaro 19: 343. See also Querengo to D'Este, 20 January 1616, in Favaro 12: 226–227; and Motta (1993, 612 n. 55).

were shown to be by Galileo. This is the kind of defense of Galileo and the kind of critical reasoning I am talking about in this book.

Finally, it should be noted that although this book advances and justifies such an overarching, interpretive, and evaluative thesis, the treatment is not meant to be complete or exhaustive, but is rather illustrative and selective. That is, my approach privileges the point of view of critical reasoning; and although it is a crucially important mental activity and human practice, critical reasoning constitutes only a particular orientation, and I am aware that one could select other points of view from which to study the same material. Furthermore, even from the point of view of critical reasoning, such an account could cover much more material and include many more studies than are included in the following chapters. Accordingly, although Galileo's critiques of several anti-Copernican arguments are analyzed, many others are not included; in particular, his answers to Bellarmine's epistemological objection from "saving the appearances"[55] and to Urban's theological objection from divine omnipotence[56] are not explicitly analyzed, but would be especially instructive. Although several important general features of Galileo's defense of Copernicus are discussed, the deeper or finer structure of Galileo's critical reasoning is not articulated; and certainly this limitation is not meant to imply that there is not such a deeper or finer structure.[57] Although many of the major anti-Galilean criticisms or clerical apologias are discussed, no attempt has been made to be completely exhaustive: for example, the fascinating history and the subtle logic of the observational and experimental demonstration of the earth's motion are not explicitly covered in any chapter below, but only briefly discussed in this introduction above and barely mentioned in the chapters below; and it is obvious that I have just scratched the surface of the social criticism of Galileo in the previous section above and in later discussions (Sections 8.15 and 12.5).

Similarly, although I have studied the most recent literature and have taken it into account to some extent, it has not been engaged directly and explicitly, and parts of it may not have been fully assimilated.[58] This is especially true of the work of Annibale Fantoli, whose account of the original trial reaches new heights of comprehensiveness, balance, and interpretive detail; of Francesco Beretta, whose account of the legal and theological aspects attains an unprecedented level of documentation and erudition; of Antonio Beltrán Marí, whose discussion of evaluative

---

[55] See Galilei (2008, 146–167), Favaro (5: 351–370, 12: 171–172), Finocchiaro (1989, 67–86). Cf. Beltrán Marí (2006, 227–248), Blackwell (1991), Feyerabend (1985), Godman (2000), Westfall (1989, 1–30).

[56] In the *Dialogue*: Favaro 7: 485–489, Galilei (1967, 460–465; 1997, 303–308; 2008, 167–171). Cf. Beltrán Marí (2006, 412–437), Besomi and Helbing (1998, 2: 899–902), Bianchi (2000; 2001), Camerota (2004, 406–417), Finocchiaro (1980, 8–12; 1985; 1997a, 306–308), Morpurgo-Tagliabue (1981, 99–107), Speller (2008, especially 143–160, 375–396), Wisan (1984).

[57] This has been analyzed in Finocchiaro (1980, 343–431; 1997a, 309–335; 2005a, 34–45, 65–91).

[58] See, for example, the works listed in the bibliography by such authors as Artigas and Sánchez de Toca, Beltrán Marí, Beretta, Biagioli, Blackwell, Bucciantini, Camerota, Fantoli, Feldhay, Mayaud, McMullin, Renn, Shea and Artigas, and Speller. But cf. my reviews, for example: Finocchiaro (1995; 1997b; 1999b,c; 2001; 2003; 2004; 2005c; 2006a,b,c; 2007b,c; 2008c,d; 2009c,d).

issues achieves an unmatched level of synthesis, sophistication, and argumentation; and of Jules Speller, whose reexamination of the trial displays unsurpassed depth and subtlety of textual analysis as well as unsurpassed acquaintance with and utilization of the history of the historiography of the trial. However, for all such cases, I would claim that such further studies would add only incremental support or nuanced refinement to the *overarching thesis* formulated here, and not alter its substance. If so, the limitation and incompleteness just noted can be viewed more positively as signs of open-endedness and fertility.

Part I
Defending Copernicus

Chapter 1
The Geostatic World View

The intellectual roots of the Galileo affair lie in the controversy that was precipitated by Copernicus's work *On the Revolutions of the Heavenly Spheres* in 1543. The Copernican controversy may be defined in terms of the arguments in favor and against the key hypothesis that the earth moves. To properly understand and assess these arguments, especially the anti-Copernican ones, we need to understand the fundamentals of the world view which the Copernican theory replaced. This chapter aims to sketch these fundamentals.

1.1 Terminology

The world view accepted until the middle of the sixteenth century contained two main theses. One was that the earth is motionless, and so we may speak of the *geostatic* world view, or more simply of geostaticism. The other asserted that the earth is located at the center of the universe, and so we may call it the *geocentric* theory, or more simply geocentrism.

Although we know now that the geocentric view is not true, it corresponds, even today, to everyday observation and common sense intuition; and, although it has this natural appeal, its technical elaboration was the result of arduous work by some of the greatest thinkers of antiquity. Two individuals made contributions which were so important that their names became synonymous with this view of the universe. Aristotle (384–322 B.C.) was a pupil of Plato who lived in Athens during the period of classical Greek civilization; he contributed primarily by elaborating the cosmology, the physics, the general philosophical principles, and the qualitative astronomical ideas of the geostatic world view. Ptolemy lived in Alexandria in the second century A.D., at the end of the Hellenistic phase of Greek culture; he contributed primarily the mathematics and the quantitative details of the astronomical system, forging a synthesis of the observational, mathematical, and theoretical discoveries of the five intervening centuries. Thus, we may also label the old view the *Aristotelian* or the *Ptolemaic* theory of the universe. Furthermore, since the

Aristotelians acquired the nickname of Peripatetics, geocentrism was also traditionally labeled the *Peripatetic* world view.

These remarks suggest that the geostatic world view was not just an astronomical theory, but contained parts belonging to philosophy, physics, and cosmology. The explanation of its details will make this interdisciplinary mixture more obvious.

Moreover, the geocentric view was not a monolithic entity, but rather a theory that underwent 2,000 years of explicit historical development comprising five centuries before and fifteen centuries after the birth of Christ (not to speak of its prehistory). Thus, there are many versions of the theory; for example, Aristotle's and Ptolemy's versions differ not only in emphasis but also in substantive detail. The version expounded here is not a synopsis of any one work, but rather a reconstruction of the most widespread beliefs at the start of the sixteenth century, in a form useful for understanding and evaluating the Copernican Revolution and the Galileo affair.[1]

1.2 Cosmology

Let us begin with the question of the earth's *shape*. The geostatic view held that the earth is a sphere, so that its surface is not flat but round; this is, of course, true. In fact, the arguments proving this fact were known to Aristotle and can be found in his writings. Although uneducated persons or primitive peoples at the time of Aristotle or Galileo may have believed that the earth is flat, scholars had settled the question a long time earlier; thus, it should be clear that the Copernican controversy had nothing to do with the shape of the earth, but with its behavior and location.

Similarly, the maritime voyages and geographical discoveries of Columbus and others at the end of the fifteenth century and thereafter did provide additional confirmation of the earth's spherical shape; but this was only a more direct, experiential proof of the earth's roundness. Those voyages also provided new evidence about the earth's size, structure, and composition; and this evidence affected cosmological and astronomical thought.[2] But this means that the geographical discoveries may have been a factor in the Copernican Revolution, not that the issue was about the earth's shape or size.

The size and shape that became part of the dispute were the size and shape of the whole universe. The old view held that the universe was a sphere much larger than the earth, but of *finite* size, the size being slightly larger than the orbit of the

---

[1] My account has been inspired by Galileo's own *Treatise on the Sphere, or Cosmography*, a short elementary textbook of traditional geostatic astronomy which he wrote and used in the early part of his teaching career, but never published; cf. Favaro 2: 205–255. My account also relies on I.B. Cohen (1960), Kuhn (1957), Toulmin and Goodfield (1961).

[2] For example, Margolis (1987) suggests that learning about the existence of large land masses in the western hemisphere made it implausible to continue to believe that the terrestrial globe consisted essentially of a series of concentric spherical layers of the elements earth, water, air, and fire; as explained

1.2 Cosmology

outermost planet; that is, the distance from the outermost planet to the stars was about the same as the distance between one planet and another. The stars were all at the same distance from the center, attached to the surface of the *stellar sphere*; this sphere (also called *celestial sphere*) enclosed the whole universe, and outside it there was nothing physical. That is, the size and shape of the universe were defined in terms of the size and shape of the sphere on which were attached about six thousand fixed stars visible with the naked eye. This contrasts with the classical modern view that the universe is infinite; space goes on without end; stars are scattered everywhere in infinite space; and so it does not even make sense to speak of the shape, size, or center of the universe.

The finite spherical universe was based on the same set of observations that led to the belief that at the center of the stellar sphere was the motionless earth. This was the phenomenon of *apparent* diurnal motion: the earth feels to be at rest; the whole universe is seen to move daily around the earth in a westward direction; thousands of stars visible with the naked eye at night appear to undergo no change in size or brightness, but seem to be at a fixed distance from us; they appear to move in unison, so that their relative positions remain fixed; they appear to move in circles which are larger for stars lying closer to the equator and smaller for those lying closer to the poles; in short, the stars appear to move as if they were attached to a sphere that rotates daily westward around a motionless earth at the center. Given the plausible principle that what appears to normal observation corresponds to reality, one had the fundamental argument in support of the basic tenets of the geostatic world view.

In the spherical finite universe, position or location or place had an absolute meaning. The geometrical center of the stellar sphere was a definite and unique place, and so was its surface or circumference; and between the center and the circumference, various layers or spherical shells defined various intermediate positions. The part of the universe outside the earth was called *heaven* in general, and to distinguish one heavenly region from another, one spoke of different heavens (in the plural); for example, the stellar sphere was the highest heaven, which meant the most distant one from the earth, and which was also called the *firmament*; whereas the closest heaven was the spherical layer to which the nearest heavenly body (the moon) was attached, and so the lunar sphere or sphere of the moon was the first heaven. Between the lunar and the stellar spheres, six other particular heavens or heavenly spheres were distinguished; one was for the sun, and there was one for each of the other five known planets (Mercury, Venus, Mars, Jupiter, and Saturn). Details about the motion of the planets will be discussed later.

---

below, the latter thesis was an important part of the geostatic view. The connection was that if one believed that land emerged out of the oceans only in a small part of the earth's surface or on one side of the globe, then one could regard this as a minor exception to the rule that the natural place of the element earth is below the element water; but if one knows that there is another continent in the western hemisphere, this implies that land emerges out of the oceans on opposite sides of the earth, and it becomes harder to believe that the normal arrangement is or should be to have the element earth below water.

Here it is important to distinguish a *heavenly sphere* from a *heavenly body*: a heavenly sphere was one of the eight nested spherical layers surrounding the central earth, each of which was the region occupied by a particular heavenly body or group of heavenly bodies, and to each of which these heavenly bodies were respectively attached; whereas a heavenly body was a term referring to either the sun, the moon, a planet, or one of the fixed stars. The two terms are confusing because heavenly bodies were considered to be spherical in shape, and so they were spheres in their own right; but the term heavenly sphere referred only to one of the spheres concentric with the center of the universe to which the (spherical) bodies of the sun, moon, planets, and fixed stars were attached.[3]

The terrestrial region too had its own layered structure. This is related to a threefold meaning for the term *earth*. In saying earlier that the earth is a sphere, I was referring to the terrestrial globe consisting of land and oceans; this globe is a sphere, not in the sense of a perfect sphere, but only approximately because the land is above the water and is full of mountains and valleys; such an approximation is very good because the height of even the tallest mountain is insignificant compared to the earth's radius. But it was only natural to distinguish water from earth, taking the latter term to mean just land, rocks, sand, and minerals; when so understood, earth was obviously only a part of the whole globe. It was also natural to count the air or atmosphere surrounding the globe as part of the terrestrial region; and so by earth one could also mean the whole region of the universe near the terrestrial globe, up to but excluding the moon and the lunar sphere. In short, *earth* had three increasingly broad meanings: it could refer to just the solid part of the terrestrial world; or to the globe consisting of both land and oceans; or to land plus oceans plus atmosphere.

Terminology aside, the substantive point is that the earth (namely, the place where mankind lives) is not a body of uniform composition, but contains three main parts: a solid, a liquid, and a gaseous part. These three parts (earth, water, and air) were labeled *elements* to signify their fundamental importance. In regard to their arrangement, the element earth sinks in water, and so earth must extend to the central inner core of the world and must make up most of what exists below the surface; but most of the surface of the globe is covered with water, and the element water mostly surrounds the element earth. This was expressed theoretically by claiming that the *natural place* of the element earth was a sphere immediately surrounding the center of the universe, and that the natural place of the element water was a spherical layer surrounding the innermost sphere. As for air, simple observation tells us that it surrounds the spheres of the first two elements, and so its natural place was a third sphere surrounding the first two.

There was a fourth terrestrial element, which was called fire; but it required a more roundabout explanation. Just as we see earth sink in water, and water fall

---

[3] This clarification has been made by Rosen (1959; 1992), who stressed that the title of Copernicus's book refers to spheres concentric with the center of the universe and not to heavenly bodies.

1.2 Cosmology

through air, we also see flames shoot upwards through air when something is burning, currents of heat move upwards through air during hot summer days, and smoke generally rise; we also see trapped fire escape upwards in volcanic eruptions. Such observations were taken as evidence that the natural place of fire was a fourth spherical layer above the atmosphere and just below the lunar sphere.

The existence of the element fire was also derived from considerations about basic physical qualities.[4] There were two fundamental pairs of physical opposites: hot and cold, and humid and dry. The element earth was a combination of cold and dry; the element water a combination of cold and humid; the element air a combination of hot and humid; so there had to be a combination of hot and dry, and that was what constituted the element fire.

In summary, from the point of view of location in the geostatic finite universe, there were twelve natural places, each consisting of a sphere or spherical layer with a common center. The four terrestrial spheres were the natural places of the four terrestrial elements (earth, water, air, and fire). The eight heavenly spheres were the natural places of the heavenly bodies; they ranged from the lunar sphere to the stellar sphere, with six intermediate spheres for the sun and the five planets. The stellar sphere enclosed everything, while the earth was at the focus of it all.

As with position, direction had an absolute meaning in the finite universe. There were three basic directions: toward the center of the universe, called *downward*; away from the center of the universe, called *upward*; and around the center of the universe. Thus, one important way of classifying motions was in these cosmological terms: bodies could and did move toward, away from, and around the center of the universe.

Geometrically, motion could be classified as simple or mixed. Simple motion was motion along a simple line. A simple line was defined as a line every part of which is congruent with any other part. Thus, there were only two such lines – circles and straight lines; and there were two types of simple motion – straight and circular motion. Mixed motion was motion which is neither straight nor circular.

Another way to classify motions was in terms of the motions characteristic of the elements, namely, the motions that the elements undergo spontaneously. This categorization was meant to correspond with the two other classifications as follows. For example, earth and water characteristically moved straight downward, while air and fire characteristically moved straight upward. Now, since heavenly spheres and heavenly bodies moved characteristically with circular motion around the center, this meant that they were composed of a fifth element; the term *aether* or *quintessence* was used to refer to this heavenly element.

Finally, another important classification was in terms of the opposition between natural and violent motions. *Violent motion* was motion caused by some external action; *natural motion* was motion which a body underwent because of its nature, so that the cause was internal. For example, the downward motion of earth and water, the upward motion of air and fire, and the circular motion of heavenly

---

[4]Galileo, *Cosmography*, in Favaro 2: 213.

spheres and heavenly bodies were all cases of natural motion; whereas rocks thrown upward, rain blown sideways by the wind, a cart pulled by a horse, and a ship sailing over the sea were all cases of violent motion.

More fundamentally, motion was the opposite of rest. Rest was the natural state of bodies, and so all motion presupposed a force in some way. Natural motion was essentially the motion of a body toward or within its proper place; only when displaced from its proper place by some force, would a terrestrial body engage in natural motion up or down; and only if started by some mover, would a heavenly sphere rotate around the center of the universe, thus carrying its planet or stars in circular motion. On the other hand, violent motion was motion that was not toward the body's proper place, and such motion could happen only by the constant operation of a force.

From what has been said, it is apparent that earth and heaven were very different; indeed this radical difference was enshrined in an idea that needs to be made explicit and that deserves a special label. The key term is the *earth–heaven dichotomy*; but one can equivalently speak of the dichotomy between the earthly or terrestrial or sublunary or elemental region of the universe on the one hand, and the heavenly or celestial or superlunary or aethereal region on the other. We have already seen that one difference between the two regions was location, which was absolute in the finite spherical universe; terrestrial bodies occupied the central region of the universe below the moon, whereas heavenly bodies occupied the outer region from the lunar to the stellar sphere. Similarly, there was another difference in regard to natural motions; earthly bodies moved naturally straight toward or away from the center of the universe, whereas celestial bodies moved circularly around the same center.

We have also seen that the two regions differed in regard to the elements of which bodies were composed. Sublunary bodies were made of earth, water, air, and fire. On the other hand, in the superlunary region things were made of aether, or various concentrations thereof; that is, aether in low concentration made up the heavenly spheres, which were invisible; whereas, aether in a highly concentrated state generated the moon, sun, planets, and stars, which were the visible heavenly bodies.

Just as the natural places and the natural motions of the two regions obviously corresponded to each other, the elements in the two regions also corresponded to the natural places and motions. That is, the natural places and the natural motions of terrestrial bodies could be conceived as defining the essential properties of the terrestrial elements, while the natural places and motions of celestial bodies could be conceived as defining the essential properties of aether. Other differences between earth and heaven could be defined in terms of additional properties of the different elements. For example, whereas superlunary substances had no weight, sublunary bodies obviously did; or to be more exact, whereas aether was weightless, earth and water had weight (and so they were called *heavy bodies*), and air and fire had *levity* (namely the tendency to go up) and so they were called *light bodies*. Moreover, aether was intrinsically luminous, namely capable of giving off its own light; but, earthly elements were dark, namely incapable of emitting their own light; even fire did not emit an inherent light of its own, but only temporarily produced light when it was in the process of escaping from lower regions to move to its natural place just below the lunar sphere.

1.2 Cosmology

Of the many differences between earth and heaven, two deserve special attention: natural motion and susceptibility to qualitative change. Natural motion has always been regarded as an essential or defining characteristic of a physical body. This seems to have remained unchanged even by the Copernican Revolution; from this viewpoint, what changed was the type of natural motion attributed to bodies. Since the geocentric view attributed different natural motions to terrestrial and celestial bodies, it is no surprise that it included the earth–heaven dichotomy.

To elaborate, one must first understand why the geostatic universe was *not* a trichotomy, given that there were three visible kinds of natural motions (namely, downward for earth and water, upward for air and fire, and around the center of the universe for aether). The answer implied by the discussion above is that downward and upward natural motions were both straight, and so they were conceived as two minor subspecies of the same fundamental kind, namely rectilinear motion. Geometrically, there were only two lines with the property that all parts are congruent with any other part – the circle and the straight line; thus, what was common to both upward and downward natural motions (straightness) was more important than what distinguished them (toward and away from the universal center).

However, this geometrical reason was not the only justification for making the essential distinction to be the twofold one between straight and circular natural motions, rather than the threefold one among upward, downward, and around. There was also the cosmological reason that, unlike circular natural motion, straight natural motion could not be perpetual. For once a rock had reached the center of the universe, its nature would make it remain there rather than continue moving past the center, which would constitute upward and thus unnatural motion for the rock; similarly, once a fiery body had reached the region above the terrestrial atmosphere just below the lunar sphere, it had reached its natural place; it had no place to go because to continue moving would bring it into the first heavenly sphere reserved for the aethereal moon where the element fire could not subsist.

Finally, there was a theoretical reason why upward and downward natural motions could belong to the same fundamental region of the universe, but were essentially different from natural circular motion. The theory in question was the theory of change as contrariety, according to which all change derives from contrariety, and no change can exist where there is no contrariety; by contrariety was meant such oppositions as hot and cold, dry and humid. Now, up and down, together with the related pair of light and heavy, was another fundamental contrariety. Thus, a region full of bodies some of which moved naturally downward and some upward was bound to be full of all sorts of qualitative changes; and indeed observation obviously revealed that the terrestrial world is full of birth, growth, decay, generation, destruction, weather and climatic change, and so on. On the other hand, the circular natural motion of the heavenly bodies was thought to have no contrary; consequently, the heavenly region lacked an essential condition for the existence of change. Add to this the belief that the opposition between hot and cold and between dry and humid belonged only within the four terrestrial elements, and one could claim that the region of aether lacked any of the proper conditions for change. And observation confirmed that claim, too, because no physical or organic or chemical changes are easily detected in the heavens, and none were said to have ever been seen; the only essential phenomenon in the heavens was motion, but all heavenly motion was regular and involved the rotation of concentric spheres, which thus remained in place, so that there was not even change of place; what changed was only the relative position of the various bodies attached to these celestial spheres.

This analysis clarifies how natural motion and qualitative change provided the basis for the earth–heaven dichotomy. There were many differences between earth and heaven, but two interrelated differences were crucial: in the terrestrial world bodies moved naturally with straight motion and underwent qualitative change, whereas in the celestial region things moved naturally with circular motion and were not subject to qualitative change.

To summarize the discussion so far, the Aristotelians and Ptolemaics believed that the earth was spherical, motionless, and located at the center of the universe; that the universe was finite, was bounded at the outer limit by the stellar sphere, and was structured into a series of a dozen nested spheres, all inside the stellar sphere and surrounding the central sphere of the solid earth; that there was a fundamental division in the universe between the earthly and the heavenly regions; and that these regions consisted of bodies with very different properties and behavior, such as different natural places, natural motions, elemental composition, and possibilities for qualitative change. Two things must now be added to this general cosmological picture: the details of the physics of the motion of terrestrial bodies and the astronomical details of the heavenly bodies. Let us begin with the former.

1.3 Physics

In the terrestrial region, the natural *state* of bodies was rest. To be more exact, it was rest at the proper place, depending on the elemental composition of the body: at the innermost core for the element earth; just above that for water; above water for air; and above air for fire. This meant that, whereas no cause was sought to explain why a body rested at its proper place, when a body was in motion or at rest outside its proper element, then an explanation was required.

The explanation for why a body was in motion could be that it was going to rest at its proper place; this was the case of natural motion like rocks and rain falling or smoke rising though air. Or the explanation could be that the body was being made to move by an external agent; this was the case of violent motion, for example, a cart pulled by a horse, a boat sailing over the water, rain blown by the wind, or weights being lifted from the ground to the top of a building. But both natural and violent motions required a force; the only difference was that in natural motion the force was internal to the body, whereas in violent motion the force was external. For example, falling bodies fell because of their inherent tendency to go to their natural place if they were not already there; this internal force was termed *gravity* and was measured by the weight of an object. On the other hand, for a sailboat, the wind was obviously the external force, and for a cart, the horse.

Sometimes "violent motion" was equated with "forced motion," but in such cases it was understood that by "forced motion" one meant motion caused by an *external* as distinct from an *internal* force. Since all motion was forced, the term "forced motion" was sometimes regarded as redundant if taken to mean caused motion, and it was found informative only if taken to mean externally-caused motion. That is, the term *force* was ambiguous and could mean either any cause of motion or an external cause of motion; this may generate some confusion, but the context usually clarifies the meaning.

All motion, then, whether natural or violent, was caused by a force, whether internal or external. But another condition was required by all motion, namely resistance. That is, motion was the overcoming of resistance. This was so partly because all space was filled and there was no vacuum or void, so that whenever a body moved it could move only through some medium, be it air, water, oil, molasses, sand, or soil. Even the heavenly region, interplanetary and interstellar space, was not devoid of matter; it was filled with aether.

Moreover, it was argued that if there were no resistance to overcome, then a force (however small) would make a body move instantaneously, namely with infinite speed; and this was an absurdity since it meant that the body would occupy different places at the same time, indeed many different places at the same time. This argument depended on the idea that speed is inversely proportional to resistance, for this idea would provide the justification of why motion without resistance would be instantaneous; that is, not only was resistance required for motion to occur, but motion was correspondingly slower with greater resistance and faster with lesser resistance.

This quantitative relationship between speed and resistance was taken seriously for the extreme case of zero resistance and used as just indicated in the above argument. But the relationship was not taken equally seriously for the other end of the spectrum, namely, for very strong resistance. That is, when the resistance was very strong, rather than saying that a given force would cause some motion, perhaps at very slow speed, it was held that there was a threshold for motion to occur at all; the force had to be sufficient to overcome the resistance in the first place, and if that was the case then the speed was inversely proportional to the resistance. Here, the typical example was that of a single man trying to pull a large ship into dry dock by himself; it is clear that he would not be able to move the ship at all, not even 100 times slower than a team of the 100 men required to accomplish the task.

The relationship between force and speed (when the resistance is constant) could also be expressed quantitatively. The formula was that at constant resistance, the speed is directly proportional to the force. Here the paradigm example was the fall of heavy objects through a fluid like water: heavier objects sink faster than lighter ones, and do so more or less in proportion to their weight; and weight in this case is the (internal) force.

Combining the two relationships one obtained the formula that, given that the force can overcome the resistance, the body moved at a speed which was directly proportional to the force and inversely proportional to the resistance: *speed = constant × (force/resistance)*.[5]

These ideas were plausible and largely in accordance with observation, except for situations like free fall through air and violent projectile motion. For free fall, the Aristotelian theory implied that a lead ball fell much faster than a rock, so that when both were dropped from the same height the lead would reach the ground earlier than the rock; and for a given object, its speed of fall should not increase indefinitely with time because its maximum value depended only on its fixed weight and the fixed resistance of the air. The problem of projectiles involved the motion of such things as arrows shot from bows, and the question was what was the force making them move after the projectiles left the ejector. The Aristotelians were aware of these problems and tried to solve them, but their solutions were found to be increasingly inadequate. The discussion of these problems provided one line of development in the rejection of the old physics and the construction of the new one.[6]

1.4 Astronomy

Let us now go on to the main astronomical details of the geostatic world view. I have already mentioned that the whole universe outside the earth moved around it daily in a westward direction. This phenomenon was called the *diurnal motion* and was regarded as directly observable. The observation is that all heavenly bodies appear to revolve westward around us; this is most obvious for the case of the sun, whose rising in the east and setting in the west causes the cycle of night and day; the moon is also easily seen to do the same; and at night each fixed star appears to follow the same westward trajectory as the previous night.

Since the universe was spherical, this diurnal motion was conceived as the daily rotation of a sphere around a line, called the *axis* of diurnal rotation, which went through the north and the south celestial *poles* (N and S, in Fig. 1.1);[7] this line also intersected the earth's center and two points on its surface, the north and the south poles of the earth (N' and S'). From an observational viewpoint, the celestial poles were the two points in the heavens which appeared to be motionless; the north celestial pole seemed motionless to observers in the earth's northern hemisphere, and the south celestial pole seemed motionless to observers in the southern hemisphere;

---

[5]This formula is a modern way of expressing the combination of the two relationships, and not one which was available to the ancients or medievals; their concepts of quantity and proportion were such that they thought only magnitudes of the same kind could be compared.

[6]For more details on the physics of the geostatic world view, see, for example, I.B. Cohen (1960, 22–35), Franklin (1998), Lindberg (1992, 58–62, 290–307), Toulmin and Goodfield (1961, 93–103).

[7]This diagram is adapted from Kuhn (1957, 31–36) and from the *New Columbia Encyclopedia* (Harris and Levey 1975, 883).

1.4 Astronomy

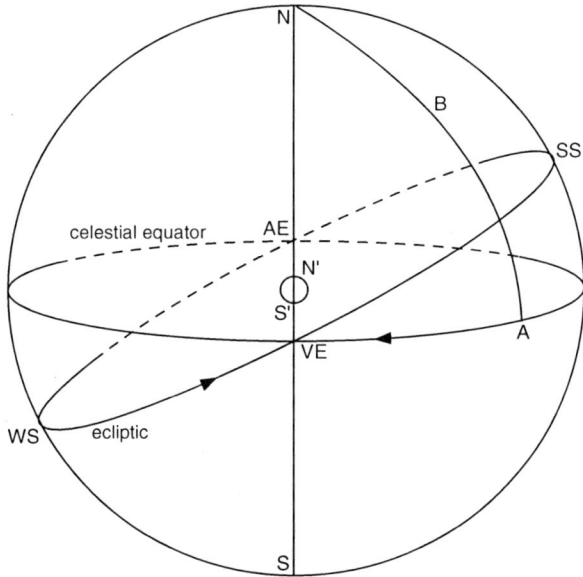

Fig. 1.1 Celestial Sphere in the Geostatic World View

and the circular paths of the fixed stars appeared to be centered at the respective poles. On the surface of the celestial sphere, midway between the poles was a great circle of special importance, called the *celestial equator*; it too had a terrestrial counterpart (the earth's equator), which could be defined as the intersection of the plane of the celestial equator with the earth's surface, or as the great circle on the earth's surface halfway between the north and south terrestrial poles.

One reason for the importance of the celestial poles and equator was that they yielded a fixed frame of reference to define the position of the heavenly bodies, and correspondingly the position of points on the earth's surface. One could measure the angular position of a star north or south of the celestial equator, which was called *declination* (AB, in Fig. 1.1); similarly, the angular distance from the terrestrial equator of a point on the earth's surface was called *latitude*. For each declination or latitude one could conceive a plane parallel to the equator whose intersection with the surface of the two spheres generated circles (called *parallels*) that became smaller as one moved toward a pole. The east–west position of a star (B) required first the drawing of a *meridian*, namely, a great circle (partially shown as NBA) through the star and the poles; then one would measure the *ascension*, namely, the angular distance from this meridian to some particular meridian (for example, A to VE); it was analogous for positions on the earth's surface, except that this east–west angular distance was called *longitude*.

There were two kinds of heavenly bodies, called *fixed stars* and *wandering stars*. A fixed star was a heavenly body that moved daily around the earth in such a way that its position relative to most other heavenly bodies did not change; for example, its declination remained constant, and so did its angular distance from any other fixed star. A wandering star was a heavenly body that not only moved daily around the earth but also changed its position relative to other heavenly bodies; that is, the wandering stars were those heavenly bodies which, besides undergoing the diurnal motion, appeared to move in other ways (to be discussed presently). There were only seven wandering stars, which were also called planets; indeed the word *planet* originally meant literally "wandering star." Because wandering stars were often called simply planets, fixed stars were often called simply stars.[8]

Thousands of fixed stars were visible on a clear night with the naked eye; they were catalogued both in terms of apparent brightness (called *magnitude*) and in terms of shapes or patterns formed by groups of stars close to each other (called *constellations*). The naked eye could be trained to distinguish six magnitudes; stars of the first magnitude were the brightest, and those of the sixth magnitude were the faintest. The brightest star was named Sirius or the Dog Star; it was located near the equator and was part of the constellation of Canis Major. A star of the second magnitude was especially important because it was so close to the north celestial pole that, for practical purposes (such as navigation), it could be regarded to be the pole; it was called Polaris or the North Star and was part of the constellation of Ursa Minor.

Both the sun and moon were planets because they moved ("wandered") in relation to the fixed stars. Because of their brilliance and their relatively large size, they were called the two *luminaries*. The other known planets were named Mercury, Venus, Mars, Jupiter, and Saturn. We now know that there are at least two other planets (Neptune and Uranus) circling the sun in orbits beyond Saturn, but they were unknown not only to the ancients but even to Copernicus and Galileo; so they played no role in the Copernican Revolution.

The most basic point about the planets was that, out of the thousands of heavenly bodies, there were seven which circled the earth westward once a day like all others, but did not do so in unison with them; these seven bodies also revolved slowly eastward, so that from day to day their position shifted. That is, whereas a fixed star revolved around the earth in such a way that after 24 hours it returned to the same position it had before, after 24 hours a planet did not quite return to the earlier position but had fallen behind somewhat, being located slightly eastward. This can be seen most easily for the case of the moon by observing its position on succeeding nights at midnight; relative to the fixed stars, it appears to move eastward. The planets seemed to behave as if their motion were a combination of two circular motions in opposite directions: they circled the motionless earth westward with the universal diurnal motion, and in addition they simultaneously moved slowly eastward.

The planets moved eastward at different rates. For example, the moon took about a month to return to the same position relative to the fixed stars; the sun took 1 year; Mars about 2 years; and Saturn about 29 years. Thus, the planets moved not only relative to the earth and the fixed stars, but also relative to each other; each planet had its own distinctive motion, besides the universal diurnal motion. Since

---

[8] That is, though one broad meaning of the word *star* was synonymous with the term *heavenly body*, one narrow meaning of *star* was identical to the term *fixed star*; the point is that the term *fixed* was often dropped when the context made it clear that one was indeed referring to fixed stars.

1.4 Astronomy

the westward diurnal motion was common to all, when one spoke of planetary motions one usually referred to the distinctive individual motions of the planets. Note that, while all the individual planetary motions were eastward, this direction was opposite to that of the diurnal rotation, which was westward.

The planetary motion of the moon, which took about a month, was the most readily observable one since it was connected with the cycle of its phases; a full moon is easily seen and the period from one full moon to the next is an obvious unit of time that can be used as the basis for a calendar. The planetary motion of the sun was also easy to observe since it is related to the cycle of the seasons of the year; hence, it was called the *annual motion*. Because of its crucial importance, I shall discuss it in some detail.

Everyone can easily observe that in the course of a year the sun slowly moves in a north–south direction. Sometimes it rises near due east and sets near due west, which is to say that it is seen on the celestial equator; this happens around March 21, which is the time of the *vernal equinox*; it also occurs around September 23, the time of the *autumnal equinox*. Sometimes it rises and sets about $23.5°$ north of due east and due west respectively (namely north of the celestial equator); this happens in the northern hemisphere around June 22, the time of the *summer solstice*. Sometimes it rises and sets about $23.5°$ south of due east and due west respectively (south of the celestial equator); this occurs in the northern hemisphere around December 22, the time of the *winter solstice*. One can also observe from a given location on the earth's surface the elevation above the horizon of the sun at noon; in the course of a year this elevation changes daily and ranges about $47°$, being highest around June 22 and lowest around December 22 (in the northern hemisphere).

This annual northward and southward motion of the sun indicates that its position relative to the fixed stars changes along a north and south direction since, as stated earlier, the fixed stars remain at a constant distance from the celestial equator. In short, the declination of the sun changes by about $47°$ during a year, while the declination of a fixed star does not change; so this north–south motion of the sun is part of its wandering among the fixed stars.

Though this apparent solar motion was the one most easily observed, it was not exactly identical to its planetary motion mentioned earlier; for the latter was eastward, whereas the former was northward and southward. The two were related as follows. The sun's eastward revolution in its planetary orbit did not take place in the plane of the celestial equator but in a plane inclined to it by $23.5°$. The point was that the sun's motion among the fixed stars was not *exactly* eastward, but *mostly* eastward; its trajectory was slanted north and south. The sun moved eastward and southward for 6 months, and eastward and northward for the other 6 months. This can be made clear by means of a diagram, but before explaining it, let us mention a simple kind of observation to detect the sun's eastward motion.

The difficulty in observing the sun's eastward motion among the fixed stars stems from the fact that they cannot be seen when it is visible. What one can do is to observe some star located near the celestial equator and rising in the east soon after the sun sets in the west; this means that the sun and star are diametrically opposed, or about $180°$ apart. Observe the position of the same star just after sunset about a month later; it will be seen to be not just rising, but high in the sky and about 30° west of its previous position; that means that the sun is now only about 150° away, which is to say that sun has moved eastward about 30° closer to the star. About 6 months after the first observation, the star will appear and immediately set in the west just after sunset. Twelve months later, the star will again rise in the east when the sun sets in the west.[9]

The planetary motion of the sun may be pictured as in Fig. 1.1. Imagine a large sphere surrounding a small one at its center, and let the small sphere represent the earth and the large one the stellar sphere. On the stellar sphere, picture a great circle lying in an horizontal plane to represent the celestial equator, and also a vertical line perpendicular to the equatorial plane and going through the center to represent the axis of diurnal rotation; this axis intersects the stellar sphere at two points, the north celestial pole (N) and the south celestial pole (S). Now imagine looking at the large sphere from above the north celestial pole, and picture the large sphere rotating clockwise around the motionless small central sphere to represent the westward diurnal rotation of the stellar sphere around the earth. Next, imagine a great circle on the stellar sphere in a plane cutting the equatorial one at an angle of 23.5°, to represent the sun's geocentric orbit projected onto the stellar sphere; in accordance with standard terminology, let us use the term *ecliptic* to refer to this actual orbit, or the corresponding great circle on the stellar sphere, or the plane on which they both lie. The intersection of the ecliptic and the equator on the stellar sphere defines two special points, called the vernal equinox (VE) and the autumnal equinox (AE); and halfway around the ecliptic between the equinoxes are two other special points, the summer solstice (SS) at the northern end, and the winter solstice (WS) at the southern end; these four points thus divide the ecliptic circle into four equal quadrants. Now, imagine the sun moving counterclockwise around the ecliptic at a rate that makes it traverse the whole circumference in 1 year; then the sun will be at VE around March 21, at SS around June 22, at AE around September 23, and at WS around December 22.

Let us now combine the clockwise rotation of the whole stellar sphere with the counterclockwise revolution of the sun along the ecliptic. The result is that the sun in reality moved in a helical path which in 1 year looped clockwise around the earth about 365 times (days of the year), but which in any one day corresponded almost but not quite to one of the parallels on the stellar sphere. I say "almost but not quite" first because the parallel circle was not completely traversed by the sun, but fell short by about 1° (1/360 of a circle, which approximately equals 1/365 of a year); and second because the end of the daily path rises northward or drops southward relative to the beginning of the same daily path by 1/4 of a degree on the average (namely 23.5° every 3 months, or 23.5° every 90 days).

The ecliptic was important not only because it represented the yearly eastward path of the sun among the stars, but also because it was used to define a frame of reference, distinct from the equatorial one mentioned earlier. For example, one could draw a line perpendicular to the center of the ecliptic (called the axis of the

---

[9]This example is adapted from Galileo, *Cosmography*, in Favaro 2: 214.

ecliptic); one could then speak of the poles of the ecliptic as the points where its axis intersected the celestial sphere; one could define the position of a star in terms of its angular distance from the ecliptic toward one of the poles; and one could also plot the position of a body in terms of east–west position along the ecliptic.

This ecliptic frame of reference was especially important for the other six planets because they are never seen to wander much away from the ecliptic; that is, planets are always observed to be somewhere inside a narrow belt extending 8° above and below the ecliptic. This was the result of the fact that the individual circular paths of the planets took place in planes which, while not identical with the ecliptic, intersected it at small angles no larger than 8°. This narrow belt on the stellar sphere along which the planets revolved was called the *zodiac*. It was subdivided into 12 equal parts of 30° each, and each part happened to be the location of a group of stars that seemed to be arranged into a distinct pattern. These twelve patterns were the constellations of the zodiac and were named: Aquarius, Pisces, Aries, Taurus, Gemini, Cancer, Leo, Virgo, Libra, Scorpio, Sagittarius, and Capricorn. The sun, moon, and other planets were at all times found somewhere in one of these constellations, and they moved from one constellation to the next in the order just listed. This order corresponded to what we have called an eastward direction (from the viewpoint of terrestrial observation) or counterclockwise (in connection with the pictorial diagram just described); the key point, however, was that the order of the signs of the zodiac was a direction of motion opposite to that of the diurnal rotation.

When projected onto the stellar sphere, the eastward motion of the planets could be described in terms of great circles on the surface of that sphere, all of which were within the zodiac and intersected one another at small angles. But the planets were not believed to be attached to the stellar sphere like the fixed stars; unlike the fixed stars, the planets were not regarded to be equidistant from the earth. The fact that the planets appeared to move relative to the fixed stars and that this motion took place at different rates for different planets implied that each planet was attached to its own sphere which rotated eastward at its own rate while being carried westward daily by the diurnal rotation of the stellar sphere. There was no direct way to measure the sizes of the various planetary spheres or orbits (namely the distances of the various planets from the earth), but the relative determination was done on the basis of the length of time required for a given planet to complete one circular journey among the stars. The principle used was that the bigger ones of these nested planetary spheres rotated at slower rates, and the smaller ones at faster rates; that is, the bigger the orbit, the slower the period of revolution. This principle was combined with the observation that the periods of revolution ranged from 1 month for the moon to 1 year for the sun and 29 years for Saturn. The result was that in order of increasing distance from the earth, the planets were most commonly arranged as follows: moon, Mercury, Venus, sun, Mars, Jupiter, and Saturn. Thus, as mentioned earlier, between the stellar sphere and the earth, there were seven other nested spheres, whose rotation carried the corresponding planets in their own individual eastward orbits, while they were all being carried in a westward daily whirl by the diurnal rotation of the stellar sphere.

One last topic about the planets must now be discussed to complete but also to complicate this picture of the geocentric universe. Careful observation revealed that no planet moved at a uniform rate in its orbit, but that its speed appeared to vary. Moreover, though the sun and moon always moved eastward in their planetary revolutions, periodically the other five planets were seen to slow down, stop, and reverse course and briefly move westward relative to the fixed stars; this reversed movement was called *retrogression* or *retrograde* planetary motion. Finally, during retrogression planets appeared brighter, as if they were nearer the earth than at other times. These observations meant that a planet could not be simply attached to a rotating heavenly sphere, for in that case neither the distance nor the direction of revolution nor the speed should change. The device most commonly used to explain retrograde motion and variation in brightness and speed was a mechanism consisting of *deferents* and *epicycles*. To understand this, it is best at first to disregard the nested spheres. Let us refer to Fig. 1.2.

A deferent was defined as a geocentric circle whose circumference (ABCD) rotated around the earth (E). An epicycle was defined as a circle (FGHI) whose center (A) lay on the circumference of the deferent, and whose circumference rotated in the same direction as the deferent. The planet was located on the circumference of the epicycle. Thus, when the rotation of the epicycle carried the planet on the far side (F) of the epicycle from the earth, its distance was the sum of the radii of the deferent and the epicycle; whereas when the epicyclic rotation carried the planet on the near side

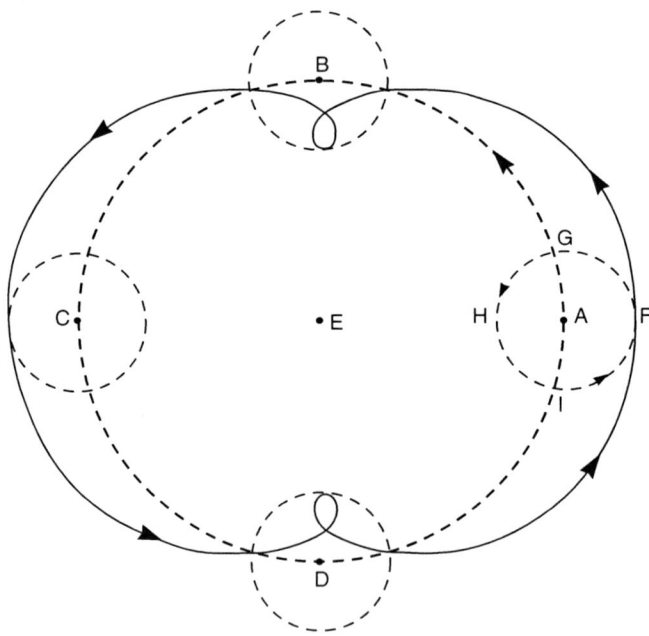

Fig. 1.2 Ptolemaic Explanation of Retrograde Planetary Motion

1.4 Astronomy

(H) of the epicycle from the earth, its distance was the difference between the two radii. Thus, in its geocentric revolution, the distance of the planet from the earth changed by an amount equal to the diameter of the epicycle. This difference accounted for the variation in brightness.

Moreover, the planet's motion was the result of its motion along the epicycle and the motion of the center of the epicycle along the deferent. Thus, when the planet was on the far side (F), its speed was the sum of the deferent speed and the epicycle speed; then its speed was faster than its average speed. But when the planet was on the near side (H) of the deferent, its speed was the difference between the two; in this case, if the epicycle speed was greater than that of the deferent, the planet appeared to move backwards (clockwise or westward). Retrograde planetary motion then resulted.

For each planet, the relative sizes of deferent and epicycle and their relative rates of rotation could be adjusted so that their combination yielded mathematically the observed details about retrogression and changes in brightness and speed. For example, if the planet was observed to retrogress twice while revolving through its complete orbit once, then the epicycle was assumed to rotate twice as fast as the deferent; this yielded a path which in reality was looped (Fig. 1.2); but from the earth (E) the loop was not seen, and instead the planet would appear brighter and retrogressing near B and D.

This framework of deferents and epicycles could also be combined with the framework of nested spheres mentioned earlier. One could conceive a planetary sphere not as a mere spherical surface on which the planet was attached, but as a spherical shell of considerable thickness, having an inner and an outer surface; the thickness of the shell could then be adjusted so that the inner surface of the sphere corresponded to the minimum distance between the planet and the earth, and the outer surface to the maximum distance; and within the shell, the planet could perform its epicyclic motion while the whole shell rotated around the earth.

The framework of deferents and epicycles was a very powerful instrument for the analysis of planetary motion. There was much more that an astronomer could do besides adjusting the relative sizes and speeds of a deferent and its epicycle. For example, one could add a second epicycle on the first epicycle; one could make the center of the deferent different from the center of the earth, in which case the deferent was called an *eccentric*; and one could even make the center of the deferent move in some way, perhaps in a small circle around the earth's center. For many centuries before Copernicus, such calculations, adjustments, and refinements involving deferents, epicycles, and eccentrics constituted the primary theoretical and mathematical task of planetary astronomy. This enhanced the power of the theory, but it also rendered the whole system increasingly more complicated. Moreover, the relationship between the framework of deferents, epicycles, and eccentrics and the system of nested spheres became increasingly unclear. However, the geostatic–geocentric system of Aristotle, as elaborated by the Ptolemaic system of deferents, epicycles, and eccentrics yielded plausible explanations and useful predictions; in short, it worked. For about 2,000 years no one was able to devise anything better, or even equally good. All this changed with Copernicus.

Chapter 2
The Copernican Controversy

If the geostatic world view provides the intellectual background for the Copernican controversy, and the latter provides the conceptual background for the Galileo affair, then the exposition of the geostatic world view (Chapter 1) must now be followed by an exposition of the Copernican controversy. In accordance with our stress on critical reasoning, this controversy will be described in terms of the arguments in favor and the arguments against the key claim of the Copernican system, the thesis that the earth moves. Now, it is obvious that this key Copernican thesis is the contradictory of the main Ptolemaic claim that the earth stands still. Thus, as a first approximation, the arguments in favor of Copernicanism are also arguments against the geostatic world view, and the arguments against the Copernican system are also arguments in favor of the geostatic thesis. Moreover, if we define an *objection* to some position to be an argument against that position, then we can also say that the focus of this chapter will be the arguments and the objections on both sides of the Copernican controversy.

2.1 Copernicus's Innovation

In 1543 Copernicus published his epoch-making book *On the Revolutions of the Heavenly Spheres*. In it he elaborated the details of a theory that may be sketched as follows.

The earth was still spherical and the universe was still finite and spherical; the fixed stars were still attached to the stellar sphere and equidistant from the center. But the stellar sphere was motionless and did not revolve around the earth with westward diurnal rotation; instead, the diurnal rotation belonged to the earth, though its direction was eastward in order to result in the observational appearance of the whole universe rotating westward. To stress this feature of the earth's rotation, the Copernican world view may be labeled *geokinetic*.

The earth was also given a second motion, an orbital revolution around the sun with a period of 1 year, and also in an eastward direction. That is, the annual motion was shifted from the sun to the earth (with the direction remaining unchanged), thus making the earth a planet, rather than the sun. This terrestrial orbital revolution meant that the earth was located off center, the center being instead the sun; to stress this feature, the Copernican world view may be labeled *heliocentric*.

The moon remained a body which circles the earth eastward once a month. The other five planets continued to be planets, but their orbits were centered on the sun rather than on the earth. Around the sun there thus revolved six planets in the order: Mercury, Venus, Earth, Mars, Jupiter, and Saturn.

What Copernicus did was to update an idea that had been advanced in various forms by the Pythagoreans, by Aristarchus, and by other astronomers in ancient Greece, but had been almost universally rejected; that is, the idea that the earth moves by rotating on its own axis daily, and by revolving around the sun once a year. In a sense, Copernicus's accomplishment was to give a *new* argument in support of this *old* idea that had been considered and rejected earlier. His theory was not primarily based on new observational evidence, but was essentially a novel and detailed reinterpretation of available data. He demonstrated in quantitative detail that the *known* facts about the motions of the heavenly bodies (especially the planets) could be explained *more simply* and *more coherently* if the sun rather than the earth is assumed to be at the center, and the earth is taken to be the third planet circling the sun yearly and spinning daily on its own axis.

For example, there are thousands fewer moving parts in the geokinetic system since the apparent daily motion of all heavenly bodies around the earth is explained by the earth's axial rotation, and thus there is only one thing moving daily (the earth), rather than thousands of stars. Thus, insofar as simplicity depends on the number of moving parts, the geokinetic system is simpler than the geostatic.

A similar point can be made in regard to the number of directions of motion. Fewer are needed in the Copernican than in the Ptolemaic system: in the geostatic system there are *two* opposite directions, but in the geokinetic system all bodies rotate or revolve in the same direction. In the geostatic system, while all the heavenly bodies revolved around the earth with the diurnal motion from *east to west*, the seven planets (moon, Mercury, Venus, sun, Mars, Jupiter, and Saturn) also simultaneously revolved around it from *west to east*, each in a different period of time. But in the geokinetic system there is only one direction of motion since, for example, if the apparent diurnal motion from east to west is explained by attributing to the earth an axial rotation, then the direction of the latter has to be reversed (west to east); whereas, if the apparent annual motion of the sun from west to east is explained by attributing to the earth an orbital revolution around the sun, then the same direction has to be retained.[1]

A third reason for the greater simplicity of the geokinetic system was that it had a single uniform pattern in the relationship between size of orbit and period of revolution, namely, the bigger the orbit, the slower the period of revolution. But in the geostatic system this pattern was only partially valid because, although the planetary motions did occur in accordance with it, the diurnal rotation of the stellar sphere

---

[1] For more details, see Kuhn (1957, 160–165), Finocchiaro (1997a, 132–133, 245).

2.1 Copernicus's Innovation

broke the uniform pattern insofar as it was the largest sphere and yet revolved at the fastest rate.

With regard to explanatory coherence, this concept means the ability to explain many phenomena in detail by means of one's basic principles, without having to add artificial and ad hoc assumptions. For Copernicus, the basic principles referred to the earth's motion, while the explained phenomena were primarily the various known facts about the motions and the orbits of the planets. But in the geostatic system, the thesis of a motionless central earth had to be combined with a whole series of unrelated assumptions in order to explain what is observed to happen.[2]

The best example of explanatory coherence is the Copernican explanation of retrograde planetary motion and of planetary variation in brightness. These phenomena are explained without the Ptolemaic ad hoc postulation and construction of epicycles, as needed to fit the observations; instead they are direct consequences of the relative motion between the earth and the other planets. The explanation of retrograde motion is that when the earth and another planet reach points in their orbits which are on the same side of the sun and are thus at the minimum distance from one another, their different speeds make the other planet seem to move backward (westward). If the other planet is a superior one (namely, one with an orbit larger than the earth's), this happens when the planet's apparent position on the celestial sphere is opposite to that of the sun; the earth's faster eastward motion leaves behind the other planet, which thus appears to move westward relative to the fixed stars. For example, in Fig. 2.1, while the earth moves through points B–I, K–M along the smaller orbit, the superior planet moves along its bigger orbit through a corresponding set of points comprising a shorter distance due to its slower speed; but the apparent position of the planet against the background of the fixed stars changes in the order P–S, S–W, W–Z.[3] If the other planet is an inferior one (namely, one with a smaller orbit), the phenomenon occurs when its apparent position on the celestial sphere is near that of the sun; the earth's slower motion enables the other planet to overtake the earth relative to the fixed stars, thus generating the appearance that the planet is moving in a direction opposite to that of the sun; since the latter always appears to move eastward relative to the fixed stars, the planet appears to move westward.

Despite these advantages of the geokinetic theory from the points of view of simplicity and explanatory coherence, as a *proof* of the earth's motion, Copernicus's argument was far from conclusive. Notice first that his argument is a hypothetical one. That is, it is based on the claim that *if* the earth were in motion *then* the observed phenomena would result; but from this it does not follow with logical necessity that the earth is in motion; all we would be entitled to infer is that the earth's motion offers an explanation of observed facts. Given the greater simplicity and coherence just mentioned, we could add that the earth's motion offered a simpler and more

---

[2] For more details on explanatory coherence, see Lakatos and Zahar (1975, 368–381), Millman (1976), Thomason (1992). But note that their terminology is different.
[3] This diagram is adapted from a passage in Galileo's *Dialogue*; cf. Favaro 7: 371 and Galilei (1967, 343).

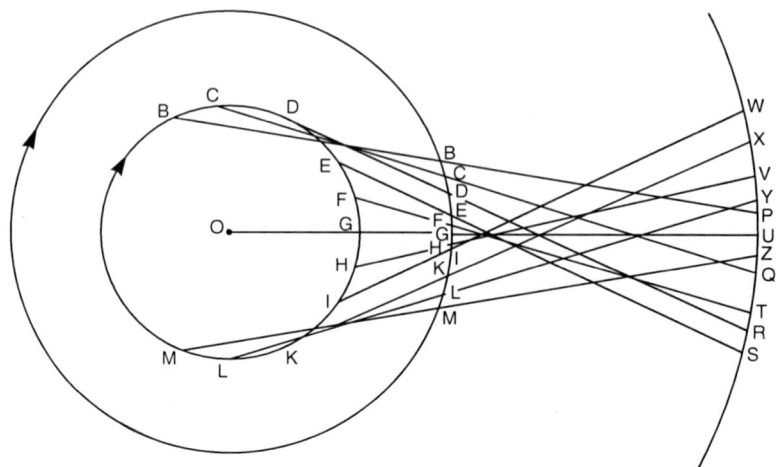

Fig. 2.1 Copernican Explanation of Retrograde Planetary Motion

coherent explanation of heavenly phenomena. This does provide *two reasons* for preferring the geokinetic idea, but they are not decisive reasons. They would be decisive only in the absence of reasons for rejecting the idea. In short, one has to look at counterarguments, and there were plenty of them.

2.2 The Anti-Copernican Arguments

The arguments against the earth's motion can be classified into various groups, depending on the branch of learning or type of principle from which they stemmed. In fact, these objections reflected the various traditional beliefs which contradicted or seemed to contradict the Copernican system. Thus, there were epistemological, philosophical, theological, religious, physical, mechanical, astronomical, and empirical objections.

Let us also distinguish two sets of issues, each applying to both the ancient and the Copernican view. The ancient view contained two main parts, the geostatic thesis that the earth is motionless, and the geocentric thesis that the earth is located at the center of the universe. These are independent of each other because there is no contradiction in holding that the earth is devoid of any motion but is located slightly off the center of the universe; this is what some ancient thinkers speculated in order to account for some of the specific details of the apparent motions of the heavenly bodies. Conversely, there is no incompatibility in holding that the earth is located at the center of the universe but moves by performing a simple daily axial rotation; in fact, this is precisely the kind of compromise position which other thinkers conceived of. Similarly, in regard to the Copernican view, the earth's daily axial rotation and its annual orbital revolution are distinct, in the sense that one could admit terrestrial rotation but deny orbital revolution; this could be done by letting the earth rotate at the center of the universe. On the other hand, if one lets the earth revolve annually around the sun then it would be absurd to deny terrestrial rotation because this would mean that the apparent daily motion of the heavens was actual, and thus we would have a situation where the earth was going around the sun once a year while the whole universe was revolving around the earth once a day.

It is instructive to begin with the empirical objections to underscore the fact that the opposition to Copernicanism was neither all mindless nor simply religious; however, to set the stage for the empirical details, it is best to begin with an argument which is empirical in the sense of involving observation and sense experience, but which does so in such a way that what we really have is an epistemological objection.

The argument was aptly called the *objection from the deception of the senses*. To understand the deception involved, note that Copernicus did not claim that he could either feel, see, or otherwise perceive the earth's motion by means of the senses. Like everyone else, Copernicus's senses told him that the earth is at rest. Therefore, if his theory were true, then the human senses would not be reporting the truth, or would be lying to us. But it was regarded as absurd that the senses should deceive us about such a basic phenomenon as the state of rest or motion of the terrestrial globe on which we live. In other words, the geokinetic theory seemed to be in flat contradiction with direct sense experience and to violate the fundamental epistemological principle that under normal conditions the senses are reliable and provide us with one of the best instruments to learn the truth about reality.[4]

One could begin trying to answer this difficulty by saying that deceptions of the senses are neither unknown nor uncommon, as shown, for example, by the straight stick half immersed in water that appears bent, or by the shore appearing to move away from a ship to an observer standing on the ship and looking at the shore. However, the difference was that these perceptual illusions involve relatively minor and secondary experiences, whereas to live all one's life on a moving globe without noticing it would be a gigantic and radical deception; moreover, it was added, the former illusions are corrigible, since we have other ways of discovering what really happens, whereas there is no way of correcting the perception of the earth being at rest. This difficulty may be labeled an epistemological objection because the real issue is whether the earth's motion ought to be (directly) observable, and whether the human senses ought to be capable of directly revealing the fundamental features of physical reality.

This general empirical objection is in a sense the reverse side of the coin of the fundamental advantage of the geostatic system. (The same applies, of course, to all the other anti-geokinetic objections, which may thus be easily turned into pro-geostatic arguments.) The most basic and important argument in favor of the geostatic view was taken from direct observation, which testifies to the correctness of the geostatic thesis: our visual experience reveals that the heavenly bodies move around the earth every day, a point which is most easily observable for the sun, whose rising in the east

---

[4] This argument is stated and criticized in Favaro 7: 273–281, Galilei (1967, 247–256; 1997, 212–220).

and setting in the west generates the cycle of night and day; further, according to our kinesthetic sense the earth is felt to be at rest; the argument here was simply that the earth must be standing still because our sense experience shows this.

The other empirical objections to Copernicanism were more specific, and were based primarily on effects in the heavens which ought to be observed in a Copernican universe, but which in fact were not. These specific empirical difficulties may therefore be also called the astronomical objections.

The *objection from the earth–heaven dichotomy* argued that if Copernicus were right then the earth would share many physical properties with the other heavenly bodies, especially the planets, since the earth would itself be a planet, the third one circling the sun. However, as we saw in the last chapter, it was widely believed that whereas the heavenly bodies were weightless, luminous, changeless, and made of the element aether, the earth was dark, subject to constant changes, and made of rocks, water, and air (which had positive or negative weight). Now, before the invention of the telescope this belief had considerable empirical support.

The *appearance of the planet Venus* was the basis of another objection.[5] For, if the Copernican system were correct, then this planet should exhibit phases similar to those of the moon but with a different period; however, none were visible (before the telescope). The reason why Venus would have to show such phases stems from the fact that in the Copernican system it is the second planet circling the sun, the earth is the third, and these two planets have different periods of revolution. Therefore, the relative positions of the sun, Venus, and the earth would be changing periodically and so would the amount of Venus's surface visible from the earth: when Venus is on the far side of the sun from the earth, its entire hemisphere lit by the sun is visible from the earth, and the planet should appear as a disk full of light (like a full moon, though much smaller); when Venus comes between the sun and the earth, none of its hemisphere lit by the sun is visible from the earth, and the planet would be invisible (as in the case of a new moon); and at intermediate locations, when the three bodies are so positioned that the line connecting them forms a noticeable angle, then different amounts would be visible, giving Venus an appearance ranging from nearly fully lit, to half lit, to a crescent shape.

The *apparent brightness and size of the planet Mars* involved another problematic issue.[6] In the Copernican system this planet revolves in the next outer orbit (M1–M2 in Fig. 2.2) after the earth's orbit (E1–E2). Since they also revolve at different rates, they are relatively close to each other when their orbital revolutions align both on the same side of the sun (E1 and M1, or E2 and M2), and relatively far when they are on opposite sides of the sun (E1 and M2, or E2 and M1). This variation in distance between the earth and Mars is considerable; according to some estimates,

---

[5]Cf. Favaro 7: 362, Galilei (1967, 334–335; 1997, 236–237; 2008, 243–244).
[6]Cf. Favaro 7: 357–60, Galilei (1967, 334; 1997, 235–236; 2008, 243).

2.2 The Anti-Copernican Arguments

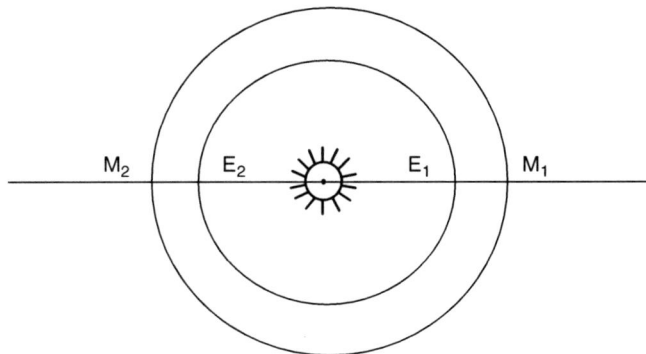

Fig. 2.2 Copernican Variations in Appearance of Mars

it was supposed to be eightfold.[7] This change in distance would cause a corresponding variation in the apparent size of Mars when seen from the earth, and an even greater change in brightness, since the intensity of light varies as the square of the distance. Now, the difficulty was that, although Mars did indeed exhibit a noticeable change in brightness with periodic regularity, this change was not nearly as much as it should be; further, there was practically no variation in apparent size (before the telescope).[8]

The last of the empirical astronomical arguments to be discussed here was based on the fact that observation revealed no change in the *apparent position of the fixed stars*;[9] this is commonly known as the objection from stellar *parallax*, a term that denotes a change in the apparent position of an observed object due to a change in the location of the observer. At its simplest level, the apparent position of a star may be thought of as its location on the celestial sphere, which in a sense is its position relative to all the other stars (also located on that sphere); or, from the viewpoint of the Copernican system it may be conceived as measured by the angular position of the star above the plane of the earth's orbit (equivalent to the plane of the ecliptic). Now, if the earth were revolving around the sun, then in the course of a year its position in space would change by a considerable amount defined by the size of the earth's orbit; therefore, a terrestrial observer looking at the same star at 6-months intervals would be observing it from different positions, the difference being a distance equal to the diameter of the earth's orbit; consequently, the same star should

---

[7]Cf. Favaro 7: 349, Galilei (1967, 321; 1997, 225–226; 2008, 236).

[8]The reason why the observed variation in apparent brightness of Mars presented a serious difficulty for the Copernican system but not for the Ptolemaic system was that in the latter the relevant quantities (distance, epicycle, and so on) could be adjusted to correspond to the actual observations, whereas in the former the variation could be derived from other elements of the system, because of its greater coherence mentioned above.

[9]Cf. Favaro (7: 385–386, 404–407), Galilei (1967, 358–359, 377–380; 1997, 247–248, 264–267).

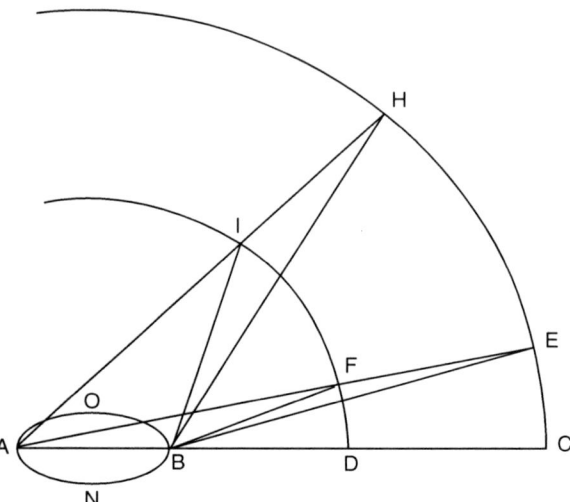

Fig. 2.3 Copernican Parallax of Fixed Stars

appear as having shifted its position either on the celestial sphere, or in terms of its angular distance above the plane of the earth's orbit. It follows that if Copernicanism were correct, we should be able to see stellar parallaxes with a periodic regularity of 1 year. However, none were observed.

For example, in Fig. 2.3,[10] let ANBO represent the earth's orbit; line AB a diameter of the earth's orbit; ABC a line in the plane of that orbit; CEH a portion of the celestial sphere; H a fixed star whose position, when observed from point B, may be defined in terms of the angle HBC. However, 6 month later, when star H is observed from point A in the earth's orbit, the star's position may be defined in terms of angle HAC; and this angle (HAC) is smaller than the previous one (HBC). That is, when observed at 6-months intervals, the same star H would appear to shift its position, appearing sometimes higher and sometimes lower above the plane of the ecliptic.

It should be noted that the first three of these empirical astronomical objections were not answered until Galileo's telescopic discoveries, and that the stellar parallax was not detected until much later, by Bessel in 1838. In fact, the magnitude of parallax varies inversely as the distance of the observed object; and the stars are so far away that their parallax is exceedingly small; so for about two centuries telescopes were not sufficiently powerful the make the fine discriminations required. One may then begin to sympathize with Copernicus's contemporaries, who found his idea very hard to accept. However, as mentioned earlier, there were many other reasons for their opposition. The next group may be labeled *mechanical* or *physical*, in the sense that they are based directly or indirectly on a number of principles of the

---

[10] This figure is taken from Galileo's *Dialogue*, in Favaro 7: 407, Galilei (1967, 380; 1997, 268).

2.2 The Anti-Copernican Arguments

branch of physics which today we call mechanics, and which studies how bodies move. We shall first mention four objections which hinge indirectly (though crucially) on the laws of motion, and later we shall present two others where the appeal to such physical principles is direct and explicit.

The *objection from vertical fall* began with the fact that bodies fall vertically. This is something that everyone can easily observe by looking at rainfall when there is no disturbing wind; or one can take a small rock into one's hand, throw it directly upwards, and notice that it then falls back to the place from which it was thrown; or one can drop a rock from the top of a building or tower and notice that it moves perpendicularly downwards, landing directly below. Then it was argued that this could not happen if the earth were rotating; for, while the body is falling through the air, the ground below would move a considerable distance to the east (due to the earth's axial rotation), and although the building and person would be carried along, the unattached falling body would be left behind; so that on a rotating earth the body would land to the west of where it was dropped, and it would appear to be falling along a slanted path. Since this is not seen, but rather bodies are observed to fall vertically, it was concluded that the earth does not rotate.[11]

An analogous argument was advanced by the *objection from east–west gunshots*. The relevant observation here was that, when ejected with equal force, projectiles range an equal distance to the east and to the west. This can be most easily observed by throwing a rock with the same exertion in both opposite directions in turn, and measuring the two distances; one could also use bow and arrow, so as to have a slightly better measure of the propulsive force; or one could use a gun, and shoot it first to the east and then to the west with the same amount of charge. Now, the argument claimed that on a rotating earth such projectiles should instead range further toward the west than toward the east. The reason given was that in its westward flight the projectile would be moving against the earth's rotation, which would carry the place of ejection and the ejector some extra distance to the east; whereas, in its eastward flight, the projectile would be traveling in the same direction as the ejector, due to the latter being carried eastward by the earth's rotation; therefore, on a rotating earth the westward projectiles would range further by a distance equal to the amount of the earth's motion, while the eastward ones would fall short by the same amount. Again, since observation reveals that this is not so, it supposedly follows that the earth does not rotate.[12]

Of course, today these arguments can be refuted. However, their refutation requires knowledge of at least two fundamental principles of mechanics (to which Galileo himself contributed). One is the law of conservation of momentum, or more simply the principle of conservation of motion, according to which the motion acquired by a body is conserved unless an external force interferes with it; the other

---

[11] For more details, including an answer, see Favaro 7: 164–167, Galilei (1967, 138–141; 1997, 155–159; 2008, 222–226), Chapter 6 below.

[12] Cf. Favaro 7: 193–197, Galilei (1967, 167–171; 1997, 145–147; 2008, 215–216).

is the principle of superposition which specifies how motions in different directions are to be added to each other to yield a resultant motion. The point that needs to be stressed is that, since the phenomena appealed to are indeed true, the two objections just presented raised issues about how bodies would or could move on a rotating earth, and the resolution of these issues depended on the possession of more accurate mechanical principles. The next objection raised these same issues, but also the question of what the facts of the case really were; however, to establish these facts was not so easy as it might seem.

The *ship analogy objection*[13] referred to an experiment to be made on a ship, and it then drew an analogy between the earth and the ship. The experiment consisted of dropping a rock from the top of a ship's mast, both when the ship is motionless and when it is advancing forward, and then checking the place where the rock hits the deck. It was asserted that the experiment yielded different results in the two cases: that when the ship is standing still the rock falls to foot of the mast, but that when the ship is moving forward the rock hits the deck some distance toward the back. Then the moving ship was compared to a portion of land on a rotating earth, and a tower on the earth was regarded as the analogue of the ship's mast. From this it was inferred that, if the earth were rotating (eastward), then a rock dropped from a tower would land to the west of the foot of the tower, just as on a ship moving forward it falls toward the back; however, since the rock can be observed to land at the foot of the tower, they concluded that the earth must be standing still.

This objection partly involves the empirical issue of exactly what happens when the experiment is made on a moving ship. If the experiment is properly made, the result will be that the rock still falls at the foot of the mast. However, it is easy to get the wrong result due to extraneous causes, such as wind and the rocking motion which the boat is likely to have in addition to its forward motion. Therefore, it is not surprising that there were common reports of the experiment having been made and having yielded anti-Copernican results (Chiaramonti 1633, 339). Nor is it surprising that when Galileo tried to refute the objection, although he disputed the results of the actual experiment, claiming to have performed it,[14] he emphasized a more theoretical answer in terms of the principles of conservation and superposition of motion.[15] These principles are needed to determine what will happen to the horizontal motion the rock had before it was dropped from the mast of the moving ship, and how it is to be combined with the new vertical motion of fall it acquires (cf. Chapter 6).

The last one of the indirectly mechanical objections to be discussed here is *the argument from the extruding power of whirling*,[16] or, as we might say today, the *centrifugal-force argument*. The basis of this objection was the fact that in a rotating system, or in motion along a curve, bodies have a tendency to move away from the center of rotation or of the curve. For example, if one is in a vehicle traveling at a

---

[13] Cf. Favaro 7: 167–175, Galilei (1967, 141–149; 1997, 159–170; 2008, 225–233).
[14] See "Galileo's Reply to Ingoli," in Favaro 6: 545–646, and in Finocchiaro (1989, 184–185).
[15] See the *Dialogue*, in Favaro 7: 170–171, and in Galilei (1967, 144–145; 1997, 163–165).
[16] Cf. Favaro 7: 214–244, Galilei (1967, 188–218; 1997, 171–212), and Chapter 5 below.

2.2 The Anti-Copernican Arguments

high rate of speed, whenever the vehicle makes a turn one experiences a force pushing one away from the center of the curve defined by the turn: if the vehicle turns right, one experiences a push to the left, and vice versa. Or one could make the simple experiment of tying a small pail of water at the end of a string and whirling the pail in a vertical circle by the motion of one's hand; now suppose a small hole is made in the bottom of the pail; as the pail is whirled one will see water rushing out of the hole always in a direction away from one's hand. Then the argument called attention to the fact that, if the earth rotates, bodies on its surface are traveling in circles around its axis at different speeds depending on the latitude, the greatest speed being about 1,000 miles per hour at the equator. This sounds like a very high rate of speed, which would generate such a strong extruding power that all bodies would fly off the earth's surface, and the earth itself might disintegrate. Since this obviously does not happen, it was concluded that the earth must not be rotating.

This objection raised issues whose resolution involved the correct laws of centrifugal force. At the time, however, these laws were not known, and so this objection was felt to be very strong. Next, we come to the objections according to which the conflict with physical principles was so explicit that the earth's motion seemed a straightforward physical impossibility.

The *natural-motion argument*[17] claimed that the earth's motion (whether of axial rotation or orbital revolution) is physically impossible because the natural motion of earthly bodies (rocks and water) is to move in a *straight* line toward the center of the *universe*. The context of this argument was a science of physics which, as discussed earlier, contrasted natural motion to violent motion, which postulated three basic types of natural motion, and which attributed each type to one or more of the basic elements: circular motion around the center of the universe was attributed to aether; straight motion away from the center of the universe was ascribed to the elements air and fire; and straight motion toward the center of the universe was given to the elements earth and water. Thus, unlike natural circular motion which can last forever, straight natural motion, especially straight-downwards, cannot be everlasting since once the center (of the universe) is reached the body will no longer have any natural tendency to move. Now, the terrestrial globe on which we live is essentially the collection of all things made of the elements earth and water, which have collected at the center (of the universe) or as close to it as possible; therefore, this whole collection cannot move around the center (in an orbital revolution as Copernicus would have it), because such a motion would be unnatural, could not last forever, and would in any case be overcome by the tendency to move naturally in a straight line toward the center; further, for the same reasons, once at the center the whole collection could not even acquire any axial rotation.

The Copernican system was also deemed physically impossible because it was in direct violation of the principle according to which *every simple body can have one and only one natural motion.*[18] This principle was another aspect of the laws of

---

[17]Cf. Favaro 7: 57–62, Galilei (1967, 32–38; 1997, 83–90).
[18]Cf. Favaro 7: 281–289, Galilei (1967, 256–264).

motion of Aristotelian physics discussed earlier, whereas Copernicanism seemed to attribute to the earth at least three natural motions: the revolution of the whole earth in an orbit around the sun, the rotation of the earth around its own axis, and the downwards motion of parts of the earth in free fall.

Just as the last two objections are essentially unanswerable as long as one accepts the two principles of traditional physics just mentioned, they are easily answerable by rejecting these two principles. However, rejecting them is easier said than done since, to be effective, the rejection should be accompanied by the formulation of some alternatives. In short, what was really required was the construction of a new science of motion, a new physics, which Copernicus did not provide. In fact, the alternatives were such cornerstones of modern physics as the law of inertia, the law of gravitational force, and the law of conservation of (linear and angular) momentum. For example, according to the law of inertia, the natural motion of *all* bodies is uniform and rectilinear; and according to the law of gravitation, all bodies attract each other with a force that makes them accelerate toward each other, or diverge from their natural inertial motion in a measurable way. Thus, the earth's orbital motion becomes a forced motion under the influence of the sun's gravitational attraction; the axial rotation of the whole earth becomes a type of natural motion in accordance with conservation laws; and the downwards fall of heavy bodies near the earth's surface becomes a forced motion under the influence of the earth's gravitational attraction.

Finally, there were theological and religious objections. One of these appealed to the authority of the Bible and may be labeled the *biblical or scriptural objection*.[19] It claimed that the idea of the earth moving is heretical or at least erroneous because it conflicts with many biblical passages which state or imply that the earth stands still. For example, Psalm 104:5 says that the Lord "laid the foundations of the earth, that it should not be removed for ever"; and this seems to say rather explicitly that the earth is motionless. Other passages were less explicit, but they seemed to attribute motion to the sun, and thus to presuppose the geostatic system. For example, Ecclesiastes 1:5 states that "the sun also riseth, and the sun goeth down, and hasteth to the place where he ariseth." And Joshua 10:12–13 asserts: "Then spake Joshua to the Lord in the day when the Lord delivered up the Amorites before the children of Israel, and he said in the sight of Israel, 'Sun, stand thou still upon Gibeon; and thou, Moon, in the valley of Ajalon'. And the sun stood still, and the moon staid, until the people had avenged themselves upon their enemies."

The biblical objection had greater appeal to those (like Protestants) who took a literal interpretation of the Bible more seriously. However, for those (like Catholics) less inclined in this direction, the same conclusion could be reinforced

---

[19] For more details see Galileo's "Letter to Castelli" and *Letter to the Grand Duchess Christina*, in Favaro (5: 281–288, 309–348), Finocchiaro (1989, 49–54, 87–118), and Galilei (2008, 103–145). See also Chapters 4 and 9 below.

2.2 The Anti-Copernican Arguments

by appeal to the *consensus of Church Fathers*;[20] these were the saints, theologians, and churchmen who had played an influential and formative role in the establishment and development of Christianity. The argument claimed that all Church Fathers were unanimous in interpreting relevant biblical passages (such as those just mentioned) in accordance with the geostatic view; therefore, the geostatic system is binding on all believers, and to claim otherwise (as Copernicans did) is erroneous or heretical.

A third theological-sounding objection was based crucially on the idea that God is all-powerful, and it may be labeled the *divine-omnipotence argument*. One of its most famous proponents was Pope Urban VIII; it was, in fact, his favorite anti-Copernican objection. A version of the argument is stated, without criticism, at the end of Galileo's *Dialogue*, and this formulation got him into trouble with Church authorities and played a role in the trial of 1633, as we shall see later (Chapter 7). Here it may be useful to quote a version found in a book by Agostino Oreggi, the pope's official theologian. Apparently reporting a discussion between Galileo and Urban, Oreggi states that the pope,

> having agreed with all the arguments presented by that most learned man [Galileo], asked if God would have had the power and wisdom to arrange differently the orbs and stars in such a way as to save the phenomena that appear in heaven or that refer to the motion, order, location, distance, and arrangement of the stars. If you deny this, said His Holiness, then you must prove that for things to happen otherwise than you have presented implies a contradiction. In fact, God in his infinite power can do anything which does not imply a contradiction; and since God's knowledge is not inferior to his power, if we admit that he could have done so, then we have to affirm that he would have known how. And if God had the power and knowledge to arrange these things otherwise than has been presented, while saving all that has been said, then we must not bind divine power and wisdom in this [particular] manner. Having heard these arguments, that most learned man [Galileo] was quieted, thus deserving praise for his virtue no less than for his intellect.[21]

This statement of the objection is instructive because it makes it clearer than the Galilean statement that the argument is not purely theological, but also raises issues of a logical, methodological, and epistemological nature. Moreover, Oreggi's statement makes it more obvious that some versions of the argument may very well be essentially correct and unanswerable; for example, part of the argument seems to suggest that scientific knowledge of the physical world is contingently true rather than necessarily true. On the other hand, the argument was also

---

[20] For more details, see Bellarmine's letter to Foscarini and Galileo's "Considerations on the Copernican Opinion" and *Letter to the Grand Duchess Christina*, in Favaro (12: 171–172, 5: 351–370, 5: 309–348), Finocchiaro (1989, 67–69, 70–118), and Galilei (2008, 109–167). Again, see also Chapters 4 and 9 below.

[21] Oreggi (1629, 194–195), quoted in Sosio (1970, 548–549, n. 1). Cf. Fantoli (2003b, 230), Mayaud (2005, 3: 652–653), Speller (2008, 375–380).

taken to suggest a general skeptical doubt about physical theories, as well as a specific difficulty for the Copernican theory; and these suggestions are controversial and questionable.[22]

In summary, the idea updated by Copernicus was vulnerable to a host of counterarguments and to considerable counterevidence. The earth's motion seemed epistemologically absurd because it flatly contradicted direct sense experience, and thus undermined the normal procedure in the search for truth. It seemed empirically and astronomically untrue because it had astronomical consequences that were not seen to happen. It seemed a physical impossibility because it was thought to have consequences that contradicted the most incontrovertible mechanical phenomena, and because it directly violated many of the most basic principles of the available physics. And it seemed religiously heretical, erroneous, or suspect because it conflicted with the words of the Bible and with the biblical interpretations of the Church Fathers and because it could be taken to undermine belief in an omnipotent God.

Copernicus was aware of many of these difficulties.[23] He realized that his novel argument did not conclusively prove the earth's motion, and that there were many counterarguments of apparently greater strength. This was an important reason why he delayed publication of his book until he was almost on his deathbed, although his motivation was complex and is not yet completely understood and continues to be the subject of serious research (e.g., Goddu 2006).

2.3 Responses to Copernicanism

In light of the many objections, a common response to Copernicanism was to regard the earth's motion as a mere instrument of calculation and prediction, rather than a description of physical reality; this may be labeled the *instrumentalist* interpretation of Copernicanism and was popularized by an anonymous foreword

---

[22]As mentioned, Galileo does not explicitly criticize the divine-omnipotence objection at the end of the *Dialogue*, where he gives a statement of it; cf. Favaro 7: 488–489, Galilei (1967, 464; 1997, 306–308). However, it should be noted that other passages in the book provide partial and indirect discussions of various points related to this objection; such is the case for the comparison of human and divine understanding (Galilei 1997, 107–116) and for the possibility of a miracle explanation of the tides (Galilei 1997, 285–286). Moreover, he (Galilei 1890–1909, 7: 565–566) does explicitly criticize a version of the objection advanced by Morin (1631, 31–32). For more details on the interpretation, analysis, and evaluation of the divine-omnipotence argument and of Galileo's attitude toward it, see Beltrán Marí (2006, 412–437), Besomi and Helbing (1998, 2: 899–902), Bianchi (2000; 2001), Camerota (2004, 406–417), Finocchiaro (1980, 8–12; 1985; 1997a, 306–308), Morpurgo-Tagliabue (1981, 99–107), Speller (2008, especially 375–396), Wisan (1984).
[23]He discussed some of these objections in Book One of his great work; see, for example, Copernicus (1992, 7–50).

preceding Copernicus's own preface in the printed *Revolutions*. This foreword was written and inserted without his approval or knowledge by one of the editors supervising the book's publication – Andreas Osiander. It is unlikely that Copernicus would have endorsed this interpretation since it is clear from the book that, although he was aware of the difficulties, he treated the earth's motion as a description of physical reality (capable of being true and false), and not as a mere instrument of calculation and prediction (limited to being more or less convenient); in short, Copernicus subscribed to a *realist* interpretation of the geokinetic theory, in accordance with the doctrine of epistemological realism.

The unauthorized foreword soon became public knowledge among experts, but many scholars adopted the instrumentalist interpretation as the only way out of the difficulties.[24]

One different response was that of Danish astronomer Tycho Brahe (1546–1601). He decided to collect new data by means of systematic naked-eye observations and the construction of new instruments. The scope, range, accuracy, and precision of his observations were unprecedented. On the basis of his observational data, Tycho constructed a new theory different from both the Ptolemaic and the Copernican ones. In the Tychonic system, the earth was still motionless at the center of the universe; the stellar sphere still had the westward diurnal motion around the earth; and the sun still had the eastward annual motion around the earth. But the planets revolved in orbits centered at the sun, so that the system was to that extent heliocentric; but the sun carried the whole solar system around the motionless earth. Moreover, the nested planetary spheres were abolished since they no longer fit properly in the new arrangement; for example, some of the orbits of the heavenly bodies intersected, so that the spheres would have had to interpenetrate one another.[25]

One of Tycho's assistants inherited his data, and analyzed them in a deeper and more systematic and sophisticated manner. He was the German mathematician and astronomer Johannes Kepler (1571–1630). He rejected Tycho's compromise and was committed to the Copernican system, in part for aesthetic and metaphysical reasons. But Kepler also had a strong empirical orientation, and so he spent his life analyzing Tycho's observational data. The result was the strengthening of some key Copernican theses, such as the earth's motion, and the refinement or revision of others, such as the replacement of circular by elliptical orbits. In fact, Kepler discovered that the planets revolve around the sun in elliptical rather than circular orbits, with the sun located at one of the two foci of these ellipses.[26]

---

[24]On the instrumentalist response to Copernicus, see Barker and Goldstein (1998), Goddu (1990), Westman (1972; 1975b; 1975c; 1975d; 1980; 1987; 1990; forthcoming). On related issues in the longer history of the controversy between instrumentalism and realism, see Duhem (1908; 1969), Jardine (1984, 225), Lloyd (1978), Morpurgo-Tagliabue (1947–1948; 1981), Popper (1963, 99 n. 6).

[25]For more details on Tycho, see Mosley (2007).

[26]Of course, Kepler is an important and interesting figure in his own right. Cf., for example, Bucciantini (2003), Field (1988), Jardine (1984), Koestler (1959, 225–424), Kozhamthadam (1994), Voelkel (2001a,b), Westman (1972; forthcoming).

Unfortunately, for reasons that remain unclear or controversial,[27] Galileo never did pay the proper attention to Kepler's writings and so ignored or neglected the elliptical nature of planetary orbits. Nor did Galileo think much of the instrumentalist interpretation of the Copernican theory. Still less did he find acceptable Tycho's compromise combining of geostatic and heliocentric elements. However, Galileo did devise his own response to the Copernican controversy; did develop a Copernican world view; and did find a way to defend Copernicus. The explicit details of Galileo's defense of Copernicus will occupy us in the next four chapters.

---

[27]But see the new evidence and arguments in Bucciantini (2003).

Chapter 3
Galileo's Stances Toward Copernican Astronomy

We have seen that the astronomical observational objections presented a serious obstacle to the acceptance of the key Copernican thesis of the earth's motion. Galileo was impressed as much as anyone by the power of these anti-Copernican arguments. Nevertheless, he was able to find a way out of these difficulties. This was a long, arduous, and tortuous process. To understand and assess how Galileo defended Copernicus in this regard, we have to examine how his attitude toward Copernican astronomy and astronomical observation evolved in the course of his life, especially in his earlier career. To this we now turn.

3.1 Historical Testing of Methodological Models

The aim of this chapter is to examine the evolution of Galileo's Copernicanism from the point of view of the methodological problem of the appraisal of scientific theories. That is, I plan to determine whether Galileo's Copernicanism confirms or disconfirms a number of specific claims about the factors that influence and determine the acceptability of a set of guiding assumptions in science. These claims correspond to criteria of appraisal gleaned from the work of such authors as Thomas Kuhn, Paul Feyerabend, Imre Lakatos, Ludwig Fleck, and Larry Laudan.[1] Moreover, some time ago a group of scholars led by Laudan formulated these claims in a relatively clear and precise manner and devised a method of testing them empirically by means of research on the history of science (Donovan et al. 1988, 3–44).

To be more specific, I shall be analyzing the evolution of Galileo's attitude toward Copernicanism in order to ascertain whether or not "the acceptability of a

---

[1] Kuhn (1970; 1977), Feyerabend (1981), Lakatos (1978), Fleck (1979), Laudan (1977; 1984).

set of guiding assumptions is judged largely on the basis of"[2] such criteria as the following: (1) empirical accuracy, (2) factors other than empirical accuracy, (3) the success of its associated theories at solving problems, (4) the success of its associated theories at making novel predictions, (5) its ability to solve problems outside the domain of its initial success, (6) its ability to make successful predictions using its central assumptions rather than assumptions invented for the purpose at hand, (7) factors other than simplicity, (8) aesthetic criteria, (9) factors other than consistency, (10) its relations to other well-established beliefs, (11) its relation to nonscientific beliefs, and (12) factors other than its practical applications. Because the focus here is Galileo's Copernicanism and its utilization as a case study for the historical testing of such philosophical theses, no elaborate analysis of their precise meaning is required here. Nevertheless, such testing will be impossible to carry out without a number of conceptual clarifications.

3.2 Conceptual Clarifications

Notice, to begin with, that, even without any reference to the background literature, the theses as formulated above are making a distinction between theories and guiding assumptions. In fact, items (3) and (4) speak of "associated theories" in relation to guiding assumptions. This corresponds to the standard and conventional distinction between sets of ideas that are more, and sets that are less, general and comprehensive from a methodological point of view; also more or less long-lasting from a historical point of view; and more or less elusive from the point of view of textual documentation. Despite the lack of terminological uniformity in the literature, there is widespread agreement about the importance of the distinction, and so let us note that the "guiding assumptions" under discussion here correspond to Kuhn's "paradigms," Lakatos's "research programmes," Feyerabend's "global theories," and Laudan's "research traditions." It is useful to note, then, that this distinction is a presupposition of the present investigation. It should be added that Copernicanism may be taken, and has been widely taken, as a paradigm example of a set of guiding assumptions, as something that almost provides an ostensive definition of the term; therefore, the full conceptual analysis of the notion, besides being outside the scope of this book, is unlikely to affect its outcome, since that analysis would have to incorporate the basic facts about this "pre-analytically intuitive" example (cf. Laudan 1977). Finally, by contrast to the Copernican set of guiding assumptions, in the present context we may give the following Galilean examples of specific theories associated with it: his theory of the tides, his theory of the lunar surface, his theory of sunspots, his theory of Jupiter's satellites, his theory

---

[2] Donovan et al. (1988, 15). In stating the criteria that follow, I am paraphrasing this passage and another one in Laudan et al. (1986, 164–165).

3.2 Conceptual Clarifications

of the phases of Venus, his theory of the conservation of motion, his theory of falling bodies, his theory of natural motion, and his theory of neutral motions.

It might seem that, just as the semi-technical meaning of guiding assumptions involves us in presupposing the distinction just mentioned, the specific meaning of "acceptability" would require us to presuppose another distinction, that between acceptance and pursuit. However, this is not so. The cluster of theses to be tested do indeed relate to the question of how "the acceptability of a set of guiding assumptions is judged," and there is no question that the distinction between acceptance and pursuit is a valuable one in some contexts. In the present inquiry, however, if one checks the passages of the works from which these theses are gleaned, it becomes clear, not only that the distinction is not being uniformly maintained, but also that some of the appraisal criteria are being put forth explicitly as criteria for pursuit rather than acceptance. This is relatively obvious for such criteria as aesthetics (number 8) and the relationship between guiding assumptions and nonscientific beliefs (number 11). The point is even more true for the case of factors other than empirical accuracy (number 2). For the main one of the alternative factors being referred to is the progressiveness of a set of guiding assumptions, either in the sense of total progress made or of rate of progress, and the proponent of this criterion explicitly advances it as relating to pursuit rather than to acceptance.[3]

There are other reasons why in the present context the term "acceptability" is meant to include "pursuit," as well as acceptance in the narrow sense. Both acceptance and pursuit are special cases of appraisal, and our present problem is precisely the appraisal of guiding assumptions. Moreover, though the distinction is generally plausible, among the primary authors whose theses are being tested, Laudan is the only one who emphasizes it; and I believe the spirit of the project of testing methodologies by history is to try to include the views common to all these authors.

Finally, there is a more historical reason why it is safe to say that the word "acceptability" is not meant literally, or at least not technically, in the formulation of these theses, but it is taken to refer indiscriminately to either acceptance strictly understood, or pursuit, or other cognitive stances. The reason is that if we do not assume the distinction in question, then we can use our historical evidence to determine whether or not it is valuable to distinguish these two cognitive stances, whether or not the distinction sheds light on the historical material, for example, whether or not the evolution of Galileo's Copernicanism is best divided into two corresponding stages, and whether his motivating reasons are best grouped into two and only two corresponding sets. The alternative possibilities are the existence of a continuum of cognitive stances, the existence of more than two but a specifically identifiable number of attitudes, and the existence of only one positive stance, namely commitment, by contrast to the pre-conversion attitude of rejection.

All of this means that we have a double amount of work, or at least that our efforts will have to be two-sided. We cannot merely assume that Galileo exhibited toward Copernicanism the well defined and unproblematic stance of acceptance,

---

[3] Laudan 1977, 107–114, 119–120.

and then search his works merely to determine what were his reasons and the criteria on the basis of which he accepted it. Instead, we shall have to ascertain, and indeed to define and identify, the various cognitive stances Galileo adopted, keeping an open mind to the various possibilities mentioned above, and simultaneously we shall have to identify, define, and ascertain his reasons and his operative criteria.

This second task is facilitated by the fact that our list of 12 items provides us with clues of what to look for. But this is not to say that we are assuming that Galileo's motivating reasons were of one or more of the twelve types in this cluster. Obviously one must be open to the possibility that none of these twelve would apply. Thus our assumption is merely that one or more of the twelve types of criteria may apply to the case of Galileo's Copernicanism. If so, we shall try to determine which one or which ones. If not, we shall try to define what those different operative criteria are.

Before we proceed to that sort of testing, we need to make another clarification. We need to clarify the individual meaning and the mutual interrelationships of the 12 criteria. To be sure, since our chief aim here is the historical testing of methodological theories, we cannot engage in any elaborate conceptual analysis, intertheoretic comparison, and examination of the internal coherence of these theories and of their relationship to other philosophical ideas and cultural trends. To invoke a meta-methodological analogy, that would be like judging a given set of guiding assumptions in science on the basis of one criterion, when the issue was to examine it on the basis of another. The essential criterion embodied by our self-imposed task of testing methodologies by history is akin to the "empirical accuracy" mentioned in our cluster of theses. Nevertheless, this would be impossible without a minimal amount of clarification of the meaning of these criteria.

The first criterion in the list, empirical accuracy, does not require any special comment here. But what are the factors other than empirical accuracy, to which criterion number (2) refers? Abstractly speaking, any one of the other ten factors could be subsumed under this general description. An examination of the references to the primary authors from which this criterion is gleaned reveals that the following are specifically meant: effectiveness at solving conceptual problems (Laudan 1977, 68), power to anticipate novel facts (Lakatos 1978, 69), comparative worth vis-à-vis alternatives (Lakatos 1978, 69), and progressiveness in the sense of either total progress or rate of progress (Laudan 1977, 107). But the first two of these can obviously be subsumed, respectively, under criterion (3), success at solving problems, and criterion (4), success at making novel predictions; and comparative worth refers to comparative problem-solving success and comparative predictive success, and so it can also be subsumed under criteria (3) and (4). Thus, the primary factor being specifically referred to in criterion (2) is progressiveness. For short, in the following discussion I shall call this criterion either "non-empirical factors" or "progressiveness."

Item number (3) is the success of associated theories at solving problems. The only thing to note here is the existence of conceptual and of anomalous problems, in addition to empirical problems. This criterion may be labeled, for short, "general problem-solving success." If need be, we may also speak of empirical problem-solving success, conceptual problem-solving success, and comparative problem-solving success.

3.2 Conceptual Clarifications

Number (4) speaks of the success of associated theories at making novel predictions, and it may be called simply "predictive novelty."

The next criterion, number (5), reads: the ability to solve problems outside the domain of initial success. This would seem to be a special case of the general problem-solving success, previously mentioned, and it may be labeled "external problem-solving success."

The sixth factor (6) is the ability to make successful predictions using central assumptions rather than assumptions invented for the purpose at hand. No special emphasis on novel predictions, as distinct from explanations, is meant here. So what is intended is the avoidance of ad-hocness. I shall label it either "anti-ad-hocness," or less awkwardly, "explanatory coherence."

Criterion number (7) refers to factors other than simplicity. Since the historical question of whether the acceptability of guiding assumptions is judged on the basis of factors other than simplicity is essentially the same as the question of whether such acceptability is judged on the basis of simplicity, no harm will normally result if we speak simply of simplicity.

There is nothing problematic about number (8), aesthetic criteria, or, more simply, aesthetics.

The next criterion (9) speaks of factors other than consistency. Here, consistency means primarily internal self-consistency. Moreover, again since the presence of this item is equivalent to the absence of consistency, and the question of the non-occurrence of this item is equivalent to the question of the occurrence of consistency, nothing will be lost by speaking of just consistency or self-consistency, and so this will be our short label.

Item number (10) involves the relation to other well-established beliefs. What the primary authors have in mind here is the relation between a set of guiding assumptions and other doctrines which the proponents of this set believe to be rationally well-founded (Laudan 1977, 50–54).

Number (11) is analogous, and it speaks of the relation to nonscientific beliefs, such as metaphysical, ethical, theological, or logical ideas. Since there is no presumption that the latter are irrational, I think it is best to refer to the previous criterion as "scientific external coherence," and to the present one as "general external coherence."

The last criterion, number (12), refers to factors other than its practical applications. For reason analogous to two previous cases, (7) and (9), in our testing we may think of the negation of this negation. That is, we may think simply of "practical application."

To summarize, our general question is whether the acceptability of a set of guiding assumptions is judged largely on the basis of empirical accuracy, progressiveness, general problem-solving success, predictive novelty, external problem-solving success, explanatory coherence (namely anti-ad-hocness), simplicity, aesthetics, self-consistency, scientific external coherence, general external coherence, or practical application. And our particular question is whether Galileo's Copernicanism was based largely on one or more of these factors. Now, what sorts of answers can we expect? That is, we cannot yet proceed to the testing, for to do so would be blind mechanism or uncritical empiricism. The fact is that only a certain range of answers may turn out to be warranted by the historical evidence. This fact is due to the various logical and quasi-logical mutual interrelationships, such as implication and exclusion, among the twelve factors (as we have just seen); but it is also due to the exact import of one other crucial term used in the formulation of the theses, namely the word "largely" (to which we now turn).

This word obviously does not mean "only." Nor does it mean "primarily." Presumably it does not mean "normally." It probably means "commonly," or "frequently," or "to a large extent." But how large is large?

Next, it is beyond the scope of this book even to try to give an exhaustive account of the various ways in which the various criteria may coalesce into groups. I shall only mention some clear examples of consistent subclusters, some clear examples of inconsistent ones, and some cases of unclear relationships.

It is obvious that it may conceivably turn out to be the case that guiding assumptions are appraised largely on the basis of predictive novelty and explanatory coherence; or largely on the basis of general problem-solving success and external problem-solving success; or largely on the basis of practical applications, nonscientific beliefs, and aesthetics. This yields three consistent subclusters. I believe it is equally obvious that our historical testing could not possibly reveal that guiding assumptions are appraised largely on the basis of empirical accuracy and aesthetics; or largely on the basis of empirical accuracy and nonscientific beliefs; or largely on the basis of empirical accuracy and simplicity. But here we must be careful.

I do not think it would be impossible for guiding assumptions to be appraised in large measure on the basis of some combination of, for example, empirical accuracy plus aesthetics. Such a combined large measure could not, however, consist of equal parts, because I believe that a large measure of one would exclude a large measure of the other; but we might have a combined large measure consisting of a large measure of empirical accuracy plus a little bit of aesthetics. I believe that a similar tension exists between simplicity and empirical accuracy on the one hand, and between the criterion of nonscientific beliefs and empirical accuracy on the other. That is, historical testing may be said to presuppose a whole series of assumptions, each derivable from one of these apparently inconsistent subclusters; for example, I am assuming that the acceptability of guiding assumptions is *not* judged largely on the basis of aesthetics *alone*; or of nonscientific beliefs *alone*; or even of aesthetics *and* empirical accuracy; or of nonscientific beliefs *and* empirical accuracy.

Finally, it is relatively unclear whether guiding assumptions could be judged largely on the basis of, for example, well-established beliefs (number 10) and nonscientific beliefs (number 11). Let us recall that well-established beliefs are meant to be beliefs deemed to be well-established by the proponents of the given set of guiding assumptions, and let us take religious beliefs as a special case of nonscientific beliefs. Now, the existence of God is a paradigm example of a religious belief, yet it is a socio-historical fact that many believers deem it to be a well-established claim. It follows that to inquire into the relationship between a set of guiding assumptions and the existence of God involves judging the acceptability of the former on the basis of both of these criteria, numbers (10) and (11). But then the question arises, why these particular two should be listed separately, if they are not meant to be mutually exclusive. Now, suppose we change the criterion of nonscientific beliefs to read: the relation of a set of guiding assumptions to beliefs that are not well-established, in the sense of not regarded as well-established. This emendation would make the two criteria properly different, but the cost would be the likelihood that the criterion of nonscientific beliefs so defined would become non-operative and non-existent. We could try changing the criterion of purportedly well-established beliefs into one of actually well-established beliefs. This might offer the proper contrast to nonscientific beliefs, but only at the risk of making it historically and empirically devoid of content.

3.3 Historiographical Considerations

These procedural considerations put us in a better position to conduct our historical testing. They give us a better idea of what it is we are looking for, what the various possibilities are, and what is unlikely or impossible to happen. However, they only relate to issues of a meta-methodological, conceptual, or generally philosophical sort. But that is only one side of our methodological situation, the other side being the historical data and evidence. This material has to be collected before it can be analyzed for the purpose of our test. So a number of historical and historiographical considerations are needed.

It might seem that, since we are dealing with only one scientist, it would be relatively easy to locate the evidence. However, Galileo's Copernicanism is an entity that spans his entire career, and hence we are faced with a large collection of writings, analogous to other historical episodes which may involve large numbers of scientists, but where the individual contributions are limited to such documents as a single book or article. On the other hand, in this particular case we are fortunate because of the existence of a critical edition of his collected works, which is not easily surpassable in scholarly quality. Yet the twenty one tomes of this collection are a rather overwhelming and forbidding domain.

The next thing to do, logically if not chronologically speaking, is a critical survey of the secondary literature. This, of course, is itself another labyrinth. Moreover, a philosopher has to be careful not to take this new domain as his new primary object of study and so forget the original Galilean one. The fruitful way to proceed here is to examine the secondary literature critically and judiciously with an eye toward collecting the relevant evidence, rather than toward accepting any one particular thesis or even any one orientation or approach. In speaking of the relevant evidence, I mean that which is *directly* relevant, for I am willing to admit that anything in Galileo's complete works is *indirectly* relevant. Here, of course, I am assuming that it is possible to distinguish direct from indirect evidence.

If we briefly reflect on how it is possible to actually make this distinction, I believe we can come to the identification of another assumption needed in our investigation, namely something about the content of Copernicanism. In fact, I believe one draws the distinction between direct and indirect evidence on the basis of another distinction, that between explicitness and implicitness. That is, directly relevant evidence is that in which Galileo discusses or mentions Copernicanism explicitly. But what does explicit mean? Partly there is a linguistic criterion for this, namely the occurrence of the word Copernicus and cognate terms. But that is obviously not enough. Equally explicit are writings where Galileo discusses any of the following topics: the state of rest or motion and the location in the universe of the earth and of the other heavenly bodies, the arrangement of such bodies, and their origin. Copernicanism must be assumed to be at least a set of propositions dealing with these topics. Such an assumption is, like all similar ones, subject to revisions in the course of the investigation.

Finally, even when dealing with direct, explicit evidence, we need another distinction, namely between relatively important or significant and relatively unimportant or insignificant. To some extent this can be based on an almost quantitative criterion, which involves a correlation between the length of explicit writings and their significance. This is sufficient to make us discard, as being relatively unimportant, documents like Galilean letters where the name Copernicus is only incidentally mentioned. Notice, however, that the correlation is far from perfect, and sometimes very brief references and statements are significant.

When all these steps are taken and these assumptions made, we arrive at the following set of documents, as constituting our data base, to be searched, interpreted, evaluated, and analyzed to try to get answers to the questions posed earlier: (A) the early work on *De motu*, composed about 1589–1592; (B) the *Treatise on the Sphere, or Cosmography*, whose time of composition ranges from 1586–1587 to 1602–1606; (C) the letter to Mazzoni, dated 30 May 1597, in which Galileo criticizes an anti-Copernican argument; (D) the famous letter to Kepler, dated 4 August 1597, in reply to the latter's gift to Galileo of his just-published *Mysterium cosmographicum*; (E) Tycho Brahe's letter to Galileo, dated 4 May 1600, which was never answered; (F) *The Sidereal Messenger*, Galileo's first scientific publication of 1610, where he announces the first telescopic discoveries; (G) the letter to Giuliano de' Medici, Tuscan ambassador to Prague, dated 1 January 1611, where Galileo announces his discovery of the phases of Venus; (H) some of Galileo's notes from the years 1611–1612, in which he uses numbers from Ptolemaic tables in his analysis of the behavior of Jupiter's satellites; (I) the letter to prince Cesi, head of the Lincean Academy, dated 30 June 1612, and discussing the reality of epicycles and eccentrics; (J) the *History and Demonstrations Concerning Sunspots*, written in 1612 and published in 1613, widely regarded as Galileo's most explicit commitment ever to Copernicanism; (K) the "Letter to Castelli," dated 21 December 1613, containing Galileo's first considered defense of Copernicanism from theological objections; (L) the letter to Baliani, dated 12 March 1614, containing one of the most explicit endorsements of Copernicanism Galileo ever wrote, as well as a summary of his position on the matter; (M) the two letters to Dini, dated 16 February and 23 March 1615, written at the height of the earlier Inquisition proceedings; (N) the "Considerations on the Copernican Opinion," written in 1615, and which may be considered, at least as a first approximation, as Galileo's reply to Cardinal Bellarmine's anti-Copernican arguments contained in the latter's letter to Foscarini; (O) the *Letter to the Grand Duchess Christina*, written in 1615, and containing Galileo's classic defense of Copernicanism from the Biblical objection; (P) the "Discourse on the Tides," dated 8 January 1616, written at the request of Cardinal Orsini, and containing Galileo's first explicit statement of two physical arguments in support of the geokinetic thesis; (Q) the well known *Assayer* of 1623; (R) the "Reply to Ingoli," written in 1624, containing Galileo's first considered reply to almost all astronomical and physical objections to Copernicanism; (S) the classic *Dialogue* of 1632, which led to the famous trial the following year; (T) the defendant's last deposition (dated June 21) during the Inquisition trial of 1633, in which, under the formal threat of torture, he answered questions about his real attitude toward Copernicanism; (U) the letter to Liceti, of January 1641; and (V) the letter to Rinuccini, dated 29 March 1641.[4]

It may seem superfluous to list and briefly identify all these documents, as we have just done. However, it is useful to have an overview of the evidence, before undertaking the task of analysis, interpretation, and evaluation; it provides a way of orienting oneself. Moreover, this list may be taken as a documentary definition of the historical entity on the basis of which we are testing methodological theories of appraisal. This is, of course, a working definition, and so it is revisable in the light of further analysis.

3.4 Periodization

Before proceeding to any textual analysis, I believe it is possible to distinguish four subsets of documents, corresponding to four periods. It would be premature at this point to say whether or not to these four stages there correspond four distinct cognitive stances and four types of rationales. While to determine the latter we need more analysis, the fourfold classification is based on relatively material considerations and on the occurrence of events whose significance is unquestionable both biographically and historically. I am almost inclined to say that this fourfold classification is a point of this inquiry which is not theory-laden, but I shall weaken the claim and say that it is the first and firmest historical interpretation (or theoretical claim, if you will) of this investigation.

The four stages are as follows. The first is a 24 year period (1586–1609) and goes up to Galileo's construction of the telescope; it contains relatively sparse references to Copernicanism. The second is very intense and spans the 7 years from the telescope in 1609 to the anti-Copernican Decree of the Index of 1616. The third

---

[4] The references, all to Galilei 1890–1909 (= Favaro 1890–1909), are as follows. (A) 1: 251–419; (B) 2: 211–255; (C) 2: 197–202; (D) 10: 67–68; (E) 10: 79–80; (F) 3: 53–96; (G) 11: 11–12; (H) 3: 521–522; (I) 11: 344–345; (J) 5: 71–250; (K) 5: 281–288; (L) 2: 33–36; (M) 5: 291–295, 297–305; (N) 5: 351–570; (O) 5: 309–348; (P) 5: 377–395; (Q) 6: 197–372; (R) 6: 509–561; (S) 7: 21–519; (T) 19: 361–362; (U) 18: 293–295; (V) 18: 314–316.

period goes from 1616 to the trial and condemnation of 1633; it is more intense than the first period, but less so than the second. The fourth stage is the post-trial period to Galileo's death in 1642.

It is tempting already to try to characterize the methodology of the evolution, but all we can do for now is to suggest various possible alternatives. The most common view is perhaps the sequence: qualified acceptance, unqualified acceptance, semi-clandestine acceptance, and secret acceptance. Others are: pursuit, acceptance, qualified acceptance, and avoidance; or nonpursuit, pursuit, qualified pursuit, and nonpursuit. Attaching to these attitudes the various alternative types of rationales constructible from our list of 12 criteria would yield a bewildering number of possibilities. But let us come down from the Platonic heaven where the latter cannot not exist to the land of the earthly, pedestrian, and concrete interpreting of texts.

My plan for the interpretation of these documents is not to give summaries of their content and interweave them into a narrative account of the development of Galileo's Copernicanism. This would be neither manageable nor especially incisive nor wholly relevant. Of course, as regards relevance I have already selected them for their relevance, but the point now is that not all parts and not all aspects of all documents in the data base are equally relevant to the theses being tested. Moreover, on a few occasions there would be no point in repeating work done elsewhere, either by others or by the present author. So, in examining the data base, not only I shall emphasize, but indeed I shall limit myself to points that strike me as especially relevant, controversial, or novel.

3.5 Indirect Pursuit (Before 1609)

Galileo's earliest reference to Copernicus is found in *De motu*, written probably in the period 1589–1592 while he was a professor at the University of Pisa, but left unpublished by him. The context is one where Galileo is arguing, against Aristotle, that when the motion of a body changes from one direction into the opposite direction, there does not have to be a state of rest at the turning point. One of Galileo's several arguments is that oscillatory motion along a straight line can be generated by combining two continuous circular motions: consider two equal circles such that the center of each is located on the circumference of the other, and such that they rotate in opposite directions at different rates; then one can adjust these rates of rotation so that there is a point on the circumference of the faster rotating circle which moves back and forth along a straight line. Galileo explicitly credits Copernicus's work *On the Revolutions* as the source of this demonstration.[5]

---

[5] See Favaro 1: 326, Galilei (1960, 97). Cf. Copernicus (1992, 125–126), *On the Revolutions*, Book 3, Chapter 4.

3.5 Indirect Pursuit (Before 1609)

This reference is important for two reasons. First, it shows that Galileo was intimately acquainted with Copernicus's masterwork at the beginning of his career. Second, it is even more important for what Galileo does *not* say: Copernican astronomy is nowhere in sight, and so he must not have been impressed by the cogency or conclusiveness of Copernicus's argument for the geokinetic thesis. Other passages in *De motu* do have a connection with the earth's rotation, but that connection is indirect and to appreciate it we must wait for other clues, which will be elaborated presently.

The first explicit discussion of Copernican astronomy is found in Galileo's letter to Mazzoni dated 30 May 1597. Jacopo Mazzoni had been a senior colleague and good friend of Galileo's during his time at the University of Pisa. Mazzoni was an eclectic philosopher with wide interests who held anti-Aristotelian ideas on the motion of falling bodies and their speed of fall; and these ideas overlapped with Galileo's own. In 1597 Mazzoni had just published a book containing a critical comparison of Plato and Aristotle. Galileo had just read the book and was writing to congratulate the old friend and to express his gratification at the fact that they seemed to agree about many things. However, the book also contained an anti-Copernican argument, and most of Galileo's letter is a lengthy analysis and refutation of that argument.

Mazzoni had argued as follows: if the earth revolves around the sun, then this off-center location would imply that terrestrial observers would not always see exactly half of the stellar sphere, but less than half at midnight and more than half at noon; the difference would be noticeable (despite the immense size of the stellar sphere), because on earth by climbing a tall mountain (such as Mount Caucasus) one can notice a difference in the visual horizon; but we always see exactly half of the stellar sphere; it follows that the earth is not located off-center revolving around the sun.

Galileo's refutation is the following: if the earth revolves around the sun, the difference in visibility of the stellar sphere between midnight and noon would be equal to that caused on earth by a mountain whose height is 1 and 1/7 miles; on earth the difference in visibility of the stellar sphere resulting from climbing such a mountain is 1° and 32 min on each side; these quantities are based on the traditional estimates of astronomical distances (which are: distance between the earth and the sun = 1,216 earth radii; radius of the stellar sphere = 45,225 earth radii; and earth radius = 3,035 miles); but the Copernican estimate of the size of the stellar sphere is much greater; so the difference in stellar horizon would be much less that 1° and 32 min; and that would be unlikely to be noticeable.

The most relevant and important aspect of Galileo's letter to Mazzoni is that it constitutes an explicit defense of Copernicanism from an astronomical objection. Moreover, the core of Galileo's reasoning is mathematical or quantitative, and so what we have here is a mathematical defense. Combining these two aspects, we might say that Galileo is exhibiting a mathematical appreciation of Copernicanism, and this is in accordance with his earlier reference to Copernicus in *De motu*, discussed above. However, it is unclear that we can describe Galileo's attitude any more precisely than is conveyed by the vague notion of what I am labeling "appreciation."

In fact, Galileo's stance is described in a similarly vague and unclear manner in the introductory part of this long letter. Its most relevant passage reads as follows:

> It has given me the greatest satisfaction and consolation to see that Your Most Excellent Lordship has written about some of those questions concerning which in the first years of our friendship we disputed together with so much merriment, and that you are now inclined to take the side which I regarded as true and you as the opposite; I suppose you did this either in order to invite argument, or to show that your powerful intellect is able to support falsehoods if you should so wish, or to preserve uncorrupted (indeed intact in their smallest parts) the truthfulness of the doctrines of that great Master under whose command appear to serve, and should, all those who undertake to investigate the truth … However, to be frank, although I was encouraged by your other conclusions, I was left confused and apprehensive when I first saw Your Most Excellent Lordship criticize so resolutely and openly the opinion of the Pythagoreans and of Copernicus on the motion and location of the earth; since [at that time] I had supported this opinion as much more probable than the other one of Aristotle and Ptolemy, I paid special attention to your criticism, insofar as I [now] have some feelings regarding this topic and others that relate to it. [Favaro 2: 197–198]

That is, now (in 1597) Galileo has some feelings towards the topic of the earth's motion and location, which he presumably did not have in the first years of his friendship with Mazzoni (in 1589–1592), when they amicably engaged in disputations, with Galileo taking the Copernican side.

It should be mentioned that this letter to Mazzoni is usually mentioned to show that by 1597 Galileo regarded "the opinion of the Pythagoreans and of Copernicus … as much more probable than the other one of Aristotle and Ptolemy." Such interpretations[6] not only overlook the past tense Galileo uses in this sentence, but also they take the sentence out of context in two ways. That is, they disregard the references to the earlier argumentative discussions between Galileo and Mazzoni; and they ignore that most of the letter elaborates a defense of Copernicanism from Mazzoni's objection that suggests a mere mathematical appreciation on Galileo's part.

More clues about Galileo's stance and rationale are found in the letter he wrote to Kepler a few months thereafter (4 August 1597). The letter was occasioned by the fact that Galileo had just received a copy of Kepler's book *Mysterium cosmographicum*, published the previous year, and he wanted to thank Kepler. Galileo said that he had only read the introduction, but planned to read the rest, and then he added:

> Indeed I shall do it so much more gladly, inasmuch as I came to Copernicus's view many years ago, and from such a position I am devising the causes of many physical effects, causes which are without a doubt inexpressible in terms of the common hypothesis. I have worked out many reasons as well as refutations of arguments to the contrary, which however I have not dared to publish, afraid of the fate suffered by Copernicus himself, our teacher, who, though he earned immortal fame with some, nevertheless became the target of ridicule and scorn with innumerable others (such is the number of fools). I would certainly dare to publish my thoughts if there were more people like you; since there are not, I shall wait on the matter. [Favaro 10: 68]

---

[6] See, for example, Bucciantini (2003, 29), Camerota (2004, 98), Fantoli (2003b, 59–60). On the other hand, other scholars display various kinds of skepticism or criticism toward this sentence, reaching conclusions that are analogous though not identical to mine; for example, Shea (1972, 113) suggests that the sentence "has no special significance," and Biagioli (1993, 100) stresses that that the letter shows that in 1597 Galileo was merely "a Copernican sympathizer but not yet a committed defender of the Copernican hypothesis."

3.5 Indirect Pursuit (Before 1609)

I take the first sentence of this passage as an expression of Galileo's being engaged in a program of physical research that fits well with Copernicanism, but not at all with the Aristotelian Ptolemaic view. We might say that he is pursuing the physical side of Copernicanism, physical by contrast with Copernicus's own astronomical motivation, or with the metaphysical flavor of Kepler's own *Mysterium cosmographicum*. Galileo must be referring to the sort of theory of motion which he had been working on for some time and part of which is recorded in *De motu*. I see no reason for reading into this first sentence a statement of commitment, adherence, or acceptance vis-à-vis Copernicanism,[7] based on the possession of an allegedly conclusive physical argument. This is not to deny that here we have an implicit reference to Galileo's geokinetic explanation of tides,[8] but we have to stress that he is saying he has many phenomena in mind and not just one, and that he is expressing certainty at most about their geostatic inexplicability and not about their geokinetic explicability. In fact, the rest of the passage, about Galileo's fear to publish, is an explicit comment on his view of the strength of his arguments in support of Copernicanism; he obviously does not think his arguments are conclusive or even strong enough to convince someone who, unlike Kepler, is not already favorably inclined.

In the second sentence of this passage, the initial clause makes an extremely important distinction. Galileo is explicitly saying that he is in possession not only of positive evidence or constructive reasons, but also of criticisms or refutations of counterarguments and objections. Thus, he is implicitly suggesting that the negative aspect of the investigation is also essential, namely that the defense of Copernicanism must include a serious and careful critical component. In fact, a few months earlier, in his letter to Mazzoni, Galileo had just conducted one exercise in such a critical defense. However, he is clear that this is just one example of many. Indeed, if one examines the historical context, one can identify other anti-Copernican arguments as the mechanical objections to the earth's motion that had recently been advanced in Tycho Brahe's *De mundi aetherei recentioribus phaenomenis* (1588) and *Epistolae astronomicae* (1596); and one can see that Tycho's work played an important negative role in Galileo's career.[9]

Now that we have a general description of the type and of the strength of Galileo's stance and rationale, let us see whether we can identify some of the constructive arguments more precisely. I have already admitted that one of them is the well-known tidal argument, but let us see whether we can identify any others. I believe one can be found in the section dealing with the earth being motionless of

---

[7] This is the usual interpretation, as one can see from Bucciantini (2003, 50, 51, 66), Camerota (2004, 98, 107), and Fantoli (2003b, 60–61). Part of their rationale is the reading of the Latin word (*venerim*) Galileo uses to describe his encounter with Copernicanism to mean "I accepted" (Santillana 1955, 11) or "I have come ... to accept" (Fantoli 2003b, 60); note that I have translated it merely as "I came to." Important exceptions are Beltrán Marí (2006, 74–76), who rightly views the letter as mostly an exaggeration; Biagioli (1993, 100) who sees the need to problematize the notions of Copernicanism and of acceptance; and Torrini (1993, 30), who stresses the methodological character of Galileo's motivation.

[8] See, for example, Drake (1978, 40–44).

[9] Here I am adopting this historical thesis from Bucciantini (2003, 49–68).

his *Treatise on the Sphere, or Cosmography*. It is well known that this *Cosmography*, which Galileo never published, was not meant to be an original contribution to knowledge, but a concise and elementary introduction to spherical astronomy for beginning students; and it is likely that he used it both in his university courses and his private tutoring. Moreover, it cannot be denied that the work is generally conservative, Aristotelian, and Ptolemaic in content. In regard to its form, however, it has some original and interesting elements (cf. Wallace 1984b). One is the methodological and epistemological introduction, where Galileo explicitly discusses the method of hypothesis, giving the following examples of hypothetical assumptions: "the sky being spherical, its moving circularly, its possessing different motions, the earth being still, and located at the center" (Favaro 2: 217). The other is the impersonal, informational, and noncommittal style which Galileo uses as he discusses the details of traditional spherical astronomy. The Copernican system is explicitly mentioned only once, and this brings us to the passage referred to above.

Unlike other sections, the one entitled "That the Earth Is Motionless" begins with an admission that this is a controversial question: "The present question is worthy of consideration since there has been no lack of very great philosophers and mathematicians who, deeming the earth to be a star, have given it motion" (Favaro 2: 223). But he is quick to add: "Nevertheless, following the opinion of Aristotle and Ptolemy, we shall adduce those reasons why one may believe that it is completely motionless" (Favaro 2: 223). Then he goes on to argue why the earth cannot have rectilinear motion, and after that he takes up the question of its possible rotation, concerning which we have the following very important passage: "But, that it may move circularly has more verisimilitude, and therefore some have believed it; they have been moved principally by their considering it almost impossible that the whole universe except the earth should experience a rotation from east to west in the period of 24 hours, and hence they have believed that it is rather the earth which undergoes a rotation from west to east during such a time" (Favaro 2: 223). He does not say whether or why he rejects this argument. Instead, in his usual impersonal style he adds that "having considered this opinion, Ptolemy argues as follows in order to destroy it" (Favaro 2: 223), and then he goes on to summarize the traditional objections from falling bodies, birds, clouds, ship, and whirling.

The argument in the passage quoted above is important because it is obviously an appeal to Galileo's own doctrine of natural and neutral motions, which he had elaborated at least as early as 1589–1592 in his *De motu*.[10] This is the theory according to which there are three basic types of motions: natural motion, or motion where the object approaches its natural place; violent motion, or motion where the object recedes from its natural place; and neutral motion, or motion which is neither natural nor violent. For Galileo, examples of neutral motions would be the rotation of a homogeneous sphere at the center of the universe or the rotation of a sphere around its center of gravity even if the sphere is located elsewhere. Finally, for him to start a body moving with neutral motion, a force as small

---

[10] Favaro 1: 304–307; cf. 2: 279.

as you like is sufficient. Given this doctrine, it presumably follows that the earth would have axial rotation even in an otherwise Ptolemaic universe.

Galileo explicitly recognized this type of argument as one favorable to Copernicus both[11] in the *History and Demonstrations Concerning Sunspots*, where he argued in support of solar rotation, and in the *Dialogue*, where he gave it at the beginning of Day II in support of terrestrial rotation. So it is not unlikely that he had made the connection even in the pre-1609 period.

Tycho's letter to Galileo, dated 4 May 1600, provides further support for my account, since Galileo never answered it. Here I should add that he also did not answer the letter dated 13 October 1597 which Kepler wrote after receiving his. Kepler was asking to be informed of Galileo's Copernican arguments and wanted him to make certain observations to try to detect stellar parallax. Although there were undoubtedly external factors that contributed to Galileo's lack of cooperation, the internal and methodological reasons can only be his lack of interest in pursuing the astronomical side of Copernicanism, and his dissatisfaction with the strength of his physical arguments.

We may summarize our results for the pre-telescopic period by saying that Galileo's attitude toward Copernicanism was one of partial, qualified, and indirect pursuit. It is easy to see that such pursuit was based largely on factors other than empirical accuracy, on factors other than simplicity, and on factors other than practical applications. More positively, it was based largely on the problem-solving success of its associated theories, primarily Galileo's theory of motion. Such an associated problem-solving success may suggest to us the ability to solve problems outside the domain of its initial success, but I do not think this was the specific motivating factor for Galileo. There is no evidence his pursuit was based on predictive novelty, aesthetic criteria, factors other than consistency, or its relation to other well-established or nonscientific beliefs. In regard to explanatory coherence, no relevant evidence has been examined yet, but some will be found below, which bears a date of 1614 but refers to this earlier period.

3.6 Full-Fledged Pursuit (1609–1616)

There is no question that the telescopic discoveries led Galileo to a significant reappraisal of Copernicanism. Less easy to ascertain is what the change was exactly, how sudden or slow it was, and what the motivating reasons were. The first piece of significant evidence here is *The Sidereal Messenger*. There are three relevant passages in this work.

One occurs in the author's dedication to the Grand Duke of Tuscany. Since this is often taken as the first published evidence that Galileo accepted Copernicanism,[12]

---

[11] See, respectively, Favaro 5: 133–135, 7: 146–147; and Galilei (2008, 97–99, 209).
[12] For example, Flora (1953, 4 n. 2), Drake (1957, 24 n. 2), Bucciantini (2003, 176). In other works, Drake (1983, 14, 223 n. 5) attributes to Galileo a weaker Copernican commitment.

it is important to be especially careful. At this point in the dedication Galileo is calling attention to the satellites of Jupiter and to the fact that he had named them Medicean stars, after the ruling family of Tuscany. The paragraph reads as follows: "Behold then, reserved for your famous name four stars, belonging not to the ordinary and less-distinguished multitude of the fixed stars, but to the illustrious order of the wandering stars; like genuine children of Jupiter, they accomplish their orbital revolutions around this most noble star with mutually unequal motions and with marvelous speed, and at the same time all together in common accord they also complete every 12 years great revolutions around the center of the world, certainly around the sun itself."[13]

The final clause of this passage is the expression often taken as evidence of commitment to Copernicanism. But that is done by translating the original Latin to read, or by interpreting it to mean:[14] "around the center of the world, that is the sun." However, I believe it is more correct to have it the way I rendered it above, which embodies a certain ambiguity. That is, a body that follows Jupiter's orbit is enclosing the sun, and in that sense is moving around it; and this is certain, with both sides of the controversy agreeing. But the clause could also mean "around the sun as center," and then we would have an acceptance of Copernicanism. At any rate, this would involve directly only the heliocentricity of planetary motions, and not necessarily the other elements of the system. So, although there is no question that he is expressing a favorable attitude toward Copernicanism, and that he is now involved with its astronomical part, any more precise definition of his stance is elusive.

This conclusion corresponds very well with the actual content of the work in which we find only two places where the status of Copernicanism is discussed explicitly. They both involve rebuttals of traditional anti-Copernican objections.

In one passage in the middle of the book, Galileo indicates he is now in the position of being able to answer the objection that the earth cannot be a planet because it is devoid of light; his telescopic discoveries about the optical properties of the moon enable him to say that the earth is not essentially different from the moon in this respect.[15] More generally, he thinks his lunar discoveries are such "that the connection and resemblance between the moon and the earth may appear more plainly";[16] that is, these discoveries show the untenability of the earth–heaven dichotomy, and so undermine the anti-Copernican objection from the earth–heaven dichotomy.

The third passage occurs at the end of the book, where Galileo explains that the ability of Jupiter's satellites to keep up with Jupiter as it revolves in its orbit allows one to refute the traditional lunar-orbit objection.[17] This was the argument that the earth cannot revolve around the sun because the moon clearly orbits the earth, and so would be left behind.

---

[13] Galilei (2008, 46); cf. Favaro 3: 56.

[14] Flora (1953, 5), Drake (1957, 24), Drake (1983, 14), Van Helden (1989, 31).

[15] Favaro 3: 75, Galilei (1989, 57; 2008, 63)

[16] Galilei (2008, 60); cf. Favaro 3: 72, Galilei (1989, 53).

[17] Favaro 3: 95, Galilei (1989, 84), (2008, 83–84).

3.6 Full-Fledged Pursuit (1609–1616)

Of course, these refutations of objections do not prove Copernicanism; they merely strengthen it. That is why the attitude expressed by Galileo toward it is such that, as we have seen, it may be described as a slightly higher degree of pursuit or favorable appraisal than any expressed earlier.

Finally, it is important to mention a piece of negative evidence, so to speak, regarding something which Galileo does *not* say or do in *The Sidereal Messenger*. In the printed book he dropped a clause which he had written down in the manuscript draft. This occurs at the end of the discussion of Jupiter's satellites, near the end of the book. The manuscript draft speaks of "the Copernican system (which above all I judge to be consonant with the truth),"[18] but the printed book lacks the parenthetical remark. This change clearly shows that in the printed book Galileo was being more cautious in his endorsement of Copernicanism.

The next significant document for this period is Galileo's letter to Giuliano de' Medici, Tuscan ambassador to Prague, dated 1 January 1611. Its main purpose was to decipher the anagram Galileo had sent him in an earlier letter. When properly transposed the anagram stated that Venus shows phases like the moon, a phenomenon Galileo had been able to observe with the help of the telescope. The attitude he displays now is not, as some scholars have alleged (Gingerich 1982), complete acceptance of the Copernican system, but rather acceptance of two specific theses in it. In the text this is as clear as his emphasis on empirical accuracy: "From this marvelous observation we have sensible and certain demonstration of two great questions, which so far have been debated by the greatest minds of the world: one is that planets are all dark (since the same thing happens with Mercury as with Venus); the other is that Venus necessarily revolves around the sun, as do also Mercury and all other planets" (Favaro 11: 11–12). He goes on to describe the change in his attitude as one from belief without, to belief with, empirical proof: "This had indeed been believed by the Pythagoreans, Copernicus, Kepler, and myself, but not sensibly proved, as done now for Venus and Mercury" (Favaro 11: 11–12). I am not sure this particular transition can be equated with that from pursuit to acceptance. More likely, it is a change from one kind of acceptance to another, the difference being the grounds for the acceptance.

In other words, we have a development from a situation in which the acceptability of an assumption was judged on the basis of factors other than empirical accuracy, to a situation in which it was judged on the basis of empirical accuracy itself. The main methodological lesson here would seem to be what may be called a pluralism of cognitive values or methodological criteria. This is not to say, however, that the state of belief or non-empirical acceptance can be extended back to before the telescope. In fact, it is important to notice that, besides distinguishing between these two attitudes, Galileo makes a still finer discrimination between his previous state of belief without empirical proof and his still earlier

---

[18] Favaro 3: 46. Cf. Favaro 3: 95, Galilei (1989, 84; 2008, 83). I thank David Wootton for bringing this evidence to my attention, although, if I understand him correctly, he wants to interpret it as somehow strengthening Galileo's commitment to Copernicanism.

stage of nonbelief, pursuit, or exploration. This is suggested when he goes on to congratulate Kepler and other Copernicans, but does not include himself: "Thus, Mr. Kepler and the other Copernicans will have reason to be proud of their having believed and philosophized correctly, although they have been, and will continue to be, regarded by all bookish philosophers as incompetents and almost fools" (Favaro 11: 11–12).

The next important milestone in our story is a letter to Prince Cesi. Its importance lies in what it reveals about the criterion of simplicity. On 20 June 1612, Cesi had written to Galileo, expressing his attraction to Copernicanism because of its doing away with epicycles and eccentrics, and asking Galileo's opinion on the problem that these seemed unavoidable for the case of the terrestrial and lunar orbits, given the periodic changes in distance between the earth and the sun, and between the earth and the moon (Favaro 11: 332–333). Ten days later Galileo replied that "we must not desire that nature should accommodate herself to what seems better arranged and ordered to us; rather it is appropriate that we should accommodate our intellect to what she has done, certain that this and nothing else is the best" (Favaro 11: 344). He then goes on to apply this principle by arguing that if by epicycles is meant orbits not encompassing the earth then we must admit their reality, examples being the revolutions of Jupiter's satellites around the planet, and the orbits of Venus and Mercury around the sun; and if eccentrics are meant relative to the earth, then the orbit of Mars encompassing the earth is a clear example, since the telescope reveals that its apparent magnitude is 60 times greater at certain places of its orbit than at others. This explicit theoretical articulation of the limitations of simplicity, together with this particular example of actual usage on his part, is completely in accordance with his behavior during the pre-telescopic period; at that time, as we have seen, Galileo did not attach enough weight to the greater simplicity of the Copernican system either to accept it or to ground his pursuit on that. This is not to say, of course, that he completely disregarded the criterion of simplicity, but only that he did not attach significant or decisive weight to it.[19]

The *History and Demonstrations Concerning Sunspots* was written in 1612 and published in 1613. As is well known, this work contains the strongest endorsement of Copernicanism that Galileo ever published, either before or after this date. What is less well known and seldom discussed is the exact form and context that this endorsement takes. It occurs at the very end, in a passage written on 1 December 1612. Galileo finishes with the topic of sunspots about three pages from the end, by stating his main conclusion that the spots are on the sun and made of some kind of volatile substances like clouds on earth.[20] Then he goes on to add that he had thought this was going to be the climax of his celestial discoveries, but that his latest observations of Saturn had just proved him wrong. In fact, he had just observed that Saturn no longer appeared as three attached bodies, which is how the planet had looked ever since the

---

[19] For more discussions of simplicity, see Favaro 7: 139–150, 349–368, 416–425; cf. Finocchiaro (1980, 113–114, 128–129, 133–134, 145–150; 1985).

[20] Favaro 5: 236; Reeves and Van Helden forthcoming.

3.6 Full-Fledged Pursuit (1609–1616)

telescope. Galileo admits his puzzlement and confusion, but goes on to make several very daring and rather precise predictions about the periodic reappearance and re-disappearance of Saturn's companions up to the summer of 1615. He does not, however, reveal the conjecture on which he says he based these bold predictions, but promises to do so later, after events would confirm or disconfirm him. He concludes his discussion as follows: "But if I do not yet doubt their return, I have reservations about the other particular events, now merely based on probable conjecture; yet whether things fall out just this way or some other, I tell you that this planet also, perhaps no less than horned Venus, agrees admirably with the great Copernican system on which propitious winds now universally are seen to blow to direct us with so bright a guide that little [reason] remains to fear shadows or crosswinds."[21]

Two main claims in the last part of this quotation deserve attention. The first is that *perhaps* Saturn's behavior too confirms Copernicanism. Despite the fact that Galileo never did reveal the theoretical basis of his prediction and the connection with Copernicanism, and despite the fact that that basis remains a puzzle for scholars,[22] the judgment is obviously based on the criterion of empirical accuracy. The second claim is that now he thinks all evidence is pointing toward Copernicanism and seems to have little doubt about its correctness. Although Galileo does not explicitly include sunspots in this evidence, the connection is obvious enough that it can easily be attributed to him; as he will argue later in Day I of the *Dialogue*,[23] sunspots contribute to the empirical undermining of the earth–heaven dichotomy, and thus to the strengthening of Copernicanism by removing the corresponding objection, and by supporting the corresponding element of the Copernican system. However, it should also be noticed that the *Sunspots* book does not contain the pro-Copernican argument from the motion of sunspots, which is one of the most powerful of those in the *Dialogue*.[24] Although at this time (1612–1613) Galileo was in a position to learn about the periodic motion of sunspots, and to formulate the geokinetic explanation of that phenomenon, he did not do this until 1629, while in the process of writing the *Dialogue* (cf. Drake 1978, 311).

In the Postscript that was added to the published version of the *Sunspots* book there is another important clue that at about this time Galileo was finding another piece of evidence which could be explained only on the basis of the earth's orbital revolution. The phenomenon involved the eclipsing of Jupiter's satellites and the variations in the duration of these eclipses. The details of the argument were never written up by Galileo, and they are extremely technical. Stillman Drake tried to identify the relevant documentary evidence and to piece together the main points.[25] However, Galileo's claim in the Postscript is as clear and unambiguous as one could wish:

---

[21] Translated in Drake (1978, 198). Cf. Favaro 5: 238, Drake (1957, 144), Reeves and Van Helden forthcoming.

[22] For some light on the matter, see Drake (1978, 198, 278).

[23] Favaro 7: 76–80, Galilei (1967, 51–55; 1997, 98–103).

[24] Favaro 7: 372–383, Galilei (1967, 345–356).

[25] Drake (1983, xix, 133–135).

But a more wonderful cause of the hiding of any of these is that which arises from various eclipses to which they are subject, thanks to the differing directions of the cone of shadow of Jupiter's body – which phenomenon, I confess to you, gave me no little trouble before its cause occurred to me. Such eclipses are sometimes of long and sometimes of short duration, and sometimes are invisible to us. These differences come about from the annual movement of the earth, from differing latitudes of Jupiter, and from the eclipsed planet's being near to or farther from Jupiter, as you shall hear in more detail at the proper time.[26]

The *Sunspots* book also contains two other passages that are relevant to understanding the character of Galileo's endorsement of Copernicanism. One is the theory that the sun rotates on its axis,[27] which represents a modification of Copernicus's original system. The other is a criticism of an argument given by Christopher Scheiner to show that Venus revolves around the sun; of course, Galileo accepts this conclusion, but he finds several faults with Scheiner's attempt to ground it on the alleged observation of a transit of Venus across the solar disk, rather than on the phases.[28] Both passages show an attitude that might be called piecemeal or non-totalistic: he does not hesitate to modify various elements of the Copernican system, or to reject proposed contributions by other supporters, as the case requires. This suggests, it seems, a greater commitment to certain procedures than to any specific physical or cosmological theses; for the case of solar rotation the procedure in question would seem to be related to empirical accuracy, while for Scheiner's argument about Venus it would seem to involve correct reasoning.

Finally, methodological duty obliges me to note the existence of a somewhat problematic statement by Galileo in this context, more specifically in the course of criticizing Scheiner's argument. This occurs when Galileo faults him for having failed to block effectively certain ways of evading the argument for the heliocentricity of Venus's orbit. The passage reads: "Just as for those very knowledgeable in astronomical science it sufficed to have understood what Copernicus writes in *De Revolutionibus*, in order to ascertain themselves of the revolution of Venus around the sun and of the truth of the rest of his system, so for those whose understanding is below the average it was necessary to block the above mentioned retreat-routes; I see that Apelles [Scheiner] touched upon only two of them, and it seems to me that even these are not completely knocked down."[29]

Here I should say that Galileo is exaggerating the strength of Copernicus's own arguments. I feel justified in this criticism by the fact that this Galilean judgment does not correspond to his previous behavior, either his actual research practice or his explicit reflections. Of course, if this were not so then I would have to accept this judgment as typical, and dismiss or criticize the conflicting ones as deviant for some reason or other. As things stand, however, we may partly use this passage as an example that we are not trying to paint a picture of a Galileo who is infallible or

---

[26] Translated in Drake (1978, 208). Cf. Favaro 5: 248, Reeves and Van Helden forthcoming.
[27] Favaro 5: 133–135, Reeves and Van Helden forthcoming.
[28] Favaro 5: 192–199, Reeves and Van Helden forthcoming.
[29] Favaro 5: 195; cf. Reeves and Van Helden forthcoming.

3.6 Full-Fledged Pursuit (1609–1616)

above criticism. But it also provides a clue that he was far from being completely insensitive to Copernicus's own original justification of his system, a point that will reappear presently.

The evolution of Galileo's Copernicanism comes to a climax in 1614, as we can see in his letter to Baliani dated March 12. In it we find for the first time a genuine expression of certainty, together with a summary of his reasons, as well as a reasoned rejection of Tycho's theory. This crucial passage is the following: "As regards Copernicus's opinion, I really hold it as certain, and not only because of the observations of Venus, of sunspots, and of the Medicean planets, but because of his other reasons, and because of many other particular reasons of mine, which seem conclusive to me … In Tycho's opinion I still find all the very great difficulties which make me abandon Ptolemy, whereas in Copernicus I find nothing which gives me the least scruple, and least of all the objections which Tycho makes to the earth's motion in certain letters of his" (Favaro 12: 34–35).

There is no question that we have here an endorsement of Copernicanism stronger than any we have seen up to this date, and than we find anywhere else subsequently. And it is clear that Galileo uses labels that are epistemically loaded: "certain" (*sicura*) to characterize the position, and "conclusive" (*concludenti*) to describe the supporting arguments. Nevertheless, I do not think we have here a qualitative jump from what came immediately before; it is a change of degree. Moreover, the notions of certainty and conclusiveness (as used informally in common parlance) do not seem to me to be discrete, but to admit of degrees within them. Finally, Galileo's words must be balanced against the following.

First, this certainty and conclusiveness may be no more literally true than is Galileo's assertion in the same passage that he finds nothing in Copernicus which gives him "the least scruple." For we have seen that solar rotation represents a departure from Copernicus; and, as the later *Dialogue* clearly explains,[30] Galileo never accepted Copernicus's "third motion," according to which the terrestrial axis was supposed to rotate with an annual period in order to compensate for the orbital revolution; and, as the *Dialogue* also shows,[31] Galileo never refuted the objection from stellar parallax, but rather accepted the difficulty, and suggested a way of testing for it. Second, this is a private letter, and there is no analogous expression in any published work. Third, Galileo was not at this time writing the work on the "System of the World" which he had promised in *The Sidereal Messenger*; now, although there were many causes for this delay, including external ones, one contributing factor may have been that he did not have yet all the arguments and evidence required to be really sure, as he was claiming in this letter; hence the crucial words here are perhaps merely that – words.

Despite all these provisos, the endorsement neither can nor should be ignored. So let us look at the motivating reasons. Notice that there are three groups of reasons: first, the observational ones depending on the telescope and involving the phases of

---

[30] Favaro 7: 424–426, Galilei (1967, 398–399).
[31] Favaro 7: 404–416, Galilei (1967, 377–389; 1997, 264–281).

Venus, sunspots, and Jupiter's satellites; second, Copernicus's own reasons; and third, Galileo's own "other particular reasons." The first group were the chronologically latest ones which had been accumulating in the last several years since the telescope, and which seem to have caused a qualitative change in Galileo's attitude, from something like indirect pursuit or qualified rejection to something like direct pursuit or qualified acceptance; they relate intimately to the criterion of empirical accuracy. The second group of reasons were never, not even in the *Dialogue*, discussed by Galileo in any great detail; in regard to their methodological character, however, we do know, as we saw above, that they were not for Galileo simplicity considerations, at least not in any simple sense of the notion of simplicity. I believe that they reduce primarily to the criterion of explanatory coherence or anti-ad-hocness.[32] The important point for us now is that Galileo was not insensitive to this sort of consideration, however insufficient he may have regarded it as a basis of acceptance, or even as a sole basis of pursuit; moreover, in view of his qualms about the principle of simplicity simply understood, notice that this conception of Copernicus's own reasons is also different from the simplicity interpretation. Finally, Galileo also had a third group of reasons, and these he had had for a very long time; collectively, they reduced, as we have seen already, to the progressiveness and problem-solving success of his theory of motion, a theory quite compatible with Copernicanism, and quite at odds with the Aristotelian-Ptolemaic system. Specifically they included such arguments as the one from tides and the one from neutral motion. Since Galileo claims he has "many" such arguments, it is interesting to speculate what else he had in mind.

One clue to what they might be is in Galileo's remark about the Tychonic system. He says he has the same basic difficulties with it as with the Ptolemaic system. These can only be problems stemming from his theory of motion, or what today we would call dynamical difficulties. They are sketched and summarized at the beginning of Day II of the *Dialogue*.[33] Two points are especially pertinent here. These objections apply with equal force to both the Tychonic and the Ptolemaic versions of the geostatic system, and so the often-heard criticism that Galileo is guilty of neglecting the Tychonic system is without foundation. Or we may say that Galileo was not insensitive to the need to appraise theories in the light of rival alternatives, and his behavior shows that he appraised them not just vis-à-vis empirical data, but also on a comparative basis. Second, those anti-geostatic objections are explicitly labeled as probable and not conclusive by Galileo, and this helps us to resolve one last question about the present passage.

Here he judges the strength of his reasons as "conclusive." The actual expression and punctuation he uses ("which seem conclusive to me") are ambiguous, since the conclusiveness could be referring to either his third group of reasons individu-

---

[32] The argument here would be essentially the one given by Lakatos (1978, 168–192), whose thesis may be accepted when limited to this specific issue and qualified in this manner. See also Sections 2.1 and 6.2 of this book.

[33] Favaro 7: 139–150, Galilei (1967, 114–124; 1997, 128–142; 2008, 201–213).

3.6 Full-Fledged Pursuit (1609–1616)

ally, or his third group of reasons as a whole, or all his three groups of reasons collectively. I believe that the last one is meant, as suggested by the just mentioned passage in the *Dialogue*. But one difficulty now remains. It is this. Since the tidal argument is obviously included in the third group, among Galileo's physical reasons, and since this argument is not discussed in the same passage at the beginning of Day II but in Day IV of the *Dialogue*, could he not here be attributing conclusiveness to the tidal argument? It is conceivable that he might, but in fact he is not. First, it can be argued that the tidal argument is presented in the *Dialogue* as an inductive, probable, hypothetical, non-necessitating argument,[34] but this thesis is insufficient here since the *Dialogue* was written after the anti-Copernican decree of 1616, and hence Galileo had external motives for such a presentation. So we need to examine the tidal argument as it exists in the version written by him in January 1616, 2 months before the Decree of the Index. At any rate, this "Discourse on the Tides" is one of the relevant documents in its own right.

An important but often overlooked fact about the "Discourse on the Tides" is that it is not just about tides, but also about winds. In other words, we have not one but two arguments for the earth's motion, the first based on the tides, the second on prevailing easterly winds, the so-called trade winds. This immediately suggests that neither one is considered to be absolutely conclusive, for the argument from the trade winds would be superfluous if the tidal argument had that quality, and vice versa. It is true that in the about 20 pages of this essay, only about two at the end are devoted to the wind argument, but that only means that the topic of tides involves more details and that the argument has more complications. So if length were at all relevant that might actually weaken it by exposing it to more potential difficulties or errors. Moreover, degrees of strength may be a function of length, but absolute conclusiveness is not because it would lead to the absurdity of labeling the tidal argument both absolutely conclusive and also nine times stronger than the wind argument.

This suggestion is reinforced by frequent explicit remarks on Galileo's part. For example, after listing several possible causes of the motion of water in general, he introduces the connection between tides and the earth's motion with the following words:

> When I examine these and other facts pertaining to this cause of motion considered last, I would be greatly inclined to agree that the cause of tides could reside in some motion of the basins containing seawater; thus, attributing some motion to the terrestrial globe, the movements of the sea might originate from it. If this did not account for all particular things we sensibly see in the tides, it would thus be giving a sign of not being an adequate cause of the effect; similarly, if it does account for everything, it will give us an indication of being its proper cause, or at least of being more probable than any other one advanced till now.[35]

This judgment is typical of the rest of this essay, but we shall limit ourselves to a concluding note on the last page of the essay, which is simply too revealing to be ignored:

---

[34] Cf. Finocchiaro (1980, 6–24); and Chapter 9.
[35] Finocchiaro (1989, 122); cf. Favaro 5: 381.

I could propose many other considerations if I wanted to delve into finer details. Many, many more could be advanced if we had abundant, clear, and truthful empirical reports of observations made by competent and diligent men in various places of the earth; for by comparing and collating them with the assumed hypothesis we could decide more firmly and ascertain more correctly the things that pertain to this very obscure subject. At the moment I only claim to have given something of a sketch, suitable at least for stimulating students of nature to reflect on this new idea of mine. I hope, however, that it does not turn out to be delusive, like a dream which gives a brief image of truth followed by an immediate certainty of falsity. This I submit to the judgment of intelligent investigators.[36]

My conclusion is that there is no doubt the certainty expressed by Galileo in the letter to Baliani was an inductive, practical, not absolute kind of certainty; it was based on the practical conclusiveness of all the arguments taken together, physical, telescopic, and Copernicus's original ones. Certainly he did not think that any one individual argument or piece of evidence was absolutely conclusive.

This analysis also fits very well with another series of relevant remarks in another key document of the period, Galileo's *Letter to the Grand Duchess Christina* (1615). The main purpose of this letter, as we shall see later (Sections 4.5 and 9.3), is to defend the Copernican system from the scriptural objection. However, in the letter's introductory part, to set the stage for the discussion, Galileo describes his attitude toward Copernicanism. These descriptions deserve to be analyzed here.

To begin with, Galileo tells us, "in my astronomical and philosophical studies, on the question of the constitution of the world's parts, I hold that the sun is located at the center of the revolutions of the heavenly orbs and does not change place, and that the earth rotates on itself and moves around it."[37] This is a clear and explicit statement of endorsement, but the strength and nature of this endorsement must be inferred from other statements. There are two sets of relevant statements: those that are meant to clarify his relationship to Copernicus, and those that are intended to explain how Galileo's cosmological position relates to his own astronomical discoveries.

By the latter it is obvious that Galileo is referring to such things as the lunar mountains, Jupiter's satellites, the phases of Venus, and sunspots. Although these were questioned at first, he now regards their existence and main features as conclusively proved, for he notes with pride that "then it developed that the passage of time disclosed to everyone the truths I had first pointed out."[38] By contrast, about the Copernican hypothesis Galileo says that "I confirm this view not only by refuting Ptolemy's and Aristotle's arguments, but also by producing many for the other side, especially some pertaining to physical effects whose causes perhaps cannot be determined in any other way, and other astronomical ones dependent on many features of the new celestial discoveries; these discoveries clearly confute the Ptolemaic system, and they agree admirably with this other position and confirm it."[39] The key notion here is that of confirmation. He seems to regard the Copernican position as confirmed. What does this mean?

---

[36] Finocchiaro (1989, 133); cf. Favaro 5: 395.
[37] Finocchiaro (1989, 88), or Galilei (2008, 111); cf. Favaro 5: 310–11.
[38] Finocchiaro (1989, 88), or Galilei (2008, 110); cf. Favaro 5: 310.
[39] Finocchiaro (1989, 89), or Galilei (2008, 111); cf. Favaro 5: 311.

That it does not mean conclusively proved is shown by Galileo's understanding of his relationship to Copernicus. On the next page of the letter's introductory part, in the context of discrediting some of his opponents who thought that the geokinetic idea was Galileo's invention, he clarifies that "Copernicus was its author, or rather its reformer and confirmer."[40] The same terminology of confirmation is used. Thus, it is obvious that Galileo thinks he is doing more of the same of what Copernicus did. There is no claim of a breakthrough from Copernicus's mere confirmation to his own strict demonstration. There is, of course, a strengthening of the position, which Galileo describes with the words that now "one is discovering how well-founded upon clear observations and necessary demonstrations this doctrine is."[41] He does not say that the earth's motion is now clearly observed and necessarily demonstrated, but that it is well founded. The necessary demonstrations referred to must be those that prove the truth of his celestial discoveries mentioned earlier. There is no problem, of course, about a long and complex probable proof, such as that supporting Copernicanism was for him at that time, consisting partly of segments that are necessary demonstrations or clear observations, because the final conclusion would be only as weak as the weakest supporting subargument. And Galileo clearly realizes this since, apropos of Copernicus, one remark he makes is that parts of his work too consist of clear observations and necessary demonstrations. That is, rather than getting involved in biblical interpretation, Copernicus "always limits himself to physical conclusions pertaining to celestial motions, and he treats of them with astronomical and geometrical demonstrations based above all on sensory experience and very accurate observations."[42]

3.7 The Post-1616 Period

My critical examination of the evidence is far from complete yet. So far I have not even touched upon any of the documents pertaining to the theological controversy and the Inquisition proceedings. Far from being irrelevant to the present inquiry, they constitute a classic test case for a specific methodological issue, namely whether or not the acceptability of a scientific theory is significantly judged on the basis of its relation to nonscientific beliefs. However, I shall deal with this evidence in a different way as contrasted to the direct analysis of documents carried out above. The problem is this.

We have seen Galileo moving from an attitude of partial pursuit to one of practical or tentative acceptance of Copernicanism. Yet the public Decree of the Index and the private, personal warning by Bellarmine in the name of the Inquisition in 1616 forbade him to hold, support, or defend the truth of the earth's motion; and he promised to obey this prohibition.[43] And from 1616 to the election of pope Urban

---

[40] Finocchiaro (1989, 89), or Galilei (2008, 112); cf. Favaro 5: 312.
[41] Finocchiaro (1989, 90), or Galilei (2008, 113); cf. Favaro 5: 312.
[42] Finocchiaro (1989, 91), or Galilei (2008, 114); cf. Favaro 5: 313.
[43] These developments are discussed in more detail in Chapter 7.

VIII in 1623 he apparently did obey it, and one could say that he did not even pursue Copernicanism then. From 1624 to the trial of 1633, his work on the *Dialogue* (which was published in 1632) may be taken to constitute at least some kind of pursuit, perhaps more. Then after the condemnation of 1633, his work on the *Two New Sciences* may be taken to constitute the same kind of partial pursuit he was engaged in before the telescope. Are we then to picture the evolution of Galileo's Copernicanism as a kind of parabola, such that his commitment to it first slowly increased until 1616, on the basis of such criteria as explanatory coherence, problem-solving success, and empirical accuracy, and then it gradually declined on the basis of pressure from the Church?

The problem with this interpretation is that in the present context such Church pressure would have to be translated into nonscientific beliefs held by Galileo which led him to certain kinds of judgments about Copernicanism. Now, there is no question that the Church pressure was real, and that in 1613–1616 he felt the need to write extensively to defend Copernicanism from religious objections. However, these very writings show that he held the view that the acceptability of scientific ideas should not be judged at all on the basis of their relation to religion. Indeed his writings constitute a classic source for this principle.

So should we adopt the view that Galileo never really judged the status of Copernicanism on the basis of its religious heterodoxy, but that he merely said so on some occasions, uttering words to that effect, but without meaning them? Hence such words and actions would be the understandable behavior of a practical man, but not those of a man of science, of a scientist qua scientist.

The difficulty with this interpretation is that it would beg the question in the present context. For here we are trying to use historical evidence to decide whether or not certain methodological principles are acceptable, and it is not clear why we should take Galileo's judgments at their face value up to 1616, but not so thereafter.

One underlying issue here is whether the problem stems from the fact that after 1616 there is a contradiction between Galileo's words and his deeds, or at least between certain series of his words and deeds and certain other series. The problem would then be analogous to the one we encountered earlier when we criticized and dismissed a specific Galilean assertion about the decisiveness of Copernicus's own proofs. We did so on the basis of the weight of the greatest portion of the evidence. Perhaps we could proceed the same way on the science-religion question. But I cannot help feeling that the cases are different.

It may be that the way out of this quandary lies with the word "largely" in the methodological principle in question, which attributes a "large" role to the relation of scientific theories and nonscientific beliefs, such as religious ones. Perhaps we could say that after 1616, and especially after 1633, Galileo's interest in and work on Copernicanism did indeed wane, due to his belief that he must bow to the authority of the Church, but that this had a small effect on him, as shown by such works as the *Assayer, Dialogue,* and *Two New Sciences.* The significance of such a role might perhaps be compared to that of simplicity. That is, given the small weight Galileo attached to simplicity considerations, we can say it is false that he judged Copernicanism "largely on the basis of simplicity"; similarly, given that he attached a small weight to

Church authority in scientific matters, it is false that he judged it "largely on the basis of its relation to nonscientific beliefs." To make this interpretation stick, then one would have to show on the basis of the texts and the documents that the effect of his submission to Church authority on his scientific practice in regard to Copernicanism was "not large." This seems a viable and defensible thesis.

3.8 Conclusion

In summary, this chapter has examined the evolution of Galileo's Copernicanism in order to determine whether he judged its acceptability largely on the basis of empirical accuracy, progressiveness, general problem-solving success, predictive novelty, external problem-solving success, explanatory coherence, simplicity, aesthetics, self-consistency, well-established beliefs, nonscientific beliefs, and/or practical applications. The result has been that his various judgments of the acceptability of the Copernican system fall primarily into three periods. First, there is the pre-telescopic stage, when his cognitive stance may be described as partial, indirect, and implicit pursuit or qualified rejection. Then we have the full-blown middle period from 1609 to 1616, which constitutes a qualitative change from the preceding one; which may be described as full-fledged, direct, and explicit pursuit or qualified acceptance; and which contains varying and increasing degrees of endorsement or commitment, up to a stance that may be labeled tentative or practical acceptance. Thirdly, there is a problematic post-1616 stage, which I have only briefly discussed, and during which the main relevant evidence involves his relationship with the Church.

During the first period, he judged Copernicanism largely on the basis of its progressiveness and problem-solving success in the physics of motion, and its explanatory coherence in the astronomical field. During the second stage, he judged it largely on the basis of these criteria, as well as empirical accuracy. And after 1616, he judged it largely on the basis of these four criteria plus its relationship to his religious beliefs. At no time did he judge its acceptability largely on the basis of predictive novelty, or of external problem-solving success, or of simplicity, or of aesthetics, or of self-consistency, or of well-established beliefs, or of practical applications. Thus five criteria emerge as the ones which may be said to have played a large role, both individually and collectively, in Galileo's attitude toward Copernicanism: progressiveness, problem-solving success, explanatory coherence, empirical accuracy, and religious beliefs. As here listed, these are not, however, ranked in order of importance; the analysis of their relative significance would require further work.

Finally, I should mention two other methodological principles for which this chapter provides an overwhelming amount of historical evidence, although admittedly I did not explicitly or deliberately devise my test with them in mind. One principle involves the question of whether or not, during a change in guiding assumptions (i.e., during a scientific revolution), these assumptions change abruptly and totally. My evidence from Galileo's Copernicanism shows that they do not, but that his attitude toward it changed slowly, gradually, and in a piecemeal fashion; in short, judiciously, so to speak.[44] The second principle relates to the question of whether during a scientific revolution it is enough to provide constructive evidence and reasons for the new set of guiding assumptions, or whether it is also necessary to provide a critical defense of the new view from objections based on the old. Clearly the evidence presented shows that constructive evidence is insufficient and that a critical defense is necessary.[45] These conclusions are, I believe, so important that one cannot fail to mention them despite the fact that they are unintended consequences of the original test design.

---

[44] This conclusion also corresponds to an interpretation of Galileo elaborated and defended by Drake (1978); cf. also Chapter 6. For an alternative view, see Camerota (2004, 98, 253, 259, 283).

[45] This conclusion is, of course, a part of the *overarching thesis* elaborated in this book; cf. last section of the Introduction.

Chapter 4
Galilean Critiques of the Biblical Objection

We have seen that, besides the astronomical observational difficulties, another serious obstacle to the acceptance of the Copernican system stemmed from the theological biblical objections. Galileo himself, however, attached no significant force or great relevance to the biblical objections, as he did for the case of the observational difficulties. But many other people found the biblical objections unanswerable and decisive, and these people included highly competent and otherwise progressive minded astronomers and mathematicians, such as Tycho Brahe and Christopher Clavius. Thus Galileo did not get involved in the theological controversy until his views and his person were attacked as heretical and theologically erroneous. But when he did get involved, his defense of the Copernican system from theological objections was intellectually as cogent and forceful as his defense of heliocentrism from the astronomical objections. Moreover, although opposed by the ecclesiastic establishment, Galileo was inspired, encouraged, and supported by several maverick professional Catholic theologians who wrote similar and complementary critiques of the scriptural objections. Notable among them were Carmelite friar Paolo Antonio Foscarini and Dominican friar Tommaso Campanella.

4.1 Preliminary Considerations

The structure of this chapter is that of a discussion of the interaction between natural philosophy or science and biblical exegesis or hermeneutics in regard to four crucial authors whose most relevant works were written in 1615–1616, namely around the time of the first phase of Galileo's trial. Before we proceed, this aim needs to be clarified in several ways.

One clarification pertains to the expression *natural philosophy or science*. I share the misgivings expressed by many scholars[1] that in using such words as *science*, *scientific*, and *scientist* one must be careful to avoid anachronism, and that an easy

---

[1] See, for example, Ross (1962), Wilson (1996; 1999), Osler (1998).

way of avoiding it is to speak of philosophy and cognate terms, although of course the latter must be understood in the sense of *natural* philosophy; otherwise, we run into the difficulty stemming from the fact that the seventeenth-century gap between natural philosophy and first or metaphysical philosophy is perhaps greater than that between seventeenth-century natural philosophy and twentieth-century natural science. Nevertheless, such sound anti-anachronistic advice should not go too far and end up falling prey to what might be called linguistic chauvinism. For not only the Latin word *scientia* has a very long history, but even the word *scientist* has a longer history (or pre-history) than commonly thought. That is, while it may be true that the English word *scientist* was first used by William Whewell in 1834 (cf. Ross 1962, 71–72), the Italian word *scienziato* goes much further back: it appears several times even in some texts directly relevant to the present topic, namely in Federico Cesi's letter to Galileo of 12 January 1612 and in Campanella's letter to pope Urban VIII of 10 June 1628;[2] and their usage is not at all idiosyncratic, as one can see from the fact that Galileo used it in a crucial passage of the *Two New Sciences*,[3] and so did also Vincenzio Viviani in a 1690 letter to Antonio Baldigiani that is an important source in the history of the subsequent Galileo affair.[4] Thus, in contexts such as the present one I will not always feel obliged to use the full locution *natural philosophy or science*, and may speak of just *science* or *scientific*; in so doing, I will feel justified to some extent by more than just brevity's sake.

If this first clarification amounts to a recognition that two things (terms) may be associated in the present context, the next point requires that two things be distinguished within the natural philosophy (or science) under scrutiny. The two developments I would want to distinguish are Galileo's telescopic discoveries and the re-evaluation of Copernicanism they implied. These Galilean discoveries were the mountains of the moon, the satellites of Jupiter, the stellar composition of the Milky Way and nebulas, the phases of Venus, and sunspots. They were made in the period 1609–1613 and described publicly in *The Sidereal Messenger* (1610) and in the *History and Demonstrations Concerning Sunspots* (1613). The cosmological implications of these discoveries were of course a much more controversial matter, and the formulation on which I would like to focus is the one which I believe was favored by Galileo himself: "the demonstrations for the earth's motion are much stronger that those for the other side";[5] that is, the Copernican theory is much more likely to be true than either the Ptolemaic or the Tychonic theory. This re-evaluation involves two main elements: a realistic, non-instrumentalist interpretation of Copernicanism, and a claim of its superiority over all geostatic alternatives. It is useful to distinguish the observational discoveries from the theoretical re-evaluation not only because obviously one could accept the former and not the latter, but also because the scriptural controversy was elicited primarily (although not exclusively) by the latter.

---

[2] Cesi to Galileo, 12 January 1612, in Favaro 12: 128–31, at 130.54; Campanella (1628, 218, 222, 224).
[3] In Favaro 8: 212, Galilei (1974, 169; 2008, 349); cf. also Section 5.12.
[4] In Favaro (1887, 153–155), Finocchiaro (2005b, 90–92).
[5] In Favaro 5: 354. Cf. Finocchiaro (1989, 73; 1997a, 5), Camerota (2004, 352).

4.1 Preliminary Considerations

Third, it is useful to distinguish not only between *exegesis* and *hermeneutics*, but also between them (especially the latter) and what I shall call *meta-hermeneutics* or *meta-scriptural methodology*. That is, in accordance with standard terminology, the label exegesis should be used to refer to the practice of textual interpretation, namely to particular interpretations of particular biblical texts; and the label hermeneutics should be taken to mean the theory of textual interpretation, i.e., principles and methods of biblical interpretation. However, we also need a label to denote the study of questions about the role (if any) of Scripture in the search for truth about nature; and this is what I am calling *meta-hermeneutics* or *meta-scriptural methodology*. Accordingly, hermeneutics may be taken to refer to principles like the following: all scriptural statements should be interpreted literally if the literal interpretation implies no absurdities (of various sorts); a scriptural statement should be interpreted as accommodating itself to popular language and beliefs if the literal interpretation implies absurdities, such as contradicting most other relevant scriptural statements (this is a version of the so-called principle of accommodation, and it is the *converse* of the preceding principle); a scriptural statement about natural phenomena should not be interpreted literally if the literal interpretation contradicts a scientific truth that has been conclusively demonstrated (this has been called the principle of priority of demonstration[6]); a scriptural statement about natural phenomena should be interpreted literally if the literal interpretation does *not* contradict any conclusively demonstrated truths (this is the *converse* of the last principle, and it could be called the principle of the *necessity* of demonstration). However, it would encourage confusion to regard the following as being principles of scriptural interpretation: "it should be said that our [scriptural] authors did know the truth about the shape of heaven, but that the Spirit of God, which was speaking through them, did not want to teach men these things which are of no use to salvation"[7] (this is a principle advanced by St. Augustine in *De Genesi ad litteram*, and quoted by Galileo in the *Letter to the Grand Duchess Christina* [1615] and by pope Leo XIII in the encyclical *Providentissimus Deus* [1893]); in Scripture, "the intention of the Holy Spirit is to teach us how one goes to heaven and not how heaven goes"[8] (which Galileo in the same letter attributes to Cardinal Cesare Baronio); and "Sacred Scripture ... does not instruct men in the truth of the secrets of nature ... its intention is now only to teach us the true road to eternal life," which is a formulation advanced by Foscarini (1615a, 30–31). These are versions of a principle that has been given such labels as the principle of scriptural limitation, or the principle denying the scientific authority of Scripture;[9] this is an epistemological, methodological, and perhaps even theological principle about the role of Scripture in

---

[6] McMullin (1998, 294; 2005c, 93).

[7] Augustine, *De Genesi ad litteram*, ii, 9, 20, as translated in Finocchiaro (1989, 95) and in Galilei (2008, 118). Cf. Favaro 5: 318, Leo XIII (1893, paragraph 18, p. 334).

[8] Finocchiaro (1989, 96), or Galilei (2008, 119); cf. Favaro 5: 319.

[9] Respectively, McMullin (2005c, 95), Finocchiaro (2005b, 266).

scientific inquiry, about its lack of a role to be more precise, but it is not a principle that says anything about how scriptural statements ought to be interpreted. Of course, there are relationships between the hermeneutical principles of the first sort and this methodological principle; moreover, in the present context we are interested in both; but these are no reasons to conflate scriptural hermeneutics with meta-scriptural methodology.[10]

The next clarification may be centered on the notion of *interaction* between science and Scripture. One reason for the complexity of this interaction is that there are really four main things that are interacting and not just two, since as we have just seen, the "science" in question can be either the Galilean telescopic discoveries or the Galilean re-evaluation of Copernicanism, and the "Scripture" can refer to either scriptural exegesis as such or meta-scriptural methodology. A second reason is that in such interactions the causal and historical influence could go in both directions, and so it is important to be on the lookout for the possibility not only that scientific developments affected scriptural ones, but also for the reverse possibility that scriptural developments may have had an effect on scientific ones. A third reason for the complexity is that the interactions to be discussed may be not only those that have actually taken place, but also those that could have occurred, as well as those that ought to have happened. In other words, in regard to the eight possible interactions (among *four* things, in *two* directions), what one wants to do is not only to understand or interpret what exactly influenced or produced what and how, but also to evaluate or assess the propriety or desirability or beneficial character of the historical developments so interpreted.

One final caveat is in order. The four authors who are the subject of our discussion are relevant and important for two distinct (although interrelated) reasons. One is that they had interesting, important, and influential ideas about our topic. A second reason is that they were practically involved in the events of those crucial years, 1615–1616. In other words, the actions as well as the thought of our four authors are part of our

---

[10] My distinction here is similar to one made by Beretta (2005b, 244) when he argues that in the 1616 condemnation of Copernicanism, "the issue is not primarily concerned with hermeneutics – that is, the principles of biblical exegesis – but with criteriology – that is, inquiry about the validity of knowledge, establishing criteria by which truth can be distinguished from error." For Beretta (2005b), the main principle involved was one about the hierarchy of disciplines, theology and natural philosophy in particular: Church officials were allegedly upholding, and Galileo and his supporters were denying, the priority of theology over natural philosophy; this was a principle that was supposedly an essential part of scholastic criteriology and that was explicitly promulgated by the Fifth Lateran Council of 1513, which used it to condemn the thesis of the mortality of the soul. However, I hesitate to adopt such a term as "criteriology," which I find too opaque, ambiguous, and unconventional; moreover, although the hierarchy of theology and philosophy was an important issue, and although as we will see below that Galileo discusses it explicitly in the *Letter to the Grand Duchess Christina*, and although Beretta deserves credit for having called attention to this issue, his emphasis on it is too one-sided. Finally, I see a serious difficulty with the analogy which Beretta tries to articulate between the Aristotelian thesis of the mortality of the soul and the Copernican thesis of the earth's motion: whereas the latter is a question of natural philosophy, the former is not, but is treated in such other branches of philosophy as metaphysics, rational psychology, etc.

story. Of course, thought and action are related, but they do not always correspond. For example, at the level of thought, Foscarini held, as mentioned above, a principle of scriptural limitation, which can also be viewed as a principle of non-interference by scriptural authority in scientific investigation; yet his act of publishing his ideas on the subject was a major factor in bringing about the Catholic Church's interference in the form of the anti-Copernican decree of 1616. In this chapter, my emphasis will be on the thought of these four authors, with only a little sprinkling about their actions, but a full account would have to examine the latter more extensively.[11]

4.2 Copernicanism and Scripture

Scriptural criticism of the Copernican theory was immediate, as we now know from the censure of *De revolutionibus* (1543) written by Giovanni Maria Tolosani in 1546–1547.[12] Indeed such criticism even antedated the publication of Copernicus's masterpiece: in an incidental remark in 1539, Martin Luther criticized Copernicanism as incompatible with the biblical passage in Joshua 10: 12–13;[13] and in 1541 the second edition of Georg Joachim Rheticus's *Narratio prima* had a preface that quoted a letter by a friend (Achilles Pirmin Gasser) suggesting that heliocentrism "can be judged heretical."[14] It is also well known that the criticism continued, for scriptural objections were usually included in discussions of the status of heliocentrism.[15] However it was not until Galileo's telescopic discoveries in 1609–1613 that the problem became a crisis. I believe the key reason for this crisis was that these discoveries entailed a major reassessment of Copernicanism:[16] they suggested that the earth's motion and its a-centric "heavenly" location could now be regarded as real possibilities and not merely convenient instruments of astronomical calculations and predictions, and indeed as more likely to be true than the alternatives, although of course they did not provide a conclusive demonstration of the physical truth of the Copernican theory.

In fact, three months after March 1610, when Galileo's *Sidereal Messenger* left the printing press, Martin Horky published *A Very Short Excursion Against the Sidereal Messenger* (Favaro 3: 127–145). A few months after that, Ludovico delle Colombe compiled an essay "Against the Earth's Motion" that included theological objections; it circulated widely, but was left unpublished (Favaro 3: 12, 251–290).

---

[11] Part of this will be found in Chapter 7 below.
[12] Garin (1975, 280), Granada (1997), Lerner (2002), Rosen (1975).
[13] Blackwell (1991, 23); cf. Luther, *Tischreden* (Weimar, 1912–1921, IV, no. 4638).
[14] Lerner (1999, 69; 2005, 11).
[15] See, for example, Grant (1984, 61–62), Lattis (1994, 106–144), Lerner (1999; 2005), Howell (2002, 39–136), Kelter (1995; 2005).
[16] See Wallace (1984a, 282–298), Finocchiaro (1980, 3–45; 1997a, 5, 28–38). Cf. Favaro 5: 354, Finocchiaro (1989, 73), Camerota (2004, 352), Galilei (2008, 151).

The following year, Francesco Sizzi published in Venice his *Dianoia astronomica, optica, physica* (1611), objecting on scriptural grounds to Galileo's discovery of the satellites of Jupiter.[17] In 1612, Giulio Cesare Lagalla, professor of philosophy at the University of Rome, published in Venice a book *On the Phenomena in the Orb of the Moon*, disputing Galileo's lunar discoveries (Favaro 3: 13, 309–399).

By the summer of that year Galileo was worried enough that he asked Cardinal Carlo Conti for advice on whether Scripture really favors Aristotelian natural philosophy. Conti was an influential churchman in Rome and replied promptly in two thoughtful letters.[18] His views can be summarized as follows: The Aristotelian doctrine of heavenly unchangeability *contradicts* Scripture and the common opinion of Church Fathers; but it will take some time to determine whether the new discoveries establish heavenly changeability since, for example, some will try to explain sunspots in terms of swarms of small planets circling the sun. The scriptural contradiction of the Aristotelian thesis of the eternity of the universe is even more obvious. In regard to the earth's motion, if one is talking about straight motion downwards, there is no difficulty with Scripture, as Johannes Lorinus has argued.[19] If one is talking about the Pythagorean or Copernican circular motions, it is less conforming to Scripture, although passages attributing stability to the earth could be interpreted as attributing perpetuity to it (as argued by Lorinus), and although Diego de Zúñiga has interpreted Job 9: 6 as referring to the earth's motion. If one is referring to the sun and heavens not moving, then the scriptural passages stating the opposite could only be interpreted as accommodating the popular manner of speaking, but such an interpretation should not be adopted "without great necessity."

On 2 November 1612, in a private conversation Dominican friar Niccolò Lorini attacked Galileo for being inclined to heresy by believing ideas, such as that the earth moves, which contradict Scripture, although on November 5 Lorini wrote Galileo a letter of apology (Favaro 11: 427). In the fall of 1613, Ulisse Albergotti published in Viterbo (Central Italy) a book, *Dialogue ... in Which It Is Held ... That the Moon Is Intrinsically Luminous ...*, containing biblical criticism of Galileo's theories.[20]

In December 1613, incited by Cosimo Boscaglia, special professor of philosophy at the University of Pisa, the grand duchess dowager Christina of Lorraine, mother of grand duke Cosimo II, questioned Galileo's disciple Benedetto Castelli

---

[17] See Favaro 3: 12, 201–250; cf. Gebler (1879a, 39), Müller (1911, 86–87).

[18] Carlo Conti to Galileo, 7 July and 18 August 1612, in Favaro 11: 354–355, 376; the cardinal apparently even encouraged his brother Conte Conti, who was living in Parma and had some ideas on the connection between *Genesis* and astronomy, to write to Galileo (cf. Conte Conti to Galileo, 11 April 1614, in Favaro 12: 47–48); the cardinal had earlier received a gift copy of *The Sidereal Messenger* (Conti to Galileo, 11 April 1610, in Favaro 10: 311–312). McMullin (2005b, 173) states that Carlo Conti was "prefect of the Holy Office" (namely chairman), which if true would be an extremely interesting and important fact; but I have been unable to confirm McMullin's claim, after consulting Favaro (20: 426), Gebler (1879a, 40), Santillana (1955, 26–27), Fantoli (2003b, 116, 126), Mayaud (1997, 45), Camerota (2004, 261, 264), Pesce (2005), and Beltrán Marí (2006, 198).

[19] Here Conti attributes this judgment to "Lorino's comment on the first verse of Ecclesiastes" (in Favaro 11: 354), which has been identified as Lorinus (1605, 215; 1606, 27) by Favaro (11: 354 n. 1) and Mayaud (1997, 44 n. 15).

[20] Favaro 11: 598–599, Drake (1957, 190).

4.2 Copernicanism and Scripture

about the compatibility of Galileo's ideas with Scripture. Castelli gave satisfactory answers, but informed Galileo of the incident. So on 21 December 1613, Galileo wrote a long letter to Castelli giving a multi-faceted refutation of the scriptural objection to Copernicanism, including a discussion of the passage from Joshua 10: 12–13, which had been advanced as especially troublesome.[21] This may be regarded as the formal beginning of Galileo's trial and the Galileo affair.

Exactly a year after Galileo's Letter to Castelli, on 21 December 1614 at the Church of Santa Maria Novella in Florence, Dominican friar Tommaso Caccini preached a sermon against mathematicians in general and Galileo in particular, because their beliefs and practices allegedly contradicted the Bible and were thus heretical.[22] Some have claimed that Caccini, besides explaining that the biblical passage on the Joshua miracle contradicts the earth's motion and thus renders belief in it heretical, also discussed the suggestive verse "Ye men of Galilee, why stand ye gazing up into heaven?" (Acts 1:10).[23]

On 6 February 1615, Christopher Scheiner sent Galileo, together with a courteous letter, a copy of a book (*Disquisitiones mathematicae de controversiis et novitatibus astronomicis*) written by one of his disciples (Johannes Locher), in which the proponents of the earth's motion were violently attacked.[24]

The following day, on 7 February 1615, Lorini sent the Roman Inquisition a written complaint against Galileo, enclosing his Letter to Castelli as incriminating evidence.[25] A month later, on 20 March 1615 Caccini gave a deposition to the Inquisition in Rome, charging Galileo with suspicion of heresy, based on the content of the Letter to Castelli and the *Sunspots* book, and on hearsay evidence of a general sort (allegedly known to everyone in Florence) and of a more specific type (involving two individuals named Sebastiano Ximenes and Giannozzo Attavanti).[26] In the process, Caccini mentioned German Jesuit Nicolaus Serarius's discussion of the Joshua passage to the effect that Copernicanism is heretical. This judgment was found in a work published by Serarius in 1609–1610, but he had also published in 1612 a book that referred extensively to Cardinal Robert Bellarmine and advanced the same views as the cardinal on biblical hermeneutics.[27]

The issue was now in the hands of the Inquisition and the trial of Galileo had begun in earnest. The events of this contentious story, which is far from over even today, will be traced later (Chapters 7 and 8); here we will focus on the more conceptual and philosophical issue of the logical, theological, epistemological, and methodological content and import of the scriptural objection to Copernicanism.

---

[21] Favaro (11: 605–606, 5: 281–288), Finocchiaro (1989, 47–54), Galilei (2008, 103–109).

[22] Favaro 12: 123, 19: 307.

[23] Fabroni (1773–1775, 1: 47 n. 1). For the mythological character of this claim, see Finocchiaro (2005b, 115), and Section 12.7.

[24] Martin (1868, 42–43), Camerota (2004, 338–342).

[25] Favaro 19: 297–298, Finocchiaro (1989, 134–135).

[26] Favaro 19: 307–311, Finocchiaro (1989, 136–141).

[27] Blackwell (1991, 39–40, 113 n. 4).

4.3 Ingoli

We have seen that Galileo's discoveries and arguments triggered many objections and criticisms by both clergymen and laymen, in oral and written comments, both published and unpublished. It should be added that these objections were not just religious, theological, and scriptural, but involved considerations ranging from the astronomical and physical to the epistemological and methodological. Of such critiques, one of the most complete, intellectually ambitious, and historically significant was an essay by Francesco Ingoli entitled "Disputation on the Location and Rest of the Earth Against the System of Copernicus" (1616).[28] Although it was not published at the time (but rather only in 1891),[29] it circulated widely: 2 years later Kepler wrote a reply to Ingoli, and 8 years later so did Galileo in a lengthy essay that may be regarded as a first draft of the *Dialogue on the Two Chief World Systems*.[30]

Ingoli was a well connected clergyman who had probably been commissioned by the Inquisition to write an expert opinion on the controversy.[31] It is likely that he wrote his essay in January 1616 and that it provided the chief direct basis for the recommendation by its committee of consultants that Copernicanism was philosophically untenable and theologically heretical.[32] However, it is a well documented fact that soon after the Index's anti-Copernican decree of 5 March 1616, on May 10 Ingoli was formally appointed consultant to the Congregation of the Index; that on 2 April 1618 he presented to this congregation a report for the correction of Copernicus's *Revolutions*; that his report was accepted; that the Index's decree of 15 May 1620 corresponded essentially to Ingoli's report; and that he was responsible also for the banning of Kepler's *Epitome of Copernican Astronomy* by the Index in February 1618.[33]

What interests us here is Ingoli's scriptural objection. First of all, it is worth repeating that Ingoli's biblical arguments occur in the context of a comprehensive presentation of all available anti-Copernican arguments. He advances a total of twenty-two arguments: by his own classification, thirteen are "mathematical," by which he means that they involve technical details of astronomy; five are "physical," by which he means that they involve principles about the motion, natural places, or optical properties of bodies. Many of the objections are adapted from Tycho Brahe's *Epistolarum astronomicarum libri* (1596), which Ingoli explicitly mentions. At least one, involving parallax, is alleged to be original with Ingoli himself.

---

[28] Ingoli (1616). Cf. Ingoli (1618a, b), Favaro (1891, 149–184), Bucciantini (1995).

[29] Cf. Favaro (1891, 165–172), Favaro 5: 400. In light of this, the chronology in Howell's (2002, 199) discussion would have to be revised.

[30] For Kepler's reply, see Kepler (1618); for Galileo's, see Finocchiaro (1989, 154–197).

[31] Brandmüller and Greipl (1992, 444–445), Bucciantini (1995, 86 ff), Lerner (1999, 87 n. 51), Beltrán Marí (2006, 342–348).

[32] Favaro 19: 320–321, Finocchiaro (1989, 146–147).

[33] Bucciantini (1995, 87–88), Gingerich (1981), Ingoli (1618a, b), Kepler (1619), Lerner (2004, 30–35), Mayaud (1997, 56–69).

4.3 Ingoli

Four of the objections are "theological." Of these, two involve common Catholic beliefs not directly traceable to Scripture: the doctrine that hell is located at the center of the earth and is most distant from heaven; and the explicit assertion that the earth is motionless in a hymn sung on Tuesdays as part of the "liturgy of the hours" of the "divine office" prayers regularly recited by priests.[34] The other two theological arguments are scriptural. One attempts to undermine specifically the Copernican arrangement of bodies on the basis of Genesis 1: 14, "And God said, Let there be lights in the firmament of the heavens to divide the day from the night." Ingoli did not think that the Copernican central location of the sun was compatible with the scriptural location in the firmament.[35]

The second scriptural objection is directed specifically against the earth's motion and is based on the Joshua passage. But Ingoli is aware that this anti-Copernican passage is just an example and that many other such passages could be appealed to. It is also noteworthy that Ingoli gives a pre-emptive answer to a possible rebuttal stemming from a non-literal interpretation of Scripture. The passage is worth quoting:

> There are infinitely many theological arguments that can be advanced against the earth's motion based on Sacred Scripture and the authority of Church Fathers and scholastic theologians; but, out of so many, I shall mention two[36] that appear to me to be stronger. One is from Joshua, chapter 10, where Scripture says that in answer to Joshua's prayers: "the sun stood still in the midst of heaven, and hasteth not to go down for a whole day. And there was no day like that before it and after it, that the Lord hearkened unto the voice of a man."[37]

> Replies which assert that Scripture speaks according to our mode of understanding are not satisfactory: both because in explaining the Sacred Writings the rule is always to preserve the literal sense, when that is possible, as it is in this case; and also because all the Fathers unanimously take this passage to mean that the sun which was truly moving stopped at Joshua's request. An interpretation which is contrary to the unanimous consent of the Fathers is condemned by the Council of Trent, Session IV, in the decree on the edition and use of the Sacred Books. Furthermore, although the Council speaks about matters of faith and morals, nevertheless it cannot be denied that the Holy Fathers would be displeased with an interpretation of Sacred Scriptures which is contrary to their common agreement.[38]

Ingoli is grounding his rejection of a nonliteral interpretation of the Joshua passage on two rules. The first is that scriptural statements must be interpreted literally (and not accommodatingly) unless it is impossible to do so (due to some internal scriptural inconsistency, or a conclusive philosophical demonstration to the contrary, etc.). The other is that scriptural interpretations are allowed if and only if they correspond

---

[34] Favaro 5: 411; cf. Blackwell (1991, 62), Howell (2002, 200–201). I thank Kenneth Howell and Father Mariano Artigas for clarifications about the meaning of this reference by Ingoli.

[35] Bucciantini (1995, 89), Favaro 5: 407–408.

[36] The other one is the one I mentioned earlier involving the hymn recited by priests on Tuesdays as part of the "liturgy of the hours" of the "divine office."

[37] Joshua 10: 13–14, King James Version. So far this passage corresponds to Favaro 5: 411, lines 14–19, and the translation is my own. For the next several lines (19–28), I quote the translation given by Blackwell (1991, 63).

[38] The second paragraph of this quotation is from Blackwell (1991, 63); cf. Favaro 5: 411, lines 19–28. Cf. also Bucciantini (1995, 90), Howell (2002, 200).

to the unanimous agreement of the Church Fathers. Given these principles, it would seem impossible for a Copernican to circumvent them. But there are two loopholes. One is implicit in Ingoli's formulation of the first rule, which is not a categorical statement but a conditional one. The second loophole is explicitly mentioned by Ingoli himself when he indicates that perhaps these principles apply only for questions of faith and morals. Let us examine the logic of the situation more closely.

First of all, it is instructive to reconstruct Ingoli's argument in a deeper, more elaborate, and more precise manner partly because such a reconstruction turns out to be essentially identical with the argument advanced by the Inquisition consultants in their report of February 1616. This report reads in part: "Propositions to be assessed: (1) The sun is the center of the world and completely devoid of local motion. Assessment: All said that this proposition is foolish and absurd in philosophy, and formally heretical since it explicitly contradicts in many places the sense of the Holy Scripture, according to the literal meaning of the words and according to the common interpretation and understanding of the Holy Fathers and the doctors of theology."[39] For the purpose of this discussion, I ignore differences between the heliostatic and geokinetic theses and use the label "Copernicanism" to refer to one or both of them.

My reconstruction is as follows: (a) Copernicanism is heretical because (b) it is contrary to Scripture, which is so for two reasons. First, (c) it is obvious that Copernicanism contradicts the literal meaning of Joshua, etc., but (d) the literal meaning of Joshua, etc., should *not* be rejected; for (e) the literal meaning of Scripture should not be rejected unless it implies absurdities, and (f) the literal meaning of Joshua, etc., implies no absurdities since (g) it is Copernicanism that is foolish and absurd, and (h) the literal meaning of Joshua, etc. is that the sun (rather than the earth) moves. Similarly, (i) it is obvious that Copernicanism contradicts the patristic interpretation of Joshua, etc., but (j) the patristic interpretation of Joshua should not be rejected, because (k) scriptural interpretations of the Church Fathers should not be rejected unless they imply absurdities, etc.

This reconstruction enables us to better understand the first loophole. For the two rules explicitly stated by Ingoli, and presupposed by the consultants, are (e) and (k). The philosophical absurdity of Copernicanism explicitly stated by the consultants and hinted at by Ingoli, namely proposition (g) here, is what grounds the non-absurdity of the literal and patristic interpretations of Joshua and other relevant scriptural passages, which in turn allows the application of the rules. However, the argument does not really need such a strong claim as (g); it would have been enough to claim that Copernicanism had not yet been "demonstrated," namely proved beyond any reasonable doubt, for then the opposite (earth's rest and sun's motion) would have been possible and non-absurd. Finally, the main part of the argument is the one that justifies the intermediate proposition (b), and the heresy-claim (a) is a further conclusion that may or may not be drawn; it was not in fact drawn by the cardinal-inquisitors when they considered the consultants' recommendation, and it is not included in the anti-Copernican decree of 5 March 1616;

---

[39]Finocchiaro (1989, 146); cf. Favaro 19: 320–321.

4.3 Ingoli

that decree only declared that the false doctrine of the earth's motion was altogether contrary to Scripture.[40]

As regards the second loophole, in Ingoli's reference to the Council of Trent, let us examine Bellarmine's attempt to close it. The Council of Trent (Fourth Session, 8 April 1546) had issued the following decree, which seems to limit the authority of Scripture to questions of faith and morals: "Furthermore, to check unbridled spirits, it decrees that no one relying on his own judgment shall, in matters of faith and morals pertaining to the edification of Christian doctrine, distorting the Holy Scriptures in accordance with his own conceptions, presume to interpret them contrary to that sense which holy mother Church, to whom it belongs to judge of their true sense and interpretation, has held and holds, or even contrary to the unanimous teaching of the Fathers, even though such interpretations should never at any time be published" (Schroeder 1978, 18–19).

On the other hand, in his famous letter to Foscarini (12 April 1615), Bellarmine claimed that scriptural statements about astronomical phenomena are matters of faith "by reason of the speaker," even if they are not so "by reason of the topic":

> ... not only the Holy Fathers, but also the modern commentaries on Genesis, the Psalms, Ecclesiastes, and Joshua, you will find all agreeing in the literal interpretation that the sun is in heaven and turns around the earth with great speed, and that the earth is very far from heaven and sits motionless at the center of the world ... Nor can one answer that this is not a matter of faith, since if it is not a matter of faith 'as regards the topic,' it is a matter of faith 'as regards the speaker'; and so it would be heretical to say that Abraham did not have two children and Jacob twelve, as well as to say that Christ was not born of a virgin, because both are said by the Holy Spirit through the mouth of the prophets and the apostles.[41]

At the beginning of this letter, Bellarmine mentions Galileo by name and makes it clear that the letter is indirectly addressed to him as well as to Foscarini. Thus, Galileo became acquainted with it and wrote down a series of notes which he did not publish at the time, but which were later edited under the title of "Considerations on the Copernican Opinion." The logical weakness of Bellarmine's argument just quoted may be seen from the answer Galileo gives: "The answer is that everything in Scripture is 'an article of faith by reason of the speaker,' so that in this regard it should be included in the rule of the Council; but this clearly has not been done because in that case the Council would have said that 'the interpretation of the Fathers is to be followed for every word of Scripture, etc.,' and not 'for matters of faith and morals'; having thus said 'for matters of faith,' we see that its intention was to mean 'for matters of faith by reason of the topic'."[42]

---

[40] Favaro 19: 323, Finocchiaro (1989, 149), Galilei (2008, 177).

[41] Finocchiaro (1989, 67–68), or Galilei (2008, 147); cf. Favaro 12:172. Bellarmine was probably adopting this discussion from Thomas Aquinas (article 6 of question 1 of Part II of the Second Part of *Summa theologica*), since Bellarmine makes a distinction and gives an example that are identical to Aquinas's; this important connection has been insightfully suggested by Mayaud (2005, 6: 349–350), although the details of his particular interpretation and the general tenor of his work are questionable and excessively apologetic.

[42] Finocchiaro (1989, 84), or Galilei (2008, 164); cf. Favaro 5:367. See also Mayaud (2005, 6: 349–350).

However, as already indicated, Bellarmine prevailed at the time. This may be taken to be an illustration of the fact that political power often prevails over logical rationality in the short run.[43]

4.4 Foscarini

In 1615, Galileo received the unexpected support of a clergyman named Paolo Antonio Foscarini (1580–1616), who published a book containing a theological defense of Copernicanism. Foscarini was the provincial head of the Carmelites in Calabria and had an ambitious agenda of works in philosophy and theology that tended to be encyclopedic in scope. For the Lent of 1615, he had been invited to preach at the church of Traspontina in Rome. On his way there from Calabria, he stopped in Naples in January 1615 to supervise the publication of a booklet written in the form of a letter to the general of the Carmelite order.[44]

The book is entitled *Letter on the Opinion, Held by Pythagoreans and by Copernicus, of the Earth's Motion and Sun's Stability and of the New Pythagorean World System*.[45] Foscarini is explicit that his aim is to give a theological defense of the earth's motion, that is a defense of the geokinetic proposition from the objection that it is contrary to Scripture (13).[46] He is equally clear that he is in no position to mount the following direct defense: that since Copernicanism is physically true, and since two truths cannot contradict each other, Copernicanism is not contrary to Scripture; this defense is not feasible because the earth's motion has not been proved with certainty (12–13). Foscarini is also at pains to repeat frequently the assertion that Copernicanism is probable or likely true, indeed more probable than the Ptolemaic system, and that this probability is largely the result of Galileo's telescopic discoveries (10, 13, 56).

---

[43] On the other hand, as discussed in Sections 8.13 and 8.17, Galileo would prevail in the long run. In fact, the error in Bellarmine's thinking, which Galileo is exposing here, is precisely the one implicitly acknowledged in 1893 by Pope Leo XII in his encyclical *Providentissimus Deus*, and explicitly admitted in 1979–1992 by Pope John Paul II in his rehabilitation of Galileo.

[44] For more biographical information, see Basile (1983; 1987), Boaga (1990), Caroti (1987).

[45] Foscarini (1615a). Cf. Foscarini (1635; 1641; 1661; 1663; 1710; 1811; 1853; 1991b; 1992; 1997; 2001). Many editions of Foscarini's book lengthen the title by adding the clause "in which it is shown that the opinion agrees with, and is reconciled with, the passages of Sacred Scripture and theological propositions which are commonly adduced against it" (Blackwell 1991, 217). But this was an editorial addition in the 1635 Strasbourg edition and emphasizes an aspect of the book that is admittedly important but not the only important one.

[46] Foscarini (1615a, 13). In this section of this chapter, further references to Foscarini (1615a) will be given in parenthesis by citing just the page number, as done here.

4.4 Foscarini

To show that Copernicanism is not contrary to Scripture, Foscarini gives several arguments.[47] One of his most important ones is based on the principle of accommodation, which he takes to be uncontroversial and universally practiced: "whenever Sacred Scripture attributes to God or to some creature anything which is otherwise known to be problematic or improper, then it is interpreted and explicated in one of the following four ways" (19); these amount to saying that Scripture is speaking metaphorically or analogically, or is accommodating itself to the common or popular manner of speaking, thinking, perceiving, describing, or believing. Foscarini (19–29) illustrates this principle with scriptural statements that attribute to God physical attributes such as walking and hands and emotional states like anger and regret; also with statements that attribute to the earth ends and foundations; and with those that speak of light and night and day having been created before anything else, of the 6 "days" of creation, and of the sun and moon as the two great luminaries. However, Foscarini is careful to formulate the conclusion of this argument by saying that "if the Pythagorean opinion were otherwise true, then it could easily be reconciled with the passages of Sacred Scripture that appear contrary to it … by saying that there Scripture speaks in accordance with our manner of understanding, with the appearances, and with our point of view" (29–30); and indeed such a conditional and relatively weak conclusion is all that follows from the principle of accommodation as stated by Foscarini, which is contingent on a scriptural attribution that is otherwise known to be literally incorrect. Thus, by means of the principle of accommodation Foscarini does not show, and does not pretend to show, that Copernicanism is indeed compatible with Scripture, but only that if we knew that Copernicanism were true then we could reinterpret geostatic statements in Scripture.

Another key argument is based on the principle of limited scriptural authority.[48] Paraphrasing various scriptural passages,[49] Foscarini claims that "Sacred Scripture … does not instruct men in the truth of the secrets of nature … because [God] has already allowed and decided that the world be occupied with disputations, quarrels,

---

[47]This multiplicity of arguments is perhaps what gives some scholars the impression that "Foscarini's *Letter* … contains a loose collection of theological and hermeneutical arguments, some of which do not always agree … Foscarini gives every appearance of inconsistency unless one interprets his *Letter* not so much as a tightly knit defense of Copernicanism as a panoply of arguments for facilitating its acceptance" (Howell 2002, 197–198). In my view, Foscarini's various arguments are distinct, but not inconsistent with each other, and they all justify the same conclusion, that Copernicanism is compatible with Scripture; it is this conclusion that gives coherence and unity to the *Letter*.

[48]In speaking of the principle of limited scriptural authority, I am adopting McMullin's terminology; he uses such labels as "principle of limitation" (McMullin 1998, 298) and "principle of scriptural limitation" (McMullin 2005c, 95), whereas in the past I have used such longer and clumsier expressions as the principle that Scripture is not a scientific authority, or the principle denying the scientific authority of Scripture, or the principle of the nonscientific authority of Scripture.

[49]Ecclesiastes (1: 13; 3: 11; 8; 9), 1 Corinthians 4: 5.

and controversies and be subject to uncertainty in everything[50] (as stated in Ecclesiastes), and that the answer will only come at the end …. Thus, its intention is now only to teach us the true road to eternal life" (30–31). The conclusion he reaches is that "so consequently we see how and why from the passages already mentioned we cannot derive any certain resolutions in such subjects, and how with this principle we can easily avoid the hits from the first and second group of passages and from any other allegation derived from Sacred Scripture against the Pythagorean and Copernican opinion" (34). This seems a more direct line of reasoning in support of his claim that Copernicanism does not contradict Scripture, for Foscarini is saying that since Scripture is not an authority on the secrets of nature, scriptural allegations about the earth's rest and sun's motion do not entitle us to infer that the earth is motionless and the sun moves, and so we are in no position to assert the earth's rest on scriptural grounds, and hence the conflict with the Copernican opinion evaporates. In other words, the earth's motion is not contrary to Scripture because Scripture is not a philosophical (or scientific) authority, and so scriptural assertions that the earth is motionless do not entail that the earth is really motionless.[51]

A third argument involves what Foscarini calls the principle of "extrinsic denomination" and the passage in Joshua 10: 12–13. This is the passage where Joshua prays to God to stop the sun and prevent it from setting so that the Israelites can have more time of daylight to finish winning a battle against the Amorites; God did the miracle and the sun stood still for a whole day. The principle states that "many times one says commonly and most properly that a motionless agent moves not because it really moves but by *extrinsic denomination*, namely because with the motion of the subject that receives its influence and action, what also moves is some property which the agent causes in the subject" (35). Applied to the Joshua miracle, we get the following analysis: if the earth moves and the sun stands still, sunlight would still move over the earth's surface, and so it would be proper to say, by extrinsic denomination, that the cause of this moving sunlight itself moves. The earth's motion can thus be reconciled with the Joshua passage.

Most of the rest of Foscarini's *Letter* consists of arguments attempting to show that various specific scriptural passages that have been alleged to be contrary to Copernicanism can be reconciled with it for various reasons and in various ways.

---

[50] Sentences such as this last clause have led Howell (2002, 196–198) to attribute to Foscarini a general skepticism about human knowledge. By contrast, I take them as an indication at most of a fallibilist position; but I think Foscarini's main point is not even fallibilism, but rather that knowledge about nature must be acquired by human beings through their own efforts.

[51] Some scholars (Boaga 1990, 186) seem to ignore this line of reasoning in Foscarini (1615a, 30–34) and portray him as accepting the philosophical primacy and authority of Scripture; they do so on the basis of a passage (Foscarini 1615a, 7–8) that seems to say that if there is a contradiction between Scripture and human reason or sense, Scripture ought to have priority; but such an interpretation neglects a qualification which Foscarini adds to the alleged contradiction, namely that the scriptural passage is so expressed that its interpretation is not subject to argument.

Foscarini's *Letter* attracted the attention of the Inquisition. By March 1615 the Inquisition had ordered an evaluation of Foscarini's *Letter*, and the consultant had written a very critical opinion.[52] Foscarini must have learned something about this censure, and so he wrote a defense of his *Letter* and sent both to Cardinal Robert Bellarmine.[53] Bellarmine replied with his famous letter of 12 April 1615, containing gracious but firm criticism.[54] Soon thereafter, Foscarini left Rome and returned home with the intention of revising his *Letter* to take such criticism into account. This revision never materialized, in part because not long after the Index's anti-Copernican decree (5 March 1616), Foscarini died on 10 June 1616 "perhaps from a heartbreak," according to one scholar's speculation (Boaga 1990, 194).

4.5 Galileo

Although encouraged by Foscarini's booklet, Galileo was also increasingly concerned with the attacks against his views, especially with the scriptural objections, and especially the criticism emanating from the pulpit. He had no way of knowing the details of the Inquisition proceedings, which were a well-kept secret, but he was able to learn of Lorini's initial complaint. So one thing Galileo decided to do was to expand his Letter to Castelli; he wrote his essay in the form of a letter to the Grand Duchess Christina.

The letter consists of a brief introductory part explaining its origin and purpose; a long central part that takes up in turn a number of distinct questions about the relationship between scriptural interpretation and physical investigation; and a brief final part in which Galileo engages in some scriptural exegesis meant to show that the earth's motion is not contrary to Scripture. In the introductory part, we are told that the letter originated from some unprovoked attacks against Galileo which charged that he was a heretic because he believed that the earth moves, and in it he plans to defend himself from this accusation. It is important to stress the apologetic and defensive character of the letter,[55] and so I quote Galileo's words: "Now, in matters of religion and of reputation I have the greatest regard for how common people judge and view me; so, because of the false aspersions my enemies so unjustly try to cast upon me, I have thought it necessary to justify myself by discussing the details of what they produce to detest and to abolish this opinion, in

---

[52] Anonymous (1615; 1882; 1991), Blackwell (1991, 253–254), Boaga (1990, 188), Kelter (1997), McMullin (2005c, 104–105).

[53] Boaga (1990, 188–189, 204–214), Foscarini (1615b; 1882; 1911–1913; 1991a).

[54] Cf. Favaro 12: 171–172, Finocchiaro (1989, 67–69), Galilei (2008, 146–148).

[55] I would thus be hesitant to speak of it as a "treatise," as McMullin (1998, 302; 2005a, 3) and Pesce (1987, 241; 2005, 89) do. This tends to distract one away from the key purpose of refuting the scriptural objection and into searching (e.g., Carroll 1997; 1999; 2001) for a generality and systematicity in Galileo's hermeneutical views which would be at odds with this main purpose.

short, to declare it not just false but heretical."[56] The apologia takes the form of the criticism of what we may call the scriptural argument against Copernicanism, and he concludes this part of the letter with the following clear statement of the objection: "So the reason they advance to condemn the opinion of the earth's mobility and sun's stability is this: since in many places in the Holy Scripture one reads that the sun moves and the earth stands still, and since Scripture can never lie or err, it follows as a necessary consequence that the opinion of those who want to assert the sun to be motionless and the earth moving is erroneous and damnable."[57]

In the central part of the letter, Galileo addresses himself to the major premise of this argument, namely the proposition that Scripture cannot err. He objects that this proposition is true but irrelevant because what is relevant is the interpretation of what Scripture says, and scriptural interpretations can indeed err. Thus the question becomes that of what interpretation, or whose interpretation, if any, is correct, and in the various sections of the letter's central part Galileo takes up, in turn,[58] literal interpretation, the interpretation by professional theologians, the interpretation in accordance with the principle of scriptural consensus, the unanimous opinion of Church Fathers, and the official interpretation of the Church (from a pronouncement of the pope speaking ex cathedra or from a decision reached by an ecumenical council). A main conclusion here is that scriptural interpretation often presupposes philosophical or scientific claims. Moreover, Galileo distinguishes between questions of faith and morals and questions about the physical universe; he points out that, though Scripture cannot err about the former, when we come to physical questions, it is not so much false as improper to say that Scripture cannot err; the reason is that it is not meant to provide scientific information, and hence it would be equally improper to say that Scripture can be wrong. A central thesis here is that Scripture is not a scientific (or philosophical) authority.

In other words, in this central part of the letter, Galileo interprets the scriptural argument against Copernicanism as essentially an argument from authority to the effect that it is erroneous to believe in the earth's motion because Scripture says so. He objects that Scripture is not a scientific authority, and therefore even if Scripture does endorse the geostatic thesis, it does not follow that it is true and the geokinetic thesis is false; that is, the reason given for the conclusion is inadequate, even if it were true. He also objects that generally speaking, to know what Scripture really says about physical questions, one has to know the scientific truth about them; this means that to know whether this reason is true, we would have to know whether the conclusion is true, or, as we might say, the argument ultimately begs the question.

The brief final part of the letter (343–348)[59] may be interpreted as a criticism of truth of the minor premise of the scriptural argument; Galileo tries to show that it

---

[56] Finocchiaro (1989, 90), or Galilei (2008, 113); cf. Favaro 5: 313.
[57] Finocchiaro (1989, 92), or Galilei (2008, 115); cf. Favaro 5: 315.
[58] Favaro 5: 315–323, 323–330, 330–335, 335–339, 339–343, respectively.
[59] Favaro 5: 343–348. In this section of this chapter, simple references to Favaro (1890–1909, vol. 5), will be given in parenthesis by citing just the page number, as done here.

4.5 Galileo

is questionable whether Scripture says that the earth stands still and the sun moves.[60] He does this by an analysis of several passages that were typically given to support the contrariety thesis. The Joshua miracle is discussed at great length. Galileo argues that this passage contradicts the *geostatic* system, whereas it could be given a literal interpretation from the Copernican viewpoint. The passage says that, in response to Joshua's prayer to prolong daylight, God ordered the sun to stop, and the sun stood still for a whole day, needed by the Israelites to defeat the Amorites. Galileo points out that in any system, to lengthen the day the diurnal motion must be stopped. Unfortunately, in the geostatic system the diurnal motion belongs to the outermost sphere in the universe called the *primum mobile*, not to the sun. The proper motion that belongs to the latter is the annual motion, which, being opposite in direction to the diurnal motion, would shorten the day if stopped, making the sun

---

[60] I believe this interpretation offers a simple and elegant solution to the problem that here Galileo apparently does what elsewhere he claims one is not supposed to, namely to use Scripture to support an astronomical theory. In the history of the Galileo affair, this problem facilitated the rise of a genuine myth which it took a long time to extirpate, that is the myth that Galileo was condemned not for being a good astronomer but for being a bad theologian; cf. Chapter 10 below. More recently, and on an altogether different level, Pesce (2000, 48–50; 2005, 1–2, 226–229) has struggled with Galileo's alleged "concordism" (the thesis that biblical statements on physical reality agree with claims established from scientific research); although Pesce's general account and mine overlap in significant ways, his view of the "concordism" problem forces him to attribute to Galileo a contradiction or incoherence, which my interpretation avoids. Similarly, McMullin (2005c, 101–102, 110–111) is right to interpret this Galilean criticism as an "ad hominem" argument, namely an argument that derives a conclusion not acceptable to an opponent from premises acceptable to him but not necessarily to the arguer; but to avoid conveying a misleading impression, one should add that this sense of "ad hominem argument" is the seventeenth-century meaning of the term, and that ad hominem arguments so construed (although they establish a merely conditional and not a categorical conclusion) are nevertheless one of the most fair-minded, effective, and sound methods of criticism; and more importantly, it must be added that such arguments are *not* ad hominem in the sense that is most common today, according to which an ad hominem argument is the fallacy of attempting to refute a conclusion by discrediting the moral character or practical situation of the arguer; cf. Finocchiaro (1980, 231–232, 368–370, 402–403, 416, 430; 2005a, 65–91, 277–291, 329–339). Analogously, Lerner (1999, 81–82) is right to protest against the attempts by recent clerical apologists (Brandmüller 1982, 1992; Poupard 1992) to try to praise Galileo as a theologian in order to demean him as a physicist and methodologist, but that is not to say that one should attribute to Galileo an "incohérence qu'il n'a pas su éviter sur le terrain exégétique" (Lerner 1999, 81); furthermore, Lerner (2005, 20) is right that Galileo held that a demonstrated physical truth could be used to interpret Scripture accordingly, but Galileo's example of such a truth is solar axial rotation (demonstrable from sunspots), not geokinetic heliocentrism (which he held to be merely probable and in any case was the very issue in question in this discussion); and Lerner (2005, 21) is also correct that "Galileo's efforts to interpret some key passages of the Scripture ... in terms of heliocentrism, were not for him a mere rhetorical or diplomatic expedient," but the reason is not that Galileo's "apparent contradiction" (Lerner 2005, 20) was real, but rather that he was engaged in the logically and epistemologically sound technique of criticizing an argument by criticizing the truth of one of its premises and by deriving unwanted consequences from its own presuppositions. Finally, Biagioli too (2003; 2006a, 219–259) struggles with the internal stresses and strains of Galileo's views and seems to have his own way out of the difficulties, although I am not sure I understand it fully due to its being laden with terminology and concepts originating from Jacques Derrida. See also Beltrán Marí (2006a, 189, 682, n. 40), Fabris (1986, 43–44), Mayaud (2005, 1: 259–62, 6: 359).

set that much sooner. It follows that if we take the Scripture literally, the miracle is physically impossible in the geostatic system, whereas if God did the miracle, he should have ordered the *primum mobile* to stop. By contrast, Galileo argues that in the geokinetic system the miracle could have happened as follows. First he refers to his own discovery that the sun is not completely motionless but rotates on its axis with a period of about a month; thus it makes sense, to begin with, to stop the sun from moving. To this Galileo adds the speculation that solar rotation probably causes the planetary revolutions, and the earth's own orbital motion is probably connected with the earth's axial rotation, all of which makes some sense because all these motions are in the same direction in the heliocentric system. Thus by stopping the sun's rotation, God could have stopped the earth's diurnal motion and thus lengthened Joshua's day.

Before we move on to more details of the central part of the letter, let me summarize my view of its overall conceptual structure. The letter as a whole amounts to a threefold criticism[61] of the argument that Copernicanism is wrong because Scripture says so: first, Scripture's saying so would not make it so; second, to know what Scripture really says about the physical universe one normally has to know what is physically true; and third, it is questionable whether Scripture does in fact say so.

---

[61] This interpretation of mine is reminiscent of, but different from, the accounts given by Howell and by McMullin. Howell (2002, 186–196) claims that Galileo had three approaches, methods, or strategies (stemming from Augustine) to answer the scriptural objection to Copernicanism; his preferred strategy was to separate the books of Scripture and nature and so make scriptural statements irrelevant to the earth's motion; if his opponents insisted on relevance, he would stress the harmony between the two books, the primacy of demonstrated physical truths, and the need for nonliteral interpretation; if they insisted on literalism, he would argue that even then Copernicus was more in accordance with scripture than Ptolemy. Analogously, McMullin (2005c) attributes to Galileo several Augustinian principles and consequent strategies, the main three of which involve accommodation (PA), scriptural limitation (PSL), and prudence (PP): "The combined implication of these three principles for the Copernican issue is clear. The Scriptures are simply irrelevant to deciding such matters as the motion of sun or earth (PSL). Further, even if PSL were to be set aside, the writers of Scripture are clearly accommodating themselves to our normal modes of speech, to what *appears* to us, when they speak of the sun as in motion or the earth as fixed (PA). Finally, even if both PSL and PA were to be set aside, ordinary prudence would counsel that on an issue where in the future a contrary demonstration could well be found, no dogmatic position should be taken now that at a later time could serve to discredit the Scriptures generally (PP)" (McMullin 2005c, 107). The similarities between the Howell-McMullin's accounts and mine are that the three logical criticisms of the scriptural argument which I am attributing to Galileo can also generate three corresponding rhetorical "strategies" depending on context and on whom he is talking to; that one of my strategies is essentially the same as one of theirs (the one I call Galileo's main criticism, aiming to show that the scriptural argument is a nonsequitur, because of the nonscientific authority of Scripture); and that my third criticism corresponds to Howell's third strategy. However, it seems to me that McMullin's and Howell's stress on a multiplicity of strategies is excessively rhetorical in the sense that it portrays Galileo as excessively concerned with influencing people and "winning" arguments, rather than primarily searching for the truth; moreover, it leads them to find tensions and incoherences in Galileo's position that are not really there; and it prevents them from appreciating Galileo's main point, namely the nonscientific authority of Scripture and the need to understand clearly, apply correctly, and justify properly this principle.

4.5 Galileo

Galileo begins the central argument of the letter by elaborating several uncontroversial points. The first is that the literal interpretation of Scripture is not always correct since, for example, some scriptural statements about God state or imply that He has eyes, ears, and so on, and we know that it is not literally true (315–316). The second point is that the literal interpretation of Scripture is incorrect when it conflicts with physical truths that have been conclusively proved (317, 320). The third is an explanation for this priority of proved scientific truths over literal scriptural meaning, and is also universally admitted; the explanation is that, whereas Scripture is the Word of God meant "to teach us how one goes to heaven and not how heaven goes,"[62] the physical universe and the human senses and mind are the *work* of God, and hence one cannot doubt the truth of physical conclusions grounded on sense-experience and conclusive arguments (316–317). From these three points, Galileo thinks it plausibly follows that the literal interpretation of Scripture is not binding when we are dealing with physical propositions that are *capable* of being conclusively proved (even if not proved yet), because this would be the more prudent policy and because what we know is a minute part of what we do not know (320–321). Galileo's own words make clear the tentativeness and prudential character of his conclusion: "I should think it would be very prudent not to allow anyone to commit and in a way oblige scriptural passages to have to maintain the truth of any physical conclusions whose contrary could ever be proved to us by the senses or demonstrative and necessary reasons."[63] This conclusion is repeated on the next page in a formulation that is of some interest because it sounds like the theological analogue of Occam's ontological razor: "Therefore, it would perhaps be wise and useful advice not to add without necessity to the articles pertaining to salvation and to the definition of the faith, against the firmness of which there is no danger that any valid and effective doctrine could ever emerge."[64]

Galileo next undertakes an explicit criticism of *theological* authority. He argues that theology is *not* the queen of the sciences because it is obvious that its principles do not provide the logical foundations of the knowledge formulated in other sciences, the way that, for example, geometry does for surveying (324–325). Moreover, theologians cannot dictate physical conclusions from the above (i.e., without actually getting involved in physical investigations), any more than a king who is not a physician can prescribe cures for the sick. Nor can theologians tell scientists to undo their own observations and proofs because this is an inherently impossible or self-defeating task (325–327). Rather, theologians can and should follow two courses. The first corresponds to already established practice: apropos of conclusively established physical truths they should strive to show that they are not contrary to Scripture by an appropriate interpretation of the latter. The second would be a rule of interdisciplinary communication. Theologians should presume

---

[62] Finocchiaro (1989, 96), or Galilei (2008, 119); cf. Favaro 5: 319.
[63] Finocchiaro (1989, 96), or Galilei (2008, 120); cf. Favaro 5: 320.
[64] Finocchiaro (1989, 97), or Galilei (2008, 121); cf. Favaro 5: 321.

scientific ideas that are not conclusively proved but are contrary to Scripture to be false and accordingly should try to give a scientific disproof of them; this is desirable because the inadequacies of an idea can be discovered more easily by those who reject it. This ingenious but plausible rule is this section's main methodological conclusion.[65]

Next, Galileo questions the traditional principle that used scriptural consensus combined with the unanimity of the Church Fathers to require acceptance of the literal meaning of physical statements (327–328). Once again he makes his fundamental distinction between physical propositions that are and those that are not capable of conclusive proof. For the latter the principle makes sense, but for the former the previous considerations suggest that it is not sound (330–331). Two new points emerge in this discussion. First, scriptural consensus is not a sign that physical statements are meant to be taken as literally and descriptively true, but rather it is the result of Scripture's desire for consistency, its appeal to common people, and the need to reflect the opinions of the time (332–334). Second, the unanimity of Church Fathers is not binding unless it is explicit, unless it is the result of reasoned discussion, and unless it refers to matters of faith and morals (335–337).

Finally, the authority of the Church herself comes under discussion. Galileo admits that she does have the power to condemn an idea as heretical (343), but he notes that "it is not always useful to do all that one can do."[66] Moreover, to make ideas heretical is not the same as making them false; indeed, "no creature has the power of making them be true or false, contrary to what they happen to be by nature and de facto."[67] At any rate the Church should not be hasty in her condemnation; he hopes that she is not "about to make rash decisions."[68] Before condemning a physical idea she should examine all the evidence and listen to all the arguments on both sides of the issue, and she should rigorously prove that her interpretation of the relevant scriptural passages is correct. For example, such a rigorous proof should use all the cautious advice elaborated by St. Augustine. To avoid potential embarrassment, it might be best to wait until the physical idea is conclusively refuted before declaring it heretical.

One of the most striking features of this central part of the letter is the negative tone[69] of its component conclusions: that the literal interpretation of Scripture is not binding in physical investigation; that theology is not the queen of the sciences; that scriptural consensus is not a sufficient condition for a literal interpretation; that the unanimity of Church Fathers is not necessarily decisive in physical questions; and that the authority of the Church should not be hastily applied.

---

[65] Finocchiaro (1989, 102), or Galilei (2008, 126); cf. Favaro 5: 327.
[66] Finocchiaro (1989, 110), or Galilei (2008, 136); cf. Favaro 5: 338.
[67] Finocchiaro (1989, 114), or Galilei (2008, 140); cf. Favaro 5: 343.
[68] Finocchiaro (1989, 114), or Galilei (2008, 140); cf. Favaro 5: 342.
[69] Thus, for example, Howell (2002, 195) is absolutely right when he says that in the *Letter to Christina* "Galileo did not attempt to prove that the Copernican theory was taught in the Bible; he only wanted to remove scriptural objections against it."

This negativity corresponds to the apologetic and critical purpose of the letter, and the general suggestion is, as mentioned earlier, a denial of the scientific (or philosophical) authority of Scripture. There is an underlying positive idea, however; that is, the principle of autonomy, according to which physical investigation can and should proceed independently of Scripture. Moreover, from the point of view of the enterprise of understanding Scripture, we get another constructive idea underlying these negative conclusions, which is the hermeneutical principle that scriptural interpretation often depends on the results of physical investigation.

A second striking theme is that of prudence and caution, which he adopts from St. Augustine and elaborates further. Galileo's explicit admonitions are, of course, against haste in condemning Copernicanism. But it would also extend to the question of accepting the theory or judging the conclusiveness of its supporting arguments.[70]

Equally striking is the theme involving the distinction between physical propositions that are and those that are not capable of conclusive proof. This is obviously the main epistemological distinction, rather than that between propositions that have and those that have not been conclusively proved. The central issue concerns the former distinction, and Galileo tries to resolve it by arguing that no physical proposition capable of conclusive proof should ever be condemned. The priority of established scientific knowledge that has already been conclusively proved, over scriptural statements, is a non-issue. From the viewpoint of this uncontroversial principle, there would be no reason to write an essay on the methodology of scriptural interpretation and physical investigation, but rather the only thing to do would have been to produce or search for the conclusive demonstration. The very fact that he writes this methodological essay indicates that he wants to advocate a (relatively) novel principle. And besides the argumentative content of the Letter and the very fact of writing it, there is a third indication of Galileo's stress on demonstrability, as distinct from demonstration. That is, when in 1636 the Letter was published for the first time by some foreign friends but with his cooperation, the stress on demonstrability was explicitly incorporated in the title of the book: *Nov-antiqua sanctissimorum patrum, & probatorum theologorum doctrina, de Sacrae Scripturae testimoniis, in conclusionibus mere naturalibus, quae sensata experientia et necessariis demonstrationibus evinci possunt, temere non usurpandis* (i.e., *New and Old Doctrine of the Most Holy Fathers and Esteemed Theologians on Preventing the Reckless Use of the Testimony of the Sacred Scripture in Purely Natural Conclusions That Can Be Established by Sense Experience and Necessary Demonstrations*).[71] Here the crucial phrase is "that can be established," which obviously is not equivalent to "that have been established."

To complete my account of Galileo's *Letter to Christina*, I now want to supplement what I have said so far with two considerations. The first involves an aspect

---

[70] For more documentation of this last point, see Finocchiaro (1980) and Chapters 3 and 9 of this book.
[71] Galilei (1636). Cf. Finocchiaro (2005b, 72–79), Garcia (2000, 2005), Motta (2000), and Section 8.4.

of the letter which so far I have largely ignored; that is, the letter is full of references to and quotations from the patristic and theological tradition, such as St. Jerome, St. Thomas Aquinas, and especially St. Augustine.[72] This aspect of the letter could be reconstructed as an argument from authority, or a series of such arguments. This is important for two reasons. First, Galileo was aware that regardless of how cogent his methodological argument was, his main conclusion (denying scriptural authority for *demonstrable* physical claims) could be taken to be so radical that its novelty needed to be toned down by trying to root it in tradition.[73] Thus, in the just quoted title of the 1636 edition of the *Letter*, the initial part of the title (especially the word *Nov-antiqua*) stresses precisely the twofold aspect of being partly radical and partly traditional. Second, the passages quoted from Augustine are so crucial that they played a significant role in the subsequent history of scriptural hermeneutics. For example, Pope Leo XIII's encyclical *Providentissimus Deus*, without even mentioning Galileo, puts forth a Galilean view of the role of Scripture in scientific investigation, and makes a crucial appeal to the same Augustinian passages that had been quoted by Galileo.[74]

---

[72] The Augustinian aspect of Galileo's *Letter to Christina* has been discussed by McMullin (1998, 1999, 2005c) and by Howell (2002, 188–189; cf. 2005). McMullin stresses that Augustine developed his hermeneutical ideas while attempting to respond to the Manichaean philosophical (or scientific) criticism of the *Genesis* account of creation; that Augustine's hermeneutics can be reconstructed in terms of a concept of the "literal meaning" of Scripture and a set of seven distinct interpretive principles; that these Augustinian principles constitute an incoherent set; but that it is valuable to point this out "not to accuse Augustine of inconsistency in what were obviously only tentative strategies on his part, but because of the consequences of tensions of this sort when Augustine's texts later reappear from Galileo's pen" (McMullin 2005c, 97). My reservations about such an approach and account are that parts of the discussions quickly become an exercise of explaining something by means of things that are more obscure and less intelligible than the original; that for example, Augustine's notion of "literal meaning" turns out to be extremely complex and in any case such as to have little to do with the ordinary meaning of "literal meaning"; that at the meta-level of the question of the proper interpretation of Augustine's own texts, I would want to question the propriety of formulating general principles that are not explicitly stated in the text under scrutiny, on the basis of Augustine's concrete interpretive practices in which such principles are allegedly implicit, for the mutual incompatibility of such principles must be regarded as the result of interpretive exaggerations or over-generalizations on the part of the interpreter, rather than tensions in the mind of the author under scrutiny; that such an approach to Augustine has an instructive similarity to the approach to Galileo's scientific methodology that tries to attribute to him general epistemological or methodological principles (apriorism, empiricism, mathematicism, positivism, anarchism, etc.) on the basis of concrete Galilean analyses of concrete scientific problems; and finally, that my point is that in such cases the critic must formulate interpretations that are nuanced enough to be able to describe the complexities of the situation under scrutiny without over-simplification; cf. Finocchiaro (1980, 145–166; 1997a, 335–356).

[73] I believe McMullin (1998, 299) and Howell (2002, 197) go too far in denying historical novelty to Galileo's hermeneutical efforts, just as other scholars have gone too far in the opposite direction of attributing him novelty. My point is that if he was not advocating some significant novelty then the issue would not have been controversial and he need not have written his essay; and that if his proposals were perceived as too novel, then his effort was doomed; thus I portray Galileo's position as partly novel and partly traditional, and self-consciously so.

[74] For more details on this, see Finocchiaro (2005b, 263–266).

4.5 Galileo

To get a flavor of Galileo's appeal to St. Augustine, the two best passages are these. One is Augustine's version of the principle of nonscientific authority of Scripture: "it should be said that our authors did know the truth about the shape of heaven, but that the Spirit of God, which was speaking through them, did not want to teach men these things which are of no use to salvation."[75] The other is Augustine's versions of the principle of the priority of demonstrated physical truth: "whenever the experts of this world can truly demonstrate something about natural phenomena, we should show it not to be contrary to our Scriptures."[76]

The second supplement regards the fact that it is undeniable that in the *Letter to Christina* there are several passages that appear to be inconsistent with the principle that Scripture is not an authority in astronomy or natural philosophy. Ernan McMullin has stressed this inconsistency and focused on three passages, which he sees as asserting a "principle of priority of Scripture":[77]

---

[75] Augustine, *De Genesi ad litteram*, ii, 9, 20, as translated in Finocchiaro (1989, 95), or Galilei (2008, 118). Cf. Favaro 5: 318, Leo XIII (1893, paragraph 18, p. 334).

[76] Augustine, *De Genesi ad litteram*, i, 21, 41, as translated in Finocchiaro (1989, 101), or Galilei (2008, 126). Cf. Favaro 5: 327, Leo XIII (1893, paragraph 18, p. 334).

[77] McMullin (1998, 308–311; 2005c, 107–110). Sylla (1991, 218–223) advances a similar view. McMullin's purpose should not be misunderstood, for he is at pains to point out that "the tension between the principles he [Galileo] sprinkles throughout the *Letter to the Grand Duchess* as ways to defuse the challenge to Copernicanism is worth the extended attention devoted to it here not so much because it implies a logical failure on Galileo's part but for what it tells us about the exegetic strategies potentially available at the time for dealing with the sort of crisis that was in the making in Rome" (McMullin 2005c, 111). According to McMullin, Galileo's flirtation with the principle of priority of Scripture is the result of two things.

The first is Galileo's complete and unqualified acceptance of the principle of priority of demonstration, which McMullin (2005c, 108) formulates as asserting that "where the literal (in the sense of normal) reading of a scriptural text conflicts with a philosophical [i.e., "scientific"] demonstration, a different interpretation must be sought." The alleged connection is that given the priority of demonstration, and given the existence of such a "demonstration" of Copernicanism, then one strategy for its theological or scriptural defense would be to present this demonstration and appeal to this principle, without having to appeal to any other principles such as the principle of limited scriptural authority; but note that such a strategy presupposes possession of a "demonstration," and indeed McMullin (2005c, 109–110) thinks that "at the time he composed the *Letter* he was evidently optimistic about the chances of demonstrating the Copernican claims." The difficulty with this interpretation is that around 1615, Galileo was aware that he possessed a conclusive, demonstrative proof of only parts of Copernicanism (such as the heliocentricity of Venus's orbit and the changeability of the heavenly region), but not of the crucial geokinetic claim, in regard to which he felt its status to be less than demonstrated. This is shown by the probabilistic language and qualifications contained in his "Discourse on the Tides" (Finocchiaro 1989, 119–133; cf. Favaro 5: 377–395); by the fact that in the *Letter to Christina* he describes his defense of Copernicanism as amounting to a process of "confirmation," in a sense distinct from "demonstration" (Galilei 2008, 110–113; Finocchiaro 1989, 88–89; Favaro 5: 310–312); and by the fact that in the "Considerations on the Copernican Opinion" the way he formulates his over-all general claim is by means of the comparative judgment that the pro-Copernican arguments are much stronger than the geostatic ones (Galilei 2008, 151; Finocchiaro 1989, 73; Favaro 5: 354). But aside from these, there is the very fact that the *Letter to Christina* is full of epistemological, methodological, theological, hermeneutical, and exegetic considerations, but does not contain a proof of the earth's motion; if he believed he had such a proof, the thing to do would have been to follow the "direct" strategy, as I called it above in my discussion of Foscarini; that is, Galileo would have simply produced the conclusive demonstration.

[1] Even in regard to those propositions which are not articles of faith, the authority of the same Holy Writ should have priority over the authority of any human works composed not with the demonstrative method but with either pure narration or even probable reasons; this principle should be considered appropriate and necessary inasmuch as divine wisdom surpasses all human judgment and speculation.[78]

[2] Some physical propositions are of a type such that by any human speculation and reasoning one can only attain a probable opinion and a verisimilar conjecture about them, rather than a certain and demonstrated science; an example is whether the stars are animate. Others are of a type such that either one has or one may firmly believe that it is possible to have, complete certainty on the basis of experiments, long observations, and necessary demonstrations; examples are whether or not the earth and the sun move, and whether or not the earth is spherical. As for the first type, I have no doubt at all that, where human reason cannot reach, and where consequently one cannot have a science, but only opinion and faith, it is appropriate piously to conform absolutely to the literal meaning of Scripture. In regard to the others, however, I should think, as stated above, that it would be proper to ascertain the facts first, so that they could guide us in finding the true meaning of Scripture; this would be found to agree absolutely with demonstrated facts, even though prima facie the words would sound otherwise, since two truths can never contradict each other.[79]

[3] In the learned books of worldly authors are contained some propositions about nature which are truly demonstrated and others which are simply taught; in regard to the former, the task of wise theologians is to show that they are not contrary to Holy Scripture; as for the latter (which are taught but not demonstrated with necessity), if they contain anything contrary to the Holy Writ, then they must be considered indubitably false and must be demonstrated such by every possible means.[80]

Now, if such assertions were expressions of the principle of the general priority of Scripture, they would seriously undermine the apologetic purpose of the *Letter*, which is to say the plan to refute the scriptural objection to Copernicanism by arguing (among other criticisms) that the objection is a nonsequitur because scriptural statements

---

Second, McMullin tries to exploit Galileo's tendency to speak as if he accepted the Aristotelian ideal of science as demonstration, according to which "demonstrations" (in a suitably deductive and apodictic sense of the term) are necessary in science. From this ideal, together with the principle of priority of demonstration, it would follow that science has nothing to fear from Scripture, for scientific demonstrations of natural phenomena will always take precedence over scriptural assertions to the contrary; this is quite consistent with giving Scripture priority outside the domain of scientific demonstration. My difficulty with this account is that Galileo's attraction toward the demonstrative ideal is not his only tendency, and that other inclinations were at work, and in fact one could interpret his career as a movement further and further away from that Aristotelian ideal and toward a fallibilist and probabilist one. Moreover, even if one equates scientific knowledge with "demonstrated" propositions, such scientific demonstrations have to be searched for and discovered, and the search for scientific demonstrations cannot avoid using and dealing with arguments that have less than a demonstrative force, and such scientific research would have much to fear from the principle of priority of Scripture. Thus to attribute Galileo such a strategy would be a quite uncharitable attribution of methodological shortsightedness, which I for one would not want to make without overwhelming evidence.

[78]Galilei (2008, 117). Cf. Favaro 5: 317, Finocchiaro (1989, 94), McMullin (1998, 308; 2005c, 109).
[79]Finocchiaro (1989, 104), or Galilei (2008, 129). Cf. Favaro 5: 330, McMullin (1998, 309–310).
[80]Finocchiaro (1989, 101–102), or Galilei (2008, 126). Cf. Favaro 5: 327, McMullin (1998, 310; 2005c, 109).

about the earth's rest and sun's motion do not entail that the earth really rests and the sun really moves. However, if these passages were regarded essentially as attempts to define more precisely the proper scope of the principle of non-scientific authority of Scripture, then such an interpretation would conform with the apologetic purpose of the *Letter*, and so would be preferable to the alternatives that undermine that purpose. I believe such an interpretation would be along the following lines.

That is, the first passage asserts that scriptural assertions have priority over other assertions in regard to historical questions; and here Galileo's talk of "pure narration" provides a crucial clue.[81] The second states that Scripture has priority over *unprovable* assertions in regard to physical and natural phenomena. The third claims that for theologians scriptural assertions have priority over *unsupported* assertions in all other writings. In other words, Scripture is a superior authority regarding (1) historical questions that depend on balancing probabilities of testimony; (2) undecidable questions about physical reality; and (3) unsupported assertions on any topic in any book. Thus, although Galileo denies the scientific (astronomical, or philosophical) authority of Scripture, he accepts its authority not only for questions of faith and morals, but also for the weighing of probable testimony in history, for undecidable questions in natural philosophy, and for questions of presumption of truth for unsupported claims. These are important nuances, complications, and qualifications in Galileo's position, but none of this undermines his criticism of the scriptural objection to Copernicanism or his principle of scriptural irrelevance in astronomical research.

4.6 Campanella

Tommaso Campanella's *Apologia* was first published in Frankfurt in 1622,[82] and although the preface indicates that it had been written several years earlier, the exact date of composition is not given. Nor have any copies of the original manuscript survived. Thus, specialists are somewhat divided. The account I find most plausible is the following.[83]

Most likely, Campanella wrote the *Apologia* just before the anti-Copernican decree of 5 March 1616 and did so at the request of Cardinal Bonifacio Caetani. Caetani was a moderate who was appointed cardinal in 1606 and member of the Congregation of the Index at the beginning of 1616, attending his first meeting on 1 March 1616. However, Caetani's request was an unofficial one. He had several

---

[81] Additionally, Galileo explicitly suggests this in his reply to Bellarmine, in "Considerations on the Copernican Opinion," in Favaro 5: 367–368, Finocchiaro (1989, 84), Galilei (2008, 148–167).
[82] Campanella 1622. Cf. Campanella (1821; 1853; 1911; 1937; 1968; 1971; 1992; 1994; 1997; 1999; 2001a, b; 2006).
[83] See Femiano (1971, 21–30), Headley (1997, 97, 169–170), Lerner (2001b, xix–liv; 2001d; 2006, ix–xxx). For other views, see Blackwell (1994, 19–24), Ditadi (1992), Firpo (1968, 21), Ponzio (1998; 2001, 5–7).

Neapolitan connections that probably served as links between him and Campanella, who was in Naples serving time in prison.

To get a proper understanding of this work, it is important to begin by examining the full title of the book: *Apologia pro Galileo, mathematico florentino, ubi disquisitur utrum ratio philosophandi quam Galileus celebrat faveat Sacris Scripturis an adversetur*. This could be translated as follows: *Apologia for the Florentine Mathematician Galileo, where One Discusses whether the Manner of Philosophizing Advocated by Galileo Conforms or Conflicts with Sacred Scripture*. The crucial phrase here is "ratio philosophandi." Michel-Pierre Lerner deserves credit for having stressed that Campanella is talking about Galileo's philosophical approach or manner of reasoning, which he renders into French as "méthode de philosopher."[84] This contrasts with the usual translations that render the phrase as: philosophical view, philosophical doctrine, scientific theory, or (just) theory.[85] The point is extremely important because Galileo's theory, doctrine, or view (whether philosophical or scientific) suggests Copernicanism or the earth's motion; this would imply that Campanella was more committed to the Copernican doctrine than he really was;[86] and it would imply that he was trying to do the same thing Foscarini had done. On the other hand, *ratio philosophandi* suggests some principle of reasoning or procedure, and so Campanella is trying to do something more general or methodological.

Campanella's general, methodological aim is also evident from other documents and passages.[87] For example, when in the course of a main argument Campanella states his conclusion, he says that "therefore I think that this manner of philosophizing should not be condemned," using a phrase (*philosophandi modum*) that leaves no doubt.[88] And in a letter to Galileo, dated 3 November 1616, Campanella states that he has sent to Rome and to him a manuscript copy of his *Apologia*, which he describes in Italian as "a discussion where it is proved theologically that the manner of philosophizing [*modo di filosofare*] you use is more in conformity with Divine Scripture than the contrary one is" (in Ditadi 1992, 238). This letter also gives a clue that Campanella's argument is specifically or primarily a theological one.

But what Galilean manner of philosophizing is he referring to? I believe this is an aspect of the manner of reasoning which Galileo uses in his discussions of astro-

---

[84] Lerner (2001b, xcv–c, 1–2; 2006, xxxiii–xxxv). Also correct, of course, is Germana Ernst's Italian translation of this phrase in the book's title as *modo di filosofare*, in Campanella (1999, 2006); but it is puzzling that in the middle of Chapter 3 (Campanella 2006, 98–99) she should translate *modum philosophandi* as *filosofia*.

[85] See respectively, Blackwell (1994, iii); Ponzio (2001, 45); Femiano (1971, 35, 43); Ditadi (1992, 117) and Firpo (1968, 27). Slightly better is "kind of philosophy" (McColley 1937, 1).

[86] See, for example, Campanella to Pope Urban VIII (10 June 1628), in Campanella (1927, 218–225); see also Headley (1997, 176).

[87] Besides the passages to be mentioned below, also revealing, although indirectly relevant, is Campanella's (n.d.) poem "Modo di filosofare"; see also Failla to Galileo, 6 September 1616, in Favaro 12: 277.

[88] Campanella (1622, 30); cf. Blackwell (1991, 79), Firpo (1968, 83), Lerner (2001b, 78–79; 2006, 98).

4.6 Campanella

nomical topics, such as one finds in *The Sidereal Messenger* (1610) and the *History and Demonstrations Concerning Sunspots* (1613), and which he reflects on and tries to justify in his critique of the scriptural objection, such as we find in the Letters to Castelli and to Christina. The most pertinent and general description of this manner of reasoning is to say that he advocates disregarding scriptural assertions in astronomical investigation. Stated as a methodological principle, it is the claim that scriptural statements about the earth's rest and sun's motion do not entail that the earth stands still and the sun moves; in other words, Scripture is not an authority in natural philosophy; or again, it is the principle of limited scriptural authority.[89] Campanella comes close to explicitly giving such a description of the Galilean manner of philosophizing when he says that "Galileo does not treat any of these subjects from a theological point of view, but rather by means of his marvelous instruments he renders previously hidden stars visible ..."[90]

In short, Campanella wants to give a scriptural argument that Scripture ought to be disregarded in physical investigation! It is also a theological argument that in such investigations it is quite proper to pay no attention to scriptural assertions. In fact, such an argument constitutes a major line of reasoning in Campanella's *Apologia*.[91]

In a central part of his book, which he calls the third assertion of the second hypothesis, Campanella stresses the fact that "in the Gospel Christ is never found to discuss physics and astronomy but only morality and the promise of eternal life" (65).[92] Correspondingly, he emphasizes two crucial scriptural passages: Ecclesiastes

---

[89] I believe my interpretation is in essential agreement with Headley's (1997, 172–177), although he speaks of the principle of *libertas philosophandi*, which is even more general than the principle of limited scriptural authority. On the other hand, here I am disagreeing with Lerner's (2001b, xcv–c; 2006, xxxiii–xxxv) description of the manner of philosophizing attributed to Galileo in the *Apologia*. Lerner seems to be saying that in that work Campanella was attributing Galileo a naïve empiricism à la Bernardino Telesio, although elsewhere Lerner (1995b) argues that in other writings Campanella's interpretation was more nuanced, and that in any case it was partly incorrect. Lerner has also objected (private correspondence) that my interpretation assumes that Campanella was acquainted with Galileo's Letters to Castelli and to Christina, which is highly unlikely, at least when he wrote the *Apologia*. However, my interpretation need only assume that Campanella was acquainted with *The Sidereal Messenger*, and we know from his letter to Galileo of 13 January 1611 that he had read that work (Favaro 11: 21–26); Campanella was clearly perceptive enough to understand this aspect (the scriptural independence) of Galileo's manner of philosophizing, and indeed in the same letter (Favaro 11: 24) he predicted that some theologians will start murmuring against it but claimed that theology itself could come to Galileo's defense.

[90] Campanella (1622, 51), my translation. Cf. Campanella (1994, 110; 2001a, 136; 2006, 173).

[91] This argument also corresponds to one advanced in various places in Campanella's *Theologia*, a monumental work in 30 volumes written in 1614–1624. Cf. Headley (1997, 168–169), Ponzio (1998, 123).

[92] Campanella (1994, 65). In this section of this chapter, further references to Campanella (1994) will be given in parenthesis by citing just the page number, as done here.

3:11, "God handed the world over to the disputes of men";[93] and Romans 1:20, "The invisible things of God come to be understood through the things which he has made" (54). And he elaborates the point with the argument: "For us to be able to do this, he gave us a rational mind, and for avenues of investigation he provided the five senses as windows to the mind ... Therefore it would have been superfluous for him, who came to redeem us from sin, to teach us what we are able and obliged to learn on our own" (65–66). Here Campanella is giving a justification of the principle of limited scriptural authority as being (1) implicit in Jesus' example in Scripture; (2) explicit in assertions of the Old and the New Testament; and (3) in accordance with plausible theological speculation.

But Campanella goes further. He argues not only that it is proper to learn about the world by using our mind and senses rather than reading Scripture, but also that it is un-Christian to prevent such learning. In what he calls the fourth assertion of the second hypothesis, Campanella holds that "anyone who forbids Christians to study philosophy and the sciences also forbids them to be Christians" (54). One reason is that "since one truth does not contradict another, as was stated by the Lateran Council under Leo X and elsewhere, and since the book of wisdom by God the creator does not contradict the book of wisdom by God the revealer, anyone who fears contradiction by the facts of nature is full of bad faith" (69). Another reason is that "from the beginning the world has been called the 'Wisdom of God' (as was revealed to St. Brigid) and a 'Book' in which we can read about all things. Hence in his Sermon 7 on the fast days of the tenth month, St. Leo says, 'We understand the meaning of God's will from these very elements of the world, as from the pages of an open book' " (57); it follows that "therefore wisdom is to be read in the immense book of God, which is the world, and there is always more to be discovered" (71). Using the metaphor of the book of nature and the theological claim that this book was authored by God and so is at least as important as the book of Scripture, Campanella is arguing that it is wrong (theologically) to prevent someone from reading that (former) book.

Similarly, in what he calls the first assertion of the second hypothesis, Campanella argues that it is not only irrational and harmful but impious "if there is anyone who chooses on his own to prescribe rules and limits for philosophers as though they were decreed in the Scriptures and who teaches that one should not think differently than he does, and who subjects and confines the Scriptures to one unique meaning either of his own or of some other philosopher" (74). What is interesting here is that this impiety looks like a description of what Galileo's scriptural critics were doing, and that the reason Campanella gives is something which he also finds in both St. Augustine and St. Thomas. Campanella's reason is that

---

[93] Campanella (1622, 13) refers to this passage simply as Ecclesiastes, Chapter 1; and this reference is glossed more specifically as Ecclesiastes 1:13 by many scholars (e.g., Firpo 1968, 20; Blackwell 1994, 54). However, the verse quoted by Campanella is actually Ecclesiastes 3:11, as one can see from Headley (1997, 173–174) and Lerner (2001b, 208–209, n. 28); the confusion is easily explained in light of the similarity of the two verses. Headley (1997, 173–176) also gives a valuable analysis of Campanella's four references to, and glosses on, Ecclesiastes 3:11.

such an impious person "exposes the Sacred Scriptures to the mockery of the philosophers and to the ridicule of pagans and heretics and thereby prevents them from listening to the faith" (74). And then he gives a long quotation from St. Thomas, that includes a quotation from St. Augustine's *On the Literal Interpretation of Genesis*, so as to be able to claim that "so says St. Thomas in agreement with St. Augustine" (76), and thus provide formidable patristic and traditional-theological credentials to his own argument.

Before concluding our examination of Campanella, it will be useful to see how he handles the Joshua passage. He denies that the miracle "would be nullified if the sun is at rest in the center of the world" (97) because "the appearances are exactly the same if either the observer or the object seen is moved" (98) and the Copernicans say that the miracle happened by stopping the earth rather than the sun; "whoever says that this happened by arresting the motion of the earth does not deny the miracle but explains it, just as the physicist does not deny that God causes the rainbow but explains how he does it and what natural and reasonable means he uses" (99). Campanella does not seem worried about the nonliteral interpretation that is needed, but about retaining the spiritual message or meaning.

Campanella ends his book with the following profound and prophetic formulation of his conclusion: "In my judgment, in agreement with what St. Thomas and St. Augustine have taught us in our Second Hypothesis, it is not possible to prohibit Galileo's investigations and to suppress his writings without causing either damaging mockery of the Scriptures, or a strong suspicion that we reject the Scriptures along with heretics, or the impression that we detest great minds … It is also my judgment that such a condemnation would cause our enemies to embrace and honor this view more avidly" (122–123).

4.7 Conclusion

In his "Disputation" (1616), Ingoli included the following statement of the scriptural argument against the Copernican system: Copernicanism is wrong because it is contrary to the literal meaning and to the patristic interpretation of scriptural passages, such as Joshua 10: 12–13, and so it is contrary to Scripture.

In his *Letter on the Earth's Motion* (1615), Foscarini criticized the scriptural objection as follows: Copernicanism is not contrary to Scripture because, although it is contrary to the literal meaning of Scripture, Copernicanism is a thesis about natural phenomena and Scripture does not aim to make claims about nature, and so scriptural statements about natural phenomena are to be interpreted accommodatingly and not literally.

In his *Letter to Christina* (1615), Galileo elaborated the following criticism. First, literal meaning, patristic interpretation, and scriptural authority are all irrelevant for demonstrable questions of natural philosophy or science, although they are relevant or binding for matters of faith, morals, history, and indemonstrable claims about nature; such irrelevance follows from the universally accepted principle of the priority of demonstration, which in turn follows from the twofold aspect of divine revelation and the asymmetries between the work and the word of God; thus, both steps of the scriptural argument (from contrariety to Scripture to falsehood, and from contrariety to literal meaning and patristic interpretation to contrariety to Scripture) are non sequiturs. Second, to determine the literal meaning of scriptural assertions about natural phenomena, one normally needs to know what is really the case about natural phenomena; and so the conclusion of the argument cannot be known to be true independently of knowing the truth of its key premise, and the argument begs the question. Third, the literal meaning of Joshua really contradicts the Ptolemaic system; whereas Copernicanism (in the modified Galilean version that attributed axial rotation to the sun) is largely compatible with the literal interpretation of Joshua; so the key premise of the argument is questionable or false.

In his *Apologia pro Galileo*, written in 1616 and first published in 1622, Campanella advanced this criticism. The scriptural objection is assuming the principle that Scripture is a scientific authority (insofar as it is trying to refute a scientific theory on the basis of Scripture). But this assumption is itself contrary to Scripture, specifically Ecclesiastes 3:11 and Romans 1:20; moreover, the assumption is generally and deeply un-Christian because is contradicts the idea that God revealed himself not only through Scripture, but also through the book of nature; finally, the assumption also contradicts the patristic tradition, as shown by the teachings of St. Augustine and St. Thomas, who regard the assumption as an abuse of scriptural authority.

We may say that Ingoli gave a sophisticated and strong statement of the scriptural objection to Copernicanism. Foscarini advanced a theological criticism of the key premise of this objection (the claim that Copernicanism is contrary to Scripture). Campanella put forth a theological criticism of the major assumption of the objection (the principle that Scripture is a scientific authority). Galileo proposed a methodological criticism of this major assumption; a scientific and textual criticism of this key premise; and an epistemological criticism of the relationship between the key premise and the conclusion. Expressed positively and constructively, Foscarini, Campanella, and Galileo provide us with theological *and* philosophical arguments justifying the claim that Copernicanism is *not* contrary to Scripture *and* that Scripture is *not* a scientific authority.

In my judgment, these arguments of Foscarini, Campanella, and Galileo are cogent and essentially valid. They are thus of perennial interest and relevance, and they have some applicability to subsequent and present-day issues, such as that of evolution versus creation. However, it is beyond the scope of this investigation to elaborate this judgment, relevance, and applicability. Instead I shall discuss some more historical questions.

Despite the cogency of these criticisms of the scriptural objection to Copernicanism, they failed to convince various officials of the Catholic Church, and so on 5 March 1616 the Congregation of the Index issued a decree declaring the Copernican doctrine of the earth's motion contrary to Scripture, condemning and totally prohibiting Foscarini's *Letter,* and banning Copernicus's *Revolutions* until

4.7 Conclusion

and unless revised in a manner to be specified later.[94] Although this anti-Copernican decree did not mention Galileo by name, it led to his condemnation as a suspected heretic in 1633[95] and later to what some called the "greatest scandal in Christendom,"[96] and contributed to the modern-times divorce between science and religion.[97]

Thus, here we have one of the greatest ironies in the history of the interaction between scriptural interpretation and natural science. On the one hand, at the intellectual level we have the conception, elaboration, and presentation of some of the best arguments ever advanced why a particular scientific theory was compatible with Scripture and why in general Scripture is not a scientific authority; on the other hand, at the institutional and organizational level, one of the world's great religions issued a formal condemnation of a key scientific theory that played a crucial role in the rise of modern science. A particularly high point in the history of thought is accompanied by a particularly low point in the history of action.

This ironical lack of correspondence is my own way of formulating an evaluation that is widely shared,[98] although I am aware that there are iconoclasts and revisionists who pretend or posture otherwise.[99] Independently of the evaluative issue, the lack of correspondence cries out for an historical explanation. The most plausible explanation, which may be gleaned from the works of Filippo Soccorsi and Ernan McMullin,[100] seems to me the following.

Copernicanism was condemned in 1616 primarily *not* because it contradicted Aristotelian philosophy and was opposed by Aristotelian professors of philosophy; *nor* because it lacked a strict and conclusive demonstration; *nor* because it aimed to be a description of physical reality and more than merely an instrument for "saving

---

[94] Favaro 19: 322–323, Finocchiaro (1989, 148–150; 2005b, 16–24), Galilei (2008, 176–178).

[95] When I say that in 1633 Galileo was condemned as a suspected heretic, I do *not* mean to imply that he was found guilty of a *formal* heresy (and that the formal heresy in question was belief in Copernicanism), since being a heretic also included other, less serious crimes, such as strong suspicion of heresy, vehement suspicion of heresy, and slight suspicious of heresy; and in fact Galileo was condemned for vehement suspicion of heresy. For some clarifications, see Finocchiaro (1989, 14–15, 363 n. 86; 2005b, 11–12, 271–74).

[96] A few days after Galileo's death on 8 January 1642, these words were uttered by pope Urban VIII to Tuscan ambassador Francesco Niccolini, as a reason to veto the plan to erect an honorific mausoleum for Galileo in the church of Santa Croce in Florence; see Niccolini to Gondi, 25 January 1642, in Favaro 18: 378–379. In that context, the phrase had an anti-Galilean connotation, but later it acquired an anti-clerical meaning.

[97] Koestler (1959); cf. Section 8.15.

[98] For some of the best examples, see Soccorsi (1947), McMullin (2005b). Cf. Section 8.14.

[99] See Duhem (1908; 1969, 104–117) and cf. Chapter 11 below; see Koestler (1959, 425–463) and cf. Section 8.15; and see Feyerabend (1985; 1987; 1988; 1993) and cf. the penultimate section of the Introduction and Section 12.5 of this book.

[100] Soccorsi (1947), McMullin (2005b). Although my reconstruction here owes much more to McMullin than to Soccorsi, and although McMullin apparently formulated his account independently of Soccorsi, the similarities between the two views are striking, and someone acquainted with Soccorsi will perceive McMullin's account as a fuller, more elaborate, and updated version of Soccorsi's view. For an account of Soccorsi, see Finocchiaro (2005b, 284–294) and Section 8.14.

the phenomena"; *nor* because it allegedly contradicted voluntarist theology and skeptical epistemology; *nor* because it was supposedly no better than the Tychonic alternative; *nor* because it undermined geocentrism and anthropocentrism; *nor* because it was allegedly advocated by Galileo with excessive zeal and imprudence; *nor* because it was (mis)perceived as implying Giordano Bruno's philosophy of the plurality of words and the infinity of the universe. Each of these factors has found serious proponents who have used it to advance an explanation in its terms;[101] and they all have some truth and played a role, some larger and some smaller. However, individually such explanations are one-sided, and even collectively they remain relatively minor and are insufficient to bring about the effect. The real explanation was that Copernicanism appeared to contradict the literal meaning and patristic interpretation of Scripture, and Church officials were unwilling or unable to disregard such literal and patristic interpretations because of the Counter-Reformation struggles over biblical interpretation and because of the personal influence of Cardinal Robert Bellarmine, who was supremely influential and held an especially conservative version of biblical literalism or fundamentalism. The two primary factors were thus a geopolitical one (Counter-Reformation) and a personal one (Bellarmine).[102] But these factors were not determinist causes in the sense that they made the outcome inevitable, and the cogent arguments of Catholics such as Galileo, Foscarini, and Campanella indicate that the controversy could have been resolved the other way and that the actual result was contingent and circumstantial.

---

[101] Just to cite two leading scholars: the first mentioned explanation was advocated by Drake (1980); the last by Koyré (1966; 1978, 136, 211 n. 30). Cf. Finocchiaro (2005b, 242; 2002) and Section 12.2.

[102] Here one might also add the factor involving the principle that theology is the queen of the sciences and the unwillingness on the part of theologians to relinquish the power resulting from this principle. But I am not sure this is a distinct third major factor because it may be regarded as a part of the other two factors.

Chapter 5
Galileo on the Mathematical Physics of Terrestrial Extrusion

We have seen that the empirical astronomical objections and the scriptural theological objections were not the only arguments against the Copernican system. Equally important, strong, and consequential were the mechanical physical objections, involving such phenomena as falling bodies, the motion of projectiles, and the extruding power of whirling. They presupposed Aristotelian physics, based on the principles that rest is the natural state and motion requires a force; and their refutation required a new physics, based on the principles of inertia or conservation of motion and the composition of motion into independent components. Galileo took the physical objections as seriously as the astronomical ones, although he conceived the refutation of the former earlier than the refutation of the latter, since the answers to the physical difficulties were implied by his earlier research into a new physics, whereas the astronomical answers were the result of the telescope.

This chapter will examine how Galileo defended Copernicus from the extremely powerful objection based on the extruding power of whirling. We will see that Galileo criticized the extrusion objection in several distinct ways, but that his major criticism was that the objection was quantitatively invalid. Now, this quantitative invalidity turns out to be a flaw that is partly mathematical, partly physical, and essentially physical–mathematical. That is, the correct understanding of the phenomenon to which the objection appeals requires the proper combination of physical and mathematical principles, namely the skill of physical–mathematical reasoning. Thus, the structure of this chapter will be that of an inquiry into the nature of mathematical reasoning by means of a Galilean case study suggesting a definition of physical–mathematical reasoning.

5.1 Introduction

What makes reasoning mathematical reasoning? I would begin by saying that, in exploring this question, we must combine philosophy and history since (to paraphrase Kant) philosophy of mathematics without history is empty, and history of mathematics without philosophy is blind. Moreover, we must follow what might be called a comparative approach, namely we must consider instances of mathematical reasoning and study their similarities and their differences with each other, as well as between them and other types of reasoning that are not mathematical.

However, I would make some further clarifications. I do not think we can equate the philosophical with the normative; in this context I would want to equate the philosophical with the theoretical, the historical with the concrete, and contrast the normative not with the historical but with the descriptive–interpretive. The upshot of my own terminological conventions would be, however, that we need to combine philosophy and history, normative evaluation and descriptive interpretation, and theoretical generalization and concrete analysis. Thus, there are at least three meta-epistemological dimensions which we need to be sensitive to in our own investigations into mathematical reasoning.

Furthermore, there is another important distinction that needs to be made in regard to normative evaluation. One normative question is, what characteristics reasoning should have in order to be classified as mathematical. This may be contrasted with the question of what characteristics reasoning does have when as a matter of historical empirical fact it is classified as mathematical. But the former question could also be contrasted to the question: what characteristics should mathematical reasoning have in order to be good, valid, or correct? It would not be proper to define the nature of mathematical reasoning in such a way that all mathematical reasoning is automatically valid, but rather we should allow the possibility that some mathematical reasoning is invalid. In other words, the defining characteristics of mathematical reasoning (whether they are normative or descriptive) are not (should not) be the same as the defining characteristics of valid mathematical reasoning.

Finally, consider the claim that the defining characteristics of mathematical reasoning should be in terms of some kind of logical structure, though the relevant conception of logic may have to be a broad one. I would want to say that this is just one possibility. What are the others? Well, the others are the other possible answers to the question: is mathematical reasoning, reasoning about some particular subject matter (like numbers, geometrical figures, functions, sets, etc.), or is it reasoning about any subject matter as long as the reasoning has certain formal properties? The logical-structure thesis is choosing the second one of these options. I believe that for a long time in the history of mathematics the first one of these options was chosen. Actually, there is a third alternative that combines both; it says that mathematical reasoning is reasoning which is about mathematical objects like numbers and which possesses certain formal properties like logical structure.

With such questions and distinctions in mind, I would like to study a piece of mathematical reasoning found in Galileo's *Dialogue*. The argument pertains to the topic of circular motion and the centrifugal effects and centripetal forces connected to it. The reasoning occurs in the context of the Copernican controversy and the discussion of the many arguments against the earth's motion. One of these objections appealed to what Galileo called the extruding power of whirling, and he tried to answer it primarily with a mathematical criticism.[1]

---

[1] Favaro 7: 214–244, Galilei (1967, 188–217; 1997, 171–212).

5.2 The Extruding Power of Whirling

The anti-Copernican argument from the extruding power of whirling claimed the following. The basis of this objection to Copernicanism was the fact that in a rotating system, or in motion along a curve, bodies have a tendency to move away from the center of rotation or of the curve. For example, if one is in a vehicle traveling at a high rate of speed, whenever the vehicle makes a turn one experiences a force pushing one away from the center of the curve defined by the turn: if the vehicle turns right one experiences a push to the left, and if it turns left one experiences a push to the right. Or one could make the simple experiment of tying a small pail of water at the end of a string and whirling the pail in a vertical circle by the motion of one's hand; now suppose a small hole is made in the bottom of the pail; as the pail is whirled one will see water rushing out of the hole always in a direction away from one's hand. Then the argument called attention to the fact that, if the earth rotates, bodies on its surface are traveling in circles around its axis at different speeds depending on the latitude, the greatest being about 1,000 miles per hour at the equator. This sounds like a very high rate of speed, which would generate such a strong extruding tendency that all bodies would fly off the earth's surface, and the earth itself might disintegrate. Since this obviously does not happen, it was concluded that the earth must not be rotating.

This objection was appealing to a phenomenon which is real and which today we call centrifugal force. So the proper refutation of this objection required a full understanding of the laws of motion in general, and of circular motion and centrifugal force in particular, and a good knowledge of the mathematical representation of physical quantities. Such knowledge was not yet available. In fact, in his critique, Galileo was groping toward these things, among others. He advances four criticisms which are distinct but interrelated.

Galileo begins with the criticism that, as usually formulated, the argument is improperly stated. That is, its crucial step (namely, its main subargument) should be stated to say that if the earth were rotating then there would now be no loose bodies on its surface, since they would have all been extruded long ago. Instead, it is usually misstated by claiming that if the earth were in rotation then we would see bodies on its surface extruded off toward the sky.

Galileo's second criticism may be summarized as follows. If a body were to be extruded from a rotating earth, the extrusion would occur along a tangent to the point of last contact with the terrestrial surface; the reason for this stems from the principle of inertia. But, because of gravity, on a rotating earth bodies would still have a tendency to move downward along the secant from the point of their position to the center of the earth. Thus, we need to do a comparison between these two tendencies; we cannot just consider the centrifugal extrusion, as the anti-Copernican argument seems to be doing. Now, the comparison shows that the downward tendency is greater than the extruding one. For example, consider Fig. 5.1,[2] where

---

[2] This corresponds to Galileo's diagram in Favaro 7: 224, Galilei (1967, 198; 1997, 184).

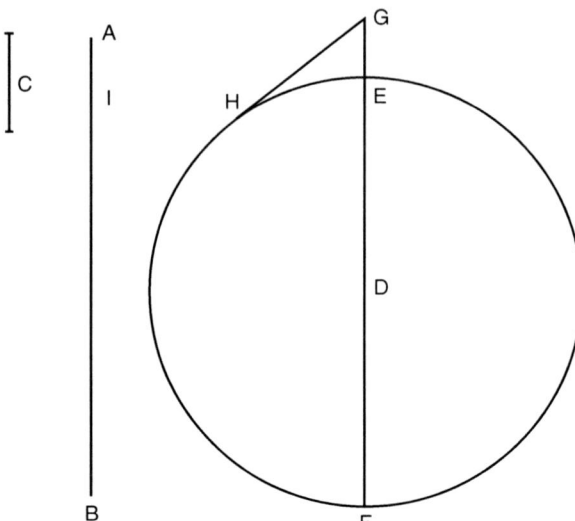

Fig. 5.1 Relationship between Tangent and Secant

the circle EHF represents the equator of the earth rotating clockwise; while a body would have the tendency to be extruded along the line HG, it would also have the tendency to fall along the line GED.

In his third criticism, Galileo tries to show that the downward tendency not only happens to exceed the tangential one, but it necessarily does so for mathematical reasons; that is, he argues that extrusion would be mathematically impossible on a rotating earth. He tries to prove this mathematical impossibility on the basis of the geometry of the situation in the neighborhood of the point of contact between a circle and a tangent, and the behavior of the external segments (called exsecants) of the secants drawn from the center of the circle to the tangent (see Fig. 5.2).[3] He argues that as one approaches the point of tangency, three things happen: (1) the ratio of an exsecant to the corresponding tangent segment tends toward zero; (2) the ratio of an exsecant to the corresponding speed of fall also tends toward zero; and (3) the ratio of one exsecant to another (at twice its distance from the point of tangency) tends to zero as well. That is, the exsecants get smaller and smaller in relation to (1) the corresponding tangent segments, (2) the speeds of fall, and (3) each other. Thus, the distances of fall required to prevent extrusion become infinitely smaller than the distances required to achieve extrusion, and they decrease more rapidly than the speeds of fall.

---

[3] This corresponds to Galileo's diagram in Favaro 7: 225, Galilei (1967, 199; 1997, 186). But I have reproduced it from MacLachlan (1977, 175).

5.3 Restatement of the Anti-Copernican Argument

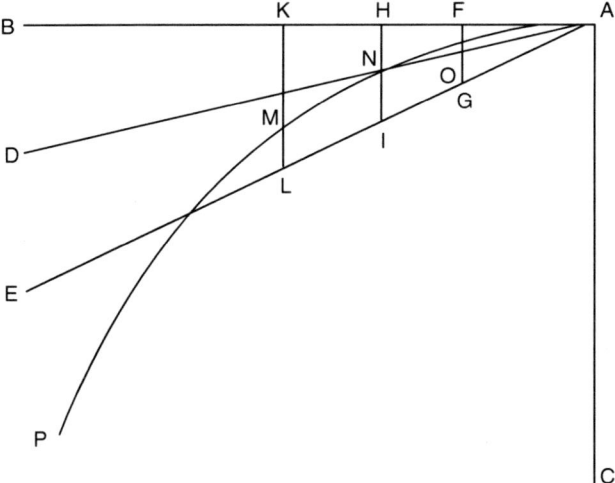

Fig. 5.2 The Neighborhood of the Point of Tangency

Galileo's fourth and last criticism is the following: it can be shown that in circular motion the extruding tendency increases with the linear speed but decreases with the radius; but on a rotating earth, the linear speed at the equator would be very small compared with the earth's radius; thus, on a rotating earth the extruding tendency would really be very small. The anti-Copernican objection ignores this aspect of the situation, and so it assumes that on a rotating earth extrusion would be more likely to happen than is the case.

There is no question that Galileo's criticism contains some insights, as well as some errors.[4] However, it is hard to distinguish one from the other, and to identify each. Nor is it easy to determine whether the errors are errors of reasoning, of fact, or of judgment. And for his errors of reasoning we need to ask whether they are mathematical, logical, neither, or both. To begin to unravel these issues, we must examine his objections more closely.

5.3 Restatement of the Anti-Copernican Argument

Recall that Galileo's first criticism charged that, as traditionally formulated, this anti-Copernican attempt to refute terrestrial rotation is misstated. He attributes its original formulation to Ptolemy, and it is worth quoting Galileo's own words:

---

[4]For other accounts, from which my own has benefited, see Boyer (1967, 245–247), Chalmers and Nicholas (1983), DiCanzio (1996, 144–145, 171–173, 353–354), Drake (1986a), Feldhay (1998), Gaukroger (1978, 189–195), Hill (1984), MacLachlan (1977), Pagnini (1964, 2: 383 n. 1, 387 n. 1, 415 n), Palmieri (2008a).

This refutation refers to the destruction of buildings and to rocks, animals, and men themselves being cast toward the heavens; but such destruction and scattering cannot happen to buildings and animals that do not already exist on the earth, nor can men be born and buildings erected on the earth unless it stands still; therefore, it is clear that Ptolemy is arguing against those who grant the earth to have been at rest for a long time (that is, while animals and masons could live on it, and palaces and cities could be built), but then suddenly set it in motion, to the ruin and destruction of buildings, animals, etc. On the other hand, if he had intended to argue against those who attribute this turning to the earth since its original creation, he would have refuted them by saying that, if the earth had always been in motion, then neither beasts nor men nor rocks could ever have been formed on it, and still less could buildings have been erected and cities founded, etc. ... [In other words] Ptolemy argues either against those who regard the earth as being always in motion, or against those who regard it to have been still for some time and then to have been set in motion. If he is arguing against the first, he should have said: 'the earth has not always been in motion because terrestrial rotation would not have allowed men, animals, or buildings to exist on the earth, and so there would have never been any of them on it.' However, he argues by saying: 'the earth does not move because the beasts, men, and buildings already found on the earth would be cast off'; hence, he supposes the earth to have been once in such a state as to have allowed beasts and men to form and live on it; this implies the consequence that once it was motionless, namely, suitable for animal life and the construction of buildings.[5]

These remarks can be interpreted as either a formal criticism or an analytical clarification of the anti-Copernican extrusion argument. The criticism is that, as ordinarily stated, the argument alleges to be proving one conclusion (that the earth is not in motion) but instead at best proves another (that the earth did not recently begin to move); that is, the argument reaches an irrelevant conclusion, namely, a proposition that is not disputed. So interpreted, the argument as originally stated is an example of a classic fallacy called *ignoratio elenchi*.[6]

But these remarks may also be taken as introducing an essential clarification in the statement of the extrusion argument; for the criticism just mentioned cannot even be stated without indicating the proper way to reformulate the argument, and so once the criticism is understood, we can immediately reformulate the argument. This is how Galileo's discussion proceeds, and so once the argument has been appropriately reformulated, he elaborates the other criticisms.

5.4 Tangential Extrusion Versus Secant Fall

Galileo's second criticism was that, to determine whether bodies would be extruded on a rotating earth, we need to compare the extruding tendency along the tangent with the downward tendency along the secant; and if the comparison is made, we realize that the downward tendency exceeds the extruding one. Here, his criticism is not as detailed as it might have been.

---

[5]Galilei (1997, 172). Cf. Favaro 7: 215–216, Galilei (1967, 189).
[6]Cf. Aristotle, *On Sophistical Refutations* 167a21.

He does suggest that, near the point of tangency, the downward tendency would be a thousand times greater than the extruding one.[7] The number 1000 is meant as a rough estimate, as a claim about the order of magnitude. At this point, Galileo could have been more quantitatively precise about the comparison. For example, he could have calculated the distance covered in one second by a body falling from rest; before his telescopic discoveries, he had done extensive research on this topic, and in another passage of the *Dialogue*[8] he gives some relevant approximate figures (i.e., 100 cubits in 5 seconds). Then he could have calculated the distance separating the extrusion tangent and the terrestrial circumference in one second of time at the equator; this can be computed from knowing the earth's radius, the rate of terrestrial rotation, and the geometry of the situation. He could have easily arrived at the number 266 as the ratio between these two distances; that is, at the conclusion that in one second free fall takes a body 266 times farther downwards than terrestrial rotation extrudes it away from the center. In fact, a few years after the book's publication, Marin Mersenne (working directly from Galileo's book) used this method of calculation and arrived at essentially this number.[9] Later in the century, Christian Huygens and Isaac Newton both arrived at more adequate solutions of the problem, corresponding to modern physics; that is, at the equator the centripetal acceleration due to gravity is 289 times greater than the centrifugal acceleration due to terrestrial rotation, which is to say that a body weighs 1/289 less at the equator than it would on a motionless earth.[10]

However, Galileo chose not to perform the Mersenne-type of calculation. Though his motivation is unclear, one likely reason is that he thought he could prove something much stronger than the contingent fact that the downward tendency due to weight happens to exceed the extruding tendency due to rotation; his stronger claim is that the downward tendency (however small it might be, as long as it not zero) can always overcome the extruding tendency (however large it might be). This, of course, is the gist of what I have called his third criticism.

5.5 Linear Versus Angular Speed

Before examining Galileo's third criticism, it is better to discuss his fourth one (namely the one which comes last in the text),[11] which is less complex. It may be reconstructed as follows.

It is true that the extruding tendency increases as the speed when the radius is constant; but when the speeds are equal, the extruding tendency decreases as the

---

[7] Favaro 7: 221, Galilei (1967, 194–195; 1997, 180).
[8] Favaro 7: 250, Galilei (1967, 223).
[9] See MacLachlan (1977, 176–178) for this account and other details.
[10] For a more modern argument along the same lines, see Chalmers and Nicholas (1983, 322).
[11] Favaro 7: 237–244, Galilei (1967, 211–217; 1997, 203–212).

radius increases; so the extruding tendency increases directly with the speed but inversely with the radius; thus, the extruding tendency remains constant when the linear speed increases as much as the radius, i.e., when equal numbers of rotations are made in equal times; hence, terrestrial rotation would cause as much extrusion as a wheel rotating once in 24 hours; so, a rotating earth would not scatter bodies toward the heavens. To show that at constant speed the extruding tendency decreases as the radius increases, one may reason thus: the extruding tendency is equal to the force needed to prevent the body from escaping along the tangent; this force is equal to that required to compensate for the tangential displacement if the body were extruded; and this compensation is smaller for greater circles. This (last claim) is so because when the linear speeds are the same (arc CE = arc BIG in Fig. 5.3),[12] the deviation from the tangent is measured by the exsecant (DE and FG), and the exsecant becomes smaller as the circle becomes larger (DE < FG).

This Galilean criticism is partly right but partly wrong. The linear speed of a body lying on the earth's equator, which is about 24,000 miles long, is about 1,000 miles per hour. The original anti-Copernican argument seems to assume that the extruding tendency varies as the linear speed, so that the speed of 1,000 miles per hour should give rise to an extruding tendency that would tear the earth apart. Galileo's key critical claim here is that this assumption is not exactly right, that it is incomplete; it fails to take into account another crucial variable, namely the radius of rotation; this is a serious flaw because the radius affects the extrusion in an inverse manner; that is, the extruding tendency decreases as the radius increases. From the viewpoint of modern physics, this claim is essentially right.

However, the manner in which Galileo derives this claim is problematic in several ways.[13] Moreover, the situation is even more complicated than he realized; that is, his critical claim (even disregarding the derivation) is itself incomplete because he failed to realize that although the extruding tendency does grow with the linear speed, the growth is in accordance with the square of this speed; in short, centrifugal force is directly proportional to the square of the linear speed and inversely proportional to the radius; thus, the extrusion is a function not simply of the angular speed (number of turns per unit time), but of the radius and the square of the angular speed.

Even so, Galileo's discussion contains evidence of his intuition that the extruding tendency depends more on angular speed than on the radius, which is one way of stating one consequence of the correct law of centrifugal force ($F = m\omega^2 R$). Before the end of the seventeenth century, this law was formulated in its modern form by Huygens (1673).

---

[12] This corresponds to Galileo's diagram in Favaro 7: 242, Galilei (1967, 216; 1997, 209).
[13] Cf., for example, Chalmers and Nicholas (1983, 323–328), Gaukroger (1978, 193–195), Pagnini (1964, 2: 415 n).

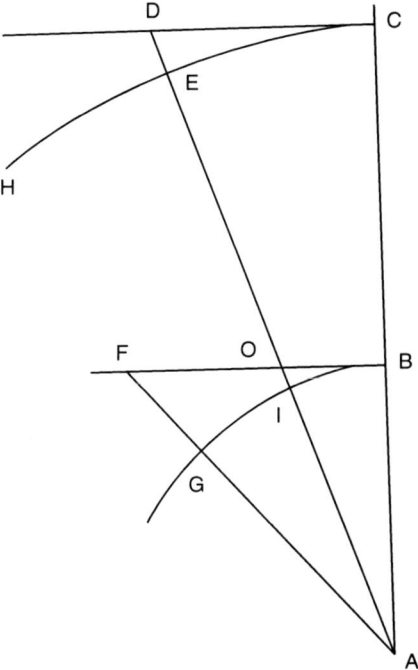

Fig. 5.3 Tangential Extrusion as a Function of Radius

5.6 Physical Processes Versus Mathematical Entities

Although this Galilean intuition would seem to be a "physical" (empirical or factual) claim, and although this claim is a crucial part of this (fourth) Galilean criticism of the anti-Copernican argument, this criticism would also seem to have a mathematical aspect (if I may be allowed to use my own – semantical or epistemological – intuition). The reason is that this criticism focuses on the quantitative aspects of the situation. When faced with the anti-Copernican claim that on a rotating earth bodies would be extruded, Galileo asks how much they would be extruded. It is this concern with asking quantitative questions that leads Galileo to engage in mathematical reasoning. When he answers that the amount of extrusion depends on two factors (linear speed and radius), rather than just one (linear speed), then we are dealing once again with matters of physical or empirical intuition or fact. Similarly, when (or if) he "intuits" that the extrusion depends more on angular speed than on the radius, this intuition presupposes a mathematical or quantitative concern, namely how much the extrusion depends on one and on the other.

This quantitative orientation – this willingness to ask quantitative questions about physical process and phenomena – is similar to the attitude Galileo displayed in the second criticism. When Galileo points out that the extruding tendency must be viewed in the context of the downward tendency, which continues to exist on a rotating earth, there is nothing mathematical or quantitative in that; but when he suggests that the important thing is to determine whether the extruding tendency is greater than the downward one, then we have a quantitative concern with how much.

However, in the earlier (second) criticism, mathematical reasoning surfaces at an earlier juncture. When Galileo goes on to represent the extruding tendency by means of the tangent line, and the downward tendency by means of the exsecant, then he has taken a step toward mathematical reasoning. In what sense? In the sense of being concerned with thinking of physical processes in terms of mathematical entities – geometrical lines in this case. This is a different aspect of mathematical reasoning than the one described before. Now we have mathematical reasoning in the sense of reasoning about mathematical entities (geometrical lines). This sense of mathematical reasoning also occurs in the fourth criticism, but there the step is taken without explicit comment, whereas in the second criticism Galileo explicitly calls attention to the fact that in representing physical processes by means of mathematical entities an important, and potentially problematic, step is being taken.

Thus both the second and the fourth criticisms exhibit mathematical reasoning in two ways. The extrusion of bodies on a rotating earth is certainly a physical process. Galileo is thinking mathematically about this physical process in the sense that he represents various aspects of it by means of mathematical entities, which in the simplest case are lines of Euclidean geometry; this is more obvious in the second than in the fourth criticism. He is also thinking mathematically insofar as he is thinking of the quantitative aspects of the physical process; and on the contrary, this is more obvious in the fourth than in the second criticism. Insofar as this kind of quantification involves numbers, and attaching numbers to physical processes, then perhaps it too involves the representation of various aspects of the physical situation by means of mathematical entities – in this case numbers, the entities of arithmetic. Thus, the two senses of mathematical reasoning perhaps reduce to reasoning about physical processes by thinking of them in terms of mathematical entities, with the difference being then determined by the branch of mathematics in question. How problematic such a representation can be is illustrated by Galileo's third criticism, to which I now turn.

5.7 Escape Extrusion Versus Orbital Extrusion

Galileo's third criticism of the extrusion objection amounts to an attempt to show the mathematical impossibility of extrusion on a rotating earth. That is, Galileo tries to show that there are mathematical reasons why the downward tendency (however small) could always overcome the extruding tendency (however large), and so bodies could not be extruded on a rotating earth.

One of the most common reactions to this Galilean argument is to say that there must be something wrong with it because the conclusion is false. The conclusion is

5.7 Escape Extrusion Versus Orbital Extrusion

the physical claim that on a rotating earth bodies could not be extruded; but in fact, if the earth rotated faster or gravity were weaker, bodies would be extruded. Of course, even if this criticism of Galileo were correct, it would only be the beginning of a critical analysis of his argument because all it would tell us is that either some of his initial premises are false or some of his principles of inference are invalid; and then we would have to reconstruct his argument in all of its complexity to determine this. However, let us first examine whether this common criticism is really correct. Is it the case that on a rotating earth bodies with weight could be extruded? I believe that, from the point of view of modern physics, Galileo's claim is indeed false in one sense, but it is true in another.

That is, two meanings of extrusion must be distinguished, and it can be shown that bodies could be extruded in one sense, but could not be extruded in another sense. The two types of extrusion are: (a) leaving the earth's surface and going into a geocentric orbit; and (b) escaping from the earth's gravitational field.

As for orbital extrusion (a), to see that it could happen as a result of (increased) terrestrial rotation, we may reason as follows: (1) when a body moves in a circle, a centripetal force is required, otherwise it would move in rectilinear uniform motion (because of inertia); (2) the centripetal force is directly proportional to the square of the linear speed and inversely proportional to the radius ($F = mV^2/R$); (3) for terrestrial bodies this force is provided by the earth's gravitational attraction, which is inversely proportional to the square of the radius, but does not depend on speed ($F = GmM/R^2$); thus, (4) as the rate of terrestrial rotation increases, the required centripetal force also increases, but the available gravitational attraction remains constant; so (5) the point would be reached when the former exceeds the latter and orbital extrusion occurs; (6) the minimum orbital speed V_o is such that $(mV_o^2/R) = (GmM/R^2)$, or more simply $V_o^2 = GM/R$; but (7) once the body is in orbit, any increase in terrestrial rotation does not affect its velocity or the required centripetal force, and so the body simply remains in orbit and does not escape from the orbit.

In regard to extrusion (b) in the sense of escaping from the earth's gravitational field, we may reason as follows: (8) to escape, a body must be given enough speed so that it will never be pulled back to earth, namely so that it can reach an infinite distance from the earth, so to speak; that is, (9) its kinetic energy must be sufficient to do the work required to move it from the earth's surface to infinity; (10) this work is equal to the body's change in potential energy; (11) setting potential energy to zero at infinity, this change equals the body's initial potential energy; that is, (12) the escape velocity is such that $(mV_e^2)/2 = GmM/R$; or, (13) in terms of orbital velocity, $V_e^2 = 2V_o^2$; that is, (14) the escape velocity is greater than the orbital velocity (by a factor of the square root of two); and so (15) the escape velocity will never be reached by just increasing terrestrial rotation.[14]

---

[14] This claim would need to be qualified if one wanted to take into account the effects of the irregularity of the earth's surface; for then a body that had reached orbital velocity could be struck some time later by a mountain that was moving at a higher velocity due to the increased terrestrial rotation, and so the body would acquire additional velocity; eventually, such additions might increase the body's velocity to the value required for escape extrusion. I thank Albert DiCanzio for this refinement.

Thus, I am not sure that Galileo's conclusion was physically wrong; it depends on which of these two physical claims we attribute to him. Therefore, the truth-value of the conclusion does not offer us a clear clue about the argument's correctness. At any rate, the argument is interesting in its own right. Let us therefore proceed to reconstruct its details. I have already suggested that Galileo has three distinct mathematical arguments. The first one is the following.

5.8 Exsecants Versus Tangents, or Achilles and the Tortoise

Consider once again Fig. 5.1, where the circle represents the equator of a sphere rotating clockwise. Galileo is saying that for a body on such a sphere, the extruding tendency can be measured by the tangential displacement that would result if the body were extruded, for example line HG.

The downward tendency can be measured by the displacement along the secant from the extruded body to the sphere's surface, for example line GE. Now, the geometry of the situation is such that a body can remain on the surface by undergoing a very small secant displacement while it undergoes a very large tangential displacement. But the secant speed, however small, is always sufficient to compensate for the tangential speed, however large. The reason for this last claim is that from the center of a circle, D, one can always draw a secant, DG, intersecting a tangent so close to the point of tangency, H, as to yield an arbitrarily small ratio between the exsecant, GE, and the tangent segment to the contact point, GH.

Galileo's point seems to be that regardless of how much greater the speed of tangential extrusion would be than the downward speed along the secant, one can always find a point on the tangent so close to the point of tangency that the body will not reach it because in the time required to reach it the body could move downward along the secant enough to remain on the surface; and in turn, this is so because at that point the exsecant distance to be covered during the same time is so much smaller than the tangent distance that the ratio between the two distances is smaller than the corresponding ratio between the two speeds. For example, suppose that the speed of tangential extrusion were 1,000 times greater than the downward secant speed; then one can find a point along the tangent so close to the point of tangency that the exsecant from there is less than 1/1,000 the tangent segment, or conversely such that the tangent segment would be more than 1,000 times greater than the exsecant; thus, the body would be able to undergo the required secant displacement and remain on the surface; and this is to say that it would never be extruded.

To prove this geometrical claim, which we may label the secant–tangent theorem, Galileo reasons as follows. Once more, let us refer to Fig. 5.1:

1. Consider two lines, BA and C, such that BA is longer than C as much as you wish.
2. The problem is to construct from the center of the circle D a secant DG such that the ratio of the tangent GH to the exsecant EG is equal to the ratio of BA to C. (For then, at any point on the tangent closer to H than G, the corresponding ratio would be even greater.)

5.8 Exsecants Versus Tangents, or Achilles and the Tortoise

3. Begin by constructing the third proportional to lines BA and C, namely, another line IA such that BA:C = C:IA; that is, take a segment within BA which is as much smaller than C as C is smaller than BA.
4. Then extend the diameter FE by an amount EG such that FE:EG = BI:IA.
5. Then from G draw the tangent GH.
6. We can prove that we have what we needed, namely that GH:EG = BA:C.
7. Proof:
7.1 FE/EG = BI/IA. (By construction, in line 4 above.)
7.2 EG/EG = IA/IA. (Identity.)
7.3 [(FE + EG)/EG] = [(BI + IA)/IA]. (By addition, from two previous lines.)
7.4 FG/EG = BA/IA. (By substitution, in previous line.)
7.5 BA/C = C/IA. (By construction, in line 3 above.)
7.6 FG/GH = GH/EG. (Proof given below,[15] in lines 7.6.1 to 7.6.9.)
7.6.1 GHD is a right triangle. (By construction, in line 5, since GH is a tangent.)
7.6.2 GH * GH + HD*HD = DG * DG (Pythagorean theorem.)
7.6.3 DG = DE + EG. (By construction.)
7.6.4 DE = HD. (Radii of the same circle.)
7.6.5 GH * GH + DE * DE = (DE + EG) * (DE+EG). (By substitution from previous three lines.)
7.6.6 GH * GH + DE * DE = DE * DE + EG * EG + 2DE * EG. (By expansion, from previous line.)
7.6.7 GH * GH = EG * (EG + 2DE). (By simplification of previous line)
7.6.8 GH * GH = EG * FG. (By substitution in previous line.)
7.6.9 FG/GH = GH/EG. (By rewriting previous line.)
7.7 BA/C = FG/GH. (From lines 7.4, 7.5, and 7.6, by the following steps.[16])
7.7.1 IA = BA * EG/FG. (From line 7.4.)
7.7.2 BA/C = (C * FG)/(BA * EG). (From lines 7.5 and 7.7.1.)
7.7.3 EG = GH * GH/FG. (From line 7.6.)
7.7.4 BA/C = (C * FG * FG)/(BA * GH * GH). (From two previous lines.)
7.7.5 (BA * BA)/(C * C) = (FG * FG)/(GH * GH). (By simplification of previous line.)
7.7.6 BA/C = FG/GH. (By simplification of previous line.)
7.8 GH/EG = BA/C. (By substitution, from lines 7.6 and 7.7.)

In order not to lose sight of the forest by focusing too much on the trees, let us pause for a moment to see what Galileo has accomplished here. Given any two lines one of which is longer than the other by any factor we wish, we can always construct a circle, tangent, and secant such that the tangent is longer than the exsecant by that same factor. The trick is to first mark at one end of the longer line a segment which is the so-called third proportional to the two original lines (namely a segment which is as much smaller than the shorter line as the shorter line was smaller than the longer line); then construct a circle whose diameter equals the longer of the two original lines minus the third proportional; then extend the diameter of this circle by an amount equal to the third proportional, thus generating an exsecant equal to

---

[15] This proof is not given in Galileo's text; he just makes the assertion (7.6).
[16] Once again, Galileo does not show these steps, though he gives explicit indication that 7.7 is based on 7.4, 7.5, and 7.6.

the third proportional; and finally draw the tangent from the outer end of this exsecant to the circumference. Such a tangent is as much longer than the exsecant as the longer of the two original lines was longer than the other.

To reverse the process, we can think of it this way. Take any circle with a tangent at some point in the circumference. If we draw a secant from the center to any point on the tangent, the tangent segment will always be longer than the exsecant. The closer we get to the point of tangency, the greater will be the factor by which the tangent segment exceeds the exsecant. In fact, and here is the crucial claim, it is always possible to choose a point so close to the point of tangency that the ratio of the tangent to the exsecant is as large as you wish. That is, using numerical terminology equivalent to Galileo's geometrical language (though not explicitly present in his text), suppose we want the tangent to be n times the exsecant; then it turns out that we must extend the diameter by $1/(n^2 - 1)$; this is also to say that the exsecant must be $1/n^2$ of the extended diameter. In other words, if the exsecant is to be $1/n$ of the tangent, it must be $1/(n^2 - 1)$ of the diameter and $1/n^2$ of the extended diameter. In nonconstructivist sounding language, we may say that: if the exsecant is $1/(n^2 - 1)$ of the diameter, then it is $1/n^2$ of the extended diameter and $1/n$ of the tangent.

These numerical formulas, which are not discussed by Galileo himself, may be proved as follows.

1. BA/C = n. (Assumption.)
2. BA/C = C/IA. (By construction of the third proportional IA.)
3. BA = nC. (From line 1.)
4. C/IA = n. (From lines 1 and 2.)
5. IA = C/n. (From line 4.)
6. BA/IA = nC/(C/n) = n^2. (From lines 3 and 5.)
7. BA = n^2 * IA. (From previous line.)
8. BA = BI + IA. (By construction.)
9. BI + IA = n^2 * IA. (From lines 7 and 8.)
10. BI = (n^2 * IA) − IA = IA * (n^2 − 1). (From previous line.)
11. IA/BI = $1/(n^2 - 1)$. (From previous line.)
12. But, EG/FE = IA/BI. (By construction.)
13. So, EG/FE = $1/(n^2 - 1)$. (From previous two lines.)
14. And, FG/EG = BA/IA (From line 7.4 in previous proof.)
15. So, FG/EG = n^2, or EG/FG = $1/n^2$. (From lines 6 and 14.)
16. And, GH/EG = BA/C. (By the relationship proved earlier, which is stated in line 7.8 of the previous proof and which may be called Galileo's tangent–secant theorem.)
17. So, GH/EG = n, or EG/GH = $1/n$. (From lines 1 and 16.)
18. So, EG = $[1/(n^2 - 1)]$ * FE = $(1/n^2)$ * FG = $(1/n)$ * GH. (From lines 13, 15, 17.)

Once again in words, this means that, for any number n however large, if we want to make the exsecant n times smaller that the tangent, we can do it by making the exsecant (n^2 − 1) times smaller than the diameter. This statement is mathematically true, and Galileo gives a correct mathematical proof for it. But, what does all

5.8 Exsecants Versus Tangents, or Achilles and the Tortoise

this mathematical reasoning have to do with the physical situation? Presumably the connection is the following: no rotational speed (however great) by the earth would cause a body to be extruded because the body's speed of fall would always be enough to counteract it; the reason for this is that the distance fallen would always be enough to compensate for the distance of extrusion; and in turn, the reason for this is that the length of the exsecant can always be made to compensate for the length of the tangent. That is, the ratio of rotational speed to falling speed can never be large enough for extrusion to occur because the ratio of tangential distance to exsecant distance can never be large enough, and this is so because for any given ratio of the tangent to the exsecant we can always find a larger such ratio by moving closer to the point of tangency. The crucial step is obviously the one that goes from a statement about geometrical tangents and secants to a statement about physical distances or speeds. This inference presupposes an identification of mathematical and physical entities.

This inferential step is crucial in two senses. It makes the passage in question more than a piece of pure mathematical reasoning; it makes it physical–mathematical, so to speak. And it is very difficult to evaluate its correctness.

In regard to the type of reasoning involved, can anything more be said than that Galileo starts with a mathematical statement about secants and tangents, and infers a physical statement about physical distances and motion? The mathematical statement is that for any arbitrarily large ratio, there is a point close enough to the point of tangency such that the ratio of the tangent to the exsecant is larger than the given ratio. Then he identifies the circle with the equator of a rotating earth, the exsecant with the distance covered by a terrestrial body in free fall, and the tangent with the distance traversed by a body during the same time if it were extruded. Then he is saying that if a body is to be extruded at some point, it has to be able to move faster along the tangent than along the secant, and sufficiently faster as to be able to traverse the longer tangent distance in the time that the body falls the shorter secant distance. This means that the ratio of the extruding tangential speed to the falling secant speed must be greater than the ratio of the tangent to the secant. However, Galileo's secant–tangent theorem establishes that as we get closer to the point of tangency or extrusion, the ratio of tangent to exsecant grows without limit. Therefore, for any finite value (however large) of the ratio of tangential speed to secant speed, there is always a point close enough to the point of tangency such that the tangential speed is insufficiently great. Thus, at that point the body will not have been extruded. Therefore, extrusion would only begin at some later point. But the tangential speed from such a later point would be subject to the same law and theorem as before, and so the body could not be extruded from that later point either. And so on. So the body will never be extruded.

When so analyzed, the argument seems like a version of the Achilles and tortoise paradox. Another similarity is that, just as we know it to be false that Achilles can never overtake the tortoise, we also know that Galileo's conclusion here is false. That is, if this Galilean argument were valid, it would apply to orbital as well as escape extrusion; but we know that orbital extrusion can happen. Thus, the key premise is true, and the essence of the conclusion is false. Thus, we also know that there must be something wrong in the reasoning. But what?

Rather than answering this question, whose answer I do not know, let me say that Galileo must have sensed that there was something wrong about his argument, for he offered another proof for the mathematical impossibility of extrusion. Let us look at that other proof. For this we will need to disregard Fig. 5.1 and consider Fig. 5.2 instead.

5.9 Distance Fallen, Distance To Be Fallen, and Speeds of Fall

In Fig. 5.2, point C represents the center of the earth; arc ANP is a portion of the earth's circumference, for example at the equator; the earth is rotating counter-clockwise; and line AB is tangent to the circumference at A.

If at point A, a body were extruded due to the earth's rotation, the body would tend to move along the tangent at a constant speed whose value would be equal to that deriving from being carried by the earth's rotation before separation. The line segments AF, FH, and HK are equal, and so they represent both equal times and equal distances since separation. However, immediately after separation, the body would start acquiring a vertically downward speed, whose value is directly proportional to time.[17] In the diagram, the segment FG represents the speed acquired in time AF; the segment HI represents the speed acquired in time AH; and the segment KL represents the speed acquired in time AK. And the time-proportionality of speed also means that FG/HI = AF/AH, that FG/KL = AF/AK, and that HI/KL = AH/AK.

The next thing to note is that in order to prevent extrusion, the body must be able to fall the distance between the tangent and the circumference, during the time that it would move a certain distance along the tangent. For example, in the time AH, it must be able to fall the distance HN. Let us also use the term "exsecant" to refer to the distance between tangent and circumference along HN and other lines parallel to it, but note that such lines are not extensions of radii and do not intersect the center. Now Galileo asks us to compare the required falling distances with the acquired speeds. As we approach point A, the point of tangency and the beginning of separation, both of these quantities approach zero. However, the exsecants diminish at a faster rate than the speeds. In fact, the acquired speeds diminish in a linear fashion, in direct proportion to the decrease in distance from the point of tangency; this is obvious from the fact that the acquired speeds increase in direct proportion to time and the distance from the point of tangency increases also in direct proportion to time. However, the required falling distances diminish as the square of the distance from the point of tangency diminishes; the

---

[17]This claim is, of course, the (correct) law of fall discovered by Galileo.

reason for this is that the required falling distances correspond to the exsecants, and for small angles the exsecants vary as the square of the distance from the point of tangency.

At this point in his argument, Galileo is appealing to a mathematical theorem. The theorem is the claim that as we get very close to the point of tangency, the ratio of an exsecant to the corresponding segment between the tangent and a line intersecting it at any acute angle however small decreases in direct proportion to the distance from the point of tangency. Therefore, as we get closer and closer to the point of tangency, that ratio approaches zero.

Once again, let us ask what this mathematical fact has to do with the physical situation. How does this mathematical fact show that the downward tendency could always overcome the extruding tendency? Galileo is identifying the exsecants with the distance which the body needs to fall to stay on the earth's surface, and the vertical segments within the acute angle with the acquired speeds of fall. But is this the proper comparison to make? Some critics have objected that Galileo is comparing apples and oranges. The more proper comparison would be between the distance a body needs to fall to stay on the surface and the distance it would actually fall. The crucial point would then be that this latter distance is also proportional to the square of the time, as Galileo himself discovered. But in the diagram, the tangent AB measures both time and distance from the point of tangency. Therefore, the actual distance fallen diminishes as the square of the distance from the point of tangency. And this rate of decrease is identical with the rate of decrease of the exsecants or distances needed to be fallen. It follows that the two quantities which it would have been proper to compare vary at the same rate, and so it is unclear that we can conclude that the centripetal tendency necessarily overcomes the centrifugal one.[18]

5.10 Exsecants Versus Exsecants

Finally, a brief word about the third argument Galileo gives for the mathematical impossibility of extrusion. This one is an easy argument to analyze and evaluate. Galileo asks us to consider the series of exsecants such that the following one is at one-half the distance from the point of tangency as the previous one. He states that as we approach the point of tangency, the ratio of the previous exsecant to the following one grows without limit. This is simply not true. The ratio of such exsecants to one another is uniformly one-fourth. The reason for this is a fact already mentioned, namely that for small angles the exsecant varies as the square of the angle. Thus we have what might be called a purely mathematical error on Galileo's part, a failure of his mathematical intuition, so to speak.[19]

---

[18] Here I am indebted to MacLachlan (1977) and Hill (1984).
[19] I have adapted this from MacLachlan (1977).

5.11 A Definition of Physical–Mathematical Reasoning

The kind of mathematical reasoning I have examined here may be labeled physical–mathematical or applied-mathematical. It is, I believe, typical of mathematical physics. Such physical–mathematical reasoning contrasts with both purely mathematical reasoning and purely physical reasoning, but it may be regarded as a special case of either mathematical reasoning or physical reasoning. Although I have focused my analysis on physical–mathematical reasoning, my discussion also provides examples of purely mathematical and purely physical reasoning. That is, the original anti-Copernican argument and Galileo's first criticism are examples of purely physical reasoning; the two proof sequences with numbered lines in Section 8 are examples of purely mathematical reasoning; on the other hand, Galileo's second and fourth criticisms, and the three parts of the third criticism provide at least five examples of physical-mathematical reasoning.

Physical–mathematical reasoning is reasoning about physical processes and phenomena such that various aspects of them are represented by mathematical entities, various mathematical conclusions are reached about these mathematical entities, and then these mathematical conclusions are applied to the physical situation. This characterization of physical–mathematical reasoning may be regarded as a definition which I have extracted from the Galilean concrete examples. The justification of my definition is that it enables us to both understand and evaluate such reasoning.

My definition presupposes various claims which I have not defended or elaborated, about the existence of mathematical entities and their distinction from physical entities, and about mathematical truth and its distinction from physical truth. In other words, at this point some questions come naturally to mind, such as, What is a mathematical entity? How does it differ from a physical entity? What is mathematical truth? How does it differ from physical truth? I believe my analysis presupposes such simple answers to these questions as the following. A physical entity is one that exists in nature, for example, the earth, the earth's rotation, the extruding power of rotation, falling bodies, and tendency to fall. Mathematical entities are those that exist in a logical system like Euclidean geometry: straight lines, circles, tangents, secants, angles, numbers, etc. Physical truths are propositions that correspond to the way things happen to be in nature. Mathematical truths are propositions that follow necessarily from the basic definitions and axioms of logical systems like Euclidean geometry. While I recognize that such claims cry out for further critical analysis, that would be beyond the scope of this investigation. Here, I would want to express my hope that these claims, while simple, are neither simple-minded nor over-simplified.

My definition is descriptive in one sense and normative in another. It is descriptive in the sense that it claims that this is what physical–mathematical reasoning is, not what it ought to be. But it is normative in the sense that such physical–mathematical reasoning can be correct or incorrect. It is incorrect when a particular application of a mathematical truth to a physical situation is incorrect, or when a given representation of a physical process by mathematical entities is incorrect, or when the mathematical conclusion being applied is mathematically incorrect. These three possibilities correspond respectively to the cases of Galileo's three arguments for the mathematical impossibility of extrusion in his third criticism. Again, it would be a long story, beyond the scope of this chapter, to elaborate on what makes "incorrect" a physical application of a mathematical truth, a mathematical representation of a physical process, and a mathematical conclusion about mathematical entities (although the last one of these three cases would be relatively unproblematic).

5.12 Galileo's Reflections on Physical–Mathematical Reasoning

There is one final question, however, which relates to the ones asked in the last few paragraphs and deserves discussion here. The question stems from the following difficulty.[20] My critical analysis of Galileo's extrusion argument and my subsequent general definition of physical–mathematical reasoning utilize and mention a distinction between mathematical entities and truth on the one hand and physical processes and truth on the other. However, Galileo (supposedly) subscribed to a Platonizing mathematical realism that tries to do away with this distinction.[21] For example, in Galileo's own memorable and eloquent words: "Philosophy is written in this all-encompassing book that is constantly open before our eyes, that is the universe; but it cannot be understood unless one first learns to understand the language and knows the characters in which it is written. It is written in mathematical language, and its characters are triangles, circles, and other geometrical figures; without these it is humanly impossible to understand a word of it, and one wanders around pointlessly in a dark labyrinth."[22] Since the philosophy in question is natural philosophy, the point is that if the book of nature is written in mathematical language, then natural phenomena (physical processes) just are mathematical entities, and mathematical truths just are physical truths. Thus, so the objection would continue, the preceding analysis is un-Galilean insofar as it analyzes Galileo's extrusion argument in accordance with a philosophy of mathematics and of science which he explicitly rejected.

To clarify this issue, I would begin by arguing that to read mathematical realism into the quoted passage is to take it out of context. The context is that of *The*

---

[20] This objection is due to Jean Dhombres, who raised it at the International Conference on Logic and Mathematical Reasoning, Mexico City, 6–8 October 1997, where an earlier version of this chapter was first presented.

[21] For some useful accounts along these lines, see Camerota (2004, 20–23, 386–387, 559–560), De Caro (1993), Galluzzi (1973), Koyré (1943; 1966; 1978), Shea (1972). These accounts are instructive not necessarily because they explicitly attribute to Galileo a conflation of mathematical and physical truth, but because they explicitly talk of his Platonism and mathematicism, and they implicitly suggest such a conflation; the most common instance of such implicit conflation is the interpretation that Galileo was certain about the truth of Copernicanism because of its mathematical simplicity. For a criticism of some aspects of such an interpretation and a clarification of such notions as Platonism, mathematicism, and realism, see Finocchiaro (1980; 1989, 7–8; 1997a, 335–356; 2005c, 554–555), as well as Chapters 3, 5, 9, 11 of this book.

[22] Galilei (2008, 183); cf. Favaro 6: 232.

Assayer, a book published in 1623 discussing the origin and nature of comets. The passage is part of a criticism of the excessive reliance on authority in general, and in particular on the authority of Tycho Brahe, whose theory of comets Galileo is criticizing. Thus, the passage is more of a plea for what I would call independent-mindedness, namely the willingness and ability to think for oneself and have a judicious attitude toward authorities, avoiding both extremes of a slavish, blind, and total acceptance on the one hand and of an uncritical and total disregard on the other. The connection is that if the book of nature is written in mathematical language, then once one learns that language, one can read the book on one's own.[23]

Secondly, if and to the extent that the remark on the book of nature written in mathematical language can be taken as an expression of mathematical realism or Platonism, it should be noted that the remark is an epistemological reflection, not an instance of concrete scientific practice, and one can raise the question whether Galileo's words and deeds correspond. However, in this regard they do not correspond.

In fact, to just mention the most striking example, let us consider his discussion in *Two New Sciences* of the fundamental law of acceleration of falling bodies.[24] Galileo first defines uniformly accelerated motion as motion whose speed increases in such a way that equal increments of speed are added in equal times. Then he assumes the postulate that a body falling along different inclined planes acquires the same increment of speed whenever the height of fall is the same. From this definition and this postulate Galileo derives mathematically two consequences: (1) the law of squares, according to which the distances fallen from rest are to each other as the square of the times; and (2) the law of odd numbers, according to which the distances fallen from rest in successively equal times are to each other as the odd numbers from unity.

Galileo is perfectly clear that all this is a piece of pure mathematical reasoning, and that observational and experimental evidence is needed if we want to claim that these are physical laws in the world that actually exists. His words and expression are not those of a mathematical realist who conflates mathematical with physical entities and truth. In the dialogue within which this pure mathematical reasoning is embedded, the most relevant exchange is that between Sagredo's expression of doubt after he hears the definition of uniformly accelerated motion, Simplicio's similar concern after the two mathematical consequences have been derived, and Salviati's response to them:

> SAGR. Although I can offer no rational objection to this or indeed to any other definition devised by any author whomsoever, since all definitions are arbitrary, I may nevertheless without offense be allowed to doubt whether such a definition as the above, established in an abstract manner, corresponds to and describes that kind of accelerated motion which we meet in nature in the case of freely falling bodies.[25]

---

[23] Here I am adopting Mario Biagioli's interpretation that "Galileo's audience for *The Assayer* was not twentieth-century historians and philosophers but early seventeenth-century courtiers. The image of the open book of nature appealed to them because of the sense of unmediated knowledge that it conveyed. True, one had to learn how to read those characters, but to learn a language was not like enslaving oneself to a philosophical system. Once that linguistic ability was acquired, the book was open and the interpretation free" (Biagioli 1993, 306–307).

[24] Favaro 8: 197–213, Galilei (2008, 335–351).

[25] Galilei (2008, 336); cf. Favaro 8: 198.

5.12 Galileo's Reflections on Physical–Mathematical Reasoning

SIMP. ... But as to whether this acceleration is that which nature employs in the case of falling bodies, I am still doubtful. So it seems to me, not only for my own sake but also for all those who think as I do, that this would be the proper moment to introduce one of those experiments – and there are many of them, I understand – which correspond in several ways to the conclusions demonstrated.

SALV. The request which you make, like a true scientist, is a very reasonable one. For this is the custom – and properly so – in those sciences where mathematical demonstrations are applied to natural phenomena.[26]

And then Galileo goes on to describe some experiments indicating that physical bodies do in fact fall in accordance with those mathematical claims. Thus, Galileo's behavior in his work on falling bodies does not exemplify the identification or conflation of mathematical truth and physical truth, which is what is sometimes meant by mathematical realism or Platonism.

Having established this much, however, there is another view, at the opposite extreme, which Galileo also rejected. It is the view that separates or divorces mathematical and physical truth in such a way that it becomes possible to say that a proposition might be mathematically but not physically true, namely true in mathematics but not in physics, true in the abstract world of mathematics but not in the concrete physical world. It is the unwillingness to attribute Galileo such a separationist view that may provide some motivation to the proponents of the Platonist or mathematical-realist interpretation to object to my account elaborated earlier. I certainly agree that Galileo cannot be attributed such a separationist view, but I would also deny that my interpretation is a separationist one. My account of physical–mathematical reasoning does distinguish mathematical from physical truth, and does attribute such a distinction to Galileo, but such a distinction is not a separation. Galileo's clarification is found in a series of methodological reflections which he found useful to insert in the middle of the passage on the extruding power of whirling in the *Dialogue*.[27]

After the (threefold) argument for the mathematical impossibility of extrusion has been stated, and after a number of technical and mathematical difficulties have been cleared up, Galileo has the Aristotelian interlocutor (Simplicio) present the following methodological objection: "These mathematical subtleties are true in the abstract, but when applied, they do not correspond to sensible physical matter; for example, mathematicians may well demonstrate with their principles that a sphere touches a plane at a single point, which is a proposition like the present one; but, when we deal with matter, things proceed otherwise; and so I want to say that all those contact angles and proportions come to nothing when we deal with sensible material things."[28]

---

[26] Galilei (2008, 349); cf. Favaro 8: 212. It should be noted that the ninth word of Salviati's speech (*scientist*) is a literal translation of the original Italian word (*scienziato*); this strengthens further the point made in Section 4.1 above that although one must be careful to avoid anachronism when using terms like science and scientist instead of natural philosophy and philosopher, one must also avoid linguistic chauvinism or provincialism.
[27] Favaro 7: 229–237, Galilei (1967, 203–210; 1997, 193–202). For other accounts of this passage, see Clavelin (1974, 404–421; 1996), Feldhay (1998), Wisan (1978).
[28] Galilei (1997, 193). Cf. Favaro 7: 229, Galilei (1967, 203).

The ensuing discussion goes on to reject the separation presupposed in this objection. Consider a bronze sphere lying on a table. One might be inclined to say that, because the weight of the sphere will flatten to some extent the part touching the table, the two will touch over an area of many points; thus, to say that a sphere touches a plane at a single point is true in the abstract (in mathematics), but not in the concrete physical world. However, a sphere is by definition a surface all points of which are equidistant from the center, and so if a portion of such a sphere is flattened then such a surface is not a sphere but another type of surface (let us call it a flattened sphere); thus, the bronze object touching the plane at several points is not a sphere but a flattened sphere; however, a flattened sphere, even in mathematics or abstractly speaking, touches a plane at several points. It follows that there is no lack of correspondence between the abstract and the concrete: a sphere touches a plane at a single point, both in abstract mathematics and in physical reality, as long as the sphere we are talking about is really a (perfect) sphere; on the other hand, a flattened sphere touches a plane at several points, regardless of whether it is an abstract or bronze flattened sphere, because that is the nature and definition of a flattened sphere.

Galileo goes on to clarify what is really meant by those who speak as if they were advocating a separationist view of truth. When they say that whereas it is abstractly true that a sphere touches a plane at a single point, physically a sphere touches a plane at several points, they should be saying that while it is true both abstractly and physically that a sphere touches a plane at a single point, and that a flattened sphere touches it at several points, in the physical world there exist flattened spheres but no perfect ones; moreover, a bronze flattened sphere is as perfect a specimen of the definition of a flattened sphere as an abstract sphere is a perfect specimen of the formal definition of a sphere.

The real methodological problem thus becomes, Galileo suggests, to determine what kind of spheres exist in the physical world. Perhaps there are not even flattened spheres in the physical world, meaning that perhaps the part of flattened surface is not a perfect plane but has an irregular shape that deviates from a plane. Or perhaps different kinds of flattened spheres are definable, and then the problem is to determine which kind exist in the physical world. But whatever is the case, clearly it is not by pure mathematical reasoning that we can determine these things. For that we need observation and physical–mathematical reasoning.

To be sure, there are two ways to proceed. We can start with a piece of abstract mathematics, i.e. a piece of pure mathematical reasoning, and then the question becomes how much of it (if any) corresponds to what part of the physical world. This was Galileo's procedure in his textual exposition of the acceleration of falling bodies in the *Two New Sciences*. Or we can start with the physical situation, engage in mathematical representation, derive some mathematical truth, and then apply it back to the physical situation. This was his procedure in the passage on the extruding power of whirling in the *Dialogue*. In either case, however, the correspondence or interrelationship between the two domains is essential. The two domains are those that have been labeled on the one hand the domain of pure mathematics, abstraction, conceptual systems, and logical deduction, and on the other the domain of the physical, material, sensible, or concrete world.

5.12 Galileo's Reflections on Physical–Mathematical Reasoning

One final point needs clarification, involving the comparison and contrast between the two Galilean responses to Simplicio's objections. Regarding the acceleration of falling bodies in *Two New Sciences*, Galileo admits the distinction between mathematical or abstract truth and physical or concrete truth and the need to test experimentally whether the latter corresponds to the former. Regarding the extruding power of whirling in the *Dialogue*, Galileo stresses the parity of truth-values in the two domains. However, the two responses are more similar than this difference would suggest.

Let us note, first, that in the former case, Galileo could have also made a comment analogous to the one he made in the latter. To see this, let us first reformulate the mathematical component of the analysis of the acceleration of falling bodies by saying that bodies moving with uniform acceleration cover distances starting from rest which are to each other as the square of the times elapsed. Given the definition of uniform acceleration, and given that the law of squares is a correct mathematical consequence of it, to experimentally test this assertion amounts to testing whether freely falling bodies in nature are moving with uniform acceleration; if the test fails, that means that freely falling bodies do not move with uniform acceleration, but with some other kind; but it is true of physical bodies as well as abstract entities that if they are uniformly accelerated then the distance traversed varies as the square of the time; and it is true of abstract entities as well as physical bodies that if they move without acceleration or with non-uniform acceleration then the distance–time relationship is different than the time-squared law.

Let us now see whether the comment made for acceleration and falling bodies would apply to the extrusion situation, or at least to the case of the sphere touching a plane. Galileo himself suggests that to question the concrete truth of this proposition does not really amount to questioning whether its truth-value is different in the physical world than it is in the world of mathematical abstraction, but rather whether the entities it is about exist in the physical world, and if not what kind of physical entities do exist and what particular truths these entities instantiate.

What Galileo seems to be saying is that we have neither separation nor identification of the two domains, but rather correspondence. We can never know in advance that a particular mathematical truth corresponds to physical reality, or that a particular physical situation is representable by a particular mathematical entity, but we can claim in advance as a matter of methodological prescription that every physical situation is representable by *some* mathematical entity. I believe this is the plausible meaning to attribute to Galileo's famous remark on the book of nature being written in mathematical language, if and to the extent that we can detach that remark from its most immediate context.

When so interpreted, the remark is a long way from the extreme mathematical realism or Platonism sometimes attributed to Galileo, for several reasons. One is that this plausible methodological advice is always open to frustration, and so we may not be able to find the right mathematical representation; and when we have not found it yet, we do not really know whether this is because we have not been resourceful enough, or because there is no mathematical representation in this case; here the methodological situation is analogous to the search for causes under the guidance of the principle that every event has a cause. Another reason is that mathematics itself ought not to be viewed as a static body of knowledge, ready-made for application (however problematic the application may be); rather, viewing mathematics as a dynamic, developing, and progressive discipline, we have to say that in a given case it may happen that the mathematical entities that would enable us to represent a physical situation may not have been invented yet, or may not have been brought into existence, so to speak.

In conclusion, Galileo's reflections on the nature of physical–mathematical reasoning, when properly contextualized (in relation to other problems under discussion, in relation to his scientific and mathematical practice, and in relation to one another), do not conflict with the definition I extracted from his extrusion argument. Not only is there no conflict, but my definition of physical–mathematical reasoning is confirmed by such contextualized reflections. Thus, this case study can claim to be Galilean in a double sense, not only insofar as it uses some examples of Galileo's physical–mathematical reasoning as material for analysis, but also in the sense that this material has been analyzed in a way he himself would have been likely to analyze it.

Chapter 6
Galilean Rationality in the Copernican Revolution

This chapter will attempt to derive some general lessons from the particular analyses of the previous chapters. The previous examples in which Galileo defended Copernicus from various kinds of objections will be combined and reworked as contributions to the Copernican Revolution. The latter episode, in turn, will be taken as an archetypical example of human rationality. The general lessons will be about the nature of rationality; the role in it of the mental activities of criticism, reasoning, and judgment, and of the intellectual traits of fallibilism, openness, fairness, and rational-mindedness; and the all-encompassing notion of critical reasoning.

6.1 The Copernican Revolution and the Role of Criticism

The Copernican Revolution is perhaps the most significant episode in science for the proper understanding and appreciation of the nature, the power, and the limitations of human rationality. The label "Copernican Revolution" may be taken to refer to the sequence of historical developments whose outcome was the replacement of the geostatic and geocentric cosmology by the geokinetic and heliocentric cosmology: the ancient view held that the earth stands still at the center of the universe, whereas according to the modern view the earth moves both by spinning around its own axis once a day and in an orbit around the sun once a year. The process lasted about 150 years, from about 1543, which is the year of publication of Nicolaus Copernicus's famous book *On the Revolutions of the Heavenly Spheres*, to the year 1687, which is the date of publication of the first edition of Isaac Newton's *Mathematical Principles of Natural Philosophy*.[1]

The episode may be said to have been earthshaking in both a literal and figurative sense. For the key development was the discovery that the terrestrial globe possesses physical motion and is not the center of the universe, and this discovery

---
[1] Kuhn (1957) remains perhaps the best synthetic account of the Copernican Revolution, although the time is ripe for updates and revisions, an excellent example being Westman (forthcoming).

was so pregnant with consequences in all areas of human culture and life that it precipitated a cultural, psychological, and social upheaval.

One of the things that makes the Copernican Revolution so relevant to human rationality stems from the fact that it involves some beliefs which today are known with certainty to be false and incorrect, and others which are now established with equal conclusiveness as being absolutely true and correct. That is, no sane person today can question the fact that the earth moves, and if mankind knows anything at all and if human knowledge encompasses any item of information, then the earth's motion is surely one of these things. Conversely, if we know anything to be false, it is the idea that the earth stands still at the center of the universe.

These epistemological facts have two implications that point in opposite directions. On the one hand, there are positive lessons to be drawn. One is that knowledge is possible because it is actual, and it is actual because we know at least one thing, namely that the earth moves and is not standing still at the center of the universe. Another positive lesson is that progress is possible because the Copernican Revolution is an instance of it, and this is so because the result was to replace ignorance by knowledge in regard to the question of the location and the behavior of the earth in the universe.

However, there are also lessons that might be called negative, and these are the ones that point in the direction of human rationality. To see this, we must focus on the fact that for thousands of years, until relatively recently in human history, almost everybody was wrong about a very fundamental matter; and the important fact is that this included not just common people but scientists and philosophers as well, that is the so-called experts. What this shows is that it is possible for everyone to be wrong, or at least for everyone to be wrong some of the time, that is for a certain time, and even for a long time; for that was certainly the case in regard to the motion of the earth up until the time of the Copernican Revolution. This opens the door for the would-be critic, no matter how radical, for he is assured that it is possible that he is the only one to see the truth about the topic in question. And so he is not afraid to claim that most people may very well be wrong in such beliefs as that there will always be rich and poor and differences of wealth and poverty are socially inevitable, or that there will always be psychological and behavioral differences between men and women and that some sexual stereotypes are natural.

Be that as it may, the point is that, since for a long time almost everyone held the false belief that the earth stands still at the center of the universe, it is always possible that almost everyone is wrong about almost anything. However, this is just one side of the lesson relevant to human rationality, the one that concerns the element of criticism. If we do not neglect the element of reasoning, then we are led to ask the following questions about the Copernican Revolution. Although factually wrong in the content of their belief, were people reasonable or unreasonable in holding this incorrect belief? In other words, let us focus on the reasoning which led pre-Copernicans to their false opinion. Was this reasoning right or wrong, correct or incorrect, valid or invalid, sound or unsound? A less well-known fact about this matter is that the reasoning of the pre-Copernicans was essentially correct; that is, pre-Copernicans were basically reasonable or rational in believing that the earth stands still at the center of the universe. This requires a rather lengthy demonstration (which we carried out in Chapters 1 and 2), but the lesson is that the Copernican Revolution does not provide us with evidence that it is possible for everyone to be *unreasonable*; on the contrary, it suggests that if we move the discussion to the level of reasoning, then mankind is essentially reasonable. In short, one lesson of the Copernican Revolution is that it is possible for almost everybody to hold a false belief, but not for almost everybody to be unreasonable.

6.2 Copernicus and Explanatory Coherence

This distinction between reasonableness and truth and between unreasonableness and falsity is related to an appreciation of the difference between the two elements of human rationality which I have labeled criticism and reasoning. In order to appreciate the next point about the Copernican Revolution we need to appreciate a third element of human rationality, namely judgment. In fact, at this point some people would begin to despair about how the transition to a geokinetic view was ever possible, and they may think that the transition was itself unreasonable or irrational.[2] However, this interpretation would be injudicious and thus lack judgment, for it is possible to show that the situation evolved in such a way that the geostatic view *became* unreasonable.

The first step in this process was the construction of a new argument in favor of the geokinetic thesis. As we have seen (Chapter 2), this was done by Copernicus when he showed that the known observed details about the motions of the heavenly bodies could be explained more simply and more coherently by postulating the earth to be the third planet revolving around the sun. Although the idea of the earth's motion was not new, the argument was. The idea was not new because the possibility of the earth moving had been considered since the ancient Greeks, but it had been rejected because of the weight of arguments and evidence against it. Further, Copernicus's argument was novel because no one had worked out the consequences of the earth's motion around the sun in full detail, and shown that the specific phenomena in the heavens could be explained this way, let alone that they could be explained with greater simplicity and coherence.

Let us recall the meaning of these concepts. The greater simplicity of the Copernican explanation referred to such features as that it utilized fewer moving parts, fewer directions of motion, and a single uniform pattern in the relationship between size of orbit and period of revolution.[3] And to say that Copernicanism had

---

[2] For some of the arguments which are often taken to imply this sort of irrationality, see Kuhn (1962; 1970), Feyerabend (1975; 1987; 1988; 1993). For a critical analysis of some of their views, see Finocchiaro (1973a, 180–200; 1973b; 1980, 180–223).

[3] Note that the concept of simplicity is itself not simple; in particular, it was not simply a matter of counting which theory used fewer epicycles, which could become a complicated business; see, for example, Price (1959).

greater explanatory coherence than the geostatic theory means that in the geokinetic theory many of the specific facts could be explained in terms of its basic postulates alone, whereas to explain the same facts the geostatic theory had to add special assumptions invented specifically for the purpose; sometimes this point is expressed by saying that the geostatic theory contained many more ad hoc elements.[4]

The next point that needs to be stressed is that Copernicus's novel argument did strengthen the geokinetic view vis-à-vis the geostatic one, but not in any decisive way. The Copernican argument is not conclusive because it is a hypothetical and explanatory argument in which the hypothesis is confirmed or corroborated, but not verified, by the observational consequences; these same observations could still be explained by the geostatic hypothesis, however less coherently, less simply, and less elegantly. Moreover, explanatory coherence was not the only, or even the chief criterion, of scientific merit; the other criteria favored the geostatic hypothesis, that is, the criteria of observation, physics, and the Bible. This is where Galileo comes in (as we have seen in Chapters 3–5).

6.3 Physics and Reasoning

At first, Galileo was primarily interested in physics and mechanics, and was working on a research program designed to understand in general how bodies move.[5] He was critical of ancient Aristotelian physics and was developing a new theory of motion more in line with the work of another ancient Greek – Archimedes. Galileo was aware of the new Copernican argument, but felt its insufficiency and the greater power of the many anti-Copernican and pro-geostatic arguments. He was initially attracted to the Copernican theory because its key geokinetic hypothesis was more in accordance with the new physics he was developing. In effect, his new physics provided him with an effective criticism of the mechanical objections to the earth's motion, and with some physical evidence in its favor. The connection can be seen most clearly and most simply for the case of the vertical fall objection.[6]

---

[4] This type of interpretation of the Copernican Revolution, though not exactly in these terms, is suggested in Lakatos and Zahar (1975) and Millman (1976), among others.

[5] The most complete and convincing account of his development is found in Drake (1978), which can be supplemented with Camerota (2004).

[6] In some cases these connections were more direct than, and occurred to Galileo chronologically prior to, the case of the vertical fall. Nevertheless I shall illustrate the point with the vertical fall objection, which he did not explicitly criticize until later when he wrote the *Dialogue*. One reason for this is that this criticism involves the most crucial principle of physics, namely what today we call the law of inertia; another reason is that this example involves critical reasoning in a much more central and vivid manner. My account of the vertical fall argument here is a digest of that found in Finocchiaro (1980, 36, 116, 192–199, 277–288; 1997a, 143–146, 155–170, 323–325). For an alternative but overlapping account, see Feyerabend (1975, 69–108; 1988, 55–109); for an appreciation and some criticism of Feyerabend, see Finocchiaro (1980, 182–200).

6.3 Physics and Reasoning

Recall that the vertical fall objection argued that the earth cannot rotate because on a rotating earth freely falling bodies would have no reason to keep up with the earth's motion, and hence during free fall they would be left behind; this in turn means that they would not be falling vertically; but it is obvious that they do fall vertically. Here the last step in reasoning can be reconstructed as an instance of "denying the consequent," namely the argument-form consisting of two premises and one conclusion such that one premise (called the major premise) is a conditional ("if-then") proposition, the other premise (called the minor premise) denies the consequent ("then") clause, and the conclusion denies the antecedent ("if") clause:

1. If the earth rotated, then bodies would not fall vertically.
2. Bodies do fall vertically.
3. So, the earth does not rotate.

Galileo begins his critique by asking us to focus our attention on the meaning of the proposition that bodies fall vertically. What does it mean? What is meant by vertical fall? Does it mean fall from the terrestrial point where the body is released to the point on the earth's surface directly below, such as the motion from the top to the base of a tower; or does vertical fall mean fall along the straight line in absolute space going from the point of release to the center of the earth? In other words, does vertical fall mean fall perpendicular to the earth's surface as viewed by a terrestrial observer, that is an observer standing on the earth's surface; or does it mean fall perpendicular to the earth's surface as viewed by an extra-terrestrial observer, that is an observer looking at the whole terrestrial globe from a fixed point at a distance. Let us call the first *apparent or relative vertical fall*, and the second *actual or absolute vertical fall*. These are indeed different.

To understand this difference, let us consider Fig. 6.1. The semicircles on the left side represent a portion of the earth's equator, and the structures labeled AB and A'B' represent towers on the earth's surface. The portions of the figure on the right side are simply highly magnified representations of the situations on the left, so that the earth's surface appears flat because the distance involved (BB') is very small. The top part (a) of the figure represents a motionless earth, whereas the middle part (b) and the bottom part (c) represent the earth undergoing axial rotation from west to east (or clockwise), as suggested by the arrows. In parts (b) and (c) of the figure, each situation exhibits two different positions of the tower: the *unprimed* position (AB) represents the tower's position at the beginning of the experiment of dropping a rock from the top of a tower and letting it fall freely; the *primed* position (A'B') represents the tower's position at the end of the experiment when the fallen rock has reached the ground. Along the vertical wall of the tower, and between towers, there are solid and dotted lines: solid lines represent apparent vertical fall, and dotted lines actual vertical fall. The lines (whether solid or dotted) between towers are drawn both as straight slanted lines and as parabolic slanted lines; here the main point to note is that they are both slanted, and although the parabolic representation is more accurate, this refinement plays no role in this discussion. In part (a) of the figure representing a motionless earth, apparent and actual vertical fall coincide. In part (b) representing a situation where the earth is rotating, apparent

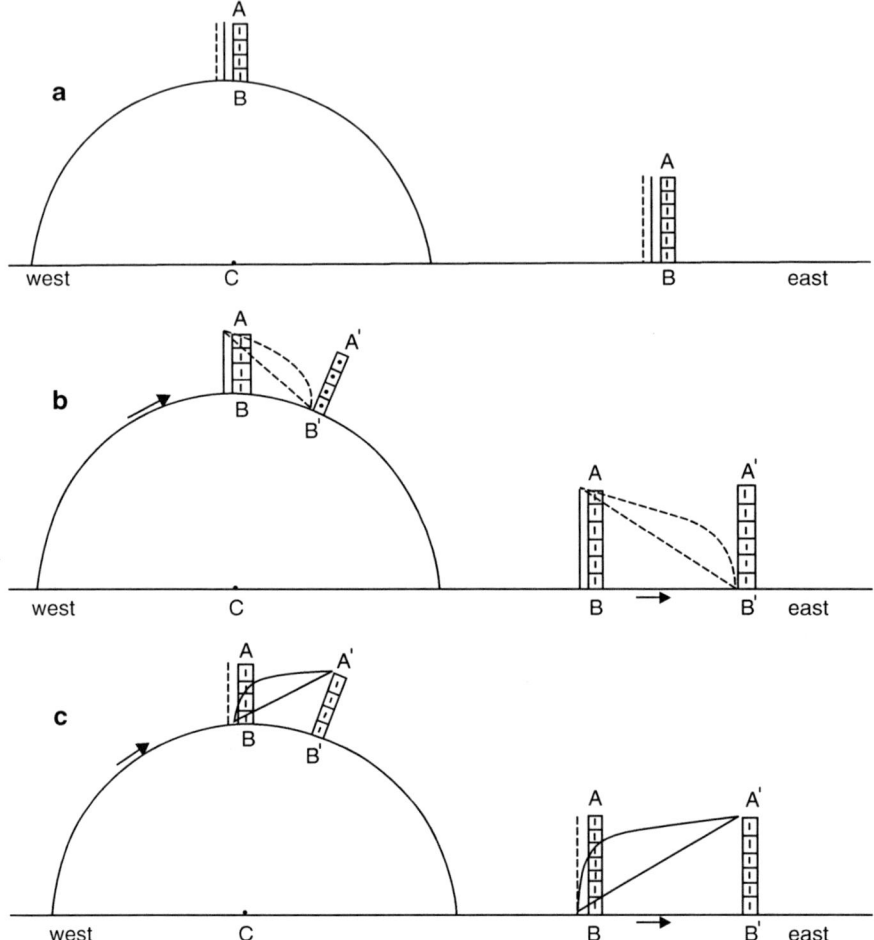

Fig. 6.1 Vertical Fall on a Motionless and on a Moving Earth

vertical fall is experienced on earth, but actually slanted fall takes place. In part (c) representing also a situation where the earth is rotating, actual vertical fall is taking place, but apparently slanted fall is experienced on earth.

With these definitions, conventions, and pictures in mind, we are now in a position to better follow Galileo's reasoning. To explain the difference between actual and apparent vertical fall, Galileo points out that although apparent and actual vertical fall would coincide on a motionless earth (Fig. 6.1a), they would *not* coincide on a rotating earth. That is, assume the earth were in rotation and a rock is dropped from the top of a tower (A); if the rock *appeared* to fall vertically to a terrestrial observer (Fig. 6.1b), then it would be seen to land at the foot of the tower (B′); but on a rotating earth the foot of the tower would have undergone some rotational motion during the time of fall, and so as viewed by an extra-terrestrial observer the

6.3 Physics and Reasoning

actual path (AB′) of the rock would be slanted toward the east; in short, on a rotating earth, apparent vertical fall would not produce actual vertical fall but rather actually slanted fall. Similarly, given the same assumption of terrestrial rotation, if the rock were to move with *actual* vertical fall (AB in Fig. 6.1c), then it would land to the west of the base of the tower because the base of the tower has moved to point B′; therefore, to a terrestrial observer the path of the falling rock would *appear* slanted westward because when the rock reached point B the terrestrial observer would have moved to A′, and so the *apparent* path would be A′B, which is not vertical; in short, on a rotating earth, actual vertical fall would not produce apparent vertical fall but rather apparently slanted fall. By contrast, as mentioned earlier, for a motionless earth, if the path were from the top to the base of the tower as seen by the terrestrial observer, then it would also be straight and perpendicular to the earth's surface for the extra-terrestrial observer; and conversely (Fig. 6.1a).

Having made such a distinction, Galileo applies it to the above mentioned argument having the form of denying the consequent. Suppose that, when the argument claims that the earth cannot rotate because bodies fall vertically, the vertical fall in question is actual vertical fall; then we would be entitled to ask how you know that bodies do actually fall vertically, for observation only reveals apparent vertical fall. In other words, it is undeniable that to us on the earth's surface bodies are seen to fall from the top to the base of a tower; and it is equally undeniable that we have no experience about how they look from an extra-terrestrial viewpoint. How could one answer that question? How could one justify that falling bodies move with actual vertical fall? It seems that one could only try an empirical justification, by basing actual vertical fall on apparent vertical fall. But to do this would presuppose that apparent vertical fall implies actual vertical fall, and in turn this implication amounts to assuming that the earth is motionless, since this is the only condition under which the implication holds. Unfortunately, the motionlessness of the earth is the very conclusion the argument is trying to prove. In short, interpreted in terms of actual vertical fall the objection from vertical fall begs the question because the premise that bodies fall vertically is either assumed gratuitously or supported circularly by reasons which in their turn presuppose the conclusion at issue.

However, perhaps the objection from vertical fall intends apparent vertical fall, in which case the minor premise of the above mentioned argument would be both true and uncontroversial. That is, the argument would now be that the earth cannot rotate because bodies *appear* to us to go from the top to the base of the tower. In such a case Galileo questions the major premise, namely the conditional proposition that if the earth were in rotation then bodies would not undergo *apparent* vertical fall. What are the grounds for asserting this conditional claim? At that time, the justification of this was based on some basic principles of Aristotelian physics: one was the principle that a body can have only one natural motion; another was the principle that the natural state of heavy material bodies is rest, and motion requires a force to sustain it (as we saw in Chapter 1). To understand the connection we have to understand the first answer the Aristotelians would give in this discussion.

They would say that if the earth were in rotational motion then falling bodies would not exhibit apparent vertical fall (which in fact they do) because if they did then they would simultaneously be moving downwards (towards the center of the earth) and horizontally (around the center); in fact, if the earth rotated and falling bodies were seen to go from the top to the base of a tower, then (as we saw above, Fig. 6.1b) in reality (to an extra-terrestrial observer) they would be slanting eastward; but this eastward slant would be the resultant of straight-downward and straight-horizontal components. Now, according to Aristotelian physics such a mixture or combination is impossible because the horizontal component of motion would be motion under the influence of no external force, and so it would have to be natural; but such a second natural motion could not co-exist with the first, downward motion.

The issue then becomes whether or not it is possible for a free-falling body to have a horizontal component of motion. This is where some of the principles of the new Galilean physics come in; they are the principle of the conservation of motion and the principle of the superposition or composition of motions. Conservation of motion is an approximation to such laws of modern physics as the law of inertia, the law of conservation of linear momentum, and the law of conservation of angular momentum. The Galilean formulation relevant to the criticism of the objection from vertical fall is the following: if a body is moving horizontally it will conserve its motion as long as it is left undisturbed. And the principle of superposition asserts that it is physically possible for a body to have more than one tendency to move, and in these cases the actual motion will be the resultant as defined by the diagonal of the corresponding parallelogram.

These principles can now be applied to answer the objection from *apparent* vertical fall. If the earth rotated, then it is possible that bodies would undergo apparent vertical fall, because in this case what would be happening is the following (Fig. 6.1b). On a rotating earth, a body before being released would be carried eastward by the earth's rotation; after being released this horizontal component of motion would be conserved because of the just-mentioned principle. The body would also start moving downwards; but this motion would not be a disturbance to the other one. Instead they would combine, by the principle of superposition, to produce the actually slanted path which would carry the body directly below the point of release, for example to the base of the tower.

Now, one last piece of reasoning was needed to complete Galileo's criticism. In that context he could not simply assert the conservation of motion without justification. But he had one ready, which reflected the way he himself had arrived at the principle. The argument is in part an empirical one and it is the following. Observation reveals that bodies which move downwards are accelerated; that is, their speed increases. We can also observe that bodies are retarded when they move upwards; that is, their speed decreases. Therefore, Galileo reasoned, if a body were moving along a path that was neither downwards nor upwards, its motion should be neutral, so to speak; that is, its speed should neither increase nor decrease in the absence of disturbances. But horizontal motion is an instance of motion which is neither upwards nor downwards. Therefore, bodies moving in a horizontal direction will conserve their speed of motion if left undisturbed.

In summary, the vertical fall objection is based on some untenable assumptions if vertical fall means apparent vertical fall; that is, this version of the argument is groundless.

On the other hand, the objection begs the question if vertical fall means actual vertical fall. It should be noted, however, that none of this, by itself, supports the earth's motion, let alone proves it; here we have simply the criticism or refutation of an argument, not a counter-argument justifying the opposite conclusion.

Eventually, Galileo did formulate several positive arguments for the earth's motion: the simplicity argument for diurnal terrestrial rotation; the argument from the law of the periods of revolutions; the simplicity argument from the heliocentrism of planetary revolutions; the coherence argument from the explanation of planetary retrogressions; the argument from the annual motion of sunspots; and the argument based on the explanation of the tides.[7] The last one is especially instructive. It involved physical and mechanical considerations, claiming that the earth's motion was the primary cause of the tides, and that there was no other way of explaining their existence. The details of the argument[8] need not be elaborated here, but it should be mentioned that there is an error in this argument in the sense that we know today that the tides are caused by gravitational attraction (primarily that of the moon, but also that of the sun). However, it should also be added that this does not invalidate Galileo's *reasoning*, and that from a historical and contextual point of view the argument had some force. Galileo did indeed have a high regard for the argument, but did not hold it to be conclusive. One reason for this inconclusiveness was that, when he first conceived the argument, he still had no answer to the astronomical and observational objections to the earth's motion. The invention of the telescope was to change all that.

6.4 The Telescope and the Role of Judgment

Galileo's attitude toward the geokinetic idea before the telescope is, in my opinion, a beautiful illustration of that element of human rationality which I call judgment. That is, we are no longer dealing primarily with questions of reasoning, for it was not merely a matter of reasoning one's way out of the various objections; nor was it a matter of criticism, for there was no question of his willingness and ability to challenge authority, as shown by the fact that he was engaged in a program of physical research which was undermining the foundations of Aristotelian physics. Rather we are dealing with questions of proportion, balance, judiciousness, and avoidance of one-sidedness and of extremes.

Galileo did indeed appreciate the novelty of Copernicus's argument, and he had begun to conceive ways of refuting the physical objections, and ways of providing

---

[7] Respectively, in Favaro (7: 139–150, 144–145, 353–354, 368–372, 372–383, 442–489), or Galilei (1967, 114–124, 118–119, 326–327, 340–345, 345–356, 416–463), or Galilei (1997, 128–142, 134–136, 232–233, 282–303).

[8] See Favaro 7: 442–489, Galilei (1967, 416–465; 1997, 282–308). Cf. Drake (1978, 33–49), Finocchiaro (1980, 74–79; 1989, 119–133), Naylor (2007), Palmieri (1998), and Chapters 3 and 9 of this book.

mechanical evidence in favor of the earth's motion. All that this meant was that physics and the criterion of explanatory coherence now favored the geokinetic idea. But direct observation was still fully on the side of the geostatic view, and Galileo could not bring himself to any one-sided disregard of sense experience.[9]

What did the telescope reveal? It made possible the observation of phenomena which enabled one to answer the general observational argument, the objection from the earth–heaven dichotomy, and most of the specific astronomical objections. In regard to the latter, the planet Mars could now be seen to vary in apparent brightness and size as required by the hypothesis of the earth's annual revolution. Similarly, the planet Venus exhibited the required phases. The earth–heaven dichotomy was undermined and the way paved for a unified view of heaven and earth: in fact, in regard to the moon, its surface could be seen to be full of mountains and valleys like the earth, and they could be seen to be nonluminous and to cast shadows from sunlight very much as it happens on the earth; the sun appeared to have on its surface dark spots that underwent changes similar to those of clouds on earth; the planet Jupiter had four moons analogous to the one circling the earth; and the phases of Venus also indicated that it was not composed out of a luminous aether, but out of some opaque nonluminous substance like the earth. Finally, the general observational argument could now be answered by saying that, besides the direct experience of the unaided senses, one should take into account the indirect observations made with the telescope; since almost all indirect observation favored the earth's motion, then according to the criterion of observation the situation was the following: at the very least it was no longer true to say that observation unequivocally favored the geostatic system, and perhaps it was possible to say that it favored the other system.

These discoveries may be said to have tipped the overall balance of evidence and argument in favor of the geokinetic and against the geostatic idea. Consequently, Galileo became increasingly outspoken about the issue, and in general an irreversible historical trend was produced which was to result in the eventual triumph of the geokinetic theory. However, the process was slow and gradual. The telescopic discoveries did not immediately decide the issue.

One reason was that at least one important astronomical objection could still not be answered, that is the argument from annual stellar parallax; even the telescope did not reveal any yearly change in the apparent position of fixed stars. Galileo was correct in arguing that stellar distances are so immense that the parallax is very minute, and therefore more powerful instruments were needed to detect it;[10] in fact, the phenomenon was first observed in 1838.

Another reason why the telescope was not immediately decisive was that for some time there were proper concerns about the legitimacy, reliability, and practical operation of the instrument. Some questioned its legitimacy on the grounds that there was no place in scientific inquiry for instruments which make us see things

---

[9] See Favaro 7: 355, Galilei (1967, 327–328; 1997, 234). Cf. Finocchiaro (1980, 128–129) and Chapter 3 of this book.
[10] Favaro 7: 385–416, Galilei (1967, 358–389; 1997, 247–281).

6.4 The Telescope and the Role of Judgment

that cannot be seen without them. Obviously this objection could not be dismissed, as we should be able to appreciate today if we compare the situation at that time with the recent issue about whether psychedelic drugs put users in contact with a deeper level of reality, or merely make them see things that are not there. Others questioned the reliability of the telescope by pointing out that Galileo had not provided a scientific explanation of how and why the telescope worked. Moreover, all empirical checks involved terrestrial observation, and there was not even one instance of a test showing that it was truthful in the observation of the phenomena in the heavens. Finally, the practical operation of the instrument required that one learn how to use it and how to avoid aberrant and deviant observations stemming from impurities of the lenses, improper lens shape, and other features of poor design.[11]

A third reason why the telescopic discoveries, although necessary and crucial, were not decisive and did not provide a conclusive defense of the earth's motion, was the existence of biblical and other theological arguments against it. For example, as we have seen, one anti-Copernican passage in the Bible was taken to be Joshua 10: 12–13, where God does the miracle of stopping the sun in its course in order to prevent it from setting at a place called Gibeon, and thus to give that region some extra daylight, needed by the Israelites to win a battle they were fighting. Understandably it took some time for Galileo to come to terms with the scriptural objection by arguing that the Bible is not a scientific authority (Chapter 4);[12] other people, of course, required an even longer time. Moreover, even though Galileo may have won all the arguments on this issue, he personally lost all the actual battles, as shown by the tragedy of the Index's prohibition of Copernicanism and the Inquisition's condemnation of Galileo (Chapter 7).

Finally, there was an epistemological issue connected with the objection from the deception of the senses which also required time for a full assimilation. The difficulty was that, although deception may be too strong a word, the earth's motion was then and remains today a phenomenon which is not observable either with telescopes or by astronauts from outer space. Do such unobservable processes have any role in science, and if so what is their role? Can they be taken seriously as descriptions of physical reality, or can they only be regarded as useful fictions, useful, that is, for the calculation, computation, and prediction of other phenomena that are indeed observable? The only point I want to make here is that to admit unobservable entities in the scientific description of the world was a giant step for mankind, to be undertaken with great caution and circumspection.

My conclusion here is that the fact that the Copernican Revolution required about 150 years to complete, and Galileo's own specific tentativeness and circumspection about the matter, both attest to the importance of judiciousness and judgment.

---

[11] For more details about the telescope controversy, see Crombie (1967), Feyerabend (1975; 1988), King (1955), Ronchi (1958), Rosen (1947), and Van Helden (1984; 1994).

[12] Galileo was forced into this type of partly theological inquiry when he began to be attacked as a heretic. His most considered analysis is found in his famous *Letter to the Grand Duchess Christina* (Favaro 5: 309–348; Finocchiaro 1989, 87–118; Galilei 2008, 109–145). For an interpretation of the latter, see Chapters 4 and 9 of this book.

6.5 Critical Reasoning

So far in this chapter, I have argued that nothing compares with the Copernican Revolution as a vivid illustration of the possibility and importance of criticism. The lesson here is both specific to science and applicable to knowledge and culture in general. The content of this lesson is that everyone may be mistaken, and everything is and ought to be open to criticism. However, man is indeed, as Aristotle declared, a rational animal, and so universal human beliefs are normally rational and reasonable, and such was certainly the geostatic belief before Copernicus and Galileo. Therefore, criticism would lack judgment if it did not recognize the importance of reasoning and the need to work at the level of reasoning. However, reasonableness and rationality are matters of degree and they are contextual; thus what is reasonable under certain conditions need not always remain so, but a change will not occur arbitrarily. Rather the change can be consummated only when one prevailing reasonable idea is shown to be less reasonable than some other new idea. Again judgment is required to insure that reasonable ideas are not discarded arbitrarily, but only in the light of more reasonable ones.

These claims about the Copernican Revolution can in turn be reinterpreted as conclusions about the nature of human rationality. For if we take the Copernican Revolution as a defining instance of human rationality in science, then our account also enables us to see that and how human rationality involves three elements: criticism, reasoning, and judgment.

Let us now connect these notions with the concept of critical reasoning, which was first defined in the Introduction (last section). As I stated there, critical reasoning may be defined as reasoning aimed at the interpretation, evaluation, or self-reflective formulation of arguments, and guided by such ideals as the principles of open-mindedness, fair-mindedness, and rational-mindedness. An argument is a piece of reasoning aiming to justify a conclusion by supporting it with reasons or defending it from objections. Open-mindedness is the ability and willingness to know, understand, and learn from the arguments, evidence, and reasons against one's own views. Fair-mindedness is the ability and willingness to appreciate the strength of arguments and reasons against one's own view, even when one is attempting to criticize or refute them. Rational-mindedness is the ability and willingness to accept the views justified by the best arguments and strongest evidence; so defined, it should not be equated with rationality in general, but is rather a particular (although important) intellectual trait that is part of rationality.

Before I elaborate this abstract conceptual framework a little further, let us see how it corresponds to the Galilean defense of Copernicus which I have elaborated in the last several chapters. For example, my account of Galileo's stances toward Copernican astronomy (Chapter 3) shows that before the telescope he was keenly aware of the strength of the anti-Copernican arguments based on the earth-heaven dichotomy, the appearance of Venus, the apparent brightness and size of Mars, and the apparent position of fixed stars. This appreciation exemplifies his open-mindedness and fair-mindedness, whereas after the telescope his critical reasoning about these

6.5 Critical Reasoning

arguments together with his rational-mindedness enabled him to seriously and actively pursue the Copernican theory. And it is important to understand that these concepts correspond not merely to Galileo's practice but also to his self-reflective pronouncements. For example, one of the most revealing reflections occurs in Day III of the *Dialogue*, in the context of the criticism of the empirical astronomical objections to the earth's motion. There he expresses amazement at "how in Aristarchus and Copernicus their [aprioristic] reason could have done so much violence to their senses, as to become, in opposition to the latter, mistress of their belief";[13] whereas he confesses that in his own case, if telescopic observation "had not joined with reason, I suspect that I too would have been much more recalcitrant against the Copernican system than I have been."[14]

Similarly, we have seen (Section 4.5) that the *Letter to the Grand Duchess Christina* provides a clear statement, an appreciative interpretation, and a nuanced evaluation of the scriptural argument against the earth's motion. My account explained the importance of seeing Galileo's essay as such an instance of critical reasoning, instead of regarding is as a "treatise" on hermeneutics.

Analogously, Galileo's critique of the argument from the extruding power of whirling (Chapter 5) is first and foremost an instance of critical reasoning. It carries to a new height the ideal of open-mindedness, for it stresses an objection to Copernicanism that is more powerful than most of those advanced by the anti-Copernicans themselves. The critique displays fair-mindedness insofar as it explicitly ensures that the anti-Copernican argument is properly stated, before it is refuted. Nevertheless, it is ultimately invalid, and rational-mindedness dictates its rejection.

Finally, the criticism of the argument from vertical fall (earlier in this chapter) is obviously a piece of critical reasoning. The analysis leads not only to rejecting vertical fall as evidence against the earth's motion (as demanded by rational-mindedness), but also to a better understanding of the objection than one finds in the proponents of geocentrism themselves (thus exemplifying open-mindedness and fair-mindedness).

Now, going back to the conceptual framework, let us note first that all those definitions take *reasoning* as an undefined primitive notion. One could define it in terms of what is an even more basic notion, namely *thinking*, and leave the latter undefined. Then we would say that reasoning is a form of thinking consisting of the interrelating of thoughts in such a way as to make some thoughts dependent on others, and this interdependence can take the form of some thoughts being based on others or some thoughts following from others.

Next, it may be useful to point out that the definition of critical reasoning repeated above is using the term *critical* to mean *evaluative*, in the general sense of either favorable and positive assessment or unfavorable and negative assessment. However, a narrower meaning of the word takes *critical* as pertaining only to negative

---

[13] Favaro 7: 355, Galilei (1997, 234; 2008, 242).
[14] Favaro 7: 355–356, Galilei (1997, 235; 2008, 242).

evaluation; this corresponds to the notion of criticism elaborated earlier in this chapter. Critical reasoning in this sense is reasoning aimed at criticism in the narrow sense, namely reasoning aimed at the refutation or negative evaluation of an argument. To bring it closer to that earlier discussion, such critical reasoning is reasoning that is aware of and open to the possibility of the falsehood or refutability of a proposition, including one's own view. Let us now give the label *principle of fallibility* to what I said above was a key lesson of the Copernican Revolution, namely that everyone may be mistaken and everything is and ought to be open to criticism. Then, critical reasoning in this narrow sense is reasoning guided by or aware of the principle of fallibility.

There is a third meaning of *critical*, which refers to self-awareness or self-reflection in general. In this sense, critical reasoning is reasoning that displays a keen awareness of what one is doing; what one is trying to prove; and what is the basis from which the argument proceeds. Reasoning guided by the principles of open-mindedness, fair-mindedness, rational-mindedness, and fallibility is critical reasoning in this sense. So is reasoning guided by judgment, namely by the ideal of judiciousness that seeks to avoid one-sidedness and extremes. In fact, I believe that open-mindedness, fair-mindedness, rational-mindedness, and fallibility are special cases of judgment so defined.

Thus, if one speaks of critical reasoning while allowing all three possible meanings of the term *critical*, then the phrase provides a single all-inclusive term. That is, at the most general level, critical reasoning is reasoning aimed at the interpretation, evaluation, or self-reflective formulation of arguments, and guided by the ideal of judgment, which ideal includes such principles as those of fallibility, open-mindedness, fair-mindedness, and rational-mindedness.

Now, these conclusions about human rationality in general can and should be applied or tested further by exploring their usefulness for the understanding and evaluation of other episodes. The Galileo affair (in the twofold sense of the original episode and the subsequent controversy) is especially instructive in this regard, not only because the subsequent controversy provides an especially relevant domain for such further testing and exploration, but also because such further testing and exploration promise to provide a comparison and contrast of the role of rationality in the Copernican Revolution and the Galileo affair. This further investigation is undertaken in the second part of this book, and it amounts to an elaboration of the second half of the *overarching thesis* formulated in the Introduction: that the proper defense of Galileo should have the reasoned, critical, open-minded, and fair-minded character which his own defense of Copernicus had.

Part II
Defending Galileo

Chapter 7
The Trial of Galileo, 1613–1633

The first part of this book has elaborated a critical interpretation of Galileo's work as a defense of the Copernican system. The Galilean defense was obviously multifaceted insofar as it involved questions of observational and theoretical astronomy, of hermeneutical and meta-hermeneutical theology, of general-qualitative and quantitative-mathematical physics, and of philosophical methodology and epistemology. More importantly, Galileo's defense was reasoned, critical, open-minded, fair-minded, and rational-minded; that is, it was focused on the critical examination of the arguments for and against Copernicus, guided by the concern to be aware of the opposite arguments, to appreciate them in their strength, but also to expose their flaws, and thus to select the conclusion justified by the better arguments. These features are reflected in the fact that before I gave an account of Galileo's replies (Chapters 3–6), I gave a lengthy account (Chapters 1–2) of the anti-Copernican objections and the traditional world view on which they were based. Equally important, I have tried to show that the Galilean defense of Copernicus was largely cogent, effective, or insightful, although I have not equated this essential correctness with complete or perfect rightness.

Now, as the discussion switches from the defense of Copernicus to the defense of Galileo, I believe something formally similar can and should be done. Thus, I will first (in this chapter) lay the foundations for why Galileo needs defending; they lie quite simply in the trial and tribulations which he had to endure from the Inquisition and other institutions and officials of the Catholic Church. That is, Galileo's defense of Copernicus did not win favor with the Church, and we need to become acquainted with the basic historical facts of the trial. Then (Chapter 8) we want to acquire a historical overview of how and why in the subsequent four centuries many people sided with the Inquisition and against Galileo, and found all kinds of faults with his original defense of Copernicus. That is, we need to become acquainted with the many objections that can and have been raised against Galileo. With that general knowledge, we will then be in the position to evaluate or assess these objections in detail. Such a defense of Galileo will be carried out primarily with regard to four major anti-Galilean criticisms (Chapters 9–12). But these replies will be sufficient to suggest that they embody a pattern; that this pattern can be applied to other cases; and that the pattern is simply the technique of critical reasoning which Galileo practiced in replying to anti-Copernican objections. By then, the analogy between Galileo's own defense of Copernicus and the defense of Galileo elaborated in this book will be complete.

As already mentioned, the aim of this chapter is to provide an account of the basic or fundamental facts of Galileo's trial from 1613 to 1633. This is necessarily a simplified account that tries to stay as close as possible to the documents and the facts, and away from controversy and complications. I do intend to avoid oversimplifications, but I firmly believe it is an error to confuse a simplification and an oversimplification. Thus, the role of this chapter in Part II of the book is analogous to that of Chapter 1 in Part I. The controversies and complications underlying this material will emerge explicitly in the next chapter as the story of the subsequent affair unfolds. It is that next chapter that by and large will define the controversy about Galileo's trial (the subsequent Galileo affair), just as Chapter 2 defined the Copernican controversy for the purpose of this investigation. It also follows that the simplified story in this chapter is not meant as a substitute for, or competitor to, the sophisticated accounts elaborated in, or derivable from, the works of such scholars as Beltrán Marí, Beretta, Camerota, Fantoli, and Speller; for my purpose here, it is sufficient that the account in this chapter be consistent with theirs.[1]

7.1 The Earlier Proceedings and the Condemnation of Copernicanism

In the years following the publication of *The Sidereal Messenger* in 1610, as support for the geostatic theory continued to dwindle, conservative scientists and philosophers began relying more and more heavily on biblical, theological, and religious arguments. These discussions became so frequent and widespread that the Medici court in Florence (where Galileo was philosopher and chief mathematician to the grand duke) must have started to wonder whether they had a heretic in their employment. Thus, in December 1613 the Grand Duchess Dowager Christina confronted one of Galileo's friends and followers named Benedetto Castelli, who had succeeded him in the chair of mathematics at the University of Pisa; she presented him with the biblical objection to the motion of the earth. This was done in an informal, gracious, and friendly manner, and clearly as much out of genuine curiosity as out of worry. Castelli answered in a way that satisfied both the duchess and Galileo, when Castelli informed him of the incident.[2] However, Galileo felt the need to write a very long letter to his former pupil containing a detailed refutation of the biblical objection.[3]

---

[1] Needless to say, while this qualification is crucial for the account in this chapter, the same does not apply for the accounts in my other chapters, especially Chapters 3–6 and 8–12. Those other chapters are intended to discuss all necessary complications and controversial issues; and as some of my notes indicate, my views there sometimes go deeper or farther than theirs, and sometimes are inconsistent with theirs (Beltrán Marí 2006; Beretta 1998, etc.; Camerota 2004; Fantoli 2003a,b; Speller 2008); and in cases of conflict I would uphold my own account against theirs.
[2] Favaro 16: 605–606, Finocchiaro (1989, 47–48).
[3] Favaro 5: 281–288, Finocchiaro (1989, 49–54), Galilei (2008, 103–109).

Recall that this objection argued that the geokinetic theory must be wrong because many biblical passages state or imply that the geostatic theory is right, one of these passages being that which describes the Joshua miracle (Joshua 10:12–13).

In this letter, Galileo suggested that the objection has three fatal flaws.[4] First, it attempts to prove a conclusion (the earth's rest) on the basis of a premise (the Bible's commitment to the geostatic system) which can only be ascertained with a knowledge of that conclusion in the first place; in fact, the interpretation of the Bible is a serious business, and normally the proper meaning of its statements about natural phenomena can be determined only after we know what is true in nature; thus, the business of biblical interpretation is dependent on physical investigation, and to base a controversial physical conclusion on the Bible is to put the cart before the horse. Second, the biblical objection is a nonsequitur, since the Bible is an authority only in matters of faith and morals, not in scientific ones, and thus its saying something about a natural phenomenon does not make it so, and therefore its statements do not constitute valid reasons for drawing corresponding scientific conclusions. Finally, it is questionable whether the earth's motion really contradicts the Bible, and an analysis of the Joshua passage shows that it cannot be easily interpreted in accordance with the geostatic theory, but that it accords better with the geokinetic view, especially as improved by Galileo's own discovery of solar axial rotation; the biblical objection is therefore groundless, aside from being question-begging and inferentially invalid.

Although unpublished, Galileo's letter to Castelli began circulating widely, and copies were made. Some of these came into the hands of traditionalists, who soon passed to the counterattack. In December 1614, at a church in Florence, a Dominican friar named Tommaso Caccini preached a Sunday sermon against mathematicians in general, and Galileo in particular, on the grounds that their beliefs and practices contradicted the Bible and were thus heretical.[5] In February 1615 another Dominican, named Niccolò Lorini, filed a written complaint against Galileo with the Inquisition in Rome, enclosing his letter to Castelli as incriminating evidence.[6] Then in March of the same year, Caccini, who had attacked Galileo from the pulpit, made a personal appearance before the Roman Inquisition. In his deposition he charged Galileo with suspicion of heresy, based not only on the content of the letter to Castelli, but also on the *Sunspots* book (1613); and he mentioned some hearsay evidence of a general sort and of a more specific type, involving two individuals named Ferdinando Ximenes and Giannozzo Attavanti.[7] The Roman Inquisition responded by ordering an examination of these two individuals and of the two mentioned writings.

In the meantime, Galileo was writing for advice and support to many friends and patrons who were either clergymen or had clerical connections.[8] He had no way of

---

[4] Cf. the discussions in Chapters 4 and 9 of this book.
[5] Favaro 12: 123, 19: 307. Cf. Fabroni (1773–1775, 1: 47 n. 1), Finocchiaro (2005b, 115).
[6] Favaro 19: 297–298, Pagano (1984, 69–70), Finocchiaro (1989, 134–135).
[7] Favaro 19: 307–311, Pagano (1984, 80–85), Finocchiaro (1989, 136–141).
[8] See, for example, Favaro (5: 291–295, 297–305), Finocchiaro (1989, 58–67).

knowing about the details of the Inquisition proceedings, which were a well-kept secret, but Caccini's original sermon had been public, and also he was able to learn about Lorini's initial complaint.[9]

Galileo also wrote and started to circulate privately three long essays on the issues. One (now known as *Letter to the Grand Duchess Christina*) dealt with the religious objections and was an elaboration of the letter to Castelli, which was thus expanded from 8 to 40 pages.[10] Another (now known as "Galileo's Considerations on the Copernican Opinion") began to sketch a way of answering the epistemological and philosophical objections, which Galileo had never done; the importance of such objections had recently been stressed in a famous letter penned by Cardinal Robert Bellarmine.[11] And the third one (the "Discourse on the Tides") was an elementary discussion of the scientific issues, in the form of a new physical argument in support of the earth's motion based on its alleged ability to explain the existence of the tides and of the trade winds;[12] this essay was an anticipation of what Galileo would later elaborate in the Fourth Day of the *Dialogue*.

He also received the unexpected but welcome support of a Carmelite friar named Paolo Antonio Foscarini, who published a book arguing in detail for the thesis that the theory of the earth's motion is compatible with the Bible.[13] Finally, in December 1615, after a long delay due to illness, Galileo went to Rome of his own initiative, to try to clear his name and prevent the condemnation of Copernicanism. He did succeed in the former, but not in the latter, undertaking.

In fact, the results of the Inquisition investigations were as follows. The consultant who examined the letter to Castelli reported that in its essence its hermeneutical views did not deviate from Catholic doctrine.[14] The cross-examination of the two witnesses, Ximenes and Attavanti, exonerated Galileo from the hearsay evidence; his utterance of heresies was found to be baseless.[15] And the examination of his work on *Sunspots* failed to reveal any explicit assertion of the earth's motion or other presumably heretical assertion, if indeed the Inquisition officials examined this book.[16] However, in the process, the status of Copernicanism had become enough of a problem that the Inquisition felt it necessary to consult its experts for a formal opinion.

---

[9] The confidentiality rule was well but not perfectly kept, and it was subject to abuse and arbitrariness, as one may gather from Beltrán Marí (2006, 41–45).

[10] In Favaro 5: 309–548, Finocchiaro (1989, 87–118), Galilei (2008, 109–145). Cf. Chapters 4 and 9 of this book.

[11] Favaro (12: 171–172, 5: 351–370), Finocchiaro (1989, 67–86), Galilei (2008, 146–167).

[12] Favaro 5: 377–395, Finocchiaro (1989, 119–133).

[13] Foscarini (1615a), Blackwell (1991, 217–251). Cf. Chapter 4 of this book.

[14] Favaro 19: 305, Pagano (1984, 68–69), Finocchiaro (1989, 135–136).

[15] Favaro 19: 316–320, Pagano (1984, 93–98), Finocchiaro (1989, 141–146).

[16] There seems to be no direct documentary evidence that the *History and Demonstrations Concerning Sunspots* was examined. The *indirect* evidence for this is a note on the back of one of the folios containing Attavanti's deposition; the note is dated 25 November 1615 and says simply: "those letters of Galileo, published in Rome under the title *Sunspots Letters*, should be looked at" (Favaro 19: 320; Pagano 1984, 98).

7.1 The Earlier Proceedings and the Condemnation of Copernicanism

On 24 February 1616, a committee of eleven consultants reported unanimously that Copernicanism was philosophically (scientifically) false and theologically heretical or erroneous.[17] In a way, much of the tragedy of the Galileo affair stems from this opinion, which even Catholic apologists seldom if ever defend nowadays.[18] Although indefensible, if one wants to understand how this opinion came about, one must recall all the traditional arguments against the earth's motion, based on empirical astronomical, mechanical physical, and epistemological considerations, as well as the scriptural theological ones (see Chapters 1–2). Moreover, one must view the judgment of heresy in the light of the two objections based on the words of the Bible and on the consensus of the Church Fathers; in the light of the traditional hierarchy of disciplines, which made theology the queen of the sciences, and which had been reaffirmed at the Fifth Lateran Council in 1513;[19] and in the light of the Catholic Counter Reformation rejection of new and individualistic interpretations of the Bible.[20] At any rate, the Inquisition must have had some misgivings about the opinion of the committee of eleven consultants, for it issued no formal condemnation. Instead two milder consequences followed.

First, the Inquisition decided to give Galileo a private warning to stop defending and to abandon his geokinetic views.[21] The warning was conveyed to Galileo within a few days by Cardinal Bellarmine, the most influential and highly respected theologian and churchman of the time, with whom Galileo was on very good terms, despite their philosophical and scientific differences. The exact content, form, and circumstances of this warning are not completely known, but they are extremely complex and a subject of great controversy. Moreover, as we shall soon see, the occurrence and propriety of the later Inquisition proceedings in 1633 hinge on the nature of this warning. For now let us simply note that Bellarmine reported back to the Inquisition that he had warned Galileo to abandon his defense of and belief in the geokinetic thesis, and that Galileo had promised to obey.[22]

The other development was a public decree issued by the Congregation of the Index, the department of the Church in charge of book censorship. On 5 March 1616 this Congregation published a decree containing four main points.[23]

---

[17] Favaro 19: 320–321, Pagano (1984, 99–100), Finocchiaro (1989, 146–147).

[18] Of course, in earlier times, there were serious attempts to justify even this conclusion. One of the most notable such justifications was written during Galileo's trial in 1632–1633, and published a few months after his condemnation. Entitled *Tractatus syllepticus* (or *Summary Treatise*), it was authored by Jesuit Melchior Inchofer. To get an idea of the care devoted to such an effort, it should be noted that the consultants distinguished between the heliostatic heliocentric thesis and the geokinetic thesis and judged the former "formally heretical" and the latter "at least erroneous in the faith." Inchofer tried to justify this differential assessment. For details, see Blackwell (2006, 45–63, 105–206).

[19] For more details on this, see Beretta (2003b; 2005b), and Chapter 4 of this book.

[20] For more details on this, see Soccorsi (1947), McMullin (2005b), and Chapter 4 of this book.

[21] Favaro 19: 321, Pagano (1984, 100–101), Finocchiaro (1989, 147).

[22] Favaro 19: 278, Pagano (1984, 223–224), Finocchiaro (1989, 148).

[23] Favaro 19: 322–323, Pagano (1984, 102–103), Finocchiaro (1989, 148–150), Galilei (2008, 176–178). Cf. Finocchiaro (2005b, 16–20).

First, it stated that the doctrine of the earth's motion is false, contrary to the Bible, and a threat to Catholicism. Second, it condemned and prohibited completely Foscarini's book; this was the work that had tried to show that the earth's motion is compatible with the Bible. Third, it suspended circulation of Copernicus's book, pending correction and revision; these corrections were eventually issued in 1620,[24] their gist being to delete or modify about a dozen passages containing either religious references or else language indicating Copernicus's realist interpretation of the earth's motion. Fourth, the decree ordered analogous censures for analogous books. Galileo was not mentioned at all.

It should be noted that this was a decree issued by the Congregation of the Index, and not a public pronouncement by the Congregation of the Holy Office or Inquisition; hence, although Catholics were still obliged to obey it, it did not carry the weight and generality of pronouncements which define the Catholic faith, just as even an Inquisition decree would not carry the authority of an official papal decree ex cathedra or of a decree issued by an ecumenical council, such as the Council of Trent.[25] Moreover, the actual wording in the decree was vague and unclear, which is a sign of having been some kind of compromise, and the exact reason or the exact offensive features of the prohibited books were not spelled out; thus the declaration that analogous works were analogously prohibited was not too informative, and was liable to great abuse. Finally, one was left in the dark concerning what type of *discussion* of Copernicanism was indeed allowed.

In view of the confusing message in this Decree of the Index, and in view of the even more confusing circumstances of Bellarmine's private warning to Galileo (which I will discuss later), it is not difficult to sympathize with Galileo's next two moves before he left Rome to return home to Florence. He obtained an audience with Pope Paul V; the precise content of their discussion is not known, but in a letter to the Tuscan Secretary of State Galileo reported that he had been warmly received and reassured, during three-quarters of an hour with the pontiff.[26] Moreover, at this time Galileo began receiving letters from friends in Venice and Pisa saying that there were rumors in those cities to the effect that he had been personally put on trial, condemned, forced to recant, and given appropriate penalties by the Inquisition (Favaro 12: 254, 257–259). Having shown these letters to Cardinal Bellarmine, Galileo was able to convince him to write a brief and clear

---

[24] Favaro 19: 400–401, Finocchiaro (1989, 200–202). Cf. Finocchiaro (2005b, 20–25).

[25] On the other hand, in issuing the decree, the Index was in part following the orders of the Inquisition, and indeed of the pope himself as chairman of the Inquisition. In fact, it was at the Inquisition meeting of 25 February 1616 that the pope ruled that the earth's motion was contrary to Scripture; ordered Bellarmine to warn Galileo; and directed the Index to decide the details of the prohibition of Copernican books (which it did at its meeting of March 1). See the documents in Favaro (19: 278, 321), Pagano (1984, 100–101, 222–223), Finocchiaro (1989, 147), Mayaud (1997, 37–41), and cf. the accounts in Beretta (2001b, 306 n. 21), Fantoli (2003b,183–187, 453–458), Beltrán Marí (2006, 321–323), Speller (2008, 79–84).

[26] Favaro 12: 247–249, Finocchiaro (1989, 151–153).

statement of what had happened and of how Galileo was affected. Thus in a document half a page long, the most authoritative churchman of his time declared the following: Galileo had been neither tried nor otherwise condemned, but rather he had been personally notified of the Decree of the Index, and of the fact that in view of this decree the geokinetic thesis could be neither held nor defended.[27]

With this certificate in his possession, Galileo left Rome soon thereafter. However, before we too leave this first phase of the affair let us pause to see where we are. Despite the fact that Galileo was not personally condemned, despite the fact that the geokinetic theory was not formally condemned by the Inquisition, despite the vagueness of the Decree of the Index, and despite the pope's and Bellarmine's personal assurances to Galileo, it seems obvious that he had lost the battle. He had been personally forbidden to defend the geokinetic theory, and to all Catholics who wanted to avoid trouble the Decree of the Index meant that they too should avoid defending it. The earth's motion had not been formally declared a heresy (as the Inquisition consultants had judged in their report), but the practical effect of what happened was about the same. Moreover, since our story continues, let us recall the three different documents that will play a crucial role later: the private, oral warning given to Galileo by Cardinal Bellarmine; the public Decree of the Index; and Bellarmine's certificate to Galileo. The propriety or impropriety of his subsequent behavior will depend on which of these three items is stressed.

7.2 The Later Proceedings and the Condemnation of Galileo

For the next several years Galileo did refrain from defending or explicitly discussing the geokinetic theory, although he did discuss it implicitly and indirectly in the context of a controversy about the nature of comets, that is, in *The Assayer* (1623).[28] Even the publication of the corrections to Copernicus's book in 1620,[29] which gave one a better idea of what was allowed and what not, did not motivate him to resume the earlier struggle. The death of both Bellarmine and Pope Paul V in 1621, and the election of Pope Gregory XV did result in some encouragement, for example when Galileo was consulted about astronomical matters by the cardinal nephew and Vatican secretary of state;[30] but those developments were not enough for a significant change.

The event that put an end to the interlude took place in 1623, when Gregory XV died and Cardinal Maffeo Barberini was elected Pope Urban VIII. Urban was a

---

[27] Favaro 19: 348, Pagano (1984, 138), Finocchiaro (1989, 153), Galilei (2008, 178). Cf. Baldini and Coyne (1984).

[28] For the indirectly Copernican aspects of *The Assayer*, see Beltrán Marí (2006, 369–381), Biagioli (1993, 267–311), Camerota (2004, 363–398), Speller (2008, 111–123).

[29] Favaro 19: 400–401, Finocchiaro (1989, 200–202). Cf. Finocchiaro (2005b, 20–25).

[30] L. Ludovisi to Galileo, 22 November 1622, in Favaro 13: 100–101; cf. Beltrán Marí (2006, 381–387).

well-educated Florentine, and in 1616 he had been instrumental in preventing the direct condemnation of Galileo and the formal condemnation of Copernicanism as a heresy. He was also a great admirer of Galileo, and in 1620 he had even written a poem in praise of Galileo.[31] He now employed as personal secretary one of Galileo's closest acquaintances. Further, at about this time, Galileo's book on the comets, *The Assayer*, was being published in Rome by the Lincean Academy, and so it was decided to dedicate the book to the new pope. Urban appreciated the gesture and liked the book very much. Finally, as soon as circumstances allowed, in the spring of 1624, Galileo went to Rome to pay his respects to the pontiff; he stayed about six weeks and was warmly received by Church officials in general and the pope in particular, who granted him weekly audiences.[32]

The details of the conversation during these six audiences are not known. There is evidence, however, that Urban VIII did not think Copernicanism to be a heresy, or to have been declared a heresy by the Church in 1616. He interpreted the Decree of the Index to mean that the earth's motion was a rash or dangerous doctrine whose study and discussion required special care and vigilance. He thought the theory could never be proved to be necessarily true, and here it is interesting to mention his favorite argument for this skepticism, namely the divine-omnipotence objection: Urban liked to argue that since God is all-powerful, he could have created any one of a number of worlds, for example one in which the earth is motionless; therefore, regardless of how much evidence there is supporting the earth's motion, we can never assert that this must be so, for that would be to want to limit God's power to do otherwise.[33] This argument, together with his interpretation of the decree of 1616, must have reinforced his liberal inclination that, as long as one exercised the proper care, there was nothing wrong with the hypothetical discussion of Copernicanism; that is to say, with

---

[31] With the revealing title of "Adulatio Perniciosa." Cf. Favaro 13: 48–49, Pieralisi (1875, 22–25)

[32] Favaro 13: 175, 182–185.

[33] For more details about this argument, see Favaro 7: 488–489, Galilei (1967, 464; 1997, 306–308; 2008, 269–271). Cf. Beltrán Marí (2006, 412–437), Besomi and Helbing (1998, 2: 899–902), Bianchi (2000; 2001), Camerota (2004, 406–417), Finocchiaro (1980, 8–12; 1985; 1997a, 306–308), Favaro 7: 565–566, Morin (1631, 31–32), Morpurgo–Tagliabue (1981, 99–107), Speller (2008, 143–160, 375–396), Wisan 1984, and Section 2.2 of this book. Urban may have had in mind the objection that God could have created a world in which the evidence suggested a moving earth despite its being motionless. There is no doubt that the exact content, structure, origin, and consequences of Urban's argument remain an open question that deserves further exploration, although a significant contribution has now been made by Speller (2008). It should also be mentioned that if Urban's argument did indeed include the objection just mentioned, this would lend support to Morpurgo–Tagliabue's thesis that Descartes's "methodic doubt is nothing but the generalization of the argument of Urban VIII" (1981, 115) and that his "*Discourse on Method* … is essentially an answer to the argument of Urban VIII" (1981, 104). This would also accord very well with the account of Descartes's development and thought advanced by Gaukroger (1995, 11–12, 292), for whom Cartesian metaphysics was a response to the methodological crisis in physics produced by the condemnation of Galileo.

7.2 The Later Proceedings and the Condemnation of Galileo

treating the earth's motion as a hypothesis and studying its consequences, its value for understanding and explaining physical reality, and its utility for making astronomical calculations and predictions.

At any rate, Galileo must have gotten some such impression during his six conversations with Urban, for upon his return to Florence he began working on a book. This was in part the work on the system of the world which he had conceived at the time of his first telescopic discoveries, but it now acquired a new form and new dimensions in view of all that he had learned and experienced since. His first step was to write and circulate privately a lengthy reply to the anti-Copernican essay written in 1616 by Francesco Ingoli.[34] This Galilean "Reply to Ingoli," as well as his earlier "Discourse on the Tides," were incorporated into the new book. After a number of delays in its writing, licensing, and printing, the work was finally published in Florence in February 1632, with the title *Dialogue on the Two Chief World Systems, Ptolemaic and Copernican*.[35]

The author had done a number of things to avoid trouble, to insure compliance with the many restrictions under which he was operating, and to satisfy the various censors who issued him permissions to print. To emphasize the hypothetical character of the discussion, he had originally entitled it "Dialogue on the Tides" and structured it accordingly. That is, it was to begin with a statement of the problem of the cause of tides, and then it would introduce the earth's motion as a hypothetical cause of the phenomenon; this would lead to the problem of the earth's motion, and to a discussion of the arguments pro and con, as a way of assessing the merits of this hypothetical explanation of the tides.[36] However, the book censors, interpreting and acting upon the pope's wishes, decided to make the book look like a vindication of the Index Decree of 1616. The book's preface, whose content must be regarded as originating primarily from the pope and the censors and only secondarily from Galileo, claimed that the work was being published to prove to non-Catholics that Catholics knew all the arguments and evidence about the scientific issues, and so their decision to believe in the geostatic theory was motivated by religious reasons and not by scientific ignorance. It went on to add that the scientific arguments seemed to favor the geokinetic theory, but that they were inconclusive, and thus the earth's motion remained a hypothesis.

Galileo also complied with the explicit request to end the book with a statement of the pope's favorite argument, namely the objection from divine omnipotence. Moreover, to make sure he would not be seen as holding or defending the geokinetic thesis (which he had been forbidden to do) the author did two things. He wrote the book in the form of a dialogue among three speakers: Simplicio, defending the

---

[34] Favaro 6: 509–561, Finocchiaro (1989, 154–197). Cf. Section 4.3 of this book.

[35] The actual title was much longer; see, for example, Finocchiaro (1980, 12–18).

[36] I realize, of course, that my interpretation here goes against the prevailing view that Galileo regarded the tidal argument as conclusive; that he therefore wanted to advertise this in the book's title; and that the Church did not want to endorse this argument. For an exposition of this view, see Shea (1972, 172–189); for some criticism, see Finocchiaro (1980, 16–18, 76–78). Cf. also Drake (1986b), MacLachlan (1990), Pitt (1992, 78–109), and Chapters 3 and 9 of this book.

geostatic side; Salviati, taking the Copernican view; and Sagredo, who is an uncommitted observer who listens to both sides and accepts the arguments that seem to survive critical scrutiny. And in many places throughout the book, usually at the end of a particular topic, the Copernican Salviati utters the qualification that the purpose of the discussion is information and enlightenment, and not to decide the issue, which is a task to be reserved for the proper authorities. Finally, it should be mentioned that Galileo obtained written permissions to print the book, first from the proper Church officials in Rome (when the plan was to publish the book there), and then from the proper officials in Florence (when a number of external circumstances dictated that the book be printed in the Tuscan capital).

The book was well-received in scientific circles. However, a number of rumors and complaints began emerging and circulating in Rome.

One complaint involved a document that had been found in the file of the Inquisition proceedings of 1615–1616.[37] It reads like a report of what took place when Cardinal Bellarmine, on orders from the Inquisition, gave Galileo the private warning to stop defending his geokinetic views. The cardinal had died in 1621, and so was no longer available to clarify the situation. The document states that in February 1616, at the same meeting with Bellarmine, the Inquisition's commissary general had given Galileo the injunction to stop holding, defending, or teaching the earth's motion in any way whatever. That is, this document states that Galileo had been given a special injunction, above and beyond what bound Catholics in general: supposedly, he had been prohibited not only to hold or defend the earth's motion as a truth, but also to hold or defend the idea in any manner as well as to teach it in any manner; in short to discuss the topic at all. The charge was then that his book of 1632 was a clear violation of this special injunction, since whatever else the book did, and however else it might be described, it undeniably contained a discussion of the earth's motion. To be sure, the document did not bear Galileo's signature, and it contradicted other genuine documents, and so it was of questionable legal validity.[38] Under different circumstances such a judicial technicality could have been taken seriously. However, too many other difficulties were being raised about the book.

One of these was that the work only paid lip service to the stipulation about a hypothetical discussion, which represented Urban's compromise; in reality, the book allegedly treated the earth's motion not as a hypothesis, but in a factual, nonconditional, and realistic manner. This was a more or less legitimate complaint on the part of the pope, but the truth of the matter is that the concept of hypothesis was ambiguous and had not been sufficiently clarified in that historical context. By hypothetical treatment Urban meant a discussion that would treat the earth's motion

---

[37] Favaro 19: 321–322, Pagano (1984, 101–102), Finocchiaro (1989, 147–148), Galilei (2008, 175–176).

[38] This only became clear and explicit much later, in the 1870s, when the trial proceedings were opened to a few scholars and were published in their entirety. Cf. Finocchiaro (2005b, 241–258) and Section 8.12 of this book.

7.2 The Later Proceedings and the Condemnation of Galileo

merely as an instrument of prediction and calculation, rather than as a potentially true description of reality. On the other hand, Galileo took a hypothesis to be an assumption about physical reality, which accounts for what is known, and which may be true, though it has not yet been proved to be true.[39]

Third, there was the problem that Galileo's book was in actuality a defense of the geokinetic theory. Despite the dialogue form, despite the repeated disclaimers that no assertion of Copernicanism was being intended, despite the non-apodictic and nonconclusive character of the pro-Copernican arguments, and despite the presentation of the anti-Copernican and pro-geostatic arguments, it was readily apparent that the pro-geostatic arguments were being criticized and the pro-Copernican ones were being portrayed favorably. And this perception led some to claim that the book was arguing in favor of Copernicanism and hence was defending it.

There were also complaints involving alleged irregularities in the various permissions to print that Galileo obtained. There were substantive criticisms of various specific points discussed in the book. There were hurt feelings about some of his rhetorical excesses and biting sarcasm. There were malicious slanders suggesting that the book was in effect a personal caricature of the pope himself.

Although some of the complaints in the last miscellaneous group were easily cleared, the sheer number in the whole list and the seriousness of some charges were such that the pope might have been forced to take some action even under normal circumstances. But Urban VIII was himself in political trouble due to his behavior in the 30 Years War between Catholics and Protestants (1618–1648). At that particular juncture the pope was in an especially vulnerable position, and thus not only could he not continue to protect Galileo, but he chose to use Galileo as a scapegoat to reassert, exhibit, and test his authority and power. The problem stemmed from the fact that in 1632 the Catholic side led by the King of Spain and by the Bohemian Holy Roman Emperor was disastrously losing the war to the Protestant side led by the King of Sweden, Gustavus Adolphus. However, religion was not the only issue in the war, which was being fought also over dynastic rights and territorial disputes. In fact, ever since his election in 1623, the pope's policy had been motivated primarily by political considerations, such as his wish to limit and balance the power of the Hapsburg dynasty, which ruled Spain and the Holy Roman Empire. And it had also been motivated by personal interest, that is, by cooperation with the French, whose support had been instrumental in his election, and who for nationalistic reasons also opposed the Hapsburg hegemony. However, in the wake of Gustavus Adolphus's spectacular victories, the Spanish and Imperial ambassadors were accusing Urban of having in effect favored and helped the Protestant cause. They mentioned such things as his failure to send the kind of military and financial support that popes had usually provided on such occasions, and his refusal to declare the war a holy war. There were even suspicions of a more direct understanding with the Protestants. Thus the pope's own

---

[39] Here, I am following in part the interpretation found in Morpurgo–Tagliabue (1981), but see also Speller (2008).

religious credentials were being questioned, and there were rumors of convening a council to depose him.[40]

Thus in the summer of 1632 sales of the book were stopped, unsold copies confiscated, and a special commission was appointed to investigate the matter.[41] The pope did not immediately send the case to the Inquisition, but he took the unusual step of appointing a special commission first. This three-member panel issued its report in September 1632,[42] and it listed as areas of concern about the book all of the above-mentioned problems, with the exception of the malicious slanders. In fact, it is from the report that we learn about these complaints that had been accumulating since the book's publication. In view of the report the pope felt he had no choice but to forward the case to the Inquisition. Or rather, this was the impression Urban wanted to convey and what he said to the Tuscan ambassador to Rome;[43] for the report can be read as leaving the question open. It is more likely that Urban was mostly manipulating the proceedings, either to cover up his own permissivism and complicity in the writing and publication of the *Dialogue*,[44] or to act on his perception that this book came close to formal heresy by its failure to treat hypothetically of the earth's motion and show the proper appreciation for divine omnipotence.[45] So Galileo was summoned to Rome to stand trial.

The entire autumn was taken up by various attempts on the part of Galileo and the Tuscan government to prevent the inevitable.[46] The Tuscan government got involved partly because of Galileo's position as Philosopher and Chief Mathematician to the Grand Duke; partly because the book contained a dedication to the grand duke; and partly because the grand duke had been instrumental in getting the book finally printed in Florence. At first they tried to have the trial moved from Rome to Florence. Then they asked that Galileo be sent the charges in writing, and that he be allowed to respond in writing. As a last resort, three physicians signed a medical certificate stating that Galileo was too ill to travel. This was true, and here it should be added that he was 68 years old, and that there had been an outbreak of the plague for the past 2 years, which meant that travelers from Tuscany to the Papal States were subject to quarantine at the border. At the end of December the Inquisition

---

[40]Ranke (1841, 2: 116–119), Pastor (1891–1953, 28: 271–321). For a more vivid account, see Redondi (1983, 288–295; 1987, 227–232). See also Miller's (2008) useful update.

[41]I do not mean to imply that *all* of the complaints had been voiced *before* the pope took this step, but only that *some* of them emerged before; in particular, I am not sure there is any way of dating exactly what I have called "the most serious complaint," involving the special injunction. However, it is certain that the ban on sales preceded the report of the special commission (Favaro 20: 571–572, 14: 391–393, 14: 397–398), and it seems to me that the first written mention of the special injunction does not occur till September (Favaro 14: 391–393, 397–398).

[42]Favaro 19: 324–327, Pagano (1984, 105–108), Finocchiaro (1989, 218–222), Galilei (2008, 272–276).

[43]Niccolini to Cioli, 18 September 1632, in Favaro 14: 391–393, and in Finocchiaro (1989, 234–237).

[44]For an elaboration of this conjecture, see Beltrán Marí (2006, 496–528).

[45]For an elaboration of this conjecture, see Speller (2008, especially 143–160, 375–396).

[46]Favaro 19: 330–336, Pagano (1984, 113–123).

7.2 The Later Proceedings and the Condemnation of Galileo

sent Galileo an ultimatum: if he did not come to Rome of his own accord, they would send some officers to arrest him and bring him to Rome in chains. On 20 January 1633, after making a last will and testament,[47] Galileo began the journey. When he arrived in Rome three weeks later, he was not placed under arrest or imprisoned by the Inquisition, but was allowed to lodge at the Tuscan embassy (Palazzo Firenze[48]), though he was ordered not to socialize and to keep himself in seclusion until he was called for interrogations (Favaro 15: 40–41).

These were slow in coming, as if the Inquisition wanted to use the torment of the uncertainty, suspense, and anxiety as part of the punishment to be administered to the old man. This was very much in line with one reason mentioned earlier by officials why Galileo had to make the journey to Rome, despite his old age, ill health, and the epidemic of the plague; that is, he had to do it as an advance partial punishment or penance, and if he did this the inquisitors might take it into consideration when the time of the actual proceedings came.

The first interrogation was held on April 12.[49] The questions did not focus on Pope Urban's complaint about the book's failure to treat the earth's motion hypothetically and to appreciate divine omnipotence, but rather on the events of 1616. In answer to various questions, the defendant claimed the following. He admitted having been given a warning by Cardinal Bellarmine in February 1616, and described this as an oral warning that the geokinetic theory could be neither held nor defended, but only discussed hypothetically. He denied having received any special injunction not to discuss the earth's motion in any way whatever, and he introduced Bellarmine's certificate as supporting evidence. His third main claim was made in answer to the question why he had not obtained any permission to write the book in the first place, and why he had not mentioned Bellarmine's warning when obtaining permission to print the book; these omissions had angered the pope and had made him feel deceived. Galileo answered that he had not done so because the book did not hold or defend the earth's motion, but rather showed that the arguments in its favor were not conclusive, and thus it did not violate Bellarmine's warning.

This was a very strong and practicable line of defense. In particular, Galileo's third point may be interpreted to suggest that the *Dialogue* was *discussing*, not defending, the earth's motion, insofar as it was a critical examination of the arguments on both sides. Moreover, just as the special injunction was news to Galileo, so Bellarmine's certificate must have surprised and disoriented the Inquisition officials. Thus, it took another three weeks before they finally decided on the next step in the proceedings. In the meantime Galileo was detained at the headquarters of the Inquisition, but allowed to lodge in the chief prosecutor's apartment.[50]

---

[47] Favaro 15: 27, 29; 19: 520.

[48] Not Villa Medici; for a clarification of the difference, see Shea and Artigas (2003, 30, 74, 106–107, 134–135, 179–180, 195).

[49] Favaro 19: 336–342, Pagano (1984, 124–130), Finocchiaro (1989, 256–262), Galilei (2008, 276–282).

[50] Favaro 15: 86–87, 94–95, 109–110.

What the inquisitors finally decided was something very close to what might be called an out-of-court settlement involving a plea-bargaining agreement. That is, they would not press the most serious charge (of having violated the special injunction), nor the charge of having violated Urban's request for a hypothetical treatment of the earth's motion and an appreciation of divine omnipotence; but Galileo would have to plead guilty to the lesser charge of having inadvertently transgressed the order not to defend Copernicanism, in regard to which his defense was the weakest; and to reward such a confession, they would show some leniency toward such a lesser violation.

The deal was worked out as follows. The Inquisition asked three consultants to determine whether or not Galileo's *Dialogue* taught, defended, or held the geokinetic theory; in separate reports all three concluded that the book clearly taught and defended the doctrine, and came close to holding it.[51] Then the commissary general of the Inquisition talked privately with Galileo to try to arrange the deal, and after lengthy discussions he succeeded.[52] Galileo requested and obtained a few days to think of a dignified way of pleading guilty to the lesser charge. Thus on April 30, the defendant appeared before the Inquisition for the second time, and signed a deposition stating the following.[53] Ever since the first hearing he had reflected about whether, without meaning to, he might have done anything wrong. It dawned on him to reread his book, which he had not done for the past 3 years since completing the manuscript. He was surprised by what he found, since the book did give the reader the impression that the author was defending the geokinetic theory, even though this had not been his intention. To explain how this could have happened, Galileo attributed it to vanity, literary flamboyance, and an excessive desire to appear clever by making the weaker side look stronger. He was deeply sorry for this transgression, and was ready to make amends.

After this deposition, Galileo was allowed to return to the Tuscan embassy. On May 10 there was a third formal hearing at which Galileo presented his defense, including the original copy of Bellarmine's certificate, repeating his recent admission of some wrongdoing together with a denial of any malicious intent, and adding a plea for clemency and pity.[54] The trial might have ended here, but was not concluded for another six weeks. The new development was one of those things that make the Galileo affair such an unending source of controversy and such rich material for tragedy.

Obviously the pope and the cardinals of the Congregation of the Holy Office would have to approve the final disposition of the case. Indeed, it was standard Inquisition practice for an official (the assessor) to compile a summary of the

---

[51]Favaro 19: 348–360, Pagano (1984, 139–153), Finocchiaro (1989, 262–276).

[52]Favaro 15: 106–107, Finocchiaro (1989, 276–277), Beretta (2001a, 571), Baldini and Spruit (2001, 683–684).

[53]Favaro 19: 342–344, Pagano (1984, 130–132), Finocchiaro (1989, 277–279), Galilei (2008, 282–284).

[54]Favaro 19: 345–347, Pagano (1984, 133–134, 135–137), Finocchiaro (1989, 279–281), Galilei (2008, 284–287).

7.2 The Later Proceedings and the Condemnation of Galileo

proceedings for the benefit of the cardinal-inquisitors.[55] So a report was written, summarizing the events from 1615 to Galileo's third deposition just completed.[56] Through a series of misrepresentations, this report left no doubt that Galileo had committed some criminal act; on the other hand, by various quotations from his confessions and pleas, the report made it clear that he was not obstinately incorrigible, but rather was sorry and willing to submit.[57] However, it did not resolve Urban's doubts about Galileo's intention, and so he directed that the defendant be interrogated under the verbal threat of torture in order to determine his intention. Moreover, the pope decreed that even if his intention was found to have been pure, Galileo had to make an abjuration and was condemned to formal arrest at the pleasure of the Inquisition, and the *Dialogue* had to be banned.[58]

Threat of torture and actual torture were, at the time, standard practice of the Inquisition,[59] and indeed of almost all systems of criminal justice in the world. Nevertheless, such an interrogation, together with the abjuration, the arrest, and the book ban were not really in accordance with the spirit or the letter of the out-of-court plea-bargaining and agreement. Thus, Galileo felt betrayed and remained always bitter about this outcome.[60]

On June 21, Galileo was subjected to the interrogation under the formal verbal threat of torture.[61] The result was favorable, in the sense that, even under such a threat, Galileo denied any malicious intention, and showed his readiness to die rather than admit that. The following day, at a ceremony in the convent of Santa Maria sopra Minerva in Rome, he was read the sentence[62] and then recited the formal abjuration.[63]

The sentence banned Galileo's *Dialogue* and stated that he had been found "vehemently suspected of heresy." This was a technical legal term which meant much more than it may sound to modern ears. Although the Inquisition dealt with other offenses such as witchcraft, it was primarily interested in two main categories of crimes: formal heresy and suspicion of heresy. Here, the term *suspicion* did not have

---

[55] Cf. Mercati (1942), Beretta (1998, 196–206), Finocchiaro (2002, 86–88; 2005b, 198–218).

[56] Favaro 19: 293–297, Pagano (1984, 63–68), Finocchiaro (1989, 281–286).

[57] Speller (2008, 285–298) deserves credit for having pointed out this twofold aspect of the report.

[58] Favaro 19: 282–283, Pagano (1984, 229), Finocchiaro (2005b, 247).

[59] Cf. Masini (1621, 120–151), Giacchi (1942), Mereu (1979, 212–228), Beretta (1998, 214–221). For more details and references, see Finocchiaro (2009a).

[60] The discrepancy between the out-of-court agreement and the actual outcome has been the subject of considerable reflection and divergent interpretations. Some view it as an actual betrayal by the officials involved; others as the result of the fact that they were split and the more rigorous faction prevailed; and still others as a planned deception that corresponded to inquisitorial practices and rules. See Beltrán Marí (2006, 559–594), Blackwell (2006, 13–26), Fantoli (2003a, 198–203; 2003b, 322–323), Speller (2008, 255–314).

[61] Favaro 19: 361–362, Pagano (1984, 154–155), Finocchiaro (1989, 286–287), Galilei (2008, 287–288).

[62] Favaro 19: 402–406, Finocchiaro (1989, 287–291), Galilei (2008, 288–293).

[63] Favaro 19: 406–407, Finocchiaro (1989, 292–293), Galilei (2008, 293–294).

the modern legal connotation, pertaining to allegation and contrasting it to proof. One difference between formal heresy and suspicion of heresy was the seriousness of the offense.[64] Another difference was whether or not the culprit, having confessed the incriminating facts, admitted having an evil intention (Masini 1621, 166–167). Furthermore, within the major category of suspicion of heresy, three main subcategories were distinguished: strong suspicion, vehement suspicion, and slight suspicion of heresy;[65] their difference depended on the seriousness of the crime. Thus, there were four main types of religious crimes, in descending order of seriousness: formal heresy, strong suspicion, vehement suspicion, and slight suspicion of heresy.

In short, it seems that "suspicion of heresy" was not merely suspicion of having committed a crime, but was itself a specific category of crime; Galileo was in effect being convicted of the third most serious offense handled by the Inquisition.

Two distinct allegedly heretical views were mentioned: the astronomical and cosmological thesis that the earth rotates daily on its axis and circles the sun once a year; and the methodological and theological principle that one may believe and defend as probable a thesis contrary to the Bible.[66] One other interesting detail about the sentence is that only seven out of the ten cardinal-inquisitors signed it; two of the three who did not were Cardinal Francesco Barberini, the pope's nephew and the Vatican secretary of state, who was the most powerful man in Rome after the pope himself, and Cardinal Gaspare Borgia, the Spanish ambassador and leader of the Spanish party, who a year earlier had threatened the pope with impeachment on account of his behavior in the 30 Years War.

It took another 6 months before Galileo was allowed to return home, to remain under arrest in his own house. He was first confined to Villa Medici in Rome, where he stayed for another 10 days.[67] Then for about 5 months he was under house arrest at the residence of the archbishop of Siena, who proved to be a very congenial and sympathetic host.[68]

---

[64]For example, a standard Inquisition manual of the time stated that "heretics are those who say, teach, preach, or write things against the Holy Scripture; against the articles of the Holy Faith; … against the decrees of the Sacred Councils and the determinations made by the Supreme Pontiffs; … those who reject the Holy Faith and become Moslems, Jews, or members of other sects, and who praise their practices and live in accordance with them …" (Masini 1621, 16–17). The same manual stated that "suspected heretics are those who occasionally utter propositions that offend the listeners … those who keep, write, read, or give others to read books forbidden in the Index and in other particular Decrees; … those who receive the holy orders even though they have a wife, or who take another wife even though they are already married; … those who listen, even once, to sermons by heretics …" (Masini 1621, 17–18).

[65]Masini (1621, 188) states that, in practice, "strong" suspicion of heresy was rarely charged, but was equated to "vehement" suspicion.

[66]Insufficient attention has been paid to the double character of the alleged heresy attributed to Galileo, and to the probabilistic character of the proscribed methodological hermeneutical principle. However, Speller (2008, 250, 332–334) has shown some appreciation. For more details, see also the Introduction and Chapters 9 and 12 of this book.

[67]Favaro 15:165, 19: 284. See Shea and Artigas (2003, 30, 74, 106–107, 134–135, 179–180, 195), to understand that this residence was at Villa Medici, and not at Palazzo Firenze.

[68]Favaro (15: 168, 170–171; 19: 284, 362, 363); cf. Finocchiaro (2005b, 56–57).

7.2 The Later Proceedings and the Condemnation of Galileo

Thus, the original Galileo affair ended and a new one began. That is, what ended was the Inquisition's trial of Galileo which started in 1613 with his letter to Castelli, refuting the biblical objection to the geokinetic hypothesis, and which climaxed in 1633 with his condemnation as a suspected heretic for defending this hypothesis and rejecting the astronomical authority of Scripture. What began then was the unresolved and perhaps irresolvable controversy about the facts, the issues, the causes, and the lessons of the original episode; this is a cause célèbre which continues to our own day and whose fascination rivals that of the original one. The subsequent Galileo affair is much more complex than the original one and so deserves a narrative chronological overview even more than the original trial. That is the task of the next chapter.

Chapter 8
The Galileo Affair, 1633–1992

The focus in the last chapter was the fundamentals of Galileo's trial. Its scope and import bear an analogy to the scope and import of Chapter 1, which covered the essentials of the geostatic world view. But note that this analogy (like any other analogy) exists in regard to things that are different: cosmological beliefs and theses on the one hand, and historical events and actions on the other. In this chapter the stress will be on the controversy generated by Galileo's trial and condemnation.

This subsequent Galileo affair is also meant to be analogous to the Copernican controversy that followed the publication of Copernicus's *Revolutions* in 1543. We have seen that the Copernican controversy centered on the crucial question of the earth's motion and was definable in terms of the arguments for and against the geokinetic hypothesis. Analogously, the subsequent controversy about Galileo's trial also has a focus, namely the question whether Galileo was justly or rightly condemned; and it also consists of arguments for and against the condemnation and the Inquisition, or alternatively of criticisms or objections to Galileo and replies to these or defenses of him. However, whereas the Copernican controversy was largely resolved a century and one-half after Copernicus, with the publication of Isaac Newton's *Mathematical Principles of Natural Philosophy* (1687), the controversy about Galileo's trial shows no signs of termination after about four centuries, although it has undergone considerable evolution. Partly for this reason, in this chapter the pro- and anti-Galilean arguments will not be presented abstractly (as the more manageable Copernican controversy was presented in chapter 2), but rather they will emerge in the course of a historical account of the subsequent Galileo affair.[1]

Moreover, given that the central purpose of this historical account is to provide an introduction to the issues (the arguments and counter-arguments *about* Galileo's

---

[1] This sketch is a summary, digest, or abridgment of the more detailed, documented, and nuanced account found in Finocchiaro (2005b); accordingly, for each section of this chapter, further information, references, and analysis are found, respectively, in the corresponding chapters of that book. Moreover, in the synoptic overview of this chapter, it will be unnecessary to give many general bibliographical references in the notes because they will be given implicitly in the body of the text, when the exposition mentions authors and dates or titles of their works; however, the usual details are found explicitly in the corresponding entries of the bibliography.

trial), this is the place where one will find a discussion of many details of the trial (such as the 1633 condemnation in Section 8.1 and the 1616 special injunction in Section 8.12) that could have been presented in the account given in the last chapter. This is like the presentation (in Chapter 2) of the anti-Copernican arguments, almost all of which could have been presented as part of the geostatic world view (in Chapter 1). There is more to this manner of exposition than personal idiosyncrasy or stylistic preference. The key point is that the greater details about the geostatic world view (of Chapter 1) provided in the discussion of the Copernican controversy (of Chapter 2) are themselves not the end of the story, that is, of the analysis; in fact, the rest of my discussion of "defending Copernicus" (Chapters 3–6) shows how much deeper one could dig into those same details. Similarly, in "defending Galileo," later chapters will show how much more there is to understand and to assess about the issues and arguments that will emerge in this chapter, even though these are themselves one level deeper or more advanced than the developments sketched in the last chapter. It is not necessary, in this context, to go into an elaborate digression about the epistemological basis and import of this manner of exposition. Suffice it to say that it reflects a particular view about the nature of knowledge, both scientific and historical. The epistemological stance reflected in this manner of exposition views knowledge as a dynamic process that has no ultimate foundations and no final ending, but consists of a series of better and better approximations to understanding various problems about various domains, as distinct from a series of hierarchically arranged propositions that exhibit a cumulative progression.

8.1 The Condemnation of Galileo (1633)

We have seen that on 22 June 1633, the Roman Inquisition concluded the trial of Galileo by pronouncing a sentence that condemned him for various transgressions. The sentence was immediately followed by the defendant's abjuration, in which he retracted previous opinions and actions. To begin to understand the controversy it generated, we must examine these defining documents more carefully.[2]

    1. The sentence contained not only a statement of the verdict, but also an account of the proceedings since 1615 and a list of penalties. The account began by describing several charges advanced against Galileo in 1615.[3] Five distinct accusations were mentioned: (1) holding the truth of the earth's motion; (2) corresponding about this doctrine with some German mathematicians; (3) publishing a book on

---

[2] In this section, I shall give line-by-line references to the Italian text of the sentence and abjuration as found in Favaro 19: 402–407. However, I shall give no specific references to the English translation (Finocchiaro 1989, 287–293; Galilei 2008, 288–294), for which this single general reference should suffice here.

[3] Favaro 19: 403, lines 15–27.

Sunspots that "explained" the truth of the doctrine; (4) answering scriptural objections against the doctrine by elaborating personal interpretations of Scripture; and (5) writing a letter to a disciple containing various propositions against the authority and the correct meaning of Scripture. No names or details were mentioned, in accordance with standard inquisitorial practice.

The account continued by relating that after those charges were made against Galileo, the Inquisition consulted its experts requesting an opinion on the Copernican doctrine.[4] For this purpose, two parts were distinguished in this doctrine: the thesis of heliocentrism or heliostaticism, and the geokinetic thesis. Both theses were judged false and absurd from a natural philosophical point of view. Theologically speaking, heliocentrism was declared formally heretical, on the grounds that it was explicitly contrary to the literal meaning and the patristic interpretation of Scripture, whereas geokineticism was determined to be at least erroneous in faith.

The sentence continued with the story of Cardinal-inquisitor Robert Bellarmine's warning and the Inquisition's special injunction to Galileo.[5] It recalled that in the earlier phase of the proceedings, the Inquisition had decided to treat Galileo with benign consideration. Consequently, at the Inquisition meeting of 25 February 1616 chaired by the pope, it was decided that Bellarmine would privately and informally warn Galileo to abandon Copernicanism; that if he refused, the Inquisition's commissary would give Galileo a formal injunction not to hold, teach, defend, or discuss it in any way whatever; and that if he did not acquiesce at the injunction, he would be arrested and prosecuted. The next day, we are told, Bellarmine gave Galileo the friendly warning; the commissary gave him the injunction; and Galileo agreed.

These statements raise many questions. Here suffice it to stress what the sentence said: that Bellarmine's warning was supposed to be something distinct from the special injunction; that the injunction was supposed to be contingent on Galileo's rejection of Bellarmine's warning, just as arrest and prosecution were to be contingent on Galileo's rejection of the injunction. The sentence did not explicitly say that Galileo rejected Bellarmine's warning, and so it is unclear whether the procedure on February 26 followed the Inquisition's orders of the day before.

Next,[6] the sentence mentioned the anti-Copernican decree issued by the Congregation of the Index. We are told that this decree declared the earth's motion false and contrary to Scripture, which is indeed an accurate statement. But we are also told that the decree prohibited books discussing the earth's motion, which is not exactly right because the decree prohibited only the assertion or defense and not the hypothetical discussion of the earth's motion.

The next development mentioned[7] occurred 16 years after the anti-Copernican decree and 1 year before the sentence itself, that is Galileo's publication in 1632 of

---

[4] Favaro 19: 403, lines 28–38.
[5] Favaro 19: 403–404, lines 39–51.
[6] Favaro 19: 404, lines 52–56.
[7] Favaro 19: 404, lines 57–68.

the *Dialogue*. The Inquisition was soon informed that this book was causing the dissemination and establishment of the geokinetic doctrine. An examination of the book was ordered, revealing that it defended the earth's motion, and so was an explicit violation of the special injunction.

This finding led the Inquisition to summon Galileo, examine him under oath, and obtain the following confession.[8] Galileo admitted that he wrote the book during the previous 10 years, after the special injunction; that he requested the imprimatur without disclosing the special injunction; that the book defended the earth's motion, in the sense that it gave the reader the impression that the arguments in favor of this false doctrine were stronger than those favoring the geostatic view; and that this transgression was due to literary vanity (the natural tendency to show off one's cleverness), rather than being intentional.

Then the document summarized Galileo's defense,[9] which hinged on Bellarmine's certificate and Galileo's denial of a malicious intention. Galileo presented a brief memorandum written by Cardinal Bellarmine, which Galileo had obtained in 1616 soon after those earlier proceedings had been concluded, in order to clarify the situation; it stated that Galileo had not been formally condemned or put on trial, but only notified of the Church's decision that the earth's motion was contrary to Scripture and so could be neither held nor defended. Galileo pointed out that Bellarmine's certificate did not contain the prohibition "to teach in any way whatever," and for this reason he eventually forgot about this part of the special injunction and felt no need to mention the injunction at all when requesting the imprimatur. All this was meant not to deny or excuse the error (of defending the earth's motion), but to show that it was due to literary vanity rather than to malicious intention.

The sentence was however quick to rebut this defense.[10] In regard to Bellarmine's certificate, it supposedly aggravated Galileo's situation because it stated that the earth's motion was contrary to Scripture, and yet Galileo had dared to defend it. Furthermore, the imprimatur obtained for the book was invalid because Galileo did not disclose the special injunction.

Regarding the question of the intention, the Inquisition was unconvinced by Galileo's denial of malice, and to resolve its doubts it conducted a "rigorous examination" of the accused.[11] Galileo apparently passed this test.

This term – rigorous examination – was the standard inquisitorial jargon for torture. Hence, this passage of the sentence generated one of the most hotly debated questions of the Galileo affair, namely whether Galileo was indeed tortured, why he was tortured (if indeed he was), and its propriety.[12] Here let me clarify that the torture could be merely threatened rather than actual. Thus, when the sentence asserted nonchalantly that Galileo was subjected to a rigorous examination, we can

---

[8] Favaro 19: 404, lines 69–83.
[9] Favaro 19: 404–405, lines 84–97.
[10] Favaro 19: 405, lines 97–101.
[11] Favaro 19: 405, lines 102–105.
[12] For more details on the torture question, see Finocchiaro (2005b, 222–258; 2009a).

take this to mean that at least he underwent an interrogation under the verbal threat of torture, although it does not necessarily imply that he was subjected to any step or degree of actual torture.

After this summary of the proceedings since 1615, the sentence assured us that the inquisitors had carefully considered the merits of the case and reached the following verdict.

The verdict[13] was that Galileo had been found guilty of "vehement suspicion of heresy." Two main errors, and not just one, were mentioned. The first involved holding a doctrine that was false and contrary to Scripture, namely the heliocentric and geokinetic theses. The second imputed error was the principle that it is permissible to defend a doctrine contrary to Scripture.

The notion of "vehement suspicion of heresy" embodies the complexity of the theological concept of heresy and of the Inquisition's anti-heretical practices.[14] Galileo was being convicted of a religious transgression that, while falling short of the most serious possible crime, was nonetheless regarded as a religious crime. The most serious such crime would have been "formal heresy"; below formal heresy there was the crime of "suspected heresy," which was in turn subdivided into three kinds: strong, vehement, and slight suspicion of heresy. There were thus four subtypes of heresy. And besides heresies, there were lesser religious crimes: beliefs or behavior could be erroneous, scandalous, temerarious, and dangerous. So Galileo was being found guilty of a relatively serious type of religious crime, clearly *not* the most serious one (formal heresy), nor even the second most-serious offense (strong suspicion of heresy), but equally obviously above the non-heretical transgressions and the slight suspicion of heresy.

The verdict was followed by a list of penalties.[15] The first was that Galileo was to immediately recite an "abjuration" of the "above mentioned errors and heresies." Second, the *Dialogue* was to be banned by a public edict. Third, Galileo would be kept under imprisonment indefinitely. Fourth, he would have to recite the seven penitential psalms once a week for 3 years. Finally, the Inquisition declared that it reserved the right to reduce or abrogate any or all of these penalties.

Although the abjuration was a penalty in the sense that it was a great humiliation for anyone to have to recant one's views, it was also a procedural step for the culprit to gain absolution of the sin of heresy.[16] The book's prohibition took effect immediately, although a year passed before it was included in a formal decree of the Index that included other books as well. The imprisonment stipulated here never did involve detention in a real prison, although it did last for the rest of Galileo's life and took the form of house arrest.[17] The last penalty in the list was a "spiritual" penance for the good of his soul.

---

[13] Favaro 19: 405, lines 117–126.

[14] Cf. Masini (1621), Limborch (1692; 1731), Garzend (1912), Genovesi (1966), Mereu (1979), Beretta (1998), Speller (2008).

[15] Favaro 19: 405–406, lines 126–137.

[16] Cf. Masini (1621), Mereu (1979), Beretta (1998), Speller (2008).

[17] For more details and references on the imprisonment question, see Finocchiaro (2009a).

The sentence ended with the signatures of seven (out of ten) cardinal-inquisitors.[18] That is, three signatures were missing. The explanation of this fact has become one of the many controversial questions in the Galileo affair.

2. As regards the abjuration, its text was relatively standardized and provided to him by the Inquisition's officials. The document begins with a multi-faceted confession. Galileo admitted having been notified that the heliocentric geokinetic doctrine was contrary to Scripture and having subsequently published a book that discussed and defended this doctrine.[19] These two admissions corresponded to what the sentence had reported.

However, Galileo also admitted having been served with the full special injunction, namely with the judicial prescription to completely abandon the doctrine and not to hold, defend, or teach it in any way whatever, orally or in writing;[20] and this admission went beyond the sentence and represented an additional confession. For the sentence reported primarily that Galileo confessed to supporting the geokinetic doctrine, but defended himself by claiming that this transgression was unintentional and by suggesting that Bellarmine's certificate cast doubt on the special injunction "not to discuss in any way whatever." In other words, the text of the abjuration made Galileo confess to violating the special injunction (besides unintentionally violating Bellarmine's warning).

Galileo then acknowledged that because of these transgressions, he had been judged to be vehemently suspected of heresy, the suspected heresies being heliocentrism, heliostaticism, geokineticism, and nongeocentrism.[21] This was followed by the admission that this verdict was right, when he spoke of "this vehement suspicion, rightly conceived against me."[22] Here we have a second new confession, which of course could not have been made during the (pre-sentencing phase of the) trial; he was also confessing that he was guilty as judged.

After these confessions, we come to the abjuration proper, where the culprit "with a sincere heart and unfeigned faith" abjured and cursed the above-mentioned errors and heresies.[23] This part of the document has led to the question whether such Galilean abjuring amounted to perjuring – another classic controversy of the Galileo affair.

The culprit then made a series of solemn promises about future thought and behavior: that he would never again hold any such beliefs; that he would denounce to

---

[18] Favaro 19: 406, lines 138–148. For some discussions of this issue, see Cantor (1864), Pieralisi (1875, 218–224), Santillana (1955, 310–311), Langford (1971, 153), Redondi (1983, 328), Beretta (2001a, 568 n. 98).
[19] Favaro 19: 406, lines 160–163.
[20] Favaro 19: 406, lines 155–160.
[21] Favaro 19: 406, lines 163–165.
[22] Favaro 19: 407, line 167.
[23] Favaro 19: 407, lines 167–168.

Church officials any heretics or suspected heretics; that he would comply with the penalties imposed on him by the judges; and that he would submit to further penalties if he failed to comply with the current ones.[24] An important aspect of these affirmations is the explicit distinction between heresy and suspected heresy; in the complex taxonomy of theology and canon law, suspected heresy may have been a lesser crime than formal heresy, but was a crime nonetheless. The sentence had declared Galileo to be a suspected heretic; the abjuration here repeated this characterization.

Whether this abjuration was extorted from Galileo during the "rigorous examination," perhaps in exchange for doing without additional "rigors," as some scholars have claimed,[25] is a controversial question that we can leave here simply as an issue to be added to our accumulating stock. However, it ought to be clear by now and it certainly is striking that in this abjuration Galileo was not only being made to retract earlier beliefs and behavior, but was also being made to plead guilty to the verdict already announced by the judges and to confess to a transgression not confessed earlier.

8.2 Diffusion of the News (1633–1651)

News of Galileo's condemnation spread quickly by means of official and unofficial channels; public and private communications; word of mouth and print media; and posters, newspapers, and books.

1. In the summer of 1633, all papal nuncios in Europe and all local inquisitors in Italy received copies of the sentence against Galileo and of his abjuration, together with orders to publicize them. Such publicity was unprecedented in the annals of the Inquisition and was never repeated in subsequent inquisitorial practice. Such a promulgation of Galileo's condemnation had been decided at the Inquisition meeting of 16 June 1633, chaired by pope Urban VIII; this was the same meeting when Galileo's trial was discussed and the pope reached a decision on its conclusion, the verdict, and the penalty.[26] The orders were transmitted by means of a memorandum signed on July 2 by Cardinal Antonio Barberini,[27] the pope's brother and one of the inquisitor-judges who had signed the sentence.

2. The orders were implemented in various ways. In university circles, professors of mathematics and philosophy were called to meetings at which a Church official read the text of the sentence and abjuration. The most memorable of these meetings was the one held in Florence on July 12, as we know from an August 27 letter by Mario Guiducci to Galileo reporting on the matter.[28]

---

[24] Favaro 19: 407, lines 170–179.
[25] For example, Genovesi (1966, 268).
[26] Favaro (19: 282–283, 360–361), Pagano (1984, 154, 229), Finocchiaro (2005b, 247).
[27] Favaro 15: 169, Finocchiaro (2005b, 27).
[28] Favaro 15: 240–242, Finocchiaro (2005b, 28–29).

3. Another common response of Church officials was to print a flyer summarizing the sentence and abjuration and to post and circulate it within their jurisdictions. The most famous of these posters was the one compiled in Latin by the nuncio to Cologne, Petrus Carafa, and printed in Liège on 20 September.[29] This was the flyer that was seen by Descartes and that led him to a fateful reorientation of his research.

4. Besides the text of the Inquisition's condemnation, another main contemporary source of information about the trial was an account written by Giovanfrancesco Buonamici, a Tuscan government official and distant relative of Galileo.[30] During the trial, Buonamici happened to be living in Rome, and so it is likely that he learned about it first-hand from Galileo himself, as well as from the many connections Buonamici had in Rome, including the Vatican secretary of state (Cardinal Francesco Barberini). After writing this account, Buonamici sent a copy to Galileo as well as to several persons in Germany, Spain, and Flanders (Favaro 14: 245–46). Buonamici's account was not merely factual, but interpretive and evaluative as well; and it stressed the psychological motivation and personal factors underlying actions, rather than the legal aspects of the case. Thus, his account complements the Inquisition's sentence both in regard to content and in regard to circulation.

5. In addition to the Church's official promulgation and Buonamici's informal transmission, there were others ways by which the news spread. Abridgments, parts, or summaries of the text of the condemnation were published in several venues: in French in Théophraste Renaudot's *Gazette* (Paris, December 1633); in Latin in Libert Froidmont's *Vesta* (Antwerp, 1634); again in French in the *Mercure françois* (Paris, 1636); and again in Latin in Jean Baptiste Morin's *Tycho Brahaeus in philolaum pro telluris quiete* (Paris, 1642). More importantly, the full text soon found its way into print: translated into French in Marin Mersenne's *Les Questions theologiques, physiques, morales, et mathematiques*[31] (Paris, 1634); in the original Italian version in Giorgio Polacco's *Anticopernicus catholicus* (Venice, 1644); and translated into Latin in Giovanni Battista Riccioli's *Almagestum novum* (Bologna, 1651).

Riccioli's publication is especially noteworthy because in his monumental work of two ponderous volumes in folio, he found a place (Riccioli 1651, 2: 496–500) to include an important collection of Latin documents on Galileo's condemnation: the anti-Copernican decree of the Index of 1616; the 1620 warning of the Index detailing the corrections to Copernicus's *Revolutions*; the Inquisition's sentence against Galileo of June 1633; Galileo's abjuration; and Cardinal Barberini's letter (of July 1633) to the inquisitor of Venice announcing Galileo's condemnation and abjuration. This small collection of documents was for more than two centuries the most complete and authoritative source of information about the trial.

---

[29] Monchamp (1893, 14–17), Favaro 19: 412–413, Finocchiaro (2005b, 30–31).
[30] Favaro 19: 407–411, Finocchiaro (2005b, 33–36). The history of this particular document involves both flawed or inauthentic, as well as correct, versions. Cf. Favaro (1902), Finocchiaro (2005b, 373 n. 43).
[31] Mersenne (1634, 214–218); cf. Pessel (1985, 385–393, 425).

6. The actions and documents just described may be said to have provided the material basis for what was to become a key cause célèbre in Western culture. At a more conceptual level, the analysis of those documents reveals two important and surprising findings. First, despite what the sentence asserts, almost everyone who read it failed to make a distinction between the prescription not to discuss the earth's motion in any way whatever and the prescription not to hold or defend it; they ignored the former (which may be labeled the special injunction) and spoke as if Galileo had been issued only the latter (which may be labeled Bellarmine's warning); and so presumably Galileo was condemned only for defending Copernicanism. Such an interpretation was advanced not only by Galileo's friends (Guiducci and Buonamici), but also by an authoritative churchman (Carafa) and by someone who was perhaps relatively neutral (Renaudot). In other words, there seemed to be an unwillingness to admit the existence of the special injunction and its role in the condemnation of Galileo, even though the text of the sentence and abjuration is clear in claiming that he was issued the special injunction and that he was condemned in part for violating it.

Secondly, there was a common tendency to incorrectly interpret the sentence as containing an official declaration that the Copernican doctrine was heretical, and not merely contrary to Scripture. This tendency was less strong and less universal that the other one to overlook the special injunction, since for example we find Buonamici explicitly distinguishing between heretical and contrary to Scripture. But the heresy interpretation was exemplified by Carafa (among others), and if this nuncio could not make such a fine but important distinction, it is easy to understand how the myth of Copernicanism as a heresy developed in the history of the Galileo affair.

8.3 Emblematic Reactions (1633–1642)

If we distinguish the phenomenon of the initial diffusion of news about Galileo's condemnation from the phenomenon of its initial reception, and if we now focus on the latter, then it is useful to begin with a period (1633–1642) that corresponds to the rest of Galileo's life after the trial and with four individual reactions of emblematic significance.

1. One of the most immediate and sensational reactions was that of Descartes. He had been on the verge of publishing a cosmological treatise entitled *Le Monde*, but decided to abort it when he heard of the condemnation in the autumn of 1633. His reason was that the geokinetic thesis was so central to his world view and so interwoven with its other parts that he could not detach that problematic thesis from the rest. And indeed this treatise was never published during his lifetime, but only posthumously in 1665.

However, he soon devised a way of detaching the earth's motion from the rest of his world view, and he published the latter without the former in his *Discourse on Method* of 1637. Later, in the *Principles of Philosophy* of 1644 he even found ways of discussing the topic. One way was to regard the earth's motion as a hypothesis rather than an assertion, which was a subterfuge that was increasingly being adopted by Catholic scholars in order to ensure that the ecclesiastic prohibitions did not bring research to a complete halt. The other way was for Descartes to adopt a version of the relativity of motion according to which the earth both moved and did not move (from different points of view), which was a way to obfuscate rather than solve the problem.

Finally, at a deeper if more controversial level, it may be argued that the most substantial Cartesian response was a creative reorientation of his thought that forced him to attempt to provide a metaphysical justification of the new heliocentric, geokinetic world system. Until the condemnation of Galileo, he had focused on mathematics and natural philosophy and had arrived at almost all of his major scientific discoveries; afterwards, he focused on first philosophy in order to justify them by embedding them in a metaphysics that provided a viable alternative to the traditional scholastic approach. This intriguing interpretation is due to Stephen Gaukroger, who has defended it with meticulous documentation and cogent arguments. But this account may also be viewed as a refinement and updating of a similar account advanced earlier by Guido Morpurgo-Tagliabue.[32] Its implication in this context would be that modern philosophy (if and to the extent that it has been a series of footnotes to Descartes) originates in the condemnation of Galileo.

2. From the point of general human interest, one of the most significant reactions to Galileo's condemnation was that of his elder daughter Maria Celeste, a nun in the monastery of San Matteo in Arcetri, within walking distance of Galileo's post-trial residence under house arrest. Anyone who reads her letters to her father cannot help being impressed by her warmth, love, intelligence, sensitivity, and unassuming eloquence. The most touching of these is perhaps the one she wrote (on 3 October 1633[33]) while her father was in exile in Siena: she had finally been able to read the text of the Inquisition's sentence and decided to assume onto herself the burden of her father's salutary penance to recite the seven penitential psalms once a week for 3 years. Of course, it was not legally or theologically possible to substitute one person's penance with that of another, but that was irrelevant. The point was to share her father's burden by taking on an equal burden herself, so that the old man would have a companion in his misery. Human nature being as it is, the effect of such company in such misery is proverbial. Unfortunately, Maria Celeste died on 2 April 1634 after a brief illness, and the old man felt devastated.

3. Let us now turn from a person whose only claim to fame is her sheer humanity to one of supreme worldly renown, power, and savoir faire, Nicholas Claude Fabri de Peiresc. This Provençal nobleman was a lawyer, politician, diplomat, and amateur natural philosopher. Like Marin Mersenne, he was also a kind of clearing house for correspondence from all parts of Europe, as well as North Africa, the

---

[32]Gaukroger (1995, 11–12, 292), Morpurgo-Tagliabue (1981, 104–105, 114–119).
[33]Favaro 15: 292–293, Sobel (1999, 312–14; 2001).

8.3 Emblematic Reactions (1633–1642)

Middle East, and Asia; his correspondence takes up 10 volumes of about 1,000 pages each. He became acquainted with Galileo in Padua in 1509–1602 while he studied law at the university there. Peiresc was in a good position to try to help Galileo because he was on very good terms with the Barberini family and had hosted at his own house Cardinal Francesco Barberini, the pope's nephew and the Vatican Secretary of State, when the cardinal was returning from Paris on a special diplomatic mission in 1625.

Acting on his own initiative and from his own conscience, Peiresc tried to obtain a pardon for Galileo. For that purpose, twice he wrote passionate letters to Cardinal Barberini, in December 1634 and again in January 1635.[34] The most memorable parts of these letters are the comparison of Galileo with Socrates and the interpretation of the *Dialogue* as a "philosophical play"[35] that discusses in an impartial and noncommittal manner both sides of a controversy. Peiresc's plea fell on deaf ears, but not for lack of argument or eloquence.

4. Regarding Galileo's own reaction, there is little need to dwell on the pain, suffering, and occasional depression he experienced as a result of the condemnation. He reached the lowest point when his elder daughter died in April 1634, a few days after receiving another blow to the effect that he should desist from making pleas for his freedom, otherwise he would be placed in a real prison. In general, however, there were enough positive developments occurring during the post-trial period that they far outweighed the setbacks and enabled him to reach a state of imperturbable serenity. This is how he described his state of mind in his reply to Peiresc in February 1635:

> I do not hope for any improvement, and this is so because I did not commit any crime. I could hope to obtain clemency and pardon if I had erred, because mistakes are the subject matter over which princes can exercise the power of reprieve and pardon; but in regard to someone who has been unjustly condemned, it suits them to maintain rigor, as a cover for the legality of the proceedings. And believe me, Your Most Illustrious Lordship, also for your own consolation, that this rigor afflicts me less than what others may think because there are two comforts that constantly assist me.[36]

And indeed, besides his innocence, Galileo had much to feel good about. Let me just highlight some of the most striking items.

During his 5-month exile in Siena under house arrest at the archbishop's residence, Galileo had been treated so well by everybody that he started writing his *Two New Sciences*. Moreover, in February 1634 an anonymous complaint was filed with the Inquisition, against Galileo and the archbishop of Siena, regarding such treatment.[37] Indeed such treatment had become common knowledge even abroad; for

---

[34]Favaro (16: 169–171, 202), Finocchiaro (2005b, 53–55).

[35]*Scherzo problematico*, in Favaro 16: 170, line 43. Cf. Campanella to Galileo, 5 August 1632, in Favaro 14: 366; Westman (1984, 334); Finocchiaro (2005b, 55, 376 n. 53).

[36]In Finocchiaro (2005b, 59); cf. Favaro 16: 215.

[37]Favaro 19: 393; cf. Pieralisi (1875, 254–261).

example, in autumn 1633, on his way from Rome to France, Gerard Marc Anthony Saint-Amant visited Galileo in Siena and later reported that he was lodged in rooms elegantly decorated with damask and silk tapestries (Favaro 15: 344, 363).

Second, no sooner had Galileo returned home at Arcetri on 17 December 1633 than the grand duke of Tuscany personally made a trip there (on Christmas day) to comfort the old man. And despite the house arrest and the Inquisition surveillance, the duke was only the first of an unprecedented stream of visitors who felt obliged to pay their respects by visiting him at Arcetri.[38] Indeed, the abbé Girolamo Ghilini was perhaps not exaggerating when he stated in a biographical dictionary published in 1647, that after the 1633 condemnation no important personage traveling through Tuscany failed to make an effort to pay a visit to Galileo, who was thus considered to be something of a major tourist attraction (cf. Madden 1863, 35).

Finally, Galileo's success as an author accelerated at an unprecedented pace. In 1634, Mersenne published in Paris *Les mecaniques de Galilée*, a French translation of an unpublished manuscript Galileo had written in Italian in his university days. In 1635, a Latin translation of Galileo's *Dialogue* was published in Strasbourg, entitled *Systema cosmicum*, edited by Elia Diodati, and translated by Matthias Bernegger. In 1636 also in Strasbourg, Galileo's *Letter to the Grand Duchess Christina*, written in 1615, was published for the first time in an edition that contained both the Italian text and a Latin translation by Diodati. In 1638 Galileo was gratified to see his *Two New Sciences* published by the Elzeviers in Leiden. In 1639, Mersenne edited and translated into French Galileo's just published *Two New Sciences* and published it in Paris under the title *Les nouvelles pensées de Galilei, mathematicien et ingenieur du Duc de Florence*. In 1640, there was a reprint in Padua of Galileo's booklet on the calculating device known as the proportional compass, entitled *Operazioni del compasso geometrico e militare*. In 1641, the Latin *Systema cosmicum* was reprinted in Lyons.

8.4 Polarizations (1633–1661)

The response by Descartes, Peiresc, Galileo's daughter, and Galileo himself in the period 1633–1642 may be regarded as a first wave of reactions to Galileo's condemnation. Then we may speak of a second wave to refer to another series of reactions that span the period 1633–1661. They involve three groups of people and institutions: Catholic states and political authorities; Protestants; and conservative Catholic intellectuals, especially the Jesuits. Ideologically, they could be subsumed, respectively, under the headings of secularism, liberalism, and fundamentalism. This second wave is important because it was instrumental in polarizing the controversy and had a formative influence on it.

---

[38]For a complete list, see Finocchiaro (2005b, 63–64).

8.4 Polarizations (1633–1661)

1. In France, in 1635 chief minister Richelieu tried unsuccessfully to have the Sorbonne endorse the condemnation of the doctrine of the earth's motion. But France had the policy that decisions of the Roman congregations were not valid unless endorsed by the Sorbonne and the Parliament of Paris. Thus, it seems that the condemnation of Copernicanism, and the consequent condemnation of Galileo stemming from it, had no formal legal standing in France.

Spain had a separate Inquisition, whose responsibilities included the endorsement of decisions of the Roman Inquisition and the publication of a separate *Index of Prohibited Books*. But Galileo's *Dialogue* was never listed in the Spanish *Index*, because the 1634 Roman decree banning it also included a pro-Spanish book which Spain did not want to ban. Hence that decree never received the proper endorsement.

In the Holy Roman Empire, the most relevant developments stem from the pro-Galilean efforts of Giovanni Pieroni, one of Galileo's former students who had an important position as a military engineer in the service of the emperor. This devoted disciple spared no time, money, or effort to arrange for the publication of Galileo's *Two New Sciences* at a time when he was exploring various possibilities for printing the manuscript, before the Elzeviers of Holland made a firm commitment. In the process, Pieroni managed to borrow a number of printing presses and to obtain the signed imprimaturs of a number of Church officials, including a Jesuit.

In Poland, king Ladislaus IV, far from endorsing the condemnation of Galileo, tried to have him pardoned. Although there is no explicit and direct evidence for this, there is circumstantial and indirect but concrete evidence. For example, in the spring of 1636 the king wrote personally to Galileo a letter that asked for some good telescopic lenses, but it seems likely that the gesture was meant primarily to signal another message, that Ladislaus continued to hold him in high esteem and was ready to help (Favaro 16: 420–21). And Galileo lost no time in replying, sending three (and not just one) pairs of telescopic lenses, relating a little about the trial and his current state. He did not explicitly ask the king to intercede, but the message was subliminally clear (Favaro 16: 458–59).

Finally, in the Venetian Republic, the revealing reaction involved its official theologian, Fulgenzio Micanzio, successor and disciple of Paolo Sarpi. After the trial Micanzio wrote Galileo almost every week, helping him in all sorts of ways. He encouraged Galileo to defy the Inquisition by publishing the *Two New Sciences* in defiance of an alleged ban on any Galilean publications. Micanzio also encouraged Galileo to accept the Dutch government's gift for his offer of the method of determining longitude at sea by observing Jupiter's satellites.

2. Within a few years after the trial, an international group of Protestants organized a liberal defense and justification of Galileo. This was accomplished by means of a Latin translation of the *Dialogue* in 1635 and a bilingual edition of his *Letter to Christina* in 1636. The moving force behind this project, its chief editor, and the translator of the *Letter to Christina* was Elia Diodati (1576–1661), a Genevan-born lawyer for the Parliament of Paris and a long-standing dear friend and strong supporter of Galileo. The translator of the *Dialogue* and the co-editor of

both works was Matthias Bernegger, an Austrian-born professor and later rector at the University of Strasbourg, who in 1612 had published a Latin translation with extensive annotations of Galileo's booklet on the proportional compass. The publisher was the Dutch firm of the Elzevier family, which had a branch office in Strasbourg. And the printing was done by the press of David Hautt in Strasbourg, which was at that time a free city federated with the Holy Roman Empire.

The 1635 volume was entitled *Systema cosmicum* and included two appendices: an excerpt from Kepler's introduction to his *Astronomia Nova* of 1609 and Paolo Foscarini's *Letter on the Copernican Opinion* of 1615, both of which argued that the earth's motion was compatible with Scripture. Because of this content, Foscarini's booklet had been explicitly condemned and banned by the anti-Copernican Index decree of 1616, as we have seen. The title page of the *Systema cosmicum* exhibited two quotations: one from Alcinous in Greek meaning, "One must be mentally free if one wants to become a philosopher"[39]; the other from Seneca in Latin meaning, "It is especially among philosophers that one must have equal liberty."[40] Finally, there was a preface by Bernegger making it sound as if the book was being published without Galileo's permission or knowledge; but this story was told in order not to cause further problems to Galileo, and there is no doubt that he participated indirectly and discreetly in the project.

The 1636 volume contained the Italian text and a Latin translation of Galileo's *Letter to Christina* of 1615. It bore a long Latin title that begins with the word *Novantiqua* and summarizes its central thesis; that is, *New and Old Doctrine of the Most Holy Fathers and Esteemed Theologians on Preventing the Reckless Use of the Testimony of the Sacred Scripture in Purely Natural Conclusions That Can Be Established by Sense Experience and Necessary Demonstrations*. The book also had an appendix with a four-page excerpt from Diego de Zúñiga's *Commentaries on Job*, suggesting a geokinetic interpretation of the biblical passage Job 9:6; this was the part of Zúñiga's book that had caused its suspension in the anti-Copernican decree of 1616. And there was a preface that, by contrast with the one in *Systema cosmicum*, was openly critical of Galileo's condemnation and explicitly praised his moral character: it called him the "new father of astronomy"; it blamed Galileo's troubles on jealous and hateful rivals and their maneuverings; and it commented on the *Letter to Christina*, claiming that it advanced sound advice on the topic and proved the depth of his Catholic piety and religiosity.

The Strasbourg liberal defense of Galileo was later echoed in England. In 1644 in the *Areopagitica*, John Milton referred explicitly to Galileo and criticized the Inquisition as an obstacle to intellectual progress. And in 1661, in the first volume of his *Mathematical Collections and Translations*, Thomas Salusbury translated into English the *Dialogue*, the *Letter to Christina*, Foscarini's letter, Kepler's introduction, and Zúñiga's passage; and in a foreword he explicitly praised Galileo and satirized the Church.[41]

---

[39] Garcia (2000, 320 n. 44); cf. Galilei (1635, title page).
[40] Garcia (2000, 320 n. 44); cf. Galilei (1635, title page).
[41] For more on Salusbury, see Finocchiaro (2005b, 78–79), Wilding (2008), and Section 12.7 below.

3. At first the condemnation of Galileo led to an expansion and intensification of the Copernican controversy, as shown by the growth of books on the topic. In fact, at least sixty books were written in the 1633–1651 period,[42] although a few remained unpublished then.[43] Most works focused on the question of the earth's motion, but some stressed the methodological and theological question whether Scripture ought to be regarded as an authority in astronomy and natural philosophy.[44]

One of the most important of the subcontroversies of this period regarded the question of the correctness, interpretation, status, and implications of Galileo's science of motion. For the high incidence of Jesuits among Galileo's critics in this subcontroversy, one might get the impression that it was an instance of Jesuit conspiracy and persecution. On the other hand, it was the pro-Galilean Pierre Gassendi who started the discussion in 1642, and so perhaps the Jesuits merely seized the opportunity. If this was the first major posthumous retrial of Galileo, there were several other minor ones that immediately followed his death.

When it emerged that there was considerable interest in Florence to build a sumptuous mausoleum for him, his enemies raised the question whether this was legitimate, in canon law, in light of his condemnation for vehement suspicion of heresy. A formal legal opinion was written, concluding that it was legal.[45] But on 25 January 1642 the pope let the grand duke of Tuscany know that it would not be pious or proper to build a mausoleum for a condemned heretic like Galileo (Favaro 18: 378–379, 379–380). The grand duke acquiesced (Favaro 18: 380, 381–382), and it took more than a century before the mausoleum project was accomplished.

Another minor posthumous skirmish involved Galileo's last will and testament. His enemies also raised the question whether someone convicted of suspected heresy had the right to have his will executed. Again, a formal legal opinion concluded in his favor.[46]

Next, in 1643 the "legitimacy" of Galileo's birth was impugned. In a biographical dictionary entitled *Pinacotheca imaginum illustrium virorum*, Janus Nicius Erythraeus (also known as Giovanni Vittorio de' Rossi) stated that Galileo's parents were not formally married at the time of his birth or conception![47] Viviani refuted

---

[42]Most of the published books are listed in the bibliography in Carli and Favaro (1896); but the following are not: Lansbergen (1633), Ward (1635), Descartes (1637), Morin (1640), White (1642), Argoli (1644), Descartes (1644), Wendelen (1644), Mousnier (1646), Hevelius (1647), Renieri (1647), Wendelen (1647), Mousnier and Fabri (1648), Le Tenneur (1649), Morin (1650), Varenius (1650).

[43]Inchofer (1635), Cavalieri (1642), Hobbes (1642), Le Tenneur (1646).

[44]Especially Inchofer (1633; 1635), Accarisius (1637), Campanella (1637), Parasin (1648). Cf. Pesce's works, especially Pesce (1987; 1991a).

[45]Favaro 19: 559–562. Cf. Nelli (1793, 2: 852), Venturi (1818–1821, 2: 324), Favaro (1891, 377–378).

[46]Favaro 18: 383 (#4204), 19: 535–537. Cf. Drinkwater Bethune (1832, 299), Brewster (1841, 113), Venturi (1818–1821, 2: 324), Favaro (1891, 378–380, 384–388).

[47]Erythraeus (1643, 279). Cf. Brucker (1766–1767, tome 4, part 2 [= vol. 5], p. 634, note e), Nelli (1793, 1: 25), Carli and Favaro (1896, 41, 91, 112), Motta (1993, 102).

the claim in his biography of Galileo, written in 1654 but not published until 1717.[48] So this myth circulated for more than a century.

Moving on to matters of greater weight, we now come to the two ponderous folio volumes of Jesuit Giovanni Battista Riccioli's *Almagestum novum* (1651), which may be regarded as the climax of the original reception of Galileo's condemnation. There is no denying the book's ambitious and encyclopedic scope. The material was subdivided into ten "books": spherical astronomy, terrestrial elements, sun, moon, eclipses, fixed stars, planets, comets and new stars, systems of the world, and general problems. Book 9, on the systems of the world, was by far the longest and covered 342 pages.[49] Section 4 of this book treated of the system of the moving earth. It contained 40 chapters and covered 210 pages. It examined scores of astronomical, physical, observational, and philosophical arguments in favor as well as against the earth's motion. By one count,[50] no fewer than 40 pro-Copernican arguments were stated and criticized, and no fewer than 77 anti-Copernican arguments were presented and developed. As a result of such a critical examination, there was no doubt for Riccioli that the earth was motionless.[51]

Then Riccioli went on to discuss the question of the role of Scripture, and elaborated explicitly and systematically a very conservative version of biblical literalism.[52] According to Riccioli, the literal meaning of biblical statements is physically true and scientifically (philosophically) correct, and it takes precedence in cases of conflict with doctrines in natural philosophy. One interesting consideration he advanced is the following holistic argument: "If the liberty taken by the Copernicans to interpret scriptural texts and to elude ecclesiastic decrees is tolerated, then one would have to fear that it would not be limited to astronomy and natural philosophy and that it could extend to the most holy dogmas; thus it is important to maintain the rule of interpreting all sacred texts in their literal sense."[53] He also examined many particular biblical passages to show they accorded with the geostatic system.

The upshot of Riccioli's monumental effort was to provide the first explicit justification of Galileo's condemnation. The justification was twofold. On the astronomical issue, Galileo was (allegedly) wrong because the evidence did not favor the Copernican system, but rather a Tychonic system (in which the earth is motionless at the center). Galileo was also (allegedly) wrong on the key methodological and theological issue: Scripture is and must be regarded to be an authority in astronomy and natural philosophy, and not merely for questions of faith and morals. The Inquisition thus acted justly and prudently insofar as it was upholding this principle and attempting to prevent the disregard of Scripture and of ecclesiastic

---

[48]Viviani (1654; 1717).

[49]Riccioli (1651, 2: 193–535).

[50]Stimson (1917, 79–84); for the critique of the pro-Copernican arguments, see Riccioli (1651, 2: 311–407); for the anti-Copernican arguments, see Riccioli (1651, 2: 408–478).

[51]Riccioli (1651, 2: 478), which corresponds to bk. 9, section 4, chapter 35.

[52]Here I rely on Pesce (1987, 266–268), Delambre (1821, 672–681). Cf. Riccioli (1651, 2: 479–495), Dinis (1989, 239–255; 2002; 2003), Baldini (1996a).

[53]Riccioli (1651, 2: 290); here quoted and translated from Delambre (1821, 1: 672).

8.5 Compromises (1654–1704)

decrees from spreading beyond astronomical subjects. Riccioli's apology for the Inquisition was not only implicit in what he said and accomplished in section 4 of book 9, but corresponded to an explicit intention on his part.[54]

8.5 Compromises (1654–1704)

Let us now deal with what may be called a third wave of reactions to Galileo's trial, covering the second half of the seventeenth century (1654–1704, to be more precise) and most significantly represented by the figures of Vincenzio Viviani (1622–1703), Adrien Auzout (1622–1691), and Gottfried Leibniz (1646–1716). This third wave produced mostly attempts at compromise.

1. Although Vincenzio Viviani was a notable mathematician in his own right, he liked to think of himself first and foremost as "the last disciple of Galileo." As such, he spent the greatest portion of his time and effort preserving Galileo's papers and legacy (in which he was largely successful) and working on a projected annotated edition of Galileo's works and a comprehensive biography of his teacher (which never materialized). In 1654 he wrote a biographical sketch in the form of a letter addressed to prince Leopold de' Medici, but it was first published posthumously in 1717.[55] It is a relatively short essay of about thirty pages, and it has all the historical value and the scholarly limitations one might expect from a biography written by one's last disciple. Viviani's account focuses on Galileo's work in astronomy and physics and on his personality. The topic of the trial is not totally avoided, but is discussed only in one paragraph that tries to be both pro-Galilean and pro-clerical (see Section 12.7 below). He accomplishes this bipartisan aim by claiming that Galileo did commit an error in the *Dialogue* insofar as he showed a preference for a world system that was contrary to Scripture, but drawing the positive lesson that divine providence wanted to remind us that Galileo was not a god but a fallible human being. Although ingenious, the account seems to have a rhetorical and mythological quality.

Viviani's only other major pronouncement was characterized by a similar spirit of compromise. It was occasioned by Leibniz's visit to Florence in 1689–1690, during which he encouraged Viviani to do something concrete to get the Church to ease or repeal its anti-Copernican and anti-Galilean decrees. Viviani decided to write to Antonio Baldigiani, a Jesuit professor of mathematics at the Roman College, who had recently been appointed consultant to both the Inquisition and the Index and with whom he had corresponded earlier (1678) about related matters.

In his letter dated 22 August 1690,[56] Viviani requested something very modest, namely that the ban on the *Dialogue* be made contingent on its being appropriately revised or "corrected" in a way analogous to how Copernicus's *Revolutions* had

---

[54]Riccioli (1651, 2: 500); cf. Stimson (1917, 79–80).
[55]Salvini (1717, 397–431). Cf. Favaro 19: 599–632, Viviani (1992).
[56]Favaro (1887, 152–55), Finocchiaro (2005b, 90–92).

been corrected in 1620. Viviani spared no argument in his attempt to persuade Baldigiani. He appealed to Baldigiani's Jesuitical pride by reminding him of the Jesuit conspiracy theory and suggesting that his action might help to refute it. He appealed to his Catholic pride by claiming that ultramontanes and Protestants tended to be Copernicans and thus took the anti-Copernican decree as one more reason for disregarding Roman decrees on other subjects; this was a point Galileo had made in the Preface to the *Dialogue* to explain his motivation, and his having done so is important even though it failed to impress the authorities. Viviani mentioned a previous action taken by Cardinal Leopold de' Medici which had received an encouraging response. Viviani appealed to the patriotism of Baldigiani, who was also a Florentine. And there was considerable flattery in what Viviani said about Baldigiani's qualities and prospects. However, nothing came out of Viviani's effort. Thus, later and a few years before he died, Viviani decided to build a public monument to Galileo in the façade of his (Viviani's) house with an inscription full of praise.[57] He had obviously given up hope for the repeal of the condemnations.

2. During the period we are dealing with (second half of the seventeenth century), a conceptually more satisfactory and indeed brilliant compromise was worked out by Adrien Auzout,[58] although practically it was no more effective or consequential than Viviani's. Auzout was commenting on a judgment attributed to Jesuit Honoré Fabri in a 1661 book published under the name of Eustachio Divini but really authored by Fabri himself. This judgment really originated from Bellarmine's 1615 letter to Foscarini, although this was not known for more than two centuries, until Bellarmine's letter was published in 1875. The judgment is nowadays a very familiar one. It states that if there were a conclusive demonstration of the earth's motion then geostatic passages in Scripture would have to be interpreted figuratively, but as long as there is no such demonstration they must be interpreted literally.

Auzout advanced an ingenious argument designed to show that if we accept the first part of the Fabri–Bellarmine judgment, it follows that the condemnation of the earth's motion was provisional, and not absolute or permanent. His reasoning is that truth is eternal, and this proposition should be applied to biblical interpretations; hence if a figurative interpretation of a scriptural passage is legitimate after the demonstration of the earth's motion, it is also legitimate beforehand. Similarly, if a literal interpretation is mandatory before a demonstration is found, it is mandatory afterwards. Auzout seemed to be suggesting that the two parts of Fabri's biconditional were inconsistent with each other, and that the proper way to resolve the inconsistency was to reject biblical literalism (in astronomy and physics).

Auzout also had a second argument to show the provisionality of the condemnation. The argument stressed the evaluation of Copernicanism as philosophically false and absurd that was explicitly contained in the Inquisition's sentence of 1633 and more

---

[57] Viviani (1701, 122), Nelli (1793, 2: 854–67), Fahie (1903, 404), Russell (1989, 382), Galluzzi (1993b; 1998).

[58] Auzout (1665, 58–66), Finocchiaro (2005b, 94–97).

8.5 Compromises (1654–1704)

cryptically in the anti-Copernican decree of 1616. He insightfully suggested that it was probably this alleged philosophical untenability that led to the evaluation of Copernicanism as contrary to Scripture. Next Auzout easily showed that at his time (1665) the earth's motion was no longer absurd philosophically. It followed that the condemnation of the earth's motion should be withdrawn now that it was deprived of its natural-philosophical grounding.

A third argument involved the concept of demonstration. Auzout distinguished what he called a reasonable demonstration from mathematical and metaphysical proofs. He claimed that in the problem at hand the relevant notion was that of reasonable demonstration, in part because otherwise astronomers would be unable to assert propositions which no one doubts in the least, such as that the sun does not circle Jupiter. Then he suggested that the earth's motion was at that time susceptible of a reasonable demonstration. It followed from Fabri's own (first) conditional claim that geostatic scriptural passages should be interpreted figuratively.

Auzout's spirit of compromise and bipartisan attitude could be elaborated from the fact that he tried to build upon two authoritative ecclesiastic judgments. The first was the conditional prediction published by Fabri and stemming from Bellarmine; the second was the philosophical evaluation of Copernicanism mentioned in the trial documents as a key reason for the condemnations of 1616 and 1633. Then from these traditional ecclesiastic judgments he tried to derive reformist conclusions critical of the Church.

3. Although Leibniz is best known for his co-invention of the calculus, his metaphysics of monads, and his contributions to dynamics, he also played a significant role as a cultural politician and diplomat. As such, one of his dreams was the ecumenical project of re-unifying the Catholic and Protestant Churches. Another one was his goal to have the Catholic Church repeal its condemnation of Copernicanism. His failure in these cultural activities should not make us underestimate the amount of effort he devoted to them, which probably exceeded the amount he devoted to his more successful achievements. In all his efforts, Leibniz was temperamentally and methodologically moderate, bipartisan, diplomatic, and ecumenical-minded. In his attempt to have the anti-Copernican censures repealed, this spirit of compromise unfortunately produced results that border on incoherence.

In an early (1679–1686) essay with the revealing title "Apologia for the Catholic Faith Based on Right Reason,"[59] Leibniz argued in favor of repealing the anti-Copernican decree on the basis of the Fabri–Bellarmine principle. However, in endorsing this principle he added a conservative qualification that could easily lead others to conclude the opposite. Leibniz's more qualified version states: if there is a conflict between a physical proposition supported by very clear reasons and a scriptural statement interpreted literally, and if the scriptural statement can be given a nonliteral interpretation that is not too strained or unprecedented and reconciles it with the physical proposition, then the nonliteral interpretation is preferable.

---

[59] In Grua (1948, 1: 30–34), Mayaud (1997, 329–330), Finocchiaro (2005b, 100–101).

Leibniz elaborated his position in a 1688 letter[60] to landgrave Ernst von Essen-Rheinfels intended to provide this German convert to Catholicism with an argument he might present to Rome. This time Leibniz added another ambivalent qualification to his conclusion, when he said that allowing scholars to maintain the truth of the earth's motion should be conjoined with requiring them to declare that Scripture does not speak improperly on the topic.

In one of two essays Leibniz wrote in 1689 while he was himself in Rome,[61] he started by articulating a concept of truth that was partly rationalist insofar as it equated truth with intelligibility, but also partly pragmatist and contextualist insofar as it equated intelligibility with contextually appropriate usefulness. The concept implied that while Tycho (the geostatic thesis) was wrong in the context of "theoretical astronomy," so was Copernicus (the geokinetic thesis) in the context of spherical astronomy as well as in common speech, but that in its context Scripture spoke properly and truthfully. The following compromise was also implied. On the one hand, the Church did not have to say that the anti-Copernican censure had been provisional; nor was a retraction on the part of censors needed; thus ecclesiastic authority and dignity would be preserved. On the other hand, theoretical astronomers were justified to maintain the truth of Copernicanism because this claim meant merely that the Copernican hypothesis was the best alternative in terms of intelligibility, intellectual adequacy, explanatory power, and simplicity, and even the censors agreed to the latter claim.

In his second Roman essay,[62] Leibniz in part reiterated his previous arguments, but also said things that could be interpreted as endorsing the scientific authority of Scripture and the condemnation of Galileo, as well as justifying the Inquisition. The ecclesiastic silence and inaction that followed might be taken as an indication that even Church officials did not want to go as far as Leibniz seemed willing to go. It is at this point that one cannot help wondering whether he was being consistent and what his real aim was. Was he really defending Galileo, the Copernican cause, or even philosophic freedom? Perhaps he was pursuing ecumenism first and foremost.

Leibniz's last word, however, was more pro Galilean. It came in 1704 in the *New Essays on Human Understanding*.[63] When discussing various categories of errors, he defined one as involving wrong measures of probability and attributed it to the anti-Copernicans of Galileo's time. After reiterating the claim that the condemnation of Copernicanism had harmed progress, he went on to argue that for all their talk of hypotheses, the anti-Copernicans had the tendency to regard only their opponents' view as hypothetical, but their own as factually and categorically true; this attitude was an example of the error of treating a received hypothesis as nonhypothetical.

---

[60] In Rommel (1847, 2: 200–202), Finocchiaro (2005b, 101–103).

[61] "Cum geometricis demonstrationibus …," in Couturat (1903, 590–593), Robinet (1988, 111–114), Ariew and Garber (1989, 90–94).

[62] "Praeclarum Ciceronis dictum est …," in Robinet (1988, 107–110).

[63] Book 4, chapter 20 of these *Essays*, in Leibniz (1923ff, vol. 6, part 6, 509–521; 1997, 509–521).

8.6 Myth-making or Enlightenment? (1709–1777)

Despite all its enlightenment about other matters, the eighteenth century was almost a golden age for the invention and diffusion of myths about Galileo's trial. And we will soon see that this judgment applies not only to minor writers but also to such iconic figures as Voltaire (1694–1778), Jean D'Alembert (1717–1783), and the French *Encyclopedia*.

1. In 1709, in a work on the history of heresy Domenico Bernini asserted that Galileo was held in an Inquisition prison for 5 years.[64] This assertion is not true, but was widely repeated for a long time.[65] This myth was not a pure fabrication because the Inquisition's sentence did speak of formal imprisonment at the pleasure of the Holy Office, whereas information about the various locations and forms of such imprisonment was harder to come by. The issue was not resolved until at least 1774–1775, when it was widely debated in Tuscany and the relevant documents were researched and found.[66]

In 1737, the pro-Galilean mythology got a boost when Galileo's body was exhumed from the original modest grave in the church of Santa Croce's bell tower and moved to a sumptuous mausoleum on the north side of the church's main aisle, across from Michelangelo's tomb.[67] This was the climax of a project that had first been conceived in 1642 immediately after Galileo's death, and that finally had been approved by the Inquisition in 1734.

In 1755, in a book on the history of astronomy published in Paris, Pierre Estève stated that Galileo had his eyes gouged out as part of his punishment.[68] This is not true. But this claim also had some basis in fact, although a weaker basis than the prison myth. What may have happened for the eye-gouging myth is that one started with the fact that toward the end of his life Galileo was tried, condemned, and punished by the Inquisition; then one considered the fact that at the end of his life he was blind; and then the imagination invented a connection between these two facts.

In 1757, the legend of Galileo's uttering *e pur si muove* made its first appearance in print. This happened in an English-language book by Giuseppe Baretti published in London and entitled *The Italian Library*. Baretti (1719–1789) was a literary critic born in Turin who lived most of his life in London. This work was essentially an annotated bibliography of books in Italian. The legend is told in a section dealing

---

[64] Bernini (1709, 615). Cf. Carli and Favaro (1896, 99), Cooper (1838, 72), Müller (1911, 455).

[65] As late as Haeckel (1878–1879, 33). Cf. Reusch (1879, 266), Müller (1911, 455 n. 2).

[66] Nelli (1793, 2: 537–538); cf. Fabroni (1773–1775, vol. 2 [1775]). For more details and references, see Finocchiaro (2009a).

[67] Cf. Gebler (1879a, 311), Fahie (1903, 405), Galluzzi (1993b, 174), Galluzzi (1998).

[68] Estève (1755, 1: 289f); already Andres (1776, 24) dismissed it as false; cf. also Redondi (1994, 97).

with natural philosophers, in which there was an entry on the *Dialogo di Galileo Galilei*.[69] Baretti had the sense to attribute these words to Galileo when he was supposedly freed from prison, whereas later versions of the myth speak of the utterance having been made at the trial after the abjuration.[70] This myth was in part based on the judgment that Galileo did not really change his mind about the earth's motion as a result of the trial;[71] he did abjure, of course, but the abjuration amounted to a mere external verbal compliance, rather than an internal mental assent.

In 1773, Angelo Fabroni published in Florence a collection of previously unknown correspondence pertaining to Galileo's telescopic discoveries of 1610, the anti-Copernican decree of 1616, and the activities of prince and later Cardinal Leopold de' Medici in 1639–1671. Fabroni had gathered the published documents mostly from the Medici archives. In a footnote, Fabroni (1773–1775, 1: 47 n. 1) claimed that Dominican friar Caccini, in his sermon of 21 December 1614, discussed the suggestive biblical verse "Ye men of Galilee, why stand ye gazing up into heaven?" (Acts 1:11).[72] It is indeed true that (as we saw in Chapter 7) on that date at the church of Santa Maria Novella in Florence, Caccini preached a sermon against mathematicians in general and Galileo in particular. But before Fabroni there is no documentary evidence that Caccini focused on this passage. On the other hand, it would have been brilliant to exploit the ambiguity of the biblical verse to use it against Galileo. So here we have a case where one's aesthetic or rhetorical imagination suggests something that should or could have happened, even if it did not.[73]

2. Few persons can be taken to represent the eighteenth century and to embody the spirit of the Enlightenment as well as Voltaire. It is perhaps not surprising that he found many occasions to comment on Galileo's trial, given Voltaire's multi-disciplinary interests and his involvement in the critique of religion and the struggle for freedom of expression. Voltaire's comments represent a mixed bag of intuition of new insights, invention of new myths, and reiteration of old myths.

Voltaire's earliest comment was included in letter number fourteen of the *Philosophical Letters*, first published in 1734. This letter had been written in 1728 during Voltaire's exile in London (1726–1729), and it was a critical comparison of Descartes and Newton.[74] Voltaire was an admirer of Newton, but was reacting to

---

[69]Baretti (1757, 52–53). However, the words *e pur si muove* appear in a painting of Galileo in prison by Spanish artist B.E. Murillo made in 1643 or 1645. Cf. Fahie (1929, 72–75, plate XVI), Favaro (1911a, b), Drake (1978, 356–358).

[70]For example, in 1847 by French artist Joseph-Nicolas Robert-Fleury; cf. Redondi (1994, 75–83).

[71]Barni (1862, 206–237) expresses the point by saying that the myth is historically false but philosophically true.

[72]The pun can be better appreciated in Latin: "Viri Galileai, qui statis aspicientes in coelum?"

[73]For the mythological character of this story, see also Section 12.7 below.

[74]See Voltaire (1877–1883, 22: 127–132; 1901, 37: 164–171).

8.6 Myth-making or Enlightenment? (1709–1777)

the anti-Cartesian English criticism of Fontenelle's eloge of Newton. The result was a surprisingly balanced interpretation and assessment. Besides such a nuanced comparison between Newton and Descartes, the letter contains a cursory one between the latter and Galileo. After mentioning Descartes' free expatriation to Holland, Voltaire remarked that by contrast, "the great Galileo, at the age of fourscore, groaned away his days in the dungeons of the Inquisition, because he had demonstrated the motion of the earth."[75]

Besides falling under the spell of the prison myth, Voltaire was explaining the condemnation as caused by Galileo's *demonstration* of the earth's motion; and by describing the Galilean accomplishment in this manner, Voltaire was contributing to the creation of another anti-clerical myth. For it is indeed true that Galileo was condemned in part for defending the earth's motion, and that his geokinetic arguments were convincing and did convince almost all open-minded or progressive thinkers. But to speak of demonstration is an exaggeration and oversimplification, and to propose it as the cause of the condemnation makes the Church's conduct seem more incoherent and irrational than it really was or needed to be.

Moreover, to speak of demonstration implies that what is being demonstrated is true, and so Voltaire's explanation amounts to saying that Galileo was condemned for having demonstrated the truth. Indeed, in 1751 in a second relevant passage, in *The Age of Louis XIV* Voltaire explicitly asserted that "the great Galileo ... asked pardon at the age of seventy for being in the right."[76] On the other hand, if Galileo's side is characterized by demonstration of the truth, it would not be hard to guess the traits that must characterize the Church's side. In 1753, in his *Essai sur les moeurs*, Voltaire (1877–1883, 12: 249) spoke of prejudice. And in 1756, in section 2 of the entry on "Newton and Descartes" of the *Philosophical Dictionary*, Voltaire added ignorance.[77]

Voltaire's explanation is, then, two-sided: Galileo was condemned because he demonstrated the truth about the place and behavior of the earth in the universe and because the Church's authorities were prejudiced and ignorant. This view turned out to be one of the most popular of the anti-clerical myths about the Galileo affair, whose appeal continues to our own day.[78]

Other aspects of Voltaire's discussions are worth noting. In more than one place,[79] he drew a comparison between Galileo and Socrates, which was an obvious but important one to make. As we have seen, the comparison had been made earlier in private by Peiresc, in his letter to the Vatican secretary of state pleading for Galileo's freedom;[80] but Voltaire's statement may very well be the first such statement in print. And in one of his last remarks on the subject, in 1770, in the entry on

---

[75] Voltaire (1901, 37: 167). I have emended this translation slightly; cf. Voltaire (1877–1883, 22: 129).

[76] Voltaire (1901, 23: 277). Cf. Voltaire (1877–1883, 14: 534; 1951, 352).

[77] Voltaire (1824, 5: 113–114); cf. Voltaire (1877–1883, 20: 122).

[78] E.g., Draper (1875) and the song "Galileo" by the Indigo Girls, popular in the late 1990s and early 2000s; cf. "Epigrafi ed offese" (1887).

[79] Voltaire (1877–1883, 12: 249; 1824, 5: 113–114); cf. Voltaire (1877–1883, 20: 122).

[80] Favaro 16: 202, Finocchiaro (2005b, 54–55).

"authority" of the *Philosophical Dictionary*,[81] Voltaire mentioned Galileo's trial under the topic of authority and made it clear that he had in mind various kinds of authorities, religious and civil, Catholic and non-Catholic. Thus the key issue was presumably the struggle between reason and authority as such (rather than religious authority), and this was a relatively novel and important interpretation of the affair, which would be echoed in the twentieth century by playwright Bertolt Brecht.

3. Strangely enough, there was no entry on Galileo in the original edition of the French *Encyclopedia*, edited by Denis Diderot and Jean D'Alembert in 17 volumes from 1751 to 1765. The oversight was remedied when the four volumes of *Supplement* were published in 1776–1777, which included an article entitled "Galileo, Philosophy of" by Italian scientist and Barnabite priest Paolo Frisi.[82] This article had originally been published in Italian in 1766 and then translated into French in 1767.[83] It was a comprehensive and reasonable account of Galileo's scientific work, if somewhat apologetic and excessive in praise. But it said nothing about the trial.

On the other hand, the original *Encyclopedia* did contain several revealing discussions of the trial. These were found in the general introduction included in volume 1 of 1751 and entitled "Preliminary Discourse" and in the entries on antipodes, astronomy, Copernicus, and Aristarchus, all authored by D'Alembert. For D'Alembert, Galileo's trial was only one of several similar cases. Two other important ones were Pope Zachary's condemnation of Bishop Virgil's belief in the antipodes in the eighth century and Cleanthes's persecution of Aristarchus in ancient Greece. In the "Preliminary Discourse" we find the following general account.

D'Alembert claimed that a major reason retarding philosophic and scientific progress was an attitude by theologians that involved several elements: an "abuse" of power vis-à-vis the people;[84] the fear of "blind reason";[85] the zealotry of "enthusiasts";[86] and the failure to appreciate that the best defense from the attacks of reason is to fight reason with reason. Another alleged factor was what might be called theological expansionism or imperialism, namely the tendency to expand the articles of faith necessary for salvation by indiscriminately adding opinions which theologians or clergymen happen to hold. Then there was a cause which D'Alembert himself called "theological despotism,"[87] that is the desire to dictate belief outside the spiritual domain of faith and morals into matters of fact and natural philosophy. He concluded by speaking of theology (or religion) having "made open war"[88] against philosophy (or science).

---

[81] Voltaire (1877–1883, 19: 501–50 2; 1824, 1: 365–366).
[82] Frisi (1777). Cf. Frisi (1756; 1775), Casini (1985; 1987).
[83] Frisi (1766); cf. Frisi (1777, 176).
[84] D'Alembert (1963, 71); cf. D'Alembert (1751, xxiii).
[85] D'Alembert (1963, 72); cf. D'Alembert (1751, xxiii).
[86] D'Alembert (1963, 72); cf. D'Alembert (1751, xxiii).
[87] D'Alembert (1963, 73); cf. D'Alembert (1751, xxiv).
[88] D'Alembert (1963, 74); cf. D'Alembert (1751, xxiv).

Here D'Alembert was creating or promulgating a conflictual image or metaphor of the relationship between science and religion in general, as well as an interpretation of Galileo's trial as epitomizing the conflict. And this view was destined to generate considerable heat but less light in the centuries to come. Despite these suggestions, we must not ignore the respectful, balanced, and nuanced tone which D'Alembert exhibited and tried to convey. In alleging the abuse of power, he stressed that such abusive theologians were few in number and warned against hastily overgeneralizing from some to all. In talking of the absurdity of religions, he explicitly excluded Catholicism (labeling it "ours"[89]). The same tone appeared when he clarified that the threat to religion came not from philosophic reason as such, which can actually help religion, but from the zealotry of those who fear it. And in advancing his warfare generalization, he qualified it by restricting it to enemies that are "poorly instructed or badly intentioned."[90] Finally, the same message is conveyed by the remarks which D'Alembert made in the article on Copernicus about the then-reigning pope, Benedict XIV. After expressing his wish that the prohibition of Copernicanism be lifted, D'Alembert said that "such a change would be worthy of the enlightened pontiff who governs the Church nowadays. Friend of the sciences and himself a scholar, he ought to legislate to the inquisitors on this subject, as he has already done for more important subjects."[91] And this brings us to the next topic, which is what happened in mid eighteenth century during the reign of this pope.

8.7 Incompetence or Enlightenment? (1740–1758)

In 1740 Prospero Lambertini (from Bologna) was elected pope Benedict XIV; he reigned until 1758. He was widely respected and liked by Catholic, non-Catholic, and non-Christian rulers, scholars, and common people. As we have just seen, D'Alembert was one of them. Another was Voltaire, who exchanged letters, compliments, and gifts with Benedict. It was during his papacy that two important events in the subsequent Galileo affair occurred: in 1744 Galileo's *Dialogue* was republished for the first time with the Church's approval as the fourth volume of the Padua edition of his collected works; and in 1758 the new edition of the *Index* dropped the prohibition against "all books teaching the earth's motion and the sun's immobility." Let us relate some of the details.

1. In 1741, Padua's inquisitor wrote to the Inquisition in Rome to ask its opinion on a projected publication by the press at Padua's seminary of Galileo's complete works, including the *Dialogue* (cf. Mayaud 1997, 130–131). The editors had promised to revise this book to make it "hypothetical"; to have the revision done by persons who were both learned and of proven Catholic faith; and to include

---

[89] D'Alembert (1963, 72); cf. D'Alembert (1751, xxiii).
[90] D'Alembert (1963, 74); cf. D'Alembert (1751, xxiv).
[91] In Finocchiaro (2005b, 123); cf. D'Alembert (1751, 174).

Galileo's abjuration and any other declaration required by the authorities. The Inquisition promptly approved the project as described (Mayaud 1997, 131–132).

However, in February 1742 the Paduan inquisitor wrote again to Rome describing some delays and difficulties encountered by the project, as well as some changes (Mayaud 1997, 135–137). The editors were now planning to leave the text of the *Dialogue* unchanged, but to add an apologetic editorial preface, as well as the Inquisition's 1633 sentence and Galileo's abjuration; moreover, they were planning to also publish the *Letter to Christina*. The preface gave a cryptic account of the trial reminiscent of the one Viviani had advanced in his biographical sketch. This time the Inquisition asked a consultant to study the question and write a recommendation. The consultant compiled a report that reads mostly like a summary of the Inquisition minutes of 16 June 1633, which had been chaired by pope Urban VIII and at which he decided how to bring Galileo's trial to a conclusion. On the basis of this report, the Inquisition approved the revised project.

But once again, other changes were in store before the edition finally appeared. In May 1742, the Paduan inquisitor again wrote to Rome with a slightly different proposal (Mayaud 1997, 144–146). The editors still did not think feasible or appropriate to change the text of the *Dialogue*, but they now suggested that they were ready and willing to make deletions and changes in the marginal postils that dotted the pages of the book and that read like a running interpretive commentary by the author about the topics being discussed by the three interlocutors. Moreover, the editors had dropped the idea of including Galileo's *Letter to Christina*; instead they were planning to include an already published essay in Italian by French Benedictine friar and biblical scholar Augustin Calmet that presumably defended the geostatic world view on the basis of Scripture. They were still thinking of reprinting the text of both the Inquisition's sentence and Galileo's abjuration. Finally, they were willing to rewrite an appropriate editorial preface.

The Inquisition resorted again to a consultant for a recommendation. In a four-page opinion, a consultant named Luigi Maria Giovasco focused on the distinction between a thesis and an hypothesis implied by the Index's decree of 1620 and generally adopted by Catholics afterwards, and applied the distinction to claim that Copernicanism was prohibited and condemned if treated as a thesis, but allowed if treated as a hypothesis. Giovasco made an attempt to clarify this distinction by associating a thesis with a family of notions such as "absolute" and "doctrinal" and a hypothesis with the cluster of "to better know the revolutions of the heavenly spheres", "more useful for contemplating … phenomena", "imagined", and speaking "problematically." The last notion is especially interesting because it is reminiscent of a concept used by Peiresc in his plea for Galileo's liberation, when he defended the *Dialogue* as a "philosophical play."[92] However, Giovasco did not distinguish between the instrumentalist notion of hypothesis and the probabilist conception, which was crucial for understanding, let alone evaluating, the Galileo affair.

---

[92] See Finocchiaro (2005b, 52–56) and Section 8.3 above.

The report cited a 1734 work by G.D. Agnani entitled *Philosophia neo-palaea* as a source where one can read the story of the affair, as well as a justification of the condemnation based on the thesis-hypothesis distinction. Besides being a useful and responsible reference, this citation may also be a confession by consultant Giovasco that he largely relied on Agnani's account rather than examining the primary sources and documents. In fact, Giovasco's report is also full of inaccuracies, such as the publication dates of Galileo's *Dialogue*, of Copernicus's *Revolutions*, and of Zúñiga's *Commentaries on Job*, and the authorship of the 1620 Index decree. The errors are so striking that one begins to sympathize with writers such as Voltaire, whose factual errors can be more easily excused by their lack of access to primary sources and documents.

Without further complications, the edition in four volumes was published by Padua's seminary in 1744. It was edited by Giuseppe Toaldo, who had conceived, nurtured, and executed the project. The *Dialogue*, with the related material, was in the fourth volume. The edition had the usual ecclesiastic imprimatur by the local Paduan officials, with one noteworthy difference. There were two, rather than just one, sets of imprimaturs: one set in volume one, applying to everything included in this edition and dated June and July 1742; and another set in volume four, applying specifically to the *Dialogue* and dated May and June 1743.[93]

The text in the body of the *Dialogue* had indeed been left intact. Only the marginal postils had been "corrected": sixteen of them were deleted, and 46 edited to qualify the earth's motion as "hypothetical" (cf. Besomi and Helbing 1998, 955–959). The Inquisition's sentence of 1633 and Galileo's abjuration preceded the text; they were printed in Latin, having been taken from Riccioli's *Almagestum novum*. Also preceding the text, in accordance with the latest approved plan, were Calmet's essay on biblical exegesis and an editorial preface by Toaldo.

Toaldo's preface of one page[94] mostly echoed Galileo himself rather than Viviani (as the earlier proposed preface had done). Thus, the published preface stated that it endorsed Galileo's own "retraction and qualification." It declared that the earth's motion was nothing but a "pure mathematical hypothesis," which was Galileo's own phrase in the preface to the *Dialogue*. It mentioned the removal or emendation of marginal postils that were not "indeterminate," which was the term used by Galileo in the full title of the book to describe the type of discussion he had aimed at.

Calmet's introduction was entitled "Dissertation on the World System of the Ancient Jews."[95] It was lengthy (twenty pages of small print), scholarly (about ten citations per page), and erudite (many biblical verses quoted in Hebrew). In the beginning section, Calmet elaborated the theme of epistemological modesty and

---

[93] Galilei (1744, vol. 1, unnumbered page following p. 601), Galilei (1744, vol. 4, unnumbered last page of book), Mayaud (1997, 120).

[94] In Galilei (1744, vol. 4); English translation in Fantoli (2003b, 353–354).

[95] In Galilei (1744, 4: 1–20). Cf. Calmet (1720; 1734), Martin (1868, 256), Fahie (1903, 427), Mayaud (1997, 122–123).

partial revelation by God through his work, which Galileo had mentioned on the last page of the *Dialogue*; formulated and discussed the principle of accommodation, which Galileo had also espoused, although it was not at all original with him; criticized with clarity and forcefulness the common abuse of reading one's own preconceptions in Scripture; and proposed a relatively novel approach that would pay more attention to the historical and intellectual context of the writers and audience.

In the central part of the essay, Calmet described in more detail his contextual, historical, and comparative approach, which led him to conclude that the biblical world view was very different from ours. This conclusion was elaborated with scholarly erudition and scriptural quotations. It claimed that the biblical world view was that of a flat earth capped by a tent-like heavenly vault. It followed, of course, that Aristotelian cosmology and the Ptolemaic system were contrary to Scripture as much as Copernicanism. Thus, if the choice was between Ptolemy (or Tycho) and Copernicus, Scripture did not favor the former any more than the latter, or conversely did not undermine the latter any more than the former. The point was that the biblical world view was scientifically (philosophically) untenable, and so one better not regard Scripture as a philosophical authority. This final conclusion had been, of course, Galileo's own view of the matter.

Calmet stated as much in the epilogue of his essay. It is very revealing that there he quoted a passage from St. Augustine that had also been quoted and capitalized upon by Galileo in his *Letter to Christina*.[96] The same Augustinian passage would be quoted and stressed by pope Leo XIII in his encyclical *Providentissimus Deus* (1893).[97] Calmet also rejected and criticized the argument stemming from Riccioli that scriptural authority had to be upheld in astronomy because otherwise its authority would dissolve in other, more spiritually relevant subjects.[98]

In light of all this, it is difficult to think of a more pro-Galilean introduction to the *Dialogue*. Even if Toaldo had printed Galileo's own *Letter to Christina*, it might have not been equally effective; for Calmet was a highly respected biblical scholar.

However, it is unclear whether this pro-Galilean statement was deliberate. For during the negotiations for the imprimatur, Calmet's dissertation was described as pro-geostatic and anti-Copernican by all involved; it was supposed to be one of several means designed to neutralize the text of the *Dialogue* (together with the revision of the marginal postils and the reprinting of the 1633 sentence and abjuration). In reality, the essay neutralized these latter documents. Thus, the question arises whether the 1744 re-edition of the *Dialogue* was a sign of incompetence or enlightenment. Pierre-Noël Mayaud (1997, 142, 159, 161), who for the first time has published the relevant documents and who has tried to analyze them critically, stresses on several occasions the "incompetence" of the officials involved. Mayaud (1997, 149 n. 48, 161) also suggests the possibility of what might be called bureaucratic overwork, when he points out, for example, that at the Inquisition's meeting

---

[96] Galilei (1636, 15), Favaro 5: 318, Finocchiaro (1989, 95), Motta (2000, 99), Galilei (2008, 118). Cf. Chapters 4 and 9 of this book.
[97] Leo XIII (1893, paragraph 18, p. 334); cf. Finocchiaro (2005b, 263–266).
[98] Cf. Finocchiaro (2005b, 79–85) and Section 8.4 above.

of 13 June 1742 when the cardinal-inquisitors gave their final approval, this project was the *tenth* case they deliberated upon in the *first part* of that meeting; and they met several times a week.

On the other hand, perhaps the tolerant and liberal climate created by Pope Benedict XIV was indirectly responsible, although there is no documentary evidence that he was directly involved in this particular episode. Thus, his enlightenment may have encouraged church officials to adopt toward the issue of the scientific (philosophical) authority of Scripture an attitude similar to that adopted toward the issue of the earth's motion. That is, one would pay lip service to the hypothetical character of Copernicanism, but then elaborate the Copernican system in any way allowed by the observational evidence and the physical theorizing; similarly, one would pay lip service to biblical literalism, but in reality develop new methods of biblical interpretation and new exegeses.

2. By contrast with his indirect involvement in the 1744 publication of Galileo's *Dialogue,* pope Benedict's involvement in the revision of the 1758 *Index* was explicit and direct. In July 1753, he issued a bull entitled *Sollicita ac provida* on the reform of the criteria for the censure and prohibition of books in the *Index*. The following year, the secretary of the Congregation of the Index proposed to the pope some additional reforms, involving the restructuring of its contents and the possibility of removing the prohibition of some books after proper correction; among these were mentioned the works of Descartes, Copernicus, and Galileo. The pope approved the secretary's proposal, and the Congregation began working on the publication of a restructured and reformed *Index*.

During such work, as part of the analysis of the Galilean question, in December 1755 the proceedings of Galileo's trial were removed from volume 1181 of the Inquisition archives (where they had been kept since the trial) and placed into a self-contained free-standing file, which began to acquire a life of its own and which would thereafter remain their location.[99] Soon thereafter, the Index commissioned one of its consultants (Jesuit Pietro Lazzari, professor of Church history at the Roman College) to make a recommendation specifically about the general prohibition of "all books teaching the earth's motion and the sun's immobility." This clause stemmed from the anti-Copernican decree of 1616 and had been included in all subsequent editions of the *Index*. Lazzari wrote a lengthy memorandum full of arguments in favor of removing from the *Index* the general anti-Copernican clause. In April 1757, with the approval of pope Benedict, the Congregation decided to drop from the forthcoming edition of the *Index* the anti-Copernican clause. Thus, its 1758 edition no longer listed this entry, although it continued to include the five previously prohibited books by Copernicus, Foscarini, Zúñiga, Kepler, and Galileo.

The main puzzle in this episode is why the repeal of the anti-Copernican censure was so partial and incomplete. Mayaud (1997, 189) finds this decision illogical in the sense of self-contradictory; that is, presumably, retaining the five mentioned geokinetic books was incompatible with dropping the general anti-geokinetic

---

[99] Beretta (1999a, 465–466), Baldini (2000, 304–306).

clause. However, although the partial repeal is problematic, there is really no strict inconsistency. The point is that the retraction was partial and incomplete, that the step taken was a small one; while it suggested other steps, it did not "logically" necessitate them. In this regard, the situation was similar to the case of the "correction" of Galileo's *Dialogue*. Moreover, there is a crucial piece of documentation which was unknown to Mayaud (or not utilized by him) which helps to explain why the step taken was small, partial, and incomplete. The document is the text of the opinion of the Index's consultant recommending the withdrawal of the general anti-Copernican clause. It has recently been discovered, transcribed, and published by Ugo Baldini.[100]

Consultant Lazzari's memorandum is 24 pages long, and strikes me as well argued, impressively nuanced, and often insightful, although of course it is not beyond criticism. Lazzari tried to show that although the prohibition was originally justified, it should now be removed. For him, it was originally justified because at that earlier time Copernicanism was generally regarded as (I.1) false, as (I.2) contrary to Scripture literally interpreted, and as (I.3) not supported by any argument with demonstrative force. Lazzari went on to argue that the general prohibition should be removed because (II) it was no longer justified and (III) it was indeed expedient to remove it. It was no longer justified because the earth's motion was now (II.1) generally accepted by astronomers and natural philosophers; (II.2) generally regarded as consistent with Scripture literally interpreted; and (II.3) supported by a demonstration. And it was practically expedient to abolish the general prohibition because keeping it (III.1a) did not do any good but (III.1b) did harm, and because removing it (III.2a) would do no harm but (III.2b) would rather do good. This last part of the argument, as befits its practical nature, was a cost-benefit analysis, so to speak. The prohibition did no good because it had become ineffective. Its harm involved such things as encouraging Catholic disregard of ecclesiastic decrees; encouraging duplicity on the part of Catholics, by way of the "hypothesis" subterfuge; and encouraging non-Catholics to extend their rejection of Catholic ideas from questions of natural philosophy to questions of faith and morals. To show that no harm would come by removing the prohibition, Lazzari argued that it is a virtue to admit one's errors and revise one's ideas – a point that had been made in Leibniz's summer 1688 letter to landgrave Ernst.

It is clear that Lazzari was not addressing the issue of the condemnation of Galileo but that of the prohibition of Copernicanism. Because of the way Lazzari approached the latter, his memorandum had no obvious implications about the former, or at least no reformist implications. Similarly, Lazzari's argument did not really have any implications about the five particular books (by Copernicus, Zúñiga, Foscarini, Kepler, and Galileo). For the consultant was as clear and forceful in arguing that the general prohibition was no longer justified, as he was in maintaining that originally it was reasonable and prudent; to remove those five books from the

---

[100] In Baldini (2000, 307–328); translated in Finocchiaro (2005b, 139–151).

Index might have tended to suggest that they did not deserve to be prohibited in that earlier time, whereas to keep them was a reminder that the original prohibition was justified. This point would later (1820) be explicitly made by Maurizio Olivieri during the Settele affair.[101]

8.8 New Criticism (1770–1797)

Benedict XIV's partial unbanning of Copernicanism was not the only effect of the anti-clerical accounts of Voltaire and D'Alembert. Another effect was the formulation of pro-clerical and anti-Galilean accounts at the end of the century by such authors as Jacques Mallet du Pan and Girolamo Tiraboschi. However, just as Benedict's liberalization was only partially due to the criticism of the *philosophes*, so Mallet's and Tiraboschi's apologias also resulted from other factors. The two main such factors were two documents that started circulating at that time.

1. Around 1770, a document about Galileo's trial began circulating, purporting to be a letter written by Galileo in December 1633 to his disciple Vincenzo Renieri.[102] It reads like an account of the trial by the victim himself. The letter turned out to be a forgery, although it took about half a century for its apocryphal character to be established. The details of its content and repercussions will be discussed later (Chapter 10) since they are intimately related with the myth of Galileo as a bad theologian. Here it deserves to be briefly mentioned in order to connect it with other developments.

This apocryphal letter portrayed the inquisitors at the 1633 trial as unconcerned not only about scientific arguments but also about the precedents of 1616, and concerned exclusively with biblical interpretation. It depicted Galileo at the trial as engaged in biblical exegesis and theological disputation. This conveyed the impression that Galileo was condemned only for some kind of theological transgression. On the basis of this letter and other evidence, some people soon drew this conclusion explicitly, adding a special twist to it. Part of this twist was due to another document, an authentic letter that was first published in 1773 by Angelo Fabroni.

2. This authentic letter was a report to Grand Duke Cosimo II written by Piero Guicciardini (1560–1626), Tuscan ambassador to Rome in 1611–1621.[103] It was dated 4 March 1616, in the midst of the earlier proceedings of the trial, while Galileo was in Rome trying to prevent the condemnation of Copernicanism. The text of the letter is filled with evaluatively overcharged language and makes it clear

---

[101] Cf. Maffei (1987, 499–500), Brandmüller and Greipl (1992, 373–3740), Finocchiaro (2005b, 214), and Section 8.10 below.

[102] Tiraboschi (1782–1797, 8 [1785]: 147–149 n), Albèri (1842–1856, 7: 40–43); translated in Section 10.4 below.

[103] Fabroni (1773–1775, 1: 53–57), Favaro 12: 241–243, Gebler (1879a, 91–93), Santillana (1955, 116–117).

that the ambassador disliked Galileo intensely. Guicciardini portrayed Galileo as unwilling to keep quiet, taking the matter too personally, passionate (or zealous), imprudent, vehement (or intense), fixated (or obsessed), aggressive, argumentative, dangerous, and a troublemaker. This image was destined to become widely accepted and may be labeled the imprudence thesis.

Guicciardini appeared to have some inside information about the proceedings, and this was to be expected since his position of ambassador gave him direct access to the pope himself, as well as to cardinals and other well-connected diplomats. However, he did not seem to understand the purpose of Galileo's trip to Rome. Some of what the letter said suggested that Galileo was trying to convince the cardinals to accept the earth's motion, as if he wanted the Church's endorsement of Copernicanism, so to speak. Guicciardini showed no awareness that Galileo might be trying to convince the Church that Copernicus's opinion should not be prohibited or condemned. As we will soon see, Guicciardini's confusion on this point would soon mislead some people into thinking that in 1615–1616 Galileo went to Rome to get the Church's endorsement of Copernicanism, and tried to get it on the basis of scriptural arguments.

3. One of these people was Jacques Mallet du Pan (1749–1800), a Swiss Protestant journalist born in Geneva. In 1784, he published in the *Mercure de France* an article entitled "Lies Printed on the Subject of the Persecution of Galileo." In his own memorable words, he tried to show that "Galileo was persecuted not at all insofar as he was a good astronomer, but insofar as he was a bad theologian" (Mallet du Pan 1784, 122). The wrong theological principle was that one could support astronomical theses by means of biblical passages. The documentary basis of Mallet's account appeared to be Galileo's apocryphal letter to Renieri and the correspondence published in Fabroni's collection of documents, especially Guicciardini's letter. Of course, Mallet did not know that the 1633 Renieri letter was a forgery.

However, Mallet's bad-theologian thesis is false, indeed the opposite of the truth, since, as we have seen (Section 4.5), Galileo preached and practiced the principle that scriptural passages should *not* be used in astronomical investigation, but only when dealing with questions of faith and morals; whereas the Inquisition was the one to uphold the opposite principle that Scripture *is* a scientific authority, as well as a moral and religious one. Despite its falsity, Mallet's bad-theologian thesis deserves a fuller examination since it proved to be long lasting; acquired the status of a myth; and represents a common and to some extent natural type of anti-Galilean criticism. It will be discussed in detail later (Chapter 10).

4. Even when Mallet's account was not adopted, it probably exerted an indirect influence by encouraging the articulation of explicitly pro-clerical apologies. I believe such to be the case for Girolamo Tiraboschi (1731–1794). Moreover, since, as we will soon see, Tiraboschi stressed Galileo's alleged imprudence in 1615–1616, his account was also probably influenced by the publication of Guicciardini's letter. However, historical connections aside, Tiraboschi's account deserves attention in its

own right, because of its content, form, and sophistication. He advanced it in two lectures delivered in 1792 and 1793 to the Accademia de' Dissonanti in Modena, which were later (1797) published in an Appendix to the last volume (no. 10) of the Rome edition of his *History of Italian Literature*.[104]

In regard to the hermeneutical question and the first phase of the original episode in 1613–1616, Tiraboschi saw a clash between two principles: on the one hand, there was the traditional principle that on all subjects, whether pertaining to faith and morals or astronomy and natural philosophy, Scripture must be interpreted literally unless there is a conclusive proof that the literal interpretation implies an obvious falsehood; on the other hand, there was Galileo's principle that the literal interpretation of Scripture should not be taken into account unless the subject pertains to faith and morals. Tiraboschi was even willing to admit that the new Galilean principle was essentially correct. But the problem was that Galileo's principle was perceived to be, and was in fact, dangerous in light of the controversy between Catholics and Protestants about the interpretation of Scripture. Moreover, in terms of the traditional principle, there was no conclusive proof that the literal interpretation of relevant scriptural passages implied a falsehood because the consensus was that Galileo's arguments for the earth's motion were not conclusive. To these two reasons, Tiraboschi added the scandal caused by Galileo's imprudent advocacy while in Rome in 1615–1616. These three reasons provided for Tiraboschi a sufficient *explanation* of the 1616 condemnation; he was also implicitly suggesting that they were plausible enough reasons so as to yield a *justification* of that condemnation.

In regard to Galileo's publication of the *Dialogue* and the second phase of the affair, Tiraboschi charged him with acting in bad faith for disobeying the special injunction and indeed for never having any intention of complying with it. And Galileo deceptively violated the prohibition to defend Copernicanism, for the book actually defended this doctrine, although the preface deceptively tried to give the impression that the work was a sort of vindication of the 1616 anti-Copernican decree, and the text contained lip service to the hypothetical character of the discussion. Finally, Galileo even mocked the ecclesiastic authorities by deceptively extorting their official imprimatur. Despite Tiraboschi's toughness here, he did *not* add a fourth element of bad faith to these three, namely that Galileo insulted the pope by impersonating him with the character of Simplicio; Tiraboschi mentioned the allegation, but dismissed it.

In regard to the details of the 1633 proceedings, Tiraboschi gave a balanced and documented refutation of the prison myth, showing that Galileo was not kept in a real prison either during the trial (despite Inquisition practice and precedent) or afterwards (despite the formal inclusion of prison among the penalties announced in the sentence). Instead the diplomatic correspondence of the Tuscan ambassador proved beyond any reasonable doubt that during the 1633 trial Galileo had been

---

[104] Tiraboschi (1782–1797, 10: 362–383); the second lecture is translated in Finocchiaro (2005b, 165–172).

treated with a consideration and kindness that were unprecedented in the annals of the Inquisition. This fact suggested a radical revision of the prevailing ideas about the trial, and we have seen that some people (i.e., Mallet du Pan) went to the opposite extreme of constructing an anti-Galilean myth. But Tiraboschi refrained from drawing any sweeping implications.

Tiraboschi's summary of the sentence is revealing. Besides mentioning the prohibition of the *Dialogue* and the abjuration, he said that the system was condemned as heretical. That is, Tiraboschi judged that the Inquisition's sentence of 1633 represented not merely the condemnation of a person and his book, but was a reaffirmation and reinterpretation of the condemnation of the Copernican doctrine. In his view, such a condemnation did not undermine the infallibility of the Church, but was a reminder of the fallibility of the Inquisition. In regard to the penalty of imprisonment, Tiraboschi interpreted it as the activation of a clause of the special injunction, the clause that stated that if he refused to comply, he would be imprisoned.

In summary, for Tiraboschi, Galileo was largely responsible for his own condemnation; he was too imprudent in how he defended his cause in Rome in 1615–1616 and in how he disobeyed the 1616 prescriptions with his *Dialogue* of 1632; he was defending the truth, but was doing so by means of violence and deception. Clearly this account of the affair was an apologia of the Inquisition; but it avoided the crudities and distortions of Mallet du Pan and explicitly contradicted his bad-theologian myth. On the contrary, Tiraboschi's account is eloquent, subtle, insightful, and well argued. This is not to say that it is entirely acceptable, but its criticism is beyond the scope of this chapter.[105]

8.9 Napoleonic Wars and Trials (1810–1821)

During the Napoleonic era, the conflictual relationship between the French emperor and the papacy produced one of the most curious and progressive episodes in the history of the Galileo affair.

1. In 1809, during one phase of this conflict, pope Pius VII excommunicated Napoleon. As a result, the emperor had the pope arrested and deported to France, and everything in Rome pertaining to papal government was ordered moved to Paris. The move required a monumental effort and an astronomical sum: the first convoy (in February 1810) used 3,239 cases and cost 179,320 Italian lire; the second (April) and third (July) convoys transported the Inquisition archives. However, a few documents were shipped separately, due to their special importance; among these were the papal bull of excommunication against Napoleon and the file of the original proceedings of Galileo's trial.

In the next few years, Napoleon approved a plan to translate the proceedings into French and publish them in an edition that would display the original Latin or

---

[105] For some relevant criticism, see Sections 8.12, 8.15, 9.2, and 12.6.

8.9 Napoleonic Wars and Trials (1810–1821)

Italian text side by side with the French translation. Work began, and the first 25 folios of the manuscripts containing nine documents were translated. But this is as far as the project proceeded, and the fall of Napoleon in 1814 left the project unfinished.

2. For the next 3 years, there was a papal representative in Paris (most of the time, Marino Marini) charged with the task of retrieving the Vatican archives. The effort was only partially successful. Some archives were retrieved and sent back to Rome, for example minutes of Inquisition meetings. Others were found, but because of their bulk and the transportation cost, they were destroyed or sold to cardboard manufacturers; for example, such was the fate for 3,600 volumes of Inquisition trial proceedings and 300 volumes of Inquisition sentences, amounting to about two-thirds of all the archives transferred to Paris.

Other documents were not found; the file of Galileo's trial was among those that were not retrieved. It had been lost during the confusion surrounding Napoleon's last one hundred days, in 1814–1815. In 1817, the papal representative returned to Rome thinking that the file had been irretrievably lost. Unexpectedly, however, in 1843 the file was returned to the Holy See by the nuncio to Vienna, who had received it from the widow of one count Blacas, who had died a few years earlier; Blacas had been one of the French ministers of the royal government with whom the papal representative had dealt, to no avail, for 3 years after the fall of Napoleon.

3. Although the Vatican file of Galileo's trial was not published as Napoleon had envisaged, this minor Napoleonic feat did contribute to the diffusion of more accurate information and better understanding. This consequence happened by way of the few persons who had had the opportunity to read the manuscript documents in 1811–1814, and by the publication of some of the documents whose French translation had been made. The climax of these developments occurred in 1818–1821 and involved Jean Delambre and Giambattista Venturi. Delambre was perpetual secretary of the French Royal Academy of Sciences and was working on a history of astronomy that appeared in 1821. Venturi (1746–1822) was a retired professor of physics who had taught at the universities of Modena and Pavia; he was publishing a two-volume documentary history of Galileo's work and its reception, the first of which had appeared in 1818 and the second in 1821. Prodded by Venturi's requests, Delambre was able to find the partial French translations of the proceedings and sent Venturi a copy.

These French texts were no substitute for the original file (which Delambre had been unable to locate), but they happened to be more revealing than one might expect from the fact that they comprised only about one-tenth of the file (25 out of 228 folios). The reason is that the first translated document (from the first five folios of the manuscript) was the Inquisition's executive summary of the proceedings from 1615 to May 1633, compiled at the end of May or beginning of June 1633. According to standard Inquisition procedure, such a summary was the basis on which the cardinal-inquisitors and the pope reached a final decision on the case being tried. Moreover, for Galileo's case, this summary happened to include a verbatim quotation of his second deposition in which he changed course from the first deposition and confessed some wrongdoing. As we have seen (Section 8.1), at that second deposition, Galileo confessed that the first deposition had prompted him to re-read his *Dialogue*; from the text of the first deposition, today we know that in it he had denied that he had received any special injunction in 1616 and that his book violated Bellarmine's warning to stop defending the earth's motion. The confession continued by stating that he was surprised to find that the book gave readers the impression that the author was defending the earth's motion, even though this had not been his intention. He attributed his error to wanting to appear clever by making the weaker side look stronger; he was sorry and ready to make amends. The second deposition ended with Galileo humbling and prostrating himself by making an emotional appeal to pity to move his judges to have compassion for an infirm old man.

In 1821, both Delambre and Venturi found ways of publicizing these French translations by discussing them and quoting long excerpts.[106] Delambre did it in the preface to his history of astronomy, Venturi in an addendum to his account of the trial at the end of his documentary history. To be sure, neither Venturi's addendum nor Delambre's preface was clear about the nature of the translated documents, and in fact their exposition confused three sets of documents: (1) the Inquisition's summary, whose French translation they had in their possession and which referred to and described other documents they did not possess; (2) the other trial documents pertaining to the year 1615 whose translations were in their possession; and (3) the other trial documents from 1615 to 1633 whose translations were not in their possession. The confusion was compounded by the fact that the just-described summary included the verbatim transcript of Galileo's second deposition. Despite the confusion, however, both authors quoted Galileo's confession in full, and so its content and tenor emerged clearly. It was a very touching, moving, and eloquent statement, comparable in pathos and poignancy to the abjuration already made available two centuries earlier.

Thus, as a result of the efforts of Napoleon, Venturi, and Delambre, by 1821 the European public had access to the text of Galileo's confession at the second deposition of the 1633 trial. There was still a long way to go to have the complete proceedings, but this was a giant leap.

8.10 The Settele Affair (1820)

In 1820 a controversy raged in Rome over whether to allow the publication of an astronomy textbook that treated the earth's motion as a fact. The Inquisition sided with the author but was opposed by the chief censor in Rome, the so-called Master of the Sacred Palace. After considerable struggle, the Inquisition prevailed in regard to the publication of this book, but two other steps took longer: it was only in 1822 that it ruled that Catholics were generally free to accept the earth's motion as a fact

---

[106] Delambre (1821, 1: xx-xxxii), Venturi (1818–1821, 2: 197–199).

8.10 The Settele Affair (1820)

in accordance with modern astronomy, and only in 1835 that Galileo's and Copernicus's books were taken off the *Index*. The highlights are as follows.

In March 1819, a clergyman and professor at the university of Rome named Giuseppe Settele published the first volume (dealing with optics) of a textbook on optics and astronomy. The second volume dealt with astronomy and was ready to be published next. But in January 1820, the Master of the Sacred Palace (named Filippo Anfossi) refused to give his approval for the publication of the second volume, on the grounds that it held the thesis of the earth's motion. Settele sought the advice of an Inquisition consultant named Maurizio Olivieri and of the Inquisition's chief legal adviser, the so-called assessor (named Fabrizio Turiozzi). With their encouragement, Settele wrote a formal appeal to pope Pius VII. In March 1820, the pope forwarded the case to the Inquisition.

In April, Assessor Turiozzi talked to the pope, who seemed to give his informal approval to Settele's appeal. In May, the assessor conveyed to Master Anfossi the pope's wish to give the imprimatur to Settele's book. Anfossi replied that he respected the pope's wishes, but could not in good conscience give the imprimatur. Instead Anfossi showed Turiozzi a book he has just edited, to which he had added an "Appendix" explaining his reasons why he could not grant his imprimatur to Settele's book.

In June, several Inquisition officials, including Assessor Turiozzi, Consultant Olivieri, and the chief prosecutor (named Libert), met with Settele to see whether he was willing to make some changes of wording in his manuscript to try to appease Master Anfossi. Settele accepted Olivieri's suggestion to replace the crucial chapter's initial phrase, "The Earth being in motion around the sun ...," with the clause "Given the earth's motion around the sun ...," although Settele said that this did not change the substance and that the book still contained a demonstration of the earth's motion. In the meantime, Consultant Olivieri finished writing an essay with a detailed refutation of Master Anfossi's view and with arguments trying to convince him to grant the imprimatur. Settele also had a long discussion with the Vatican majordomo, who was on Anfossi's side and tried to convince Settele to treat the earth's motion as a hypothesis rather than as a demonstrated fact.

In July 1820, Dominican friar Olivieri, who had been a consultant until then, was appointed commissary general of the Inquisition. This development was a significant turning point in the affair.

In August, Settele with the advice and consent of Commissary Olivieri and Assessor Turiozzi submitted to the pope a second appeal requesting a formal ruling by the Inquisition. He enclosed three documents as attachments: a copy of his first appeal to the pope; a copy of Master Anfossi's "Appendix" published in May; and a critical article from the periodical *Italian Library* published in Milan. The pope agreed.

Thus, the Inquisition started its usual proceedings. It asked one of its consultants for a formal written opinion on the matter. The consultant, Antonio Grandi, wrote an opinion, concluding that it was proper to defend the Copernican thesis that the earth moves in the way in which it was customarily defended by Catholic astronomers, and also that Settele should insert a note in his book explaining that the

Copernican system no longer suffered from the difficulties from which it suffered at the time of Copernicus and Galileo. On August 16, the cardinal-inquisitors unanimously accepted the consultant's recommendation, and the pope ratified the decision. In the meantime, Settele compiled his "Insert,"[107] and on August 23, the Inquisition approved it.

Master Anfossi did not acquiesce. In the next few days, he sent the pope a memorandum explaining his "Motives" why he thought Settele's book should not be published.[108]

Then the Inquisition went to the counterattack, intellectually and bureaucratically. Commissary Olivieri wrote a reply to Anfossi's memorandum and gave it to Assessor Turiozzi. Assessor Turiozzi officially conveyed in writing to Anfossi the decision by the Inquisition and the pope on the Settele case. Master Anfossi then started temporizing, for on the following day he wrote Turiozzi that he would give the imprimatur, but suggested that the Inquisition get two more evaluations of Settele's book by consultants; he also mentioned that he had given the pope a written memorandum explaining his objections. Turiozzi agreed to request two additional reviews.

In September, the consultants both confirmed that Settele's book did not contradict Catholic faith and morals. But now Master Anfossi formally withdrew his imprimatur; and printed a lengthy explanation (in a booklet entitled *Reasons*),[109] including objections to Settele's "Insert" note requested and approved by the Inquisition; and he maneuvered to appeal the Inquisition's ruling to the pope. Seeking to avoid a disciplinary confrontation, the pope approved the idea that if Anfossi did not want to sign his name, the book's imprimatur be granted by another authority, the pope's own deputy as bishop of Rome, the so-called vicar-apostolic.

The controversy still did not end. During the months of September and October there were some reports in the popular European press saying that the Inquisition had ruled in favor of publishing Settele's book, but that despite this approval, the book's publication was being opposed and delayed by the Master of the Sacred Palace. And indeed in November the Vatican majordomo was instructed by the pope to forward to the Inquisition Anfossi's latest (September) printed booklet against Settele and the earth's motion.

However, Olivieri had already started his reply, and sometime in November he finished writing his criticism of Anfossi's latest effort and included it in his "Summary" of the whole affair;[110] he attached all the available documents and had the whole thing printed for internal distribution and use (cf. Maffei 1987, 451–580). Anfossi immediately wrote a brief reply to Olivieri's latest criticism. Then Grandi was again consulted, and he wrote a lengthy criticism of Anfossi's latest objections.

---

[107] In Settele (1819, 130–133, note 1), Maffei (1987, 543–545), Brandmüller and Greipl (1992, 303–305).

[108] In Maffei (1987, 548–554), Brandmüller and Greipl (1992, 310–317).

[109] In Maffei (1987, 452–463), Brandmüller and Greipl (1992, 336–349).

[110] In Maffei (1987, 427–450), Brandmüller and Greipl (1992, 351–380), Finocchiaro (2005b, 199–218).

8.10 The Settele Affair (1820) 193

On November 20, the Inquisition consultants met and approved the immediate publication of Settele's book. On December 14, the Inquisition cardinals agreed that the imprimatur would be given by the vicar-apostolic, and the pope approved the decision.

Finally, on 2 January 1821, the printing of Settele's book was completed. The book carried imprimaturs and endorsements by the vicar-apostolic (Candido Frattini), as well as by the two additional consultants (Mazzetti and Ostini), and the rector of the University of Rome (Belisario Cristaldi).

Although the publication of Settele's astronomical work settled the issue involving that particular book, the more general doctrinal question had not really been settled. Thus before long, new difficulties arose.

In April 1822, Master Anfossi refused to give his imprimatur to an extract of Settele's book on astronomy by a certain Dr. D. de Crollis scheduled to be published in a Roman periodical, *Arcadian Journal*. This refusal was followed in July by Anfossi's anonymous publication of a booklet against the earth's motion. Thus, the Inquisition became re-involved and in August Commissary Olivieri got back into action by writing an opinion criticizing Anfossi's latest refusal.

On 11 September 1822, the Inquisition ruled that in the future the Master of the Sacred Palace shall not refuse the imprimatur to publications teaching the earth's motion.[111] But it postponed a decision about removing from the *Index* five particular Copernican books including Galileo's *Dialogue*; Olivieri was given the task of examining whether these five explicitly mentioned books were prohibited because of advocating the earth's motion or for some other reason. Two weeks later (September 25), Pope Pius VII ratified the Inquisition's decision to permit works teaching the earth's motion. Thus after several months of controversy, the proposed extract of Settele's book was published in the October 1822 issue of the *Arcadian Journal*; the imprimatur was formally signed by Anfossi himself, who finally chose to yield to the Inquisition and the Vatican secretary of state.

About a year later, Olivieri finished writing his account of the original reasons for prohibiting the five Copernican books explicitly mentioned in the *Index*. In November 1823, the Inquisition consultants discussed his report and decided to ask him to write out answers to various questions that came up at the meeting. In December, the Inquisition consultants discussed Olivieri's answers and decided to request the opinion of two other experts, one of whom was Bartolomeo Capellari (who would later be elected Pope Gregory XVI).

At this point the documentary trail is lost, but not the historical connection. For on 20 May 1833, while deliberating on a new proposed edition of the *Index*, Pope Gregory XVI decided that it would omit the five books by Galileo, Copernicus, Kepler, Foscarini, and Zúñiga, but that this omission would be accomplished without explicit comment. Thus, the 1835 edition of the *Index* for the first time omitted from the list Galileo's *Dialogue*, as well as the four other books.

---

[111] Settele (1820–1833, 413, 418), Favaro 19: 421, Brandmüller (1992, 187–188), Brandmüller and Greipl (1992, 426–428), Mayaud (1997, 244–245).

Many documents pertaining to the Settele affair have survived, and almost all of them have been recently published. The most important is the summary which Commissary Olivieri wrote in the fall of 1820 in preparation for the crucial Inquisition meeting of December 14. In accordance with standard Inquisition procedure (as we have seen for Galileo's own case), such a summary was the document on the basis of which the cardinal-inquisitors reached their decision. Olivieri's summary is a lengthy essay of 24 printed pages of small print, to which he appended 130 pages of documents that had accumulated during the year. Thus, it is also a guide to the documents, which are referred to and quoted in the text of the summary. It is a chronological and historical account of the Settele affair from January to November 1820. It is also a discussion of the theological, administrative, legal, scientific, and epistemological issues underlying the controversy. It is a criticism of Anfossi's reasons for withholding his approval for Settele's book; for objecting to any treatment of the earth's motion as a thesis; and for requiring that it be treated only as a hypothesis. And it is a justification of the propriety of holding and defending the geokinetic thesis in 1820. This "Summary" is now available in three editions that complement each other: one facsimile; one critical; and an English translation.[112]

Some of the more interesting, original, or plausible arguments in the summary are as follows. Olivieri argued that it was inconsistent to allow the earth's motion as a hypothesis and to regard it as heretical, for if a doctrine were heretical the Church would not place believers at risk, and so would not allow its acceptance even as a hypothesis. Consequently, given that acceptance of the geokinetic hypothesis was legitimate, it followed that the earth's motion was problematic in some other way than heresy.

Olivieri stressed that at the time of Galileo the earth's motion had been problematic as regards the mechanical consequences for terrestrial bodies that seemed derivable from it, such as the westward deviation of falling bodies, westward gunshots ranging farther than eastward ones, birds' inability to fly, and a constant easterly wind. And he suggested that in 1820 the earth's motion was no longer problematic in this regard, so the condemnation no longer applied.

Echoing Auzout and Lazzari,[113] Olivieri also argued that in 1616 the earth's motion was declared contrary to Scripture (in part) because it was philosophically false and absurd, and so once it was discovered that it was no longer such, then one was no longer forced to say that it was contrary to Scripture either. For Olivieri, this change did not imply that the earlier condemnation had been wrong because the earlier condemnation obviously applied to the earlier doctrine, and the earlier doctrine referred to an earth's motion that was problematic, and that problematic earth's motion was indeed contrary to Scripture. In other words, the condemnation of the earth's motion as contrary to Scripture did not have to be revised because it did not refer to motion per se, or as it exists in itself. What had been condemned

---

[112] Respectively, Maffei (1987, 427–450), Brandmüller and Greipl (1992, 352–380), Finocchiaro (2005b, 199–218).

[113] See respectively, Sections 8.5 and 8.7.

was the proposition that the earth moved in the sense of motion that implied all the mechanical difficulties that seemed derivable from it; and the earth's motion in this "devastating" sense was still contrary to Scripture. Correspondingly, the earth's motion theorized about by the astronomy of Settele's time was a motion freed of such difficulties, and so it was not contrary to Scripture.

Similarly, Olivieri did not question the legitimacy or correctness of Galileo's condemnation as "vehemently suspected of heresy." The commissary only proposed a reinterpretation of the (suspected) heresy in question as referring to a motion of the earth in the devastating sense of Aristotelian physics.

Finally, Olivieri stressed that one of Anfossi's greatest oversights was the failure to appreciate Benedict XIV's decision of 1757–1758. In fact, Olivieri pointed out that the documents in the archives showed that the decision was significant and deliberate because it was based on explicit discussion and on a written reasoned recommendation. Furthermore, the repeal of the general prohibition was more important than the non-repeal of the specific prohibitions of Copernicus's and Galileo's books because the general repeal was relevant to the contemporary situation, whereas the particular prohibitions reflected the historical situation at the earlier time.

Not all of these arguments are sound or acceptable, but their criticism cannot be undertaken here. Suffice it to say that while Olivieri's analysis justified a liberalization in 1820, it simultaneously justified the earlier condemnation at the time it occurred. In that sense, it was an update and reaffirmation of Lazzari's argument that had led to the partial unbanning of Copernicanism in 1757–1758. Thus, although some may admire Olivieri's balanced impartiality, his argument was Solomonic in more than one sense; it was a double-edged sword of questionable value to friends of the historical Galileo.

8.11 The Torture Question and the Demythologizing Approach (1835–1867)

In the middle of the nineteenth century, Galileo's trial started receiving an unprecedented amount of attention. Sustained discussion spread from Italy and France to England, Ireland, America, and Germany. Key issues started to be seriously debated with a critical dialogue of arguments and counterarguments. The controversy also grew more heated and bitter. The variety of topics and approaches seemed to coalesce around two themes: torture and demythologization. These developments may be highlighted as follows.

1. The question whether Galileo had been physically tortured became a cause célèbre.[114] The pro-torture case was argued by Guglielmo Libri in 1841, in the context of a generally anti-clerical account that received wide circulation throughout

---

[114] For more details and references, see Finocchiaro (2005b, 225–237; 2009a).

Europe. Libri could not of course consult the trial proceedings, which in any case were still missing as a result of the Napoleonic transfer. But he was in possession of the original manuscripts of an Inquisition trial in Novara in 1705, in which the procedure and terminology used there were the same as those for the case of Galileo, and in which it was clear that the defendant was tortured. Libri's argument was also based on the following reasons: the wording of Galileo's sentence, which uses a phrase ("rigorous examination") connoting torture; the procedural rules contained in Inquisition manuals, according to which torture was standard practice in cases like Galileo's when the defendant's intention was in doubt; the fact that after the trial Galileo was afflicted with a hernia; and the fact that Galileo's silence on the matter can be explained because Inquisition defendants were sworn to secrecy.

The other side of this issue was argued by Marino Marini, who in 1850 published a book containing a general apologia of the Inquisition based on the newly returned file of trial proceedings, to which he had access in his position of director of the Vatican Secret Archives, where the file was then kept. First, Marini showed that the term "rigorous examination" was *not* synonymous with torture, but rather referred to an interrogation conducted with the verbal threat that if the defendant did not tell the truth he would be tortured; torture followed the threat if and only if the replies were unsatisfactory; Marini supported this claim by quoting from one of the trial documents available only to himself, namely the fourth deposition held on 21 June 1633, which was the written record of Galileo's "rigorous examination."[115] Another argument Marini gave was that the manuals of Inquisition procedure stipulated that for torture to be administered a prior vote and recommendation of the consultants was required, and if need be a vote by the cardinal-inquisitors; and he assured the reader that the trial documents did not contain any minutes of such decisions. And he gave a description of the documents in the file after the May–June 1633 summary, to counter Libri's claim that the documents recording the torture might have been removed, a claim based on the French translations published by Delambre and Venturi, which contained no references to any subsequent proceedings.

Although Marini was right on the torture question, at the time his argument was compelling only if one could trust his word regarding the content of the trial proceedings, available only to himself. However, his book was such that it was obvious even to a casual reader that Marini was being selective and manipulative in his treatment of the texts. Thus, his book only added fuel to the fire and intensified demands that the manuscripts be open to scholars for consultation and published in their entirety. This was not to happen for another two decades.

But physical torture was not the only kind that was argued about. Some authors who rejected the physical-torture thesis felt inclined to claim that the Church's treatment of Galileo (from the abjuration at the conclusion of the trial through the house arrest to the end of his life) amounted to *moral* (or *psychological*) torture.

---

[115]Marini (1850, 56, 59, 61–62). Cf. Favaro 19: 361–362, Finocchiaro (1989, 286–287), Galilei (2008, 287–288).

8.11 The Torture Question and the Demythologizing Approach (1835–1867)

The moral-torture thesis became especially widespread in France, where it was advanced in 1858 by physicist Jean Biot and in 1862 by literary critic Philarète Chasles. Such a thesis was especially significant in light of the fact that these authors were also engaged in an attempt to demythologize Galileo's trial and to elaborate a circumstantialist account.

2. In fact, the other main trend in the middle of the nineteenth century involved a move away from portraying Galileo as a martyr or hero and toward depicting him as a fallible and flawed human being, mostly a moral weakling and coward. It also involved moving away from interpreting the trial in terms of grand notions such as science versus religion, philosophy versus theology, and scriptural authority versus natural reason, and toward emphasizing contingent and accidental circumstances and petty human motives. By following such an anti-heroic and circumstantialist approach, such authors may be said to have been trying to de-mythologize Galileo's trial.

The first proponent of an anti-heroic account was, I believe, Scottish physicist David Brewster, who in 1835 published an encyclopedia entry on Galileo and later included it in his book *The Martyrs of Science, or the Lives of Galileo, Tycho Brahe, and Kepler* (1841). In it, far from portraying Galileo as a martyr, Brewster depicted him as having cowardly avoided martyrdom (in 1633), thus in effect harming the cause of science and benefiting that of the Church. Brewster also portrayed him as reckless and too bold in 1613–1616, for failing to appreciate the justifiable mental inertia of his opponents.

With some unfairness toward Galileo, and without solving the problem, Brewster was raising an important issue. Certainly Brewster was right that "mind has its inertia as well as matter" (1841, 58). But the problem is how to overcome such mental inertia in the search for truth. If one opposes it, one may be judged reckless. If one acquiesces and "rejects the crown of martyrdom" (Brewster 1841, 94), one may be judged a coward. At different times, in light of different circumstances, Galileo did both, and so he earned a double dose of blame from Brewster. Far from being the "martyr" of science suggested in Brewster's title, Galileo allegedly harmed science first by alienating potential converts and antagonizing opponents, and later when he "cowered under the fear of man" (Brewster 1841, 95) and let his enemies figuratively decapitate him. Although Brewster was apparently no friend of the Inquisition, clearly he was no friend of Galileo. Thus, his account was also original by being simultaneously anti-Galilean and anti-clerical.

A very different example of an anti-heroic approach is provided by Jean Biot. In 1816, he had written the Galileo article in the *Biographie universelle*, which treated Galileo's work in general but had a few pages on the trial. However, in 1858 he published a new account of the trial based on new primary and secondary sources published since then, such as Venturi's (1818–1821) collection, Albèri's (1842–1856) critical edition of Galileo's works and correspondence, and Marini's book. Biot claimed that what he called "the crux" of the trial was personal circumstances, such as the "envy" of others and primarily Galileo's managing to sour his friendship with the pope. Biot saw himself as giving both a revisionist and a bipartisan interpretation, by contrast to those accounts that blamed the whole episode on ecclesiastic obscurantism, Galilean errors (scientific and otherwise), or the alleged conflict between science and religion.

Another noteworthy example, which combined elements of Brewster and of Biot, was given by Chasles in 1862. Dismissing the idea that the trial was caused by scientific or theological controversies or the conflict between science and religion, Chasles claimed that it derived from ordinary human malice, especially envy: "in this affair, it is more of a question of personalities and small hatreds than of theology or doctrine" (Chasles 1862, 135). He portrayed Galileo as riddled with conflicting tendencies but ultimately a coward and moral weakling; his enemies as motivated by envy and petty hatred, and engaging in what amounted to the moral assassination of an innocent man; and the society in which they lived as decadent and amoral and carrying to an excess the social art of living based on comfort, sensuality, accommodation, dissimulation, appearances, and face-saving. By way of conclusion, he described his own work as a case study of a "moral assassination" of a moral weakling in a "decadent society" (Chasles 1862, 277).

One problem that emerges from Chasles is that demythologization is easier said than done. In fact, for all his anti-heroic, circumstantialist, and demythologizing fervor, Chasles ended up substituting the old heroic portrayals with his own, when he spoke of Galileo as a "genius of enlightenment." Moreover, given Chasles's stress on portraying Galileo as a victim of moral assassination, such a victim turned into another kind of hero, as is often the case when victims are romanticized and made to wallow in their condition of victimhood.

8.12 The Documentation of Impropriety (1867–1879)

I have already mentioned that interest in consulting and publishing the Vatican file of Galilean trial documents intensified after Napoleon's plan to publish it was not completed; after the file was lost and found; and after Marini's apologetic book showed that the Church herself was not going to publish it. Eventually, during the span of about a dozen years from the late 1860s to the late 1870s, four scholars were allowed to consult and publish the file. These publications led to a critical interpretation of the trial which is epoch-making in the history of the Galileo affair and whose essentials are still largely acceptable.

The first scholar to consult and publish the proceedings was Frenchman Henri de L'Epinois, who edited a large selection in 1867; but besides being incomplete, this edition was full of errors and other imperfections, and so 10 years later he published a complete and improved edition. The second scholar was an Italian, Domenico Berti, who published his first edition in 1876 and his second improved and complete edition in 1878. The third one was an Italian priest who was the director of the Barberini Library in Rome; he did not publish his own edition, but compared the manuscripts with the editions of L'Epinois and Berti and published corrections to them. The fourth scholar was Austrian Karl von Gebler, who was

8.12 The Documentation of Impropriety (1867–1879)

granted the permission after publishing an interpretive account based on L'Epinois's first edition; then from his own personal consultation and with the benefit of L'Epinois's second edition, Gebler published his own complete edition in 1877. Thus, by 1878 there existed three essentially complete editions of the Galilean file: L'Epinois (1877), Gebler (1877), and Berti (1878).

1. The first person to appreciate the significance of the new documents was a German scholar who did not have direct access to the manuscripts. In 1870, on the basis of L'Epinois's first edition of 1867, Emil Wohlwill published the following account. The focus of the discussion was four documents. One was the minutes of the Inquisition meeting of 25 February 1616 presided by the pope, which ordered the following: Bellarmine would warn Galileo to abandon his geokinetic view; if Galileo refused, the Inquisition's commissary would give him a special injunction not to discuss the topic in any way; and if Galileo did not acquiesce at this injunction, he would be arrested. Another document was the transcript of the special injunction dated 26 February 1616, which stated that immediately after Bellarmine gave Galileo the warning, the commissary gave him the special injunction; but this document lacked any signatures, especially Galileo's. The third document was a certificate written by Bellarmine for Galileo on 26 May 1616 stating that, despite rumors to the contrary regarding various proceedings against Galileo, the only thing to have happened was that he has been warned that the earth's motion was contrary to Scripture and so could not be held or defended. The fourth was the text of the first deposition of the 1633 trial, dated April 12, which was a long document of 12 folio pages whose gist was this: Galileo denied receiving a special injunction not to discuss the earth's motion in any way whatever, and in his defense he introduced Bellarmine's certificate; he admitted receiving from Bellarmine the warning that the earth's motion could not be held or defended, but only discussed hypothetically; and Galileo claimed that his book did not defend the earth's motion, but rather suggested that the favorable arguments were inconclusive, and so did not violate Bellarmine's warning.

Wohlwill argued that the special injunction was of questionable legal validity because it contradicted the other three crucial documents: the Inquisition minutes of the day before, Bellarmine's certificate, and Galileo's first deposition. Moreover, the special injunction conflicted with Galileo's attitude and correspondence of that period, with the Index decree of 5 March 1616, with the Church's good will toward Galileo, and with the fact that the *Sunspots* book was not prohibited. Furthermore, although the 1633 sentence *claimed* that Galileo also violated the requirements of Bellarmine's certificate, there was never any serious inquiry into what this certificate implied. Wohlwill concluded that Galileo's condemnation was based on a document that was legally invalid or improper; he also conjectured that the document was forged in 1632 in order to have a justification of a pre-ordained verdict, which would have been insufficiently grounded on just Bellarmine's warning.

Wohlwill had advanced a radically revisionist account. As suggested by my labeling one of his theses a conclusion and the other a conjecture, his claim of the legal impropriety of the special injunction was powerful, solid, and well argued, whereas his forgery thesis was highly speculative. And indeed during the next decade the impropriety thesis was reinforced, while the forgery thesis was disconfirmed.

2. By fortuitous coincidence, the same year that Wohlwill published his account, an Italian named Silvestro Gherardi made a completely independent documentary contribution, one of whose implications was to reinforce the legal-impropriety thesis. Gherardi had been involved in the 1848 revolution in Rome that overthrew the papal government and established a republic, but was suppressed within a year by French military intervention. During the short-lived republic, two republican officials (one of whom was Gherardi) searched the Inquisition archives looking for material related to Galileo's trial. They found the archives organized into three main groups of material: files called "Decrees," containing minutes of meetings and summaries of resolutions; lengthy proceedings consisting of charges, depositions, sentences, and the like, called "Trials"; and indexes to the decrees and trials. They did not find the Galilean proceedings among the "Trials," because (unbeknown to them) that dossier was kept in the Vatican Secret Archives rather than in the ordinary Inquisition archives. However, by examining the "Decrees," they found that many minutes contained information about Galileo's trial. So they copied relevant passages from 32 minutes, spanning the years from 1611 to 1734, although most were from the crucial years 1616 and 1633. These documents remained unpublished for about 20 years, but eventually Gherardi published them with a critical commentary in 1870.

These documents were (and remain) invaluable because they constitute an independent and equally authoritative source of information about the trial; that is, independent of, and as authoritative as, the documents in the special file of trial proceedings. We have seen that this file was created in 1755, and we can now add that it was created by removing the appropriate documents from the relevant volume of "Trials." According to Inquisition practice, minutes of meetings were first written on loose sheets of paper during the meeting, and then they were copied onto various volumes of "Decrees" or "Trials" or both. Thus, many minutes of Inquisition meetings are found with identical wording in both the special Galilean file of trial proceedings and in Gherardi's set of minutes. A consequence of this coincidence was that the general accuracy and integrity of the file documents was reinforced. This was important because the vicissitudes of the special file (especially the Napoleonic removal and temporary loss) made the anti-clericals suspect that perhaps the documents had been tampered with.

However, due to bureaucratic sloppiness, the recording of minutes from the loose sheets used at the actual meetings onto the appropriate volumes did not happen with absolute regularity. Thus, some minutes were never recorded in the trial proceedings. A crucial example of such a document was the minutes of the Inquisition meeting of 3 March 1616. This was an important meeting because Cardinal Bellarmine reported that he had done what he had been ordered to do at the previous meeting presided by the pope the week before (25 February). Gherardi had copied the March 3 minutes in 1849 and published the text along with the rest in 1870.

The document states that Bellarmine reported that Galileo had acquiesced when warned to abandon his geokinetic opinion.

In his commentary, Gherardi pointed out that Bellarmine's report provided further evidence against the special injunction. For the cardinal was reporting that when he had given his warning to Galileo, there had been no complications and Galileo had agreed. Bellarmine's report said nothing of any Galilean refusal that would have triggered the stricter special injunction by the Inquisition commissary. The report implied that, in accordance with the papal orders of February 25, there had been no need for a special injunction.

3. Despite the two independent avenues of support, the legal-impropriety thesis was not immediately or unanimously accepted. Among the critics was one of the editors of the Vatican file, Berti, who argued that Galileo could and would have been convicted of disobedience even without the special injunction, merely for violating Bellarmine's warning not to hold or defend the earth's motion; for Galileo had not denied receiving a warning so phrased, and beginning with the second deposition he even admitted this violation (while insisting that it had been unintentional). Berti was right that in 1632–1633 Galileo was open to the charge of having violated the warning not to defend Copernicanism, but whether he could have been convicted of this charge (without the special injunction) is questionable, as some of Berti's critics soon pointed out. Such criticism was partly based on a new document that was discovered at the time and first published by Sante Pieralisi in 1875.

The new document was a letter dated 28 April 1633 by the Inquisition's commissary Maculano da Firenzuola to Cardinal Francesco Barberini, the Vatican secretary of state and a member of the Inquisition.[116] This date was between Galileo's first deposition (April 12) and the second (April 30), during which he changed his plea from innocent of any wrongdoing to guilty of an unintentional violation of the warning not to defend Copernicanism. One could not help wondering why Galileo changed his mind and conjecturing that it must have been some kind of pressure from the Inquisition. Maculano's letter to Barberini provided the evidence for this conjecture. The letter was not an official document, and so it was not in the Vatican file of Inquisition proceedings. It was found by Pieralisi in the Barberini family's archives in Rome. In this letter, Maculano conveyed the latest news to Francesco Barberini, who had not attended the last Inquisition meeting. Maculano said that, in light of Galileo's denials in the first deposition (supported by Bellarmine's certificate that was news to the Inquisition), he had obtained permission from the cardinal-inquisitors to engage in out of court negotiations with the defendant. During this meeting Maculano was able to convince Galileo to confess some wrongdoing, and Galileo was now (April 28) in the process of writing a confession to present in court at the next deposition.

---

[116] Pieralisi (1875, 197–198). Cf. Favaro 15: 106–107), Finocchiaro (1989, 276–277). This letter may now be read in conjunction with another one recently discovered and published: see Beretta (2001a, 571); cf. Beltrán Marí (2001b).

Partly exploiting the information contained in this letter, in 1878 J.A. Scartazzini articulated a position that also took into account the other recent documents and interpretations. He tried to develop his position as being intermediate between two extremes. On the one hand, there was the thesis (stemming from Wohlwill) that the condemnation of Galileo embodied a legal impropriety because it was based on the special injunction and the document recording this injunction contradicted the other documents and was probably a forgery. On the other hand, there was the view (stemming from Berti) that the special-injunction document was essentially consistent with the others; that they all proved that Galileo had received *an* injunction not to defend Copernicanism; that the *Dialogue* was a defense of Copernicanism; and that therefore he was guilty of disobeying this injunction, and the condemnation was proper in that sense. Scartazzini argued that the special injunction was crucial, at first, to get Galileo indicted and summoned to trial and, later, to get the trial started and the first interrogation defined; but after Galileo's denials in the first deposition and his introduction of Bellarmine's certificate, he was forced to confess having transgressed the prohibition to defend Copernicanism. In other words, the special injunction transcript was important in the *dynamics* of the 1633 trial: Galileo was not convicted just on account of the special injunction; but he could not have been convicted without it; the special injunction was instrumental in extracting (or extorting) from him the confession of 30 April 1633, which paved the way for the conclusion of the trial. This account was a plausible and reasonable one.

4. Scartazzini also articulated a more extreme position about the authenticity and completeness of the documents in the Vatican file. He questioned not only the integrity of the special-injunction transcript, as Wohlwill had done, but also the integrity of several other documents, especially the fourth deposition (21 June 1633) describing Galileo's "rigorous examination." The points are subtle and intriguing, and cannot be examined here.[117] Suffice it to say that his conclusion is based on considerations about where the various texts are found in the physical organization of the Vatican file, involving such questions as whether a given document is written on a loose sheet, or on a folded sheet, or on a folio that is part of a bundle. He also exploited the fact that both suspicious documents are next to folios that have been cut out from the bundle to which they belonged and no longer exist.

5. Despite the ingenuity and plausibility of Scartazzini's arguments, his and Wohlwill's forgery theses turned out to be untenable. This was proved by Gebler, who was also the third editor of the file. Gebler's own synthesis was that the trial did indeed involve a legal impropriety insofar as the special injunction used against Galileo contradicted all the other available documents and evidence. But the special injunction transcript was not a later (1632) falsification forged after the fact in order to make publication of the *Dialogue* a transgression; it was rather an inaccurate report written earlier (1616) that misrepresented what happened at the meeting when Bellarmine implemented the Inquisition's orders to give Galileo the warning.

---

[117] See Finocchiaro (2005b, 251–254) for the details.

Gebler argued that the special injunction document exhibits handwriting as well as watermarks identical to those of other 1616 (annotation) documents and different from the 1632–1633 documents; and that it is written on folios that are part of bundles containing other documents of the earlier (1615–1616) proceedings. Moreover, Gebler was able to exclude the possibility of the kind of paper shuffling imagined by Scartazzini, which enabled the would-be forger to use blank parts of existing and authentic documents to write inauthentic texts and then reshuffle those parts into inauthentic but apparently proper locations. This possibility could be excluded because Scartazzini neglected or was not cognizant of the fact that the sheets in the Galilean file have page numbers, indeed several sequences (one entered during the proceedings of 1615–1616, a second entered when the 1632–1633 proceedings started, a third entered at the end of the trial in May or early June 1633, and a fourth one entered in 1926); and the special injunction is written on folios bearing page numbers in proper sequence, whereas the shuffling conjectured by Scartazzini would have produced some disorder in the sequence.

Gebler's account represented not only a convincing confirmation and important refinement of Wohlwill's radically revisionist thesis of legal impropriety, but also an elegant refutation of Wohlwill's and Scartazzini's forgery theses. It also amounted to a judicious assimilation of the trial proceedings finally made accessible by the Vatican and published in the previous decade by L'Epinois, Berti, and Gebler himself, as well as the independent documents published by Gherardi and Pieralisi. It turned out that despite the puzzling features of the Vatican file, its obvious incompleteness, and its other limitations, by an almost miraculous set of coincidences it happened to provide within itself internal evidence of its own authenticity. Moreover, Gherardi's Inquisition minutes provided an external and independent check. Finally, I believe we may also speak of a third check on the authenticity of the file of proceedings, namely the incriminating evidence it contains; that is, the fact that it contains almost conclusive evidence that Galileo's condemnation embodied a judicial impropriety. If post-1633 apologists had really wanted to tamper with the evidence, they would have removed the evidence that conflicts with the special injunction, which was highly irregular within the Inquisition's own framework; they would not have removed the alleged evidence of actual torture (the presumed deposition of Galileo on the rack), which was standard procedure not only by the Inquisition's own rules but also by those of criminal justice systems throughout Europe and the whole world.

8.13 Theological Developments (1893–1912)

1. Around the turn of the century, new important developments once again entered the scene. The first of these involved a Church action that was *not* taken directly in regard to Galileo, but for other reasons, and yet it had consequences that affected a key issue of the Galileo affair. Recall that the "suspected heresy" for which Galileo was condemned was twofold: one was the physical proposition that the earth does

not stand still at the center of the universe but revolves annually around the sun and daily around its own axis; the other was the methodological, theological, and hermeneutical principle that Scripture is not a philosophical or astronomical authority but only an authority on questions of faith and morals. Recall also that with regard to the physical proposition, Galileo was proved right and the Church gradually, if slowly and grudgingly, admitted its error with such actions as the retraction of the general prohibition of Copernican books in 1757, as well as the approval of Settele's textbook in 1820, the approval of Copernican books in general in 1822, and the silent withdrawal of the prohibition of Copernicus's and Galileo's books in 1835.

However, with regard to the hermeneutical principle, due to the nature of the topic, developments were not as discrete, clear-cut, or easy to identify and assess. Nevertheless a few points have emerged. We have seen that a conservative reaction to Galileo's condemnation was the reaffirmation of biblical literalism, a leading example being Riccioli; and that some significant implications were derivable from the recurring discussions of Bellarmine's assertion, popularized by Fabri, to the effect that if a demonstration of the earth's motion is ever found, then the Church will revise its literal interpretation of biblical passages such as Joshua 10:12–13. And it has been briefly mentioned (but will be elaborated later, in Chapter 10) that a curious thing happened on the way to the consolidation of the novel principle that Scripture is not a scientific authority, namely Galileo was accused (by Mallet du Pan and his followers) of having *violated* it, and so he was alleged to have been condemned for being a bad theologian and not for being a good astronomer; while this myth illustrates the strange forms which the anti-Galilean animus can take and has taken, it also testifies to the fact that the denial of the scientific authority of Scripture was becoming an increasingly well-established principle.

A climax for the hermeneutical aspect of the Galileo affair occurred in 1893 with pope Leo XIII's encyclical letter *Providentissimus Deus*, for this document put forth a view of the relationship between biblical interpretation and scientific investigation that corresponded to the one advanced by Galileo in his letters to Castelli and Christina. To be sure, the encyclical did not even mention Galileo but was written in response to the controversy known as "the biblical question," which involved such issues as the nature, methods, and implications of the scientific study of the Bible and the validity of scientific criticism of Scripture.

The problem of the scientific criticism of Scripture was the reverse of what Galileo had to deal with: he was trying to defend astronomical theory from objections based on scriptural assertions, whereas Leo was discussing how to defend Scripture from attempts "to vilify its contents"[118] based on physical science. However, their respective answers hinged on essentially the same point, the denial of the scientific authority of Scripture. Not only were both Galileo and Leo asserting the same principle, but they also shared some crucial aspects of the reasoning to justify this principle. The argument is this. Natural science and scriptural

---

[118] Leo XIII (1893, paragraph 18, p. 334).

8.13 Theological Developments (1893–1912)

assertions cannot contradict each other, because both nature and Scripture derive from God. Hence, if there appears to be a contradiction, the conflict is only apparent and not real and must be resolved. This is normally done by interpreting the biblical statement in a nonliteral fashion. Now, to see why this is done, "to understand how just is the rule here formulated,"[119] we have to say that Scripture is not a scientific authority. Given this principle, the principle of accommodation also follows. Besides the formal similarity of problems, the substantive overlap of content, and the deep-structure correspondence of the reasoning, Leo's account was reminiscent of Galileo's even in its appearance. This parallelism involved the quotations from St. Augustine and how they were interwoven with the rest of the argument. In fact, Leo's two main passages from Augustine had also been quoted by Galileo in his letter to Christina: Augustine's statement of the priority of demonstrated physical truth ("whatever they can really demonstrate …, we must show to be capable of reconciliation with our Scripture"[120]); and his statement of nonscientific authority of Scripture ("the Holy Ghost … did not intend to teach men … the things of the visible universe"[121]).

It is not surprising that Leo's encyclical has been widely perceived to have been the Church's belated endorsement of the second fundamental belief for which Galileo had originally been condemned, namely that Scripture is not an authority in astronomy. As we shall see later, this interpretation was also explicitly endorsed by Pope John Paul II in 1979–1992.

2. Another significant development during this period was the elaboration by Pierre Duhem (1861–1916) of his novel, epistemological criticism of Galileo. It was published in 1908, first as a series of articles in the journal *Annales de philosophie chrétienne*, and then in book form under the title *To Save the Appearances*. However, Duhem's account deserves special attention, and so its examination will be undertaken later (Chapter 11).

Here, it is useful to discuss a development which, like Leo XIII's *Providentissimus Deus*, belongs to the domain of theology. This was a new solution to an old problem centered on the concept of heresy: what was the notion of heresy according to which the Inquisition's 1633 sentence had found Galileo "vehemently suspected of heresy"? Was it a legitimate notion? And was this notion the same concept as was presupposed by those authors who subsequently claimed that Galileo had been condemned for "disobedience not heresy"? In 1912, Frenchman Léon Garzend published in Paris a well documented work that addressed this cluster of issues, and whose full title conveyed a good idea of its content: *The Inquisition and Heresy: Distinction between Theological Heresy and Inquisitorial Heresy, with Regard to the Galileo Affair.*

---

[119] Leo XIII (1893, paragraph 18, p. 334).

[120] Augustine, *De Genesi ad litteram*, i, 21, 41; cf. Favaro 5: 327, Finocchiaro (1989, 101), Galilei (2008, 126), Leo XIII (1893, paragraph 18, p. 334).

[121] Augustine, *De Genesi ad litteram*, ii, 9, 20; cf. Favaro 5: 318, Finocchiaro (1989, 95), Galilei (2008, 118), Leo XIII (1893, paragraph 18, p. 334).

Garzend's main thesis was that there were two concepts of heresy, a theological and an inquisitorial one. The theological concept was strict and narrow and defined a heresy as a denial of a proposition (1a) explicitly (1b) revealed by God and (1c) officially proclaimed by the Church in a declaration addressed to (1d) all who have been (1e) baptized. This was the concept prevalent among modern theologians as well as those in the seventeenth century. On the other hand, there was an inquisitorial concept which was looser and broader and reflected both the practice and the manuals of the Inquisition. It broadened the concept to include the denial of propositions that (2a) could be clearly deduced from divine revelations, and/or (2b) embodied common Church teachings, and/or (2c) were clearly contained in Scripture but had not been officially proclaimed by the Church, and/or (2d) were declared articles of faith by lesser Church organs (inquisitors, bishops, popes when not speaking ex cathedra, etc.), and/or (2e) were applicable only to a particular person or group, and/or (2f) were articles of faith but were being denied by unbaptized persons.

One consequence of this thesis was that Garzend could explain a tension between two aspects of the Galilean trial documents. One aspect consists of a group of texts such as those stating that Copernicanism is false and contrary to Scripture (without ever stating that it is heretical). The other aspect consists of texts such as those stating that Galileo was found to be vehemently suspected of heresy. Garzend's explanation was that the first aspect reflects the theological concept of heresy, the second aspect the inquisitorial concept.

Another consequence of Garzend's thesis was to provide a *novel* answer to the anti-infallibility objection: this was the argument that in the trial of Galileo the Church made various errors, and therefore the doctrine of infallibility is itself erroneous. The usual answer to this objection pointed out that this doctrine remained unrefuted because the infallibility applied only to papal pronouncements ex cathedra publicly addressed to all faithful; it did not apply to pronouncements of the Inquisition, nor to papal decisions that were not ex cathedra. This answer left the residual difficulty that it was very strange that the Inquisition in general and the pope in particular (as an individual person) would arrive at a condemnation of Galileo in 1633 which appeared to involve a theological error in interpreting the 1616 decisions and in applying them to his subsequent behavior; for indeed from a strict theological point of view neither Copernicanism nor Galileo were heretical, nor was Copernicanism declared heretical in 1616, nor was Galileo guilty of formal heresy in 1633. Garzend's answer to this *residual difficulty* and his *novel* answer to the anti-infallibility objection was to point out that all this was indeed correct from the point of view of the theological concept of heresy; but in Galileo's trial the pope and the inquisitors were taking the inquisitorial point of view, and Copernicanism *was* an *inquisitorial* heresy and Galileo *was* an *inquisitorial* heretic. Thus, no error was committed, and the question of infallibility does not arise.

While the scholarly documentation provided by Garzend and his ingenuity are beyond question, I believe he also unwittingly showed something very far from his own explicitly apologetic intention. That is, he also showed that the Inquisition practices had no theological justification, or were theologically untenable.

Moreover, another unintended consequence follows from Garzend's argument that insubordination was an essential part of strict theological heresy because this heresy reduced to an intellectual error accompanied by a persistent defiance of the Church's injunctions on what to believe. It follows that inquisitorial heresy should not be labeled "disciplinary," to distinguish it from "formal" heresy, because all heresy is a violation of prescribed discipline. And this in turn undermines the apologetic line that Galileo was condemned for disobedience and not for heresy, because (even formal) heresy is ultimately disobedience in believing what the Church commands one to believe.

8.14 Tricentennial Rehabilitation (1941–1947)

The tricentennial of Galileo's death in 1942 occasioned a series of re-interpretations and re-assessments of Galileo's trial that may be regarded as a semi-official rehabilitation. The rehabilitation was not formal or official because it was not proclaimed either by the pope or by the Congregation for the Doctrine of the Faith (the new name of the Inquisition). On the other hand, it did involve authoritative persons and institutions: Franciscan friar Agostino Gemelli, president of the Pontifical Academy of Sciences and of the Catholic University of Milan; priest and Church historian Pio Paschini, President of the Lateran University, commissioned by the Academy to write a book on Galileo; and Jesuit Filippo Soccorsi, director of the Vatican Radio. Although these views were not publicized at the time as a rehabilitation of Galileo, that idea is a natural thought that comes to mind to anyone who considers them collectively. Moreover, the fact that there was no rhetoric of rehabilitation makes this development more genuine and important. Indeed it may be regarded as a preview or anticipation of Pope John Paul II's more explicit "rehabilitation" in 1979–1992, to be discussed later.

1. Gemelli elaborated his account in a lecture at the tricentennial commemoration held at the Catholic University of Milan, whose proceedings were published as a book in 1942. As president of this university, he gave the first lecture in the series and entitled it "Science and Faith in the Person of Galilei." Gemelli had no hesitation admitting that the condemnation of Galileo involved theological errors, such as the claims that Copernicanism was contrary to Scripture and that Scripture was a scientific authority. However, Gemelli also argued that Galileo's tragedy embodied the great positive lesson that faith and religion are harmonious with reason and science. For example, Galileo's "worship of the Author of the universe had a childlike simplicity" (Gemelli 1942, 16–17); moreover, "he worshipped with a self-conscious humility; he considered himself to be simultaneously privileged and ordered to discharge a great mission: to observe new works of God and reveal them to men" (Gemelli 1942, 18); although he was allegedly wrong to get involved in theological and scriptural questions and imprudent in the way he discussed them, this interference and this imprudence were "a proof of the sincerity of his Faith"

(Gemelli 1942, 22); and his trial never undermined his faith, for "he saw in the conflict the passions of men, not incompatibility of ideas; that is, he saw what we see nowadays; time has shown he was right" (Gemelli 1942, 26).

In short, for Gemelli the "suspected heretic" had become the embodiment of the harmony between science and religion, and this Galilean lesson was all the more instructive insofar as Galileo had not only preached such harmony, but also practiced it. What is more, he had continued to uphold it even after his condemnation, when an ordinary believer would have found reason to despair. But in his religious faith Galileo was no ordinary mortal. Here were the seeds of a new appreciation of Galileo. Gemelli was rehabilitating the "suspected heretic."

2. However, he was not the only clergyman viewing Galileo in such a light. Influenced by Gemelli, Paschini also developed a similar account. In 1941, Paschini had been commissioned by the Pontifical Academy of Sciences to write a book length re-assessment of Galileo's life and works. That project turned out to be more time consuming and challenging than anticipated, and it would be published only posthumously in 1964. But in 1943 Paschini found the opportunity to show that he was not idle by accepting an invitation to write a short article on Galileo. It was published in 1943 in the Roman journal *Studium* and was aimed at a general educated audience. It is important both as a preview of Paschini's later full account and as a sign of the new ecclesiastic appreciation inaugurated by Gemelli.

The article was entitled "The Teaching of Galileo: Do Not Be Afraid of the Truth."[122] This title was a double-entendre meant to convey the following messages: that Galileo was not afraid of the truth, even when it seemed to undermine traditional beliefs, including traditional interpretations of Scripture; that this is an instructive and positive lesson which we today can learn from Galileo; and that in particular we should not be afraid of the truth about the Galileo affair, even when the truth is that his condemnation was an error.

Paschini's account was full of simplifications, as was appropriate for the audience of educated nonspecialists to which it was addressed. However, these simplifications usually avoided being over-simplifications and often contained important insights, which could be elaborated and documented and which were occasionally relatively novel. One of these insightful simplifications was to say that Galileo's telescopic discoveries implied a re-assessment of the relative merits of the Ptolemaic and Copernican systems, with the consequence that heliocentrism became more well-supported than geocentrism, while still not conclusively demonstrated.

Paschini's interpretation of the 1616 warning or injunction to Galileo was revealing. Paschini formulated its content merely as the prohibition to defend Copernicanism, whereas "discussion was not prohibited."[123] Apparently he had

---

[122] Paschini (1943); translated in Finocchiaro (2005b, 281–283).
[123] In Finocchiaro (2005b, 282); cf. Paschini (1943, 96).

8.14 Tricentennial Rehabilitation (1941–1947)

been convinced by the arguments against the existence and/or propriety of a *special injunction*. Paschini admitted that Galileo did not acquiesce at the results of the 1616 proceedings and that he was looking for an opportunity to resume the fight; but rather than charging Galileo with bad faith, Paschini clarified that Galileo's plan was to do it "by legitimate means."[124] That is, in his *Dialogue*, Galileo discussed the arguments on both sides, showing that the pro-Copernican arguments were stronger than the pro-Ptolemaic ones; and Paschini added, "but it was not his fault if the arguments for the heliocentric system turned out to be more convincing."[125]

This was an important and relatively original thesis, and could form the basis for defending Galileo even from the charge of violating the warning not to defend Copernicanism. However, Paschini did not go that far, and all he seemed to want to claim was that this was a legitimate defense of Galileo's behavior. For Paschini also seemed to admit that the Inquisition had a legitimate reason for claiming that Galileo had violated the warning. And it is in light of the clash of such legitimate reasons that Paschini then gave credit to Galileo for having yielded. In such a context, his confession and abjuration then "rendered nobler and more sacred his reverence toward the authority of the Church";[126] they became acts of superior piety and religiousness. That is how Paschini was transforming Galileo into a model of religious faith, and in that sense rehabilitating him.

3. Although Paschini sketched an account of the affair portraying Galileo in the favorable light of a model of religious piety, he did not stress the ecclesiastical errors as Gemelli had done. Soon thereafter, Jesuit Soccorsi elaborated and documented an account that combined both points of view. Soccorsi's account also originated as a lecture, in a series delivered in Rome in 1942 and sponsored by the Royal Academy of Italy to celebrate the Galilean tricentennial. The account was first published (in 1946) in the authoritative and more or less official Italian Jesuit journal *La Civiltà Cattolica*. The following year, the press of the same organization published an expanded version of the article in book form. Then it reprinted the book in 1963. Finally, in 1964 the book was one of three works published by the Pontifical Academy of Sciences under the collective title of *Miscellanea galileiana*; the set included Paschini's *Life and Works of Galileo Galilei* (published then for the first time) and was meant to signal the Church's open-minded attitude, especially in the context of the second Vatican Council (1963–1965). This presentation and publication history suggest that Soccorsi's account has found wide acceptance in ecclesiastical circles and has come to be viewed as a semi-official statement.

Soccorsi began by admitting that in 1616 a fateful error was committed by various ecclesiastic institutions and persons (although not by the official Church or the pope speaking ex cathedra); the error was to believe and declare that Copernicanism

---

[124] In Finocchiaro (2005b, 282); cf. Paschini (1943, 96).
[125] In Finocchiaro (2005b, 282); cf. Paschini (1943, 96).
[126] In Finocchiaro (2005b, 283); cf. Paschini (1943, 97).

was contrary to Scripture. And he was clear that in the face of such an error the proper task was to try to explain why it happened, rather than to try to justify, excuse, or rationalize it. He then stated and criticized a number of common explanations, such as that the condemnation was due to Galileo's lack of proof, to the need of suppressing personalistic interpretations of Scripture, to psychological factors like envy and hatred, and to the Church's uncritical acceptance of the science of the time. He concluded by elaborating and defending his thesis that the intellectual root cause of the condemnation was the error of believing in the scientific authority of Scripture.[127]

After this explanation (but *not* justification) of the erroneous condemnation of Copernicanism in 1616, with regard to the 1633 proceedings Soccorsi elaborated what may be called a justification (which is also an explanation) of Galileo's retraction. For Soccorsi, Galileo's retraction was a sincere, indeed admirable, act of religious heroism. It was sincere because it did not require an internal adoption of the geostatic belief, let alone an internal reasoned conviction of it, but rather only an intention "not to impugn it publicly" (Soccorsi 1947, 53) or to observe "respectful silence" (Soccorsi 1947, 53) on the controversy; and this intention had to be grounded on appropriate reasons, such as "on the authority of the judges and on the value of the reasons that persuaded the judges themselves" (Soccorsi 1947, 52), and more generally on "an admission that certain disciplined behaviors are necessary for social reasons" (Soccorsi 1947, 53). And such an intention and such a motivation were quite possible and indeed likely in Galileo's situation. However, they were harder the stronger were the internal reasons supporting one's internal conviction and opposing the judges' belief. Galileo's supporting reasons were (correctly) regarded by him as very convincing and close to being conclusive. And this is where intellectual sacrifice was required, so that the renunciation can be seen to be admirable and even heroic.

Soccorsi's account so far is quite plausible with regard to the first aspect of Galileo's retraction, namely the confession (at the second deposition on April 30) that in his *Dialogue* he had unintentionally done something wrong, that he was sorry, and that he was ready to make amends. But there was a second aspect to his retraction, namely the abjuration. It amounted to more than renouncing public advocacy and keeping silent: it involved saying, "with a sincere heart and unfeigned faith I abjure, curse, and detest the above-mentioned errors and heresies."[128] While this could make Galileo's act insincere, immoral, and sinful, the Inquisition would perhaps fare no better; for as some critics have argued, its imposition of the penalty of the abjuration meant tempting Galileo to perjure himself, which is to say leading him into sin; and such temptation is itself sinful, indeed a graver sin.

Soccorsi tried to resolve this difficulty as follows. He first distinguished between the truth or falsity of a doctrine and its safety or dangerousness. Then he argued that Inquisition decrees on doctrines were essentially evaluations of their safety from the point of view of the Faith. Finally, of Soccorsi claimed that when Galileo said "I abjure

---

[127] Cf. McMullin (2005b), which (as indicated earlier, in Chapter 4) may be regarded as an independent, original, and updated elaboration of Soccorsi's account.

[128] In Finocchiaro (1989, 292), or Galilei (2008, 293); cf. Favaro 19: 407.

the above mentioned heresies and errors," he did not mean that he was really rejecting the propositions of Copernicus as heretical or erroneous theologically or speculatively (philosophically), but rather that he was giving deference to the authorities' judgment that they were practically unsafe or dangerous, namely harmful to religion (because they were widely, albeit incorrectly, perceived to be contrary to Scripture), and so he was abandoning their advocacy.

8.15 Secular Indictments (1947–1959)

In the middle of the twentieth century, Galileo became the target of unprecedented criticism on the part of various secular-minded thinkers. Earlier (in the penultimate section of the Introduction) I hinted at the occurrence of this phenomenon and its significance: it is a historical irony that his original antagonists have become friends, and his original friends have turned into adversaries; the content of such criticism is novel, consisting of social and cultural objections; the source seems to be the left wing of the political spectrum; and the most important such critics are literary intellectuals. Now it is time to examine the details.

1. In 1938, Bertolt Brecht wrote a play entitled *The Earth Moves*, which was slightly revised within several months and retitled *Life of Galileo*; it was performed for the first time in Zurich in 1943. At the time of its writing, Brecht had been living in exile in Denmark for several years, having left his native Germany when the Nazis gained power. He was then deeply concerned with the question of whether it was right to escape Germany and thus seek safety and fight Nazism from the outside, or whether it might have been better to remain there and continue the anti-Nazi struggle in a covert manner from within. This concern was reflected in the play's stress on Galileo's external abjuration in 1633 in order to covertly pursue his work in mechanics and publish the *Two New Sciences* in 1638.

Brecht moved to the United States in 1941, and in 1944–1947 he collaborated with actor Charles Laughton to complete a revision and English translation of the play that amounted to a second version of it. At that time Brecht was deeply affected by the construction and dropping of the first atomic bomb, and the consequent problem of the social responsibility of scientists. Entitled simply *Galileo*, it was performed for the first time in Los Angeles in 1947.

In the late 1940s Brecht left the United States and eventually moved to East Berlin, in communist East Germany. There, in 1953–1956 he revised the play into a third and last version; although this time the substantive revisions were not major, the stylistic and linguistic ones were because the second version existed only in the Laughton English-language version, and so this final revision included a translation into German of Laughton's text. Besides his previous social and political concerns, Brecht now felt such problems as the Cold War, the analogy between Soviet-style communism and the Catholic Church of Galileo's time, and the building of the hydrogen bomb (including the Robert Oppenheimer affair). Retitled *Life of Galileo*, it was first performed in Cologne in 1955.

The final version of the play has fifteen scenes and may be highlighted as follows. In the first scene, Galileo makes the longest speech; it is about the dawn of new age when astronomy will be widely discussed even in the market places. The sixth scene takes place at the Jesuit Roman College in 1616; various arguments are given against the new Copernican world view; the focus is the inappropriateness of the earth being off the center of the universe. In fact, in the eighth scene a minor character abandons the study of astronomy because, he says, common people could not tolerate their misery if the earth were not at the center. The tenth scene depicts a carnival at which common people blame Galileo for spreading a view that undermines the division between earth and heaven, top and bottom, and rulers and ruled; they also call Galileo "Bible-buster." In the twelfth scene, an Inquisition official presents many charges against Galileo to the pope, focusing on his displacement of the earth from the center and abolition of the distinction between top and bottom.

Scene thirteen has Galileo's disciples wait for the outcome of the trial, which takes place off stage; they expect and hope that he will not recant, but are devastated when he comes on scene and tells them he has abjured. His leading disciple, Andrea, cries, "Unhappy the land that has no heroes" (Brecht 1994, 98), and he inveighs against Galileo, "Wine-pump! Snail-eater! Did you save your precious skin?" (Brecht 1994, 98). Galileo's only words are, "Unhappy the land where heroes are needed" (Brecht 1994, 98).

In the climactic penultimate scene (number fourteen) a few years after the abjuration, we are at Arcetri, where Galileo is under house arrest. Andrea comes to say good-bye as he is about to leave for Holland. To his pleasant surprise, Andrea learns that Galileo has completed the manuscript of the *Two New Sciences* and wants him to smuggle a copy out of Italy. This redeems the master in the disciple's eyes because he now sees Galileo as having recanted at the trial in order to be able to do more serious work, namely to lay the foundations of the science of motion whose practical application to machines would produce immense benefits to mankind. However, Galileo does not agree with Andrea's assessment. First, Galileo confesses that he recanted because he was shown the instruments of torture and was afraid of physical pain. Moreover, he feels he has betrayed the true cause of science because he has wasted a unique opportunity. Here are the words Brecht makes him utter: "In my day astronomy emerged in the market place. Given this unique situation, if one man had put up a fight it might have had tremendous repercussions. Had I stood firm the scientists could have developed something like the doctors' Hippocratic oath, a vow to use their knowledge exclusively for mankind's benefit. As things are, the best that can be hoped for is a race of inventive dwarfs who can be hired for any purpose ... I handed my knowledge to those in power for them to use, fail to use, misuse – whatever best suited their objectives" (Brecht 1994, 109).

In the last scene, Andrea crosses the Italian frontier, thus taking the manuscript of the *Two New Sciences* to safety.

The play contains many minor historical inaccuracies. For example, Galileo's daughter is portrayed as engaged to be married to a nobleman, who breaks off the engagement when Galileo resumes his research into the dangerous topics of astronomy after the friendly Urban VIII becomes pope. Such inaccuracies can perhaps be easily dismissed as irrelevant to the purpose of the play.

But it also contains some major inaccuracies. For example, in the play the most important issue in the controversy is the centrality of the earth, not only whether or not it physically *true* that the sun rather than the earth is at the center, but also whether or not it is *proper* that this should be so. That is, for Brecht the replacement of geocentrism by heliocentrism was the source of most opposition to Galileo's ideas because abandoning geocentrism was seen as undermining an anthropocentrism that was felt essential for human life to have meaning, at least for common people. However, as a matter of fact this was not a key issue in the history of Galileo's trial, although of course it was one of the many minor issues. Instead, the most important issue was the question of the scientific (or philosophical) authority of Scripture, which Brecht does mention, but whose significance he does not perceive. Such inaccuracies are harder to dismiss as irrelevant, but still they probably do not affect the dramatic plausibility of the play.

The key issue for the dramatic plausibility of the play involves the social betrayal, of which Galileo accuses himself in the climactic scene, and which Brecht elaborated in various self-reflective pronouncements. I would argue that Galileo's self-accusation is dramatically flawed because it is out of character with his words and actions in the rest of the play, which in this regard are not too different from the historical Galileo. That is, his fight against biblical interferences in the search for truth suggests that he would have opposed social and political interference, whether advanced by authorities and institutions or by individuals or persons; he would have wanted to distinguish the question of whether a proposition is factually or scientifically true, from the question whether its truth or acceptance is socially harmful or beneficial. The same conclusion is suggested by his frequent criticism of teleological and anthropomorphic ways of thinking that reduce to arguing that something is true because it is useful and false because it is useless. It is more likely that the historical Galileo would justify himself in the manner in which the play's Andrea justifies the master's submission to authority.

This criticism is similar to the one which in more eloquent words has been made by Eric Bentley (1966b, 21). He advances it in the context of contrasting the first version of the play, which focused on Galileo producing the *Two New Sciences* after the condemnation, to the later versions, both of which introduced the self-accusation of social betrayal. For Bentley, the change from the earlier to the later versions is a change for the worse (from a dramatic and theatrical point of view).

One final point deserves discussion. Although Brecht's accusation is neither historically accurate nor dramatically plausible as a self-accusation by Galileo, it may have some other kind of validity, perhaps a philosophical one. That is, if we think of it merely as an accusation by Brecht himself against Galileo, then it may be regarded as a social-philosophical criticism of Galileo, and it may not be devoid of all validity. But that is not to say that it is valid, for the criticism would have to withstand (among others) the Galilean arguments (which we have inherited) against the theological-pastoral, teleological, and anthropocentric ways of thinking that were the traditional and historical precursors of Brecht's social philosophy. On the other hand, such a historical context suggests the possibility that some other historical agent or dramatic character might have formulated (some appropriate version of) Brecht's socialist accusation against Galileo. Perhaps, if the play's Andrea had advanced that complaint against his master, its dramatic plausibility might have been enhanced. And if that accusation could be attributed to some appropriate historical agent, even the historical accuracy of such a Brechtian account might be vindicated.

2. In 1959, novelist Arthur Koestler (1905–1983) caused a sensation with a book entitled *The Sleepwalkers: A History of Man's Changing Vision of the Universe*. In a sense, the work was part of the increasing concern with, and discussion of, the problem of cultural fragmentation that witnessed C.P. Snow deliver the same year at Cambridge University a lecture entitled "The Two Cultures and the Scientific Revolution." The two cultures to which Snow was referring were the sciences and the humanities. Koestler too was concerned with them, but was even more concerned with the cultural division between science and religion.

Koestler discussed two main themes. One of them was the interaction of science and religion from antiquity to the seventeenth century, in regard to which he advanced the thesis that in reality science and religion share a deep commonality, but in the modern age they experience a separation that is conflictual. The second main theme was the history and nature of intellectual or scientific discovery; in this regard Koestler claimed that "the history of cosmic theories, in particular, may without exaggeration be called a history of collective obsessions and controlled schizophrenias; and the manner in which some of the most important individual discoveries were arrived at reminds one more of a sleepwalker's performance than an electronic brain's" (1959, 15). Besides being important in its own right, the sleepwalker thesis helped Koestler justify his other thesis about the deep-structural unity of science and religion. The sleepwalker thesis also had an important corollary: the debunking of science. For Koestler, this criticism was unintentional, but the consequence was inescapable.

Koestler admitted having an anti-Galilean "bias" and explained it as stemming from Galileo's unfriendly behavior toward Kepler, who fit Koestler's scheme very well and whom he liked very much. But Koestler had other reasons for disliking Galileo. One was that Galileo did *not* exemplify "the unitary source of the mystical and scientific modes of experience" (Koestler 1959, 426). For Koestler, Galileo "was utterly devoid of any mystical, contemplative leanings, in which the bitter passions could from time to time be resolved; he was unable to transcend himself and find refuge, as Kepler did in his darkest hours, in the cosmic mystery. He did not stand astride the watershed; Galileo is wholly and frighteningly modern" (Koestler 1959, 363). Another reason was connected with the "resentment" Koestler (1959, 425) felt at the fact that the clash between Galileo and the Church occurred at all, when it could have been avoided; this thesis was Koestler's peculiar way of expressing the circumstantialist approach, but his circumstantialism was such that Galileo received most of the blame.

The details of Koestler's account amount to little more than a long list of Galileo's responsibilities. Regarding the earlier proceedings, Koestler claims that although by December 1615 the Inquisition had dismissed the charges against Galileo advanced by Lorini and Caccini, he went to Rome then to defend his cause, and his defense produced the Index's condemnation of Copernicanism and Bellarmine's warning to himself. There were several factors at work: Galileo's behavior was full of hubris insofar as he resented the fact that his assessment of Copernicanism should be questioned; his attitude was one of rejection of any compromise; his rhetoric was outwardly flashy and temporarily disarming, but offensive and unconvincing in the long run; and he was obsessed with the argument from tides to try to prove the earth's motion, but the argument was invalid and ineffective.

This account is ingenious and apparently plausible, but ultimately untenable. Koestler is constantly committing the straw-man fallacy, that is, attributing to the author under scrutiny silly and untenable views by interpreting relevant texts in the most uncharitable manner and most unfavorable light. For example, he misinterprets the tidal argument as conclusive rather than probable, and his criticism is anachronistic and wise after the event. He misinterprets and misjudges Galileo's technique of strengthening opponents' arguments before criticizing them, which is really sound, wise, and fair, rather than short-sighted. He views Galileo's rejection of the instrumentalist compromise as a rejection of any compromise, and fails to see that Galileo was proposing a probabilist compromise as well as the compromise of don't condemn but don't endorse Copernicanism.

Regarding the later proceedings of 1632–1633, Koestler criticizes the *Dialogue* in many ways: allegedly, it is committed to circular orbits, after Kepler had introduced ellipses; it is silent about Copernican epicycles and about the Tychonic system; and the argument from sunspots is a deliberate attempt to confuse and mislead. Moreover, Galileo managed to offend everybody: the Jesuits were alienated by his controversies with Christoph Scheiner and Orazio Grassi; he used "sharp practice" toward book censors; and the character Simplicio caricatured the pope. Finally, at the trial he alienated the judges by insulting their intelligence with his alleged lie that the book shows the Copernican arguments to be inconclusive; he confessed guilt in the second deposition because his "superman" self-image had been exposed; and his post-script suggestion to add an anti-Copernican addendum to the book was dishonorable and was contemptuously ignored by the Inquisition.

However, again, Koestler's interpretation of the *Dialogue* is a straw-man caricature; for example, he misunderstands the sunspot argument by failing to take the diurnal motion into account. He views everything Galileo did in the worst possible light. He covertly criticizes Galileo by claiming not to criticize his change of mind from first to second deposition and to instead want to explain it; but his explanation implies a very damaging but unfair psychological criticism of Galileo. Finally, Koestler's account suffers from a general and contextual incoherence; that is, if Galileo's *Dialogue* and his other astronomical work were as inadequate as Koestler claims, this fact would not have caused the divorce of science and religion which he bemoans, but the divorce of religion from sophistry, because the book would have been seen as a piece of sophistry rather than as a scientific contribution.

8.16 The Paschini Affair (1941–1979)

In 1941, to mark the tricentennial of Galileo's death, the Pontifical Academy of Sciences commissioned Msgr. Pio Paschini to write a book on Galileo's life and work and their historical background and significance. Although the manuscript was finished in 1945, the book was not published until 1964, 2 years after Paschini's death. Here I want to discuss the reasons for the delay, the reasons for the posthumous publication, and the controversy generated by such reasons.[129]

1. When Paschini (in November 1941) was first approached by the Academy about the project, he refused since he felt the topic was outside his field (sixteenth-century ecclesiastic history). However, when the Academy's president (Gemelli) made a personal and passionate plea, stressing their admiration for Paschini's previous work and their confidence in his abilities, he was unable to decline. After about 3 years' work, the manuscript was finished and Paschini submitted it to the Academy. The manuscript was given for a review of its scientific parts to Giuseppe Armellini, Director of the Vatican Astronomical Observatory, who (except for minor revisions) found it acceptable. However, other reviewers, who evaluated the historical parts, objected to many of Paschini's judgments for reasons which I shall discuss presently. Paschini was unwilling to change the substance of his judgments. So the Academy refused to publish the manuscript and forwarded the case to the Vatican Secretariat of State. The latter forwarded the case to the Holy Office. After about a year, the Holy Office decided against publication; paid a sum of money to the author for his time and efforts; and closed the case.

Apparently Paschini was never given any written evaluation reports of his manuscript, but he was able to talk to the officials involved. He then related the substance of these discussions in various letters he wrote. From these letters, we can get a glimpse at the objections raised against his manuscript. These letters are also valuable because they contain his answers to those objections. A general and crucial objection was that Paschini's book was an apologia for Galileo; Paschini defended himself cogently from this charge by elaborating the open-mindedness and objectivity which all his work displayed. Another criticism of Paschini stemmed from his alleged failure to appreciate that Galileo did not have a conclusive demonstration of Copernicanism; Paschini replied that the lack of conclusive demonstration applied with even greater force to geocentrism. Paschini was also faulted for being too harsh in criticizing the Inquisition consultants; he replied that

---

[129] As usual, more information, documentation, and analysis can be found in Finocchiaro (2005b, 318–37). However, it should be noted that many of the essential sources in this case are held at the library of the Udine Seminary, which I have not personally consulted. Instead I have relied on the efforts of several scholars, as found in their published works: *Atti del convegno di studio su Pio Paschini nel centenario della nascita, 1878–1978* (1979); Bertolla (1979), Blackwell (1998a, 361–366), Brandmüller (1992, 20 n. 27), Fabris (1986, 8–10), Fantoli (2003b, 363–366, 562–570), Lamalle (1964), Maccarrone (1979a, b), Nonis (1979), Paschini (1964a, b, 1965), Simoncelli (1992), Tamburini (1990, 128–129).

by being tough on the consultants, he was lessening the blame on the Church as a whole and on the Inquisition as an institution. Additionally, Paschini was perceived as having been too critical of the Jesuits and the Dominicans; as a priest of the regular clergy, he felt this criticism to be irrelevant, but he also stated that it was not his fault if the members of these orders had made big mistakes during the Galileo affair. It was also objected that the book relied too much on Italian sources, to which be gave the obvious reply. And he was told that his assertions could be misused to the detriment of the Church; he replied that he did not see how this could happen given that his central aim had been to set the record straight.

2. Paschini's manuscript remained unpublished for the rest of his life, and he remained silent about the matter. In his last will and testament, he named his former student Michele Maccarrone as the legal heir of his manuscript on Galileo. Soon after Paschini's death in December 1962, Maccarrone undertook an effort to have it published by approaching all the offices and institutions with which his mentor had dealt two decades earlier. He began with the Vatican deputy secretary of state, from whom he received encouragement. Then he sought an opinion from the dean of Church history at the Jesuit Gregorian University in Rome, whose judgment was favorable. Maccarrone then approached the Pontifical Academy of Sciences, which appeared interested in publishing the book on the occasion of the 400th anniversary of Galileo's birth (1564); it commissioned Belgian Jesuit Edmond Lamalle to make appropriate revisions and updates and to edit the book. Maccarrone also contacted the Holy Office, which advanced no objections to its publication but had some reservations about its lack of novelty and the timeliness of its appearance. In July 1963, he had an audience with pope Paul VI, informing him of the steps he had taken. In the autumn of 1963, the pope reconvened the Second Vatican Council, which had been started by John XXIII; some of its discussions focused on the relationship of the church and the world, the freedom of scientific research, and the condemnation of Galileo; and it soon emerged that the publication of Paschini's book was felt to be appropriate and useful in the context of these discussions.

Thus the book was published by the Pontifical Academy in October 1964, with a preface by Lamalle. Lamalle blamed the second world war for the nonpublication of the book at the time it was written. He mentioned several weaknesses in Paschini's book: for example, it was outdated by two decades; the topic was outside Paschini's specialty; the author relied too much on the National Edition of Galileo's works, on the scholarly work of Antonio Favaro, on textual analysis, and on quotations; and so the author ended up seeing history through eyes of Galileo and his disciples. Lamalle concluded his preface with a brief discussion of his editorial practices: first, it would not have been appropriate to modify the book's perspective; nor would it have been proper to write a second book in copious and extended footnotes; and he stated that "our changes, both in the text and in the notes, have been deliberately very discreet; they were limited to corrections that seemed to us to be indispensable and to a minimum of bibliographical updating" (Lamalle 1964, xiii).

Besides being published, Paschini's book also happened to be immortalized in the proceedings of the second Vatican council. During this council there were various proposals for the rehabilitation of Galileo. In March 1964, French Dominican Dominique Dubarle forwarded to the pope a request for a solemn rehabilitation of Galileo, endorsed by many scientists and academics; the pope forwarded the request to the Inquisition, which decided (May 15) that they had already acted on the matter by approving Paschini's manuscript for publication. Moreover, during the discussions of the relationship of the Church with the world and with culture, there were motions for Galileo's rehabilitation. A committee prepared a draft deploring in general Church interferences into scientific research and explicitly mentioning Galileo's case. But there seemed to be widespread opposition to mentioning Galileo's name in any official documents and to any statements in which the Church would explicitly admit errors. Finally a compromise was reached: the official statement of the Second Vatican Council, the constitution *Gaudium et spes* approved on 7 December 1965, would contain a statement deploring in general the failure to respect "the rightful autonomy of science," which failure often leads to the widespread belief "that there is an opposition between faith and science";[130] and there was a footnote to this passage that said, "Cf. Pio Paschini, *Vita e opere di Galileo Galilei*, 2 vols., Vatican City: Pontifical Academy of Sciences, 1964."[131]

Paschini's literary heir (Maccarrone) interpreted this citation as a posthumous vindication and triumph for Paschini, pointing out that the citation was unique for any twentieth century author who was not a pope. Maccarrone also argued that this reference to Paschini was a better response by the Church to the Galilean problem than to stage a re-trial of Galileo because a re-trial "would have been anachronistic and useless" (Maccarrone 1979b, 217), whereas Paschini's work represented "the superseding of the apologetic position that had prevailed in so much Catholic historiography, especially in the delicate Galileo question" (Maccarrone 1979b, 218).

There is no question that Paschini received a personal vindication, both from the indefatigable efforts of his faithful disciple Maccarrone and from the convergence of the world-historical circumstances of the Second Vatican Council. However, regarding the more general historiographical implications of the Paschini affair, Maccarrone's assessment turned out to be more ironical than prophetic. How ironic this assessment was, we will be able to appreciate after we examine what the 1964 Vatican edition of Paschini's *Galileo* did to the original manuscript, to which I now turn.

3. In 1978, a conference was held in Udine to commemorate the one hundredth anniversary of Paschini's birth, and the proceedings were published the following year. As one would expect, several contributions (though not all) focused on the history of Paschini's work on Galileo. Moreover, the occasion afforded a perfect opportunity for scholars to study the archival materials that were held in or near Udine. One set of such materials was the correspondence between Paschini and his friend Father Giuseppe Vale. In the Library of the Udine Seminary there was also the original manuscript of Paschini's book. Maccarrone, who inherited it, had donated it to the library after it was published. One of the conference participants,

---

[130] Here quoted from Blackwell (1998a, 365).
[131] Maccarrone (1979a, 93), Simoncelli (1992, 138).

Pietro Bertolla, decided to examine Paschini's original manuscript and compare it with the published book. The results were surprising.

It was already known, from Lamalle's preface, that the editor had made some changes; thus, no one was surprised that there were some changes. Nor was the number of changes surprising, about 100 in a two-volume work of more than 700 pages; for many of the changes were minor and could be easily classified under the heading of what is normally understood as "editorial."

However, many of Lamalle's emendations changed partially or wholly Paschini's judgments. These usually involved four topics. Regarding Aristotelianism and the Jesuits, the changes toned down Paschini's negative remarks and added favorable judgments. As regards Galileo's precursors and rivals, the changes had the effect of diminishing the novelty, originality, and importance of Galileo's work. With regard to Galileo's interaction with the Inquisition, the changes made this institution appear in a better light and Galileo in a worse one. Finally, in a few cases, the changes not only completely altered Paschini's judgments, but also reversed them by turning them into the opposite of what Paschini had said. Let us examine the more significant of these changes.

In his manuscript, Paschini had stated that Galileo's hermeneutics was right, and that of the Inquisition's theologians wrong. In the published book, Lamalle added qualifications to the effect that Galileo's reasons for his correct hermeneutics were weak or unsound, whereas the theologians had good reasons for their incorrect hermeneutics.

In regard to the February 1616 consultants' judgment that Copernicanism was philosophically false and absurd, in his manuscript Paschini had tried to guess the reasons by quoting an argument from Riccioli about the role of sense experience, and had commented that this argument was childish. In the printed work, Lamalle deleted this comment, and added a footnote giving a criticism of such a possible criticism and a positive appreciation of Riccioli's "childish" reasons. For this purpose, Lamalle quoted a subtle analysis by Robert Lenoble of the distinction between common sense experience and scientific experimentation.

Thirdly, in his manuscript, Paschini's final judgment on the 1616 proceedings expressed negative evaluations of the Inquisition and the Peripatetics. In the book, Lamalle replaced these with more nuanced and favorable ones. The change was from a mostly unfavorable judgment to a mostly favorable one.

Finally, on the 1633 condemnation, in his manuscript Paschini blamed it on Peripatetics and on the fact that the Inquisition allowed itself to be used and exploited by them, quoting a 1906 scholarly article to that effect. In the book, Lamalle removed this quotation from the text; put it in a note; called it untenable and obsolete; and gave instead in the text a different interpretation gleaned from Lenoble.

All the emendations were made silently, namely without any indications that the author's original text had been changed.

Since these discrepancies came to light in 1978, most scholars have agreed that Lamalle's emendations were improper; indeed they have condemned such a practice.

The problem is that Lamalle and the Pontifical Academy were pretending to be presenting and publishing the dead author's own work, even though they were reversing many of his judgments. Now, it would have been quite feasible and proper for them to publish the original manuscript intact (except for "merely" editorial corrections of typographical errors and the like), and then have a second set of notes (besides Paschini's own) in which Lamalle made all the "corrections" he wanted. A few scholars, however, have tried to explicitly defend the legitimacy of such changes. Still others aggravate the original adulteration with silence in the context of referring to Paschini's published text; that is, they quote Paschini as an authority to support their own claims, without mentioning the fact that what they are quoting is not really Paschini's own judgments but Lamalle's emendations. But we cannot pursue any further this ongoing controversy about a recent controversy (the Paschini affair, 1941–1979) that is part of the modern controversy (Galileo affair, 1633–1992) about the original controversy (Galileo's trial, 1613–1633).

8.17 John Paul II's Rehabilitation (1979–1992)[132]

1. In 1979, at a meeting of the Pontifical Academy of Sciences commemorating the centennial of Einstein's birth, Pope John Paul II gave a speech in which he paid homage to Einstein and also conducted some Academy business; but the pope spent most of his time talking about the Galileo affair. John Paul not only admitted errors on the part of ecclesiastic persons and institutions, but seemed to acknowledge some wrongdoing on their part. In fact, he spoke (section 6)[133] of Galileo having been caused "suffering"; of his treatment being an instance of unwarranted interference into the autonomy of scientific research (section 6); and of the fact that the Second Vatican Council had "deplored" such interferences. The pope also issued a call for further studies of the Galileo affair, to be guided by three goals: bipartisan collaboration between the Galilean scientific side and the ecclesiastic religious side; open-mindedness to the wrongs of one side and the merits of the other side; and validation of the harmony between science and religion. Although this third goal was in some tension with the other two, it was the one dearest to the pope's heart. In fact, besides stating the harmony thesis, he gave a plausible justification of it by arguing that Galileo himself believed that science and religion are harmonious; that he conducted his scientific research in the spirit of religious service and worship; and that he managed to elaborate important epistemological principles about scriptural interpretation, which correspond to the correct ones later clarified and formulated by the Church herself.

---

[132] For more details, see Finocchiaro (2005b, 338–357), Finocchiaro (2008a; 2009d), and Artigas and Sánchez de Toca (2008).

[133] All editions of Pope John Paul II's 1979 speech, as well as of his 1992 speech to be examined later, preserve the same section numbers and the same subdivision into unnumbered paragraphs; thus in this section my references to them will be given in parenthesis in the text by just citing the section number and, if needed, the paragraph number that can be easily supplied by the reader.

8.17 John Paul II's Rehabilitation (1979–1992)

John Paul was aware that his account did not solve all the problems of the Galileo affair, but felt it held the key for properly understanding the rest. The pope was keen on reversing the traditional interpretation of the trial as epitomizing the conflict between science and religion. He was reviving parts of the tricentennial rehabilitation of Galileo, updating it, and placing it in a more authoritative context. It is not surprising that the speech was widely reported at the time, and continued to be commonly interpreted later, as a "rehabilitation" of Galileo.

However, talk of rehabilitation is problematic because the speech was not a papal pronouncement ex cathedra, but rather a personal opinion; nor was it a formal official action by the same tribunal (the Inquisition) that had condemned the suspected heretic. This point is similar to the one made in the context of defending the Church from the objection that the errors committed in Galileo's trial refute the doctrine of Church or papal infallibility: there, one points out that the anti-Copernican and anti-Galilean condemnations were issued by the Inquisition (which is not infallible) and by popes not speaking ex cathedra; here, one now uses the same technical clarification to deny similar status of the proposed rehabilitation. On the other hand, there is no denying that the pope's speech was an important and revealing action; one might even speak of an "informal" rehabilitation, if this phrase is not a contradiction in terms.

Moreover, the harmony interpretation is more easily said than done. For if we regard science and religion not as self-subsisting entities that exist in some relationship to one another in some Platonic heaven of abstractions, but rather as concrete historical entities that interact dynamically and reciprocally in various ways, then in the Galilean controversy one can plausibly take Copernicanism to represent science and Scripture to represent religion; and if one does this, it is indeed true for Galileo there was no real incompatibility between the two, but it is equally true that for the Inquisition the apparent conflict between Copernicus and the Bible was real. Consequently, there was an irreducible conflictual element in the Galileo affair, between those who believed that there was *no* conflict between Scripture and science, and those who believed that there *was* a conflict. On the other hand, such a minimalist conflict is something very different from the eternal war between the two alleged by the Platonist conflictualists.

2. At any rate, John Paul's words were followed by action. In 1981 he appointed a commission to study the Galileo affair. It was headed by a cardinal and subdivided into four subcommittees: exegetical, cultural, scientific-epistemological, and historical-juridical. The memorandum of appointment states that its charge was not the review or revision of the trial, nor the rehabilitation of Galileo, but rather the "rethinking" of the Galileo affair. The "rethinking" of the affair was to be free and objective, unprejudiced and open-minded; but it was also to be guided by the idea and project outlined by John Paul in his 1979 speech, namely that the Galileo affair illustrates the harmony between science and religion.

After all the media talk of rehabilitation and all the journalistic hype about a retrial of Galileo that had followed the pope's original speech of 1979, such a clarification and definition of the commission's purpose were essential. But this remained unknown at the time, and thus the talk, perception, and expectation of a rehabilitation or retrial continued. Moreover, the commission's appointment repeated the Einstein-speech's equivocation between an open-minded inquiry and a validation of the harmony thesis; indeed it heightened the tension by having a subcommittee dealing with historical and juridical issues, and yet by warning it to stay away from retrials and rehabilitations. Finally, despite the commission's high profile, it must be noted that it did not include any experts in Galilean scholarship, nor any non-Catholics, nor many (only two) laymen; this reflected the Church's traditional approach to such questions, for we should recall that on the occasion of the tricentennial commemoration, the task of writing a re-examination of Galileo had been assigned to Msgr. Paschini, who was not a Galileo specialist.

During the next decade most of the persons appointed became active studying and publishing on the Galileo affair and promoting its study by the organization of conferences and the edition and sponsorship of relevant works. The Vatican Observatory started a monograph series, and by 1992 five works had been published. And the Pontifical Academy of Sciences commissioned, sponsored, and published a number of works. Let me highlight the most important of these contributions.

In 1983, Mario D'Addio (a member of the historical-juridical subcommittee) started publishing a series of articles entitled *Considerations on Galileo's Trial*, which were later published as a book; its most crucial and revealing conclusion was probably its endorsement of Gebler's thesis that the special injunction transcript was legally worthless and inadmissible at the trial, although authentic (D'Addio 1985, 51–52). In 1984, George Coyne (director of the Vatican Observatory and a member of the scientific-epistemological subcommittee) and a historian named Ugo Baldini published a monograph containing two new documents: the first was the text of Robert Bellarmine's lectures at Louvain in 1571, which explicitly argued for the anti-Aristotelian thesis of the fluidity of the heavens on biblical grounds, and thus implicitly displayed such a biblical literalism as to make him more of a conservative than previously thought; the second was a handwritten draft of Bellarmine's certificate to Galileo dated May 1616, which revealed that the cardinal had been gracious enough to revise its wording so as to make Galileo appear in a better light. In 1989, the Vatican Observatory series published a collection of essays on Galileo's trial by a distinguished (and non-Catholic) historian of science (Richard S. Westfall); the most important one argued clearly and explicitly that Bellarmine was primarily a biblical literalist and traditionalist in scientific methodology, and so neither an epistemological instrumentalist nor a hypothetico-deductivist, and consequently Duhem's interpretation was a one-sided oversimplification. And in 1992, German priest and professor Walter Brandmüller and German scholar Egon Greipl published the voluminous documents and an historical interpretation of the Settele affair, in a massive volume sponsored (and copyrighted) by the Pontifical Academy; although extremely valuable for the documentation and historical information provided, it is less so for its interpretative thesis that the Settele affair of 1820 ended the controversy.

There is no question that the Vatican Commission generated a considerable body of work in Galilean studies. Nor can there be any question that many of these works contained valuable and useful contributions. However, one may question whether such work as a whole amounted to a rethinking of the Galileo affair, let alone a retrial and rehabilitation (which it was not even supposed to do). For by and large such work amounted to a reaffirmation, repetition, and reinforcement of the thesis that science and religion can be in harmony and Galileo's work and even Galileo's trial can help us see such harmony. This had been John Paul's thesis in the Einstein-centennial speech, appropriated from various sources (chiefly Gemelli) generated during the tricentennial rehabilitation of 1942. On the other hand, the fact that the pope had publicly elaborated such a view in 1979, and that for about a dozen years afterwards there was a Vatican commission doing the same, represented an important socio-cultural fact.

3. Actually, during the period there was also a novel apologia of the Inquisition. This was a book by Walter Brandmüller whose key theme was the "right to make mistakes." It was first published in German in 1982, then translated into Spanish in 1987, and then revised, expanded, and translated into Italian in 1992. Brandmüller admitted that the Inquisition was wrong insofar as it condemned Galileo on the grounds (among others) that the earth's motion contradicts Scripture, but argued that it would be improper to blame the Inquisition since to do so would be a denial of its right to commit its own errors. There exists a right to make mistakes because the future development of knowledge cannot be predicted, and so one cannot be blamed for not knowing at any given time what is discovered afterwards. In Galileo's trial, the Church cannot be denied this right any more than Galileo can. In regard to Galileo, "although Galileo's propaganda for the theories of Copernicus was corroborated in subsequent years by the studies of Newton, Bradley, and others, nevertheless he was mistaken precisely in regard to the evidentiary value of the arguments which he advanced in favor of Copernicus" (Brandmüller 1992, 193). On the other hand, the Inquisition erred in saying that heliocentrism was contrary to Scripture. Thus, Brandmüller also liked to belabor "the paradox of a Galileo who makes mistakes in the field of science and of a curia that makes mistakes in the field of theology. Vice versa, the curia was right in the scientific field, and Galileo was right in the interpretation of the Bible" (Brandmüller 1992, 196). This was a paradox analogous to Duhem's view that, although Galileo and Kepler were right (and their opponent wrong) in physics and astronomy, Cardinal Bellarmine and Pope Urban were right (and Galileo wrong) in logic and epistemology.

Certainly one has the right to make mistakes, which cannot be denied to either Galileo or his opponents. However, Brandmüller seems to be giving a "liberal" justification of the Inquisition, but it is unclear that such liberal principles are part of Catholic doctrine. Insofar as they are not, it would follow that a Catholic could not consistently justify the Inquisition in this manner, although a liberal could. Another objection is that, in granting Galileo the right to make mistakes, Brandmüller attributes to him mistakes too cavalierly, too superficially, too anachronistically. A third difficulty with Brandmüller's "right to make mistakes" involves the question of how this right relates to the duty to admit one's mistakes. In the hands of Brandmüller and his followers, the right to make mistakes is often equated with the right to deny one's mistakes. For example, he implicitly denies that the Inquisition, the Index, and their consultants erred in claiming the earth's motion to be philosophically false and absurd.

4. In October 1992, there was a meeting of the Pontifical Academy of Sciences at which pope John Paul heard and accepted the Vatican Commission's report on the Galileo affair. The report was given in a speech by Cardinal Paul Poupard, president of the Pontifical Council for Culture, who had been a member of that commission since its inception and had later become its chairman. Poupard began by recalling that he was reporting on an episode that went back 13 years, starting with the pope's Einstein-centennial speech in 1979 and being practically organized with the appointment of the commission 2 years later. Then the cardinal went on to describe very succinctly the commission's investigations and the publications they yielded. Two substantive results were elaborated at greater length, the first pertaining to the original controversy and Bellarmine's role, the second involving the subsequent historical aftermath.

According to Poupard, Bellarmine deserves credit for appreciating the importance of asking whether Copernicanism was demonstrably true, and whether it was compatible with scriptural statements. Presumably Bellarmine realized that Galileo had not provided a conclusive proof of the earth's motion. By contrast Galileo did not realize that he lacked such proofs, and in particular that the tidal argument which he regarded as conclusive was not so. By portraying Bellarmine as a shrewd methodologist in this manner, Poupard was not only relying on and following Brandmüller's account, but also ignoring, and indeed contradicting, two other relevant studies stemming from the commission; that is the work by Baldini and Coyne on Bellarmine's Louvain lectures of 1571 and one of Westfall's *Essays on Galileo's Trial*. Whether one praises Poupard for relying on Brandmüller or blames him for discarding Westfall's interpretation and the Baldini-Coyne documents, it is perhaps more important to point out that Poupard was in the eternal predicament of non-experts who rely on specialists: often the specialists disagree, and then one can pick and choose among them to justify whatever conclusion one wants.

For his account of Galileo's theory and practice of demonstration, Poupard also relied largely on Brandmüller's work. Here I would argue that such an interpretation represents a regress backwards of several steps compared to the analysis made by Soccorsi in 1947 and Paschini in 1943 and traceable as far back as Auzout in 1665. Among other things, Poupard was also ignoring Soccorsi's point that Galileo's lack of conclusive proof was true but irrelevant, since such lack could not (and should not) have motivated a condemnation or prohibition; and he was also ignoring Paschini's point that if heliocentrism lacked demonstration, so did geocentrism, and hence the question reduced to which side had the better and stronger arguments.

However, with respect to this topic (Galileo's view of his demonstrations) the situation is different from the previous one (Bellarmine's view of the logical situation), because among the studies produced by the Vatican commission there were not any that followed the Auzout–Soccorsi–Paschini line, and so perhaps the non-expert

8.17 John Paul II's Rehabilitation (1979–1992)

Poupard may be excused for having had no choice. On the other hand, the failure to sponsor or encourage any study along these lines could be taken as a sign that the commission had not been as objective and bipartisan in its investigations as it was supposed to be. Moreover, there was one contributor to an anthology edited in 1983 by Poupard himself (as part of the work of the cultural sub-committee of the Vatican commission) who had mentioned such an alternative interpretation; in his essay on "Galileo and the Professors of the Collegio Romano at the End of the Sixteenth Century," William Wallace (1987, 59–60) had argued that Galileo was well aware that his Copernican arguments in the *Dialogue*, while strong, were not demonstrative and conclusive. Admittedly, this claim was not stressed by Wallace and was relatively secondary in the context of his work, but the existence of Wallace's essay brings home the point that in compiling his report as he did, Poupard was choosing sides, and not always the most well-documented or well-argued.

Regarding the historical aftermath, Poupard gave an account claiming that the 1633 condemnation of Galileo had been "reformed" several times: in 1741–1744 with the Church's imprimatur for the publication of the *Dialogue*; in 1757–1758, with the abolition of the general prohibition of Copernican books; in 1820–1822, with the explicit permission of books advocating the earth's motion as a thesis; and in 1835 with the removal from the *Index* of Copernicus's and Galileo's books. Poupard's account was largely an abstract of Brandmüller's interpretation, augmented by imprecisely reported dates and carelessly described events.[134] Poupard was also gratuitously extrapolating from considerations that affected the 1616 condemnation of the Copernican doctrine to those that affected the 1633 condemnation of the person Galileo.

In general, Poupard's report was an uncritical appropriation of Brandmüller's work, not only the latter's apologia about the right to make mistakes, but also his interpretation of the aftermath of the trial (especially the Settele affair), and his paradox about a scientifically wrong but theologically right Galileo and a scientifically right but theologically wrong curia. Thus, Poupard's report (together with his activities leading to it) was perhaps an attempt to reaffirm the conviction of Galileo and undo his rehabilitation by Pope John Paul II.

5. After Poupard made his report, at the same meeting (October 1992) the pope delivered his own speech. As for the case of John Paul's 1979 speech, although the Galileo affair was the main topic of a plenary session of the Academy, it was not the only one. This time, the other, more current and routine topic was the nature of complexity as studied in mathematics, physics, chemistry, and biology. In this 1992 speech, the pope, however, shrewdly made the following insightful connection between the two topics: although the facts of complexity lead to the fragmentation of knowledge and to the philosophical problem of keeping such fragmentation under control, they also lead us to appreciate the need and importance of methodological pluralism, that is the idea "that the different branches of knowledge call for different

---

[134] These errors have been pointed out by many scholars. See, for example, Finocchiaro (2005b, 352, 426 n. 68), Artigas and Sánchez de Toca (2008, 191–193).

methods" (section 12, paragraph 1). But for John Paul a key lesson from the Galileo affair is precisely methodological pluralism; for this is what Galileo was advocating with his principle that "the intention of the Holy Spirit is to teach us how one goes to heaven and not how heaven goes";[135] whereas his theological opponents were committed to a misplaced cultural monism that led them to fail to distinguish scriptural interpretation from scientific investigation, and so to illegitimately transpose one domain into the other.

In the central part of his speech dealing explicitly with Galileo, the pope began by expressing the proper thanks and appreciation to Poupard, but did not simply endorse his report. John Paul was expressing gratitude to all members of the commission and to all experts who had participated in its projects. He explicitly mentioned the publications produced by the commission and its conclusions in general, but not Poupard's report as such. To be sure, later in his speech the pontiff mentioned and endorsed some specific theses from the report, but he was not just rubber-stamping the whole report. Thus, the pope was acknowledging and declaring that the commission had finished its work, but he was drawing his own conclusions.

John Paul went on to reiterate the theme of the science-religion harmony, which he had stressed in his original 1979 speech; which had been studied by the commission; but which had hardly been touched upon in Poupard's report. And the pope added several reflections. One that "one day we shall find ourselves in a similar situation" (section 4), and so the lessons of the Galileo affair may be useful, relevant, and applicable in the future; and at the end of his speech the pope explicitly mentioned the worrisome future area of biology and genetics. Another reflection was the memorable judgment that "Galileo, a sincere believer, showed himself to be more perceptive in this regard than the theologians who opposed him" (section 5, paragraph 4). Moreover, John Paul tried to generalize the lesson from this aspect of the episode in a way different from the methodological pluralism already mentioned; this second more general lesson involved the epistemology of interdisciplinary interaction, namely that "the birth of a new way of approaching the study of natural phenomena demands a clarification on the part of all disciplines of knowledge" (section 6, paragraph 1).

The most novel part of the speech was perhaps the discussion of the pastoral dimension of the affair. On this question, Catholic authors have usually argued that, although Galileo may have been right in astronomy and even in biblical hermeneutics, he was definitely wrong from the pastoral point of view; this requires that the mass of believers not be scandalized or misled by new discoveries, and so the dissemination of truth (if not its pursuit) must be careful not to upset popular beliefs too suddenly and must be mindful of the social and practical consequences of truth. Even such a shrewd and pro-Galilean writer as Soccorsi had been sensitive to such pastoral considerations, as we have seen, although he had used them not to criticize Galileo but rather to justify his abjuration (by attributing them to Galileo's own deliberations). Instead of siding with Galileo's opponents, John Paul's solution to

---

[135] In Finocchiaro (1989, 96), or Galilei (2008, 119). Cf. Galilei (1636, 17), Favaro 5: 319, Motta (2000, 101).

the pastoral issue was to declare "that the pastor ought to show a genuine boldness, avoiding the double trap of a hesitant attitude and of hasty judgment, both of which can cause considerable harm" (section 7, paragraph 2). He was not reversing the traditional anti-Galilean solution, but rather he was denying it, and pointing out that the correct pastoral position is one of arriving at a judicious mean between the two extremes of too much conservation and too much innovation. Thus, while he was not really siding with Galileo on the pastoral issue, his rejection of the opposite side was contextually a pro-Galilean position.

The pope went on to accept some of Poupard's specific conclusions. One was the thesis about the unity of culture in Galileo's age, together with the explanation that "this unitary character of culture, which in itself is positive and desirable even in our own day, was one of the reasons for Galileo's condemnation" (section 9, paragraph 1). John Paul also seemed to endorse Poupard's reference to Bellarmine, but the pope traced Bellarmine's key point to St. Augustine, and so the endorsement was partial and apparently diluted. Similar remarks apply to the Poupard–Brandmüller thesis that the 1633 sentence was "reformed" in subsequent history, and that "the debate ... was closed in 1820" (section 9, paragraph 3). This could be taken as an instance of uncritical acceptance by the pope of an untenable and misleading thesis. But he was so cursory about it that one gets the impression that he mostly wanted to use it to add further support to his own historical cultural thesis: that the Enlightenment fabricated the myth that Galileo's trial illustrates the conflict between scientific progress and the Catholic Church, but that this conflict is a thing of the past.

In summary, in this speech the pope was acknowledging the completion of the commission's work, as reported by Poupard. He was reiterating his own earlier view that a key lesson of the Galileo affair is the harmony between science and religion. He was clearly and explicitly praising Galileo's biblical hermeneutics, thus finalizing what might be called the *theological* rehabilitation of Galileo. John Paul was placing such a theological rehabilitation in the context of a broader philosophical appreciation, one along the lines of the epistemology of interdisciplinary relations, the other in line with methodological pluralism. And he was giving an unprecedented pastoral interpretation of the affair which, while not implying that Galileo was right (and his ecclesiastic opponents wrong) along the pastoral dimension, did suggest that he was no more wrong than they were. John Paul did not explicitly endorse Poupard's report. Although he accepted some particular conclusions, in the context of the papal speech those theses lost the anti-Galilean flavor and implications they possessed in Poupard's speech.

Chapter 9
Galileo Right for the Wrong Reasons?

The chronology sketched and described in the last chapter now needs to be interpreted more systematically and evaluated more explicitly. As suggested earlier, our plan is to interpret the subsequent Galileo affair as a controversy about whether or not the Inquisition's condemnation of Galileo in 1633 was right. And by analogy to the Copernican controversy of the sixteenth and seventeenth centuries, the Galileo-affair controversy will be analyzed in terms of the arguments and objections on both sides: arguments in favor of the Inquisition's condemnation, which is to say against Galileo; and argument against the condemnation, which is to say in favor of Galileo. Now, the most fundamental of these issues raises the question that, regardless of whether Galileo was substantively or factually right in the beliefs he held, his supporting reasons or justifying arguments may not have been, and in any case must be assessed differently and separately. We begin in this chapter with this issue.

9.1 The Problem

On 22 June 1633, as we have seen, Galileo's trial came to an end with his being condemned for a crime called *vehement suspicion of heresy*. In the words of the official sentence, the defendant had been found, "vehemently suspected of heresy, namely of having held and believed a doctrine which is false and contrary to the divine and Holy Scripture: that the sun is the center of the world and does not move from east to west, and the earth moves and is not the center of the world, and that one may hold and defend as probable an opinion after it has been declared and defined to be contrary to Holy Scripture."[1] We need not read other parts of the sentence or other documents, nor do we need to examine any other events related to the trial, to recognize immediately the following theological problem: what is the concept of heresy being used here? Is it sound? And is it being properly applied to the specific opinions mentioned?

---

[1] Finocchiaro (1989, 291), or Galilei (2008, 292); cf. Favaro 19: 405.

Going by what this passage explicitly says, it would seem that heresy is being defined as holding and believing a doctrine that is false and contrary to the Bible. This seems to stipulate two conditions for heretical belief: physical falsehood and biblical contrariety. Several questions immediately arise about this notion of heresy. Are falsehood and biblical contrariety individually necessary and jointly sufficient conditions for heretical belief? Is falsehood really an independent condition, given that (as suggested by the sentence) falsehood might be deducible from biblical contrariety? Is even biblical contrariety an independent condition, given that to establish it one has to establish the meaning of biblical statements about physical reality, and to establish such a meaning one may have to know what is physically true.[2]

At any rate, if we focus on the condition of biblical contrariety, then the same passage in the sentence seems to be misapplying the concept to the second view being described as heretical, namely, that one may hold and defend as probable an opinion contrary to Scripture. In fact, the latter view is not itself contrary to the Bible, in the sense of being contradicted by biblical assertions, since the topic never comes up in Scripture.[3] Hence, however theologically defensible its heretical character may be, some other notion of heresy would obviously be needed. In other words, we have here a self-referential impropriety.

More extensive theological and historical investigations have led some scholars[4] to the conclusion that at the time of Galileo there existed two distinct concepts of heresy, a strict theological notion and an inquisitional one. The former defines a heresy as a belief that is contrary to a pronouncement of the pope speaking ex cathedra or to a conclusion officially endorsed by an ecumenical council. The inquisitional definition takes a heresy to be a belief contrary either to the letter of the Bible, or to an injunction by an inquisitor, or to a conclusion established by theologians. I shall not pursue these questions further, however, because the theological aspect of Galileo's trial is not the object of the present inquiry. Here it suffices to have briefly mentioned this major theological issue.

Before coming to the scientific-methodological component of the affair, it will be useful to narrow the problem further by saying a few words about the legal, judicial, or jurisprudential aspect of the situation. Once again the text of the sentence is a good starting point. The passage quoted above immediately raises two questions. First, when the sentence speaks of "vehement suspicion of heresy," and not of heresy per se or formal heresy, is this description an implicit admission that the available evidence incriminating Galileo was insufficient for a verdict of guilty of formal heresy? This evidence was the publication the previous year of the *Dialogue* (1632) and the ecclesiastic obligation incurred by Galileo 17 years earlier in 1616.

---

[2] This last question is not as far-fetched as it might seem, for it was asked and answered *affirmatively* by Olivieri (1840, 57–65), who was the Inquisition's commissary during the Settele affair. See also the reconstruction of Ingoli's argument in Section 4.3 and the summary of Olivieri's position in Section 8.10.

[3] Or perhaps, as Campanella argued, the Galilean principle is actually suggested by Scripture; see Section 4.6.

[4] Garzend (1912), Giacchi (1942, 402–403). Cf. Section 8.13 above.

9.1 The Problem

The alleged crime was that his publication of the *Dialogue* was a violation of this obligation. To understand the crime, therefore, one has to understand the content and nature of the 1632 *Dialogue* and of the 1616 obligation. A crucial legal problem concerns the reality, authenticity, validity, and applicability of this obligation and whether its violation implies even suspicion of heresy.

Another question we can ask about the above-quoted passage points in the direction of the same judicial issue. For the sentence says that one of Galileo's suspected heresies is "that one may hold and defend as probable an opinion after it has been declared and defined to be contrary to Holy Scripture."[5] Thus, we may ask when, how, and by whom such a declaration was made. This question also leads us back to the year 1616 and to the problem of exactly what Galileo was forbidden to do then.

The exact details of the 1616 story are extremely complicated, as we have seen (Chapter 7 and Section 8.12). Here it suffices to mention three documents, which represent the ambiguous and inconclusive resolution of Galileo's brush with the law on that earlier occasion.

The first document is a report found in the trial proceedings stating that on 26 February 1616 Galileo had been given a special injunction by the commissary general of the Holy Office and had promised to obey. It was the injunction "to abandon completely the above-mentioned opinion that the sun stands still at the center of the world and the earth moves, and henceforth not to hold, teach, or defend it in any way whatever, either orally or in writing."[6] Although there is no question that the publication of the *Dialogue* in 1632 violated this prohibition, the authenticity or admissibility of the document is open to question.[7]

The second document is the decree issued by the Congregation of the Index on 5 March 1616.[8] Without mentioning Galileo, it declares the thesis of the earth's motion false and contrary to Scripture. It suspends, until corrected, Copernicus's book *On the Revolutions of the Heavenly Spheres*. And it prohibits totally and condemns Foscarini's *Letter on the Copernican Opinion*, which had tried to show that the motion of the earth is compatible with Scripture. Since the corrections to Copernicus's book were issued in 1620,[9] and since Galileo's *Dialogue* does not discuss in any way whether the earth's motion is consistent with the Bible, from the point of view of the Index decree it is unclear what is the violation for which Galileo could be found guilty.

The third document is the certificate given to Galileo in 1616 by Cardinal Bellarmine. It unequivocally denies the rumors that in 1615–1616 Galileo had been tried, condemned, or forced to abjure. It states that he had merely been informed of the decree by the Index to the effect that the Copernican doctrine "is contrary to Holy

---

[5] Finocchiaro (1989, 291), or Galilei (2008, 292); cf. Favaro 19: 405.
[6] Finocchiaro (1989, 147), or Galilei (2008, 176). Cf. Favaro 19: 322, Pagano (1984, 101).
[7] See Section 8.12. Cf. for example, Santillana (1955, 261–274), D'Addio (1985, 51–52).
[8] Favaro 19: 322–323, Pagano (1984, 102–103), Finocchiaro (1989, 148–150; 2005b, 16–20), Galilei (2008, 176–178).
[9] Favaro 19: 400–401, Finocchiaro (1989, 200–202; 2005b, 20–24).

Scripture and therefore cannot be defended or held."[10] In light of this certificate, the question of Galileo's guilt reduces to the question of whether his book (the *Dialogue*), besides obviously discussing Copernicanism, also holds or defends the doctrine. Such a determination would require a careful analysis of the book, plus a clarification of the conceptual distinction between the act of discussing and the acts of holding and defending. I regard this question as being open even today. The evidence developed during the trial was inconclusive because, on the one hand, the three experts consulted by the Holy Office reported that in the *Dialogue* Galileo had defended the motion of the earth;[11] on the other hand, the evidence obtained by the officials in their interrogations of the defendant, including one conducted under the formal threat of torture, indicated that he had not held or intentionally defended the doctrine.[12]

These three documents suggest the following three alternatives: either Galileo unquestionably violated an invalid injunction (the special injunction from the commissary general); or he unquestionably did not violate a valid prescription (the decree of the Index); or it is questionable whether he violated an unquestionable warning (the one by Bellarmine). And these three alternatives define a major judicial issue about the trial.

Aside from, and independent of, the theological and legal issues, the scientific-methodological aspect of the trial needs to be examined. It too arises immediately out of the text of the sentence. In fact, Galileo is being blamed for two alleged errors: the first is a belief about the physical universe, which may be summarized in terms of the theory that the earth moves; the second is a belief about procedure in physical or scientific investigation, namely, the methodological principle that one may defend as probable a physical proposition that is contrary to the Bible. To use Owen Gingerich's eloquent expression (1982, 133), Galileo's trial did indeed involve both a question about the truth of nature and a question about the nature of truth. It is important to emphasize the two-sidedness of the scientific controversy because otherwise one gets the impression that the difference between Galileo and his opponents was merely material or substantive, and this would miss the fact that what was at stake was also the rules of the game.

The existence of a methodological disagreement is well worth further discussion. First, let me make explicit that by methodological disagreement I mean a dispute over questions of principle that stipulate proper procedure in physical inquiry. The notion thus includes the issue of the role of the Bible in science, which I am emphasizing, as well as other questions stressed by other scholars, such as the difference between induction and deduction and the avoidability of metaphysics in scientific inquiry.[13]

Second, one must pay attention to the methodological controversy merely to do justice to the complexities of the historical evidence; thus not only do we find a reference to the principle of autonomy from the Bible in the sentence with which

---

[10] Finocchiaro (1989, 153), or Galilei (2008, 178). Cf. Favaro 19: 348, Pagano (1984, 138).
[11] Favaro 19: 348–360, Pagano (1984, 139–153), Finocchiaro (1989, 262–276).
[12] Favaro (19: 336–347, 361–362), Pagano (1984, 124–137, 154–155), Finocchiaro (1989, 256–262, 277–281, 286–287), Galilei (2008, 276–288).
[13] See, Gingerich (1982) and Drake (1978; 1986b), respectively.

the trial of 1633 ended, but the same principle is what started the unfortunate affair in 1612–1613, when Galileo was denounced as a heretic for his Copernican beliefs and defended himself from this charge in the letter to Castelli. This earlier phase of Galileo's troubles climaxed with the problematic prescriptions of 1616 and with his attempt at a resolution of this methodological issue in the *Letter to Christina*. The evidence of this letter tends to be neglected or misinterpreted by all those who neglect the methodological disagreement.

Third, I believe that the proper emphasis on the methodological disagreement provides a dignified and effective way of criticizing two common myths about the Galileo affair. One is the anti-clerical myth that Galileo was tried and condemned by the Catholic Church for having seen or proved that the earth moves. This myth was propagated by Voltaire (as we saw in Section 8.6) and found its way into an inscription on a monument near Villa Medici in Rome.[14] Since to condemn someone for seeing or proving the truth can only be the result of blind prejudice or ignorance, this is also the myth that tries to exploit the Galileo affair to advocate an irreconcilable conflict between science and religion or at least between science and Catholicism.[15] This myth can be and has been refuted many times by pointing out two pieces of evidence: first, that Galileo received about as much support and encouragement from clergymen and from Catholic institutions as he received opposition and criticism; second, that Galileo's own attitude was such that he did not see any necessary conflict between scientific reason and religious faith, and his religiousness and piety were sincere and strong. It follows that the Galileo affair is better interpreted as an instance of a conflict between a conservative, authoritarian trend and an innovative, liberalizing one within the Catholic Church.[16] Such a conflict is indeed irreconcilable, but it is also universal and inevitable. It is universal insofar as it affects every existing institution, from the American Supreme Court to the United Nations. And it is inevitable inasmuch as the polarity of conservation and innovation is part of the inner logic of historical development. More to the point, this myth does violence to the historical fact that in Galileo's time a strong case could be made to the effect that one could see and prove that the earth does *not* move. Therefore, by merely following what may be called the principle of charity and the principle of rationality, one would come to suspect that methods of proving and of seeing must have been at issue as well.[17]

At the opposite extreme, there is the anti-Galilean myth that Galileo failed to give the conclusive demonstration of the Copernican hypothesis which he had

---

[14] Cf. Finocchiaro (2005b, 261) and Sections 10.5 and 12.7.

[15] It might not be unfair to attribute this myth to, among others, none less than Albert Einstein, who, in his otherwise enlightening Foreword to Drake's translation of the *Dialogue*, says that "a man is here revealed who possesses the passionate will, the intelligence, and the courage to stand up as the representative of rational thinking against the host of those who, relying on the ignorance of the people and the indolence of teachers in priest's and scholar's garb, maintain and defend their positions of authority" (Einstein 1953, vii).

[16] This view has also been put forth by Santillana (1960, 326).

[17] A good example of the use of the principles of charity and of rationality is Agassi (1971).

promised, which he boasted to possess, which was required then by the agreed-upon methodological norms, and which is required now by the canons of scientific proof.[18] Thus the Church was right to condemn Galileo for his *Dialogue* of 1632, which is supposedly full of logically invalid arguments, scientific errors, and even deceptive sophistries. It also follows that the Church was upholding the cause of scientific reason in its opposition to Galileo. This view is based on an untenable misreading of the main relevant documents, the *Dialogue* of 1632 and the *Letter to Christina* of 1615, and it will be indirectly and substantively criticized in my discussion of them below. For now, I merely wish to point out that this view neglects the existence of a methodological disagreement and that it is as uncharitable and injudiciously extreme as the anti-clerical myth. The anti-Galilean myth, however, suggests certain important questions that need to be asked about the scientific-methodological aspect of Galileo's condemnation. So let us go back once again to the statement of the formal sentence.

We have seen that it attributes to Galileo two transgressions, one involving a physical belief, the other involving a methodological principle. Galileo turned out to be right in both of his beliefs: the proposition that the earth moves is as conclusively established as any scientific fact, and the propriety of holding and defending as probable physical theories contrary to the Bible is an immediate consequence of the doctrine that the Bible is not a scientific authority but an authority in matters of faith and morals. Though there might be some disagreement from fundamentalists, the doctrine now enjoys official sanction from the Catholic Church and more generally in enlightened theological circles.[19]

However, there is much more to human rationality in general, and scientific rationality in particular, however, than being materially right or holding true conclusions. Obviously, we should consider whether Galileo was logically right, that is, whether his reasons were correct. Moreover, one should consider the possibility that there may be some deeper wisdom in the opposition to Galileo by the Holy Office. Perhaps the inquisitors were materially right in concluding that he should be condemned, even if their reasons were completely invalid; that is, even if Galileo was not a suspected heretic and did not violate the injunction of 1616, perhaps he deserved censure because of scientific, methodological, or logical transgressions. For example, when the inquisitors in their sentence condemned Galileo for having believed in the motion of the earth, perhaps they could be taken to mean that they were condemning him for having hastily and prematurely so believed, or for having addressed his arguments to the general public in Italian rather than to professional experts in Latin. And when Galileo was cited for his belief that one can defend as probable a theory contrary to the Bible, perhaps the inquisitors meant that it was

---

[18] This view is propounded not only by such popular writers as Koestler (1959), but also by such scholars as McMullin (1978; 1980) and Feher (1982).

[19] Scholars usually refer to Pope Leo XIII's encyclical *Providentissimus Deus*: see Langford (1971, 66 n. 31), Dubarle (1964, 25). John Paul II also spoke approvingly in his speeches to the Pontifical Academy of Sciences (1979; 1992). Cf. Sections 8.13 and 8.17.

not proper for a nontheologian like Galileo to tell churchmen how they ought to read or not to read Sacred Scripture, or that it was not proper to water down the principle that only a conclusive demonstration of a physical truth can justify the nonliteral interpretation of relevant biblical statements.

How much truth there is in these methodological charges, or to be more exact, how little truth they contain, will be see in due course. For now, I wish to stress that these are proper questions and that to ask them is a way of formulating the problem of the logical and methodological background of Galileo's trial and of its interpretation and evaluation from the point of view of reasoning, argument, and scientific method.[20]

9.2 The *Dialogue* and Its Critics

As stated above, the two main logical and methodological discussions pertinent to Galileo's trial are his *Dialogue* of 1632 and his *Letter to Christina* of 1615. I will begin with the former since, despite its greater complexity, the resulting issues are more straightforward and more directly relevant.

The *Dialogue* deals with the question of the physical reality of the earth's motion by means of a comprehensive evaluation of arguments and evidence for and against. The discussion takes the form of the statement and criticism of the arguments against the geokinetic view and the elaboration and favorable portrayal of supporting arguments. The net effect of the comparison is that the geokinetic thesis emerges as more probable than the geostatic one. Galileo does not, however, explicitly commit himself to holding or believing in the earth's motion, so that we might say that he is pursuing, rather than accepting, the idea. A number of clarifications are needed to understand properly the book's content and structure.

The fact that the book is discussing the physical reality of the earth's motion is sometimes confused with the claim that Galileo is asserting that the earth's motion is physically real. However, all that Galileo is doing is taking seriously the phenomenon of the earth's motion, rather than regarding it as a mere instrument of calculation and prediction. To use today's terminology we might say that he is indeed an epistemological realist, but this is not the same as being a committed geokineticist. Or we might use the terminology of Galileo's time and say that the *Dialogue* is a book on natural philosophy rather than on mathematical astronomy. This point is worth stressing because some scholars (e.g., Shea 1972) approach the book as a chapter in the development of Copernicanism and technical planetary astronomy. The evidence that the *Dialogue* is more philosophical than astronomical in intent includes the oft-cited passage in which Galileo explicitly discusses the difference

---

[20] Since I recognize that the problem is genuine, perhaps I should mention that I have been inspired to formulate it explicitly by Wallace (1981a, 4; 1983a,b). As far as I can tell, however, his published views of its solution come close to those which I attempt to refute below.

between what he calls philosophical astronomy and computational astronomy,[21] and the well-known fact that for the most part the book avoids discussion of the technical details of the latter. Less frequently appreciated evidence is that the book begins by discussing the geostatic argument from natural motion, which Galileo criticizes by showing that the Aristotelian idea of natural motion on which it is based is conceptually misconceived and empirically groundless.[22]

Galileo's epistemological realism is also one of the reasons that lead some to think that the *Dialogue* was meant to offer a conclusive proof of the earth's motion. This reasoning, of course, is no more valid than the previous inference to his acceptance (as distinct from pursuit) of Copernicanism. But the strict-demonstration interpretation is supported in other ways. For example, it is said that Galileo had committed himself to provide a conclusive demonstration of the earth's motion to convince the Church to throw its support in favor of Copernicanism. Given the apparent conflict between the geokinetic thesis and the Bible, a strict proof was required in view of the universally accepted principle that biblical statements are to be interpreted literally unless they conflict with a physical truth that has been conclusively established. I will show below, in my analysis of the relevant document, the *Letter to Christina*, that Galileo was not arguing that the Church should support Copernicanism but rather that it should not condemn it. I will also show that on the basis of this universally accepted hermeneutical principle he tries to justify the novel rule that biblical statements cannot be used against physical theories that are *susceptible of being* conclusively proved. Other evidence in support of the strict demonstration thesis consists of quotations from the text of the *Dialogue*, in which Galileo has one or more of the speakers say that particular arguments in favor of Copernicanism are cogent. But all that this textual evidence shows is that he thinks his book contains strong arguments in support of the earth's motion, which is not the same as offering a necessary demonstration.[23]

Consider, for example, Galileo's tidal argument, which he felt was his strongest. It tries to show that the combination of the earth's two motions (daily axial rotation and annual orbital revolution) could provide the basis for the explanation of a number of tidal phenomena; that no other available explanation has any plausibility; and that therefore the earth must move. I believe that there are indications that Galileo felt that this argument fell short of conclusive demonstration. The evidence is that in the context of this argument Galileo has two discussions of the logic of theoretical explanation, which throw doubt on whether the explanandum is uniquely explained in this way and not in another.[24] One of these discussions uses theological language to the effect that God could have created the world different from the way it is and

---

[21] Favaro 7: 368–372, Galilei (1967, 340–345).

[22] Finocchiaro (1980, 31, 33–34, 349–353). Drake (1986b), by a very different approach, reached a similar conclusion.

[23] For a discussion of this textual evidence, see Finocchiaro (1980, 3–26).

[24] Galilei (1967, 436–444, 460–465), Favaro (7: 462–470, 484–489). Cf. Finocchiaro (1980, 18–22, 139).

9.2 The *Dialogue* and Its Critics

such that the tides would result from some other cause. This theological qualification was the favorite argument of Pope Urban VIII, and Galileo was ordered to include it in his book. The important point is that he integrates the theological qualification in the scientific discussion rather than printing it separately in a final section of the book. Although such integration was maliciously held against Galileo at the trial, the rhetorical effect is to strengthen its tie to the tidal argument and thus to suggest a degree of cogency less than strict demonstration.

To support the intended conclusiveness of the tidal argument, scholars often mention that Galileo's original and preferred title of his book was *Dialogue on the Tides*. I would criticize this interpretation by arguing that to put the tides in the title would have stressed that his treatment of the forbidden idea of the earth's motion was merely hypothetical (see Finocchiaro 1980, 12–18). Stillman Drake (1986b) arrived at a similar conclusion by a completely different route, namely, that Galileo originally intended his *Dialogue* to be a mechanical explanation of the tides in terms of the earth's motion, thus serving the Church by showing the world that the anti-Copernican decree of 1616 allowed Catholics to discuss the idea and had not halted scientific progress.

Another important piece of evidence, often neglected, attesting that Galileo could not have meant the *Dialogue* to have established the earth's motion conclusively, is that it contains at least one objection which Galileo clearly does not refute but answers by outlining a research program.[25] The objection is the argument from stellar parallax, which he discusses at great length and concerning which he shows that the failure to detect an annual parallax could have causes other than its nonexistence, namely, not knowing exactly what to look for, not making careful enough observations, and not having instruments adequate to measure it. The programmatic answer is explicit, and there is no reason whatever to suppose that Galileo forgot about this piece of counterevidence.

Having clarified that in the *Dialogue* Galileo's attitude toward Copernicanism was that of an epistemological realist as opposed to a true believer, and that of a hypothetical probabilist as opposed to a strict demonstrationist, two other issues can now be examined, namely, whether the discussion is biased and whether the arguments are fallacious or sophistical. These new methodological concerns arise out of the previous ones as follows. First, the methodological critics of Galileo portray him as a Copernican zealot and attribute to him a fanatic and total commitment to the geokinetic view; they make this attribution on the basis of insufficient evidence such as his epistemological realism or his ambiguous letter to Kepler of 1597. Such evidence would merely warrant an attribution of the attitude that the Copernican idea is fruitful and worthy of pursuit (cf. Section 3.5). Second, pretending to be nice to Galileo, these critics reason that since he had such an absolute and consummate commitment to his cause, he must have thought that he had conclusive proofs for his view; then they examine the *Dialogue* and, finding that it lacks a strict demonstration, they question his scientific judgment and methodological self-awareness.

---

[25] Favaro 7: 409–416, Galilei (1967, 383–389), Galilei (1997, 270–281).

What I am arguing is that since the *Dialogue* does not offer a conclusive proof of the earth's motion, and since there is no other conclusive evidence that Galileo had an absolute commitment to this idea (but rather evidence to the contrary), it is best to attribute to him a degree of belief in it commensurate with that of his supporting arguments: one of pursuit grounded on strong or probable arguments. It is at this juncture that our would-be critic would make a third allegation, namely, that Galileo's arguments were not that good, but ranged from the fallacious to the deceptive.[26]

On this last issue, a preliminary point is often made in order to cast doubt on the scientific correctness of Galileo's attempt, in the *Dialogue*, to prove the geokinetic theory. Some say that, as a matter of historical fact, it was not until Isaac Newton's *Mathematical Principles of Natural Philosophy*, published in 1687, that there was a scientifically correct proof of Copernicanism in general and of the earth's motion in particular. There are several difficulties with this comparison. First, whether Newton deduced or induced his gravitational principle from Kepler's laws, or merely provided a hypothetico-deductive explanation of the latter from the former, in any case he assumed Kepler's laws as premises of his argument; now, if these laws contain an assertion of the earth's motion, then it is not clear why credit for proving Copernicanism should not go to Kepler or how the Newtonian proof can avoid a *petitio principii*; and if Kepler's laws do not claim that the earth revolves around the sun (but only that Mercury, Venus, Mars, Jupiter, and Saturn do), then it is unclear how Newton's conclusion can apply to the earth's motion. Second, if scientific validity is defined in terms of the reaction by the scientific community of the time, there is no essential difference between Newton's *Principia* and Galileo's *Dialogue*. The essentials of both were favorably received by the majority of progressively minded scientists; both were critically examined by the same scientists; and for both there were cases of fundamental rejection by scientists who, though neither incompetent nor irrational, were outside the scientific mainstream.

These difficulties lead some scholars to make a different comparison and to say that the proofs sought by Galileo in the *Dialogue* did not come until the eighteenth and nineteenth centuries: James Bradley's discovery of the aberration of starlight in 1729 proved that the earth has translational motion; Friedrich Bessel's discovery of stellar parallax in 1838 first proved that our planet revolves around the sun; and Jean Foucault's pendulum of 1851 demonstrated the earth's rotation on its axis. There is no question that each of these three phenomena makes a significant contribution to the case for a moving earth; but to make invidious comparisons between them and Galileo's evidence is to ignore the very important fact that they are not, even collectively, incontrovertible; that the Galilean argument is not dismissible; and that in general it is the accumulation of arguments and evidence from Copernicus's work in 1543 to Foucault's in 1851 and beyond that makes the earth's motion the indisputable fact it is today. The geokinetic explanation of the three

---

[26] See, for example, Koestler (1959), Feyerabend (1975; 1985; 1987; 1988; 1993).

9.2 The *Dialogue* and Its Critics

phenomena mentioned above could be undermined by explaining them away through a reformulation of the laws of mechanics in more complex mathematical formalism. The result would not be simple, but simplicity would become the issue, which would bring us back to Copernicus's situation, before Galileo came along. In summary, to question the scientific validity of the argument in the *Dialogue* the way these critics do is to take the first step in a slippery slope, at the end of which one finds himself in the situation of the great French mathematician and physicist Henri Poincaré, who at the beginning of the twentieth century argued that, if one wanted to do away with the conventional simplicity of the laws of mechanics, one could still hold that the motion of the earth remains an unproven fact.[27]

With this slippery slope behind us, or to be more exact beside us, we can now take up more directly the question of the inherent soundness and cogency of Galileo's arguments. Since we have already admitted that they were not in fact, nor were meant to be, strictly demonstrative, the relevant standards will be those of probable inference, plausible reasoning, and inductive evidence. It would be impossible, of course, to undertake here an exhaustive examination, and so I shall concentrate on one instructive and important example.

The argument from the tides, which, as mentioned earlier, Galileo felt to be strongest, is typically criticized as being in fact one of his worst. Reduced to its bare bones, the argument states that the earth must move because only its motion could cause the tides. Sometimes critics point out that Galileo's tidal theory is scientifically wrong, for we know today (and at least since Newton) that tides are the effect of lunar (and solar) gravitational attraction. This criticism may be summarily dismissed for the philosophical reason that it fails to show the proper appreciation of the distinction between the falsity of a premise and the impropriety of an inference and for the historical reason that it uncontextually disregards the fact that Galileo had no way of knowing what we know today. A more serious criticism of Galileo would be that his argument is, in the context, inductively worthless because its central premise is fallaciously supported and because it leads to false empirical consequences.[28] The supporting subargument is alleged to be a fallacy of equivocation in which Galileo would be confusing two different frames of reference, one relative to the earth and the other relative to the sun. The difficulty with such critiques is that they are internally incoherent in such a way that it is not clear what reasoning they are attributing to Galileo; nor are they textually accurate. Since it is possible to provide an accurate and coherent reconstruction, it is preferable to do so. Such, I believe, would be this interpretation: the earth's motion is the best explanation of the tides because the earth's axial rotation and orbital revolution would combine in such a way as to produce daily accelerations for every point on the earth, and this acceleration would provide the primary force to set up and keep ocean water oscillating; but no other cause can explain why the tides occur. The falsehood that Galileo's tidal theory allegedly implies is that there is only one

---

[27] Hujer (1967); he refers to Poincaré (1904; 1952, 111–122).
[28] Shea (1972, 173–186); cf. Finocchiaro (1980, 76–78).

high tide and one low tide a day; in fact, normally there are two. Here we simply have a careless misreading of the text, in which Galileo is at great pains to discuss the other causes besides the earth's motion which give rise to specific features of the tides. He calls the earth's motion the *primary* cause in the sense that it sets and keeps the waters in motion, not in the sense that it alone determines either the period or other facts of the tides. My conclusion here is that although Galileo's tidal argument uses some false propositions, it does not contain any obvious improper inferences and possesses considerable inductive strength.[29]

An examination of other arguments would yield similar results. We do not have to read Koestler's *Sleepwalkers* to find misunderstanding and misdirected criticisms. For example, Galileo's anti-Aristotelian arguments can be shown not to be the *petitio principii* alleged by as reputable a scholar as Alexandre Koyré.[30] And the Galilean tower argument is not the sophistry judged by the inimitable Paul Feyerabend.[31] At this point in the logic (or should I say the illogic?) of anti-Galilean criticism, the transgression may be weakened from a charge of poor reasoning to one of lack of objectivity. Two specific biases are often held against him: his failure to keep the discussion properly balanced between the two chief world systems and his neglect of Tycho Brahe's system of the universe.

According to the latter, the planets do move around the sun, but the sun and fixed stars circle the motionless earth. The neglect of Tycho's system can also be expressed as a criticism of the previous type, in such a way that Galileo is portrayed as committing the fallacy of affirming the consequent in the argument that the Copernican system is true because the telescope reveals that the planet Venus goes through phases like those of the moon (cf. Gingerich 1982, 137). To attribute this argument to Galileo, however, would be to attack a straw man, for it is nowhere found in the text of the *Dialogue*. What we do find is the nonfallacious, though admittedly weak argument, that since all planets seem to revolve around the sun, and since the earth seems to be located between the planets Venus and Mars, it too is probably a planet and circles the sun.[32] Thus it is better to regard the Tychonic neglect as giving rise to a general underlying bias, rather than as involving a specific fallacious argument. The standard criticism is that Galileo either did not realize or else contrived to conceal the fact the Tychonic view could explain all the evidence that made the Copernican system superior to the Ptolemaic system and that therefore there was no reason to prefer the Copernican to the Tychonic system.

To clarify this situation it must be first pointed out that, whatever the situation may have been at the time when Tycho first devised his system, by the time Galileo

---

[29] For other instructive appreciations of Galileo's tidal argument, see Drake (1978, 33–49), Finocchiaro (1980, 74–79), Naylor (2007), Palmieri (1998).

[30] Koyré (1966, 219–220); cf. Finocchiaro (1980, 207–223).

[31] Feyerabend (1975, 75–92); cf. Finocchiaro (1980, 192–200) and Section 6.3.

[32] Favaro 7: 346–368, Galilei (1967, 318–340; 1997, 221–244; 2008, 233–250). Cf. Finocchiaro (1980, 40).

published his *Dialogue* the superiority of the Copernican system was clear. In fact, the only serious reason for considering the Tychonic compromise was the weight of the mechanical arguments against the earth's motion, and it was precisely one of the great accomplishments of the *Dialogue* to call these into question.

Moreover, it is not really true that Galileo neglects the Tychonic possibility. It is correct to say only that he fails to mention it by name, although he does consider the relevant content and substance of Tycho's idea. The general reason for this inclusion is that Galileo's primary interest is to discuss the physical reality of the earth's motion and not the technical details of planetary astronomy. From that point of view there is no difference between the Ptolemaic and the Tychonic systems: they are both geostatic and have only one alternative, the geokinetic one. But more specific evidence from Galileo's book can also be given. For example, the argument which he regarded to be his second most powerful one was based on the apparent motion of sunspots across the solar disk. This motion is such that they curve and slant upward for half a year and downward the other half, and such that both the curvature and slant are continuously changing so as to show straight paths twice a year when the slant is greatest and no slant twice a year when the curvature is greatest.[33] In arguing that this phenomenon is best explained as resulting from the earth's annual revolution around the sun, together with a monthly rotation by the sun on its axis, Galileo explicitly discusses the geostatic explanation. He admits that one would be possible, if in addition to the diurnal and annual motions and monthly rotation, the sun were given a fourth motion whereby its inclined axis of rotation itself rotates yearly around the axis of the ecliptic. Galileo correctly points out, however, that this explanation would have two disadvantages compared with the geokinetic theory, namely, it would be less simple and more ad hoc.

Another specific example would be the several arguments Galileo gives in favor of the earth's rotation. These apply with equal force against the Ptolemaic as against the Tychonic view. The best of these arguments is based on the law governing periods of revolutions, namely, that whenever a number of bodies revolve around a common center their periods increase with the distance.[34] Galileo was very confident of the truth of this law because he had found it to hold for Jupiter's satellites, besides the previously known case of the planets. However, he argued, in the geostatic system the diurnal motion violates this law since the sphere of the fixed stars or of the *primum mobile* circles the earth the fastest (every 24 hours) even though its distance from the center is the greatest. Galileo realized that the force of his argument is merely probable, but it is obvious that it works against the Tychonic as much as it does against the Ptolemaic system.

If the charge of anti-Tychonic bias can thus be answered, its answer is needed more because of the popularity of the criticism than because of the intrinsic plausibility of the problem. In fact, the careful reader of the *Dialogue* knows that it is

---

[33] Favaro 7: 372–383, Galilei (1967, 345–356); cf. Finocchiaro (1980, 40–41, 129–130).
[34] Favaro 7: 144–145, Galilei (1967, 18–19; 1997, 134–136; 2008, 206–207). Cf. Finocchiaro (1980, 35–36, 113–114).

primarily a work on mechanics discussing the earth's motion, and from this perspective the Tychonic system does not represent a relevant physical difference from the Ptolemaic. But the other question of bias raises a potentially more damaging critique. For the other difficulty is that, although the book claims to be a balanced discussion of two conflicting views, the geokinetic position is actually favored.[35] The problem here is that a favorable presentation is very difficult to distinguish from a defense, and the 1616 admonition, despite its other ambiguities, was clear in proscribing the defense of the geokinetic idea. In short, if Galileo had really limited himself to discussing the two sides of the controversy in an objective and balanced manner, he would have run no risk, or at least a smaller risk, of getting in trouble. Given the book he did write, perhaps the Holy Office was trying to punish his one-sided stand in favor of the earth's motion. But is it true that the *Dialogue* is biased and one-sided?.

The reason why it might be held to be so is that, although Galileo does present the arguments and evidence on both sides, he criticizes the geostatic ones and presents the geokinetic ones favorably. Since this interpretation is correct, the issue is only whether it is any lapse in objectivity to engage in such an evaluative discussion. Let us be a bit more precise in clarifying the nature of the Galilean discussion. First, in criticizing the geostatic arguments he normally gives adequate statements of them, and often he elaborates and strengthens them before refuting them. If his criticism were so hasty that it led him to state the original arguments in a sloppy manner, then his objectivity might be impugned, but as things stand this problem is not present in the book. Second, the presentation of the geokinetic arguments is indeed favorable, but, as I have already argued, that is not to say that they are portrayed as being perfect or conclusive. For example, I have already stated that the argument from the law about periods of revolution is explicitly evaluated as merely probable; the argument from the motion of sunspots is weakened somewhat by formulating the geostatic alternative explanation; and even the tidal argument is qualified by means of two methodological reflections on the logic of theoretical explanation. Thus, if Galileo's discussion contains a bias, it is that the discussion indicates that the geokinetic arguments are better than the geostatic ones. But this feature of the *Dialogue* would be objectionable and constitute a bias only if the geokinetic superiority were a distortion of the logical and methodological situation, that is, only if the geostatic arguments were really the better ones and Galileo was willfully and consciously conveying the opposite impression. The issue thus reduces essentially to one's analysis and evaluation of the arguments in Galileo's book. As my account above suggests, he was fundamentally correct in his appraisal of the relative merits of the two sides. It follows that the alleged imbalance is not his fault but a consequence of the nature of the case, and objectivity demands that one do precisely what Galileo did. In short, objectivity requires the avoidance of one-sidedness, but this does not imply the dishonesty of saying that two sides are equally strong when they are not.

---

[35] This objection is made, among others, by Gingerich (1982, 133).

This situation bears an interesting and instructive resemblance to one issue raised by recent discussions in the United States, where some states and other local governments periodically try to enact laws mandating that biology textbooks should contain what is called a "balanced treatment" of both evolutionist theories and creationist or "intelligent-design" theories. My criticism here would be not to deny that objectivity demands balance, but rather to stress that balance is not something that can or should be artificially or arbitrarily produced; what is required is a judicious rather than a mindless balancing act. Just as it was judicious for Galileo not to conceal the fact that the arguments favoring the earth's motion were better, so it may be judicious for a textbook writer today not to hide the scientific worthlessness of some hypotheses by dignifying them with a discussion.

To summarize, Galileo's *Dialogue* cannot be charged with the methodological transgression of violating objectivity, any more than it can be blamed for containing fallacious reasoning, or for the failure to perceive that its case in favor of the moving earth is not a conclusive demonstration, or for exhibiting toward the favored idea an attitude of hasty commitment rather than tentative pursuit. It follows that there is no methodological justification for his condemnation, insofar as it involved the first of the two alleged or suspected heresies mentioned in the sentence. The action of the Holy Office did not embody a deeper scientific wisdom, hidden behind the theological and judicial chaff of the unfortunate affair.

One other methodological issue was explicitly raised in the sentence, which I have mentioned in my account so far but without a detailed analysis. It is the question of whether it was proper for Galileo to believe and defend as probable a physical theory contrary to the Bible, and here we may presume that his objective discussion, together with the conclusion that the evidence favors the geokinetic theory, amounts to a case of probable belief and defense. I have already said that we know today that there is no impropriety here, but we can still ask whether Galileo had good reasons for thinking that he was proceeding properly. This brings us to the *Letter to Christina*, which he had written 18 years before the trial – at a time when he was still free to defend the earth's motion from theological objections, one year before the decree of 1616 deprived him of that freedom.

9.3 The *Letter to Christina* and Its Critics

My account of the *Letter to Christina* was presented earlier (Section 4.5), in the context of my discussion of Galilean critiques of the scriptural objection to the Copernican system. That discussion, in turn, was part of my general account of how Galileo defended Copernicus from the many types of objections that, since antiquity, had convinced almost everyone to reject the earth's motion. That account of the *Letter to Christina* need not be repeated here, but it will be summarized and adapted to the present context and the second part of this book, in which the main question is whether Galileo can be defended from the many types of objections which his defense of Copernicus has elicited since the Inquisition's condemnation of 1633. Thus, now the stress will be on the criticism found in alternative accounts of the letter.[36]

One fundamental fault of many such accounts is that they fail to properly integrate the letter with the rest Galileo's activities pertaining to the Copernican controversy. They do this when they extract from the letter as one of its central conclusions the principle that a conclusive proof of a physical truth is required in order to adopt a nonliteral interpretation of relevant biblical passages; and then they are surprised when they discover that, as we have seen, nowhere did Galileo provide the conclusive proof of the earth's motion required by its being contrary to the literal meaning of many biblical passages.

In my view, on the other hand, the letter provides the meta-cognitive reflection or philosophical theory of which the *Dialogue* is the scientific practice or de facto involvement. This correspondence becomes possible by showing that the conservative hermeneutics of the above-stated principle is not so much a main conclusion arrived at in the letter, but rather one of its starting points, although only in part. That is, Galileo takes it for granted that nobody would disagree with the principle that a conclusive proof of a physical truth is *sufficient* to force a nonliteral interpretation of the Bible, but he distinguishes this principle from the converse principle that conclusive proof is *necessary* for nonliteral interpretation. He then goes on to argue that the reason why the sufficiency principle holds is such as to justify also another more novel, controversial, and relevant principle that denies the necessity principle. The explanation of the validity of the sufficiency principle is that the Bible is not an authority in physical investigation but only in matters of faith and morals, and from this we get the novel principle that biblical statements should not be used to condemn physical conclusions which, though not yet conclusively proved, are susceptible of being conclusively proved. This novel principle of autonomy justifies what Galileo does in the *Dialogue*, for all he needs is that the geokinetic thesis should be a proposition *capable* of being conclusively proved, which was indisputable at least since the telescopic discoveries, if not since Copernicus; and then he is entitled to be left alone despite the conflict with the literal meaning of the Bible. This also means that it is no part of Galileo's intention to argue that the Church should throw its support behind the Copernican theory, but rather that it should refrain from condemning it. Given this principle of autonomy we get as an immediate consequence the one held against him by the Inquisition in the sentence, namely "that one may hold and defend as probable an opinion [about the physical world] after it has been declared and defined contrary to Holy Scripture."[37]

---

[36] These anti-Galilean critiques can be gleaned from such accounts as Carroll (1997; 1999; 2001), Koestler (1959, 434–439), Langford (1971, 69–78), McMullin (1967c, 31–35; 1998; 2005c), Moss (1983; 1986; 1993, 181–211). Note that many of them were criticized indirectly or in the notes in my earlier account (Chapter 4).

[37] Finocchiaro (1989, 291), or Galilei (2008, 292); cf. Favaro 19: 405.

A second difficulty with common accounts of the letter is that they do not formulate a coherent interpretation of all its contents. In other words, they attribute to Galileo various inconsistencies and ambiguities, but these are really the result of insufficient analysis on the critic's part, rather than correct descriptions of Galilean faults. For example, sometimes critics claim that the letter contains two views that are somewhat in tension with each other: (1) the radically novel view that biblical statements are wholly irrelevant to scientific investigation, and (2) the traditional view that scientists must provide a conclusive proof before theologians develop a nonliteral interpretation of relevant biblical passages. They try to make Galileo's inconsistency understandable by saying that, although he was inclined toward the first view, nevertheless, he was unable or unwilling to cast off the traditional view, and in any case he was confident he could provide a conclusive proof (see McMullin 1967c, 34). Other times, critics attribute to Galileo a more explicit contradiction: both the view that the Bible is not a scientific authority and the view that even in scientific inquiry, biblical statements take precedence over probable physical arguments to the contrary. In this case, besides finding insufficient analysis, we also encounter quotations taken out of context.

For example, to support its interpretation that Galileo "gives Scripture strict scientific authority over physical arguments which are only probable" (Langford 1971, 73), a widely circulating book quotes three passages, one of which is the following:

> I take this to be an orthodox and indisputable doctrine, and I find it explicitly in St. Augustine when he speaks of the shape of heaven and what we may believe concerning it. Astronomers seem to declare what is contrary to Scripture, for they hold the heavens to be spherical while the Scripture calls it 'stretched out like a curtain' [Psalm 103:2]. St. Augustine is of the opinion that we are not to be concerned lest the Bible contradict astronomers; we are to believe its authority if what they say is founded only on the conjectures of frail humanity. But if what they say is proved by *unquestionable arguments*, this holy Father does not say that the astronomers are to be ordered to dissolve their proofs and declare their own conclusions to be false. Rather, he says, it must be demonstrated that what is meant in the Bible by 'curtain' is not contrary to their proofs.[38]

One difficulty here is that this author misinterprets the contrast being attributed to St. Augustine because of his reliance on Drake's translation rather than the original Italian of Galileo's letter. In fact the contrast in Drake's translation (just quoted) is between what is "proved by unquestionable arguments" and what is "founded only on the conjectures of frail humanity," whereas Galileo contrasts what is "proved with indubitable reasons" and what is "false and based only on conjecture typical of human weakness."[39] Here Drake's translation simply omitted the "false." The difference is, of course, enormous.

---

[38] Langford (1971, 72–73), quoted from Drake (1957, 197–198). Cf. Favaro 4: 330–331, Finocchiaro (1989, 104–105), Galilei (2008, 129).

[39] Finocchiaro (1989, 104), or Galilei (2008, 129); cf. Favaro 5: 331.

Moreover, this author does not quote the immediately preceding passage, where Galileo states the view whose orthodoxy the present passage tries to justify and whose content is the referent of its initial words, "I take this to be." The "this" refers to the principle of the sufficiency of conclusive proof, which is the principle mentioned at the end of the passage just preceding:

> I wish first to remark that among physical propositions there are some with regard to which all human science and reason cannot supply more than a plausible opinion and a probable conjecture in place of a sure and demonstrated knowledge; for example, whether the stars are animate. Then there are other propositions of which we have (or may confidently expect) positive assurances through experiments, long observation, and rigorous demonstration; for example, whether or not the earth and the heavens move, and whether or not the heavens are spherical. As to the first sort of propositions, I have no doubt that where human reasoning cannot reach – and where consequently we can have no science but only opinion and faith – it is necessary in piety to comply absolutely with the strict sense of Scripture. But as to the other kind, I should think, as said before, that first we are to make certain of the fact, which will reveal to us the true senses of the Bible, and these will most certainly be found to agree with the proved fact (even though at first the words sounded otherwise), for two truths can never contradict each other.[40]

Another best-selling author claims that the *Letter to Christina* attempts to perpetrate a "sleight of hand" by shifting the burden of proof such that "it is no longer Galileo's task to prove the Copernican system, but the theologians' task to disprove it" (Koestler 1959, 437). To support this criticism, this author quotes the following passage:

> Now if truly demonstrated physical conclusions need not be subordinated to biblical passages, but the latter must rather be shown not to interfere with the former, then *before a physical proposition is condemned it must be shown to be not rigorously demonstrated* – and this is to be done not by those who hold the proposition to be true, but by those who judge it to be false. This seems very reasonable and natural, for those who believe an argument to be false may much more easily find the fallacies in it than men who consider it to be true and conclusive.[41]

This passage occurs in a discussion of whether theology is the queen of the sciences, and it is merely one of six or seven different main topics in the letter. Clearly, when the question is one of the role of the interpretation of the Bible by professional theologians, it is proper to examine what exactly they are supposed to be doing. Galileo is, in fact, trying to be helpful and constructive, by saying that theologians are entitled to regard as false a physical proposition that is not conclusively proved and conflicts with the Bible, and they should be encouraged to formulate

---

[40] Drake (1957, 197); I quote from Drake's translation in order to stress that this particular difficulty with Langford's account is primarily a decontextualization. Cf. Favaro 5: 330, Finocchiaro (1989, 104), Galilei (2008, 128–129).
[41] Koestler (1959, 437), quoted from Drake (1957, 194–195), italics added by Koestler. Cf. Favaro 5: 327, Finocchiaro (1989, 102), Galilei (2008, 126).

9.3 The *Letter to Christina* and Its Critics

scientific counterarguments and to try to find counterevidence. Galileo feels that nothing but good can result from such a procedure.

Thus, the *Letter to Christina* is not filled with the sleights of hand, contradictions, tensions, and nonsequiturs just described. Instead, it is preferable, as we saw earlier (Section 4.5), to interpret it along the following lines. Its key claim is a principle attributed by Galileo to Cardinal Cesare Baronio, stating that "the intention of the Holy Spirit is to teach us how one goes to heaven and not how heaven goes."[42] This memorable formulation is the one directly applicable in the context of the letter, namely in the criticism of the biblical objection. In fact, this objection argued that the earth must be standing still because it is so stated or implied in the Bible; and Baronio's principle directly invalidates the inferential soundness of this objection. The letter also criticizes the truth of the objection's premise, by arguing that a literal reading of the Joshua passage would contradict and thus invalidate the Ptolemaic system more than the Copernican view. And we also saw that the letter charges the biblical objection with begging the question, by arguing that in order to know what the premise really says and means one has to know whether or not the conclusion is true.

Now, Baronio's principle has a corollary formulating a norm corresponding more directly to Galileo's procedure in the *Dialogue*. The corollary is that biblical assertions about physical reality ought to be disregarded in natural science, or that scientists are free to investigate physical theories that contradict biblical assertions. In the words of the Inquisition's sentence which in 1633 found that procedure to be heretical, the corollary is "that one may hold and defend as probable an opinion ... contrary to Holy Scripture."[43] Here, I am concurring with the Inquisition's implicit interpretation of the *Dialogue*, although of course I would reject its evaluation.

However, the *Letter to Christina* could not have just assumed Baronio's principle, and so one of its central purposes was to justify it. Now, it is interesting and important that Galileo attempts what might be called an orthodox justification, namely one based on orthodox ideas. These stem primarily from St. Augustine. Galileo accepts Augustine's stress on prudence. At the substantive level, the key premise of Galileo's argument is Augustine's traditional principle that if a biblical assertion contradicts a physical claim that has been conclusively proved, the latter is to be given priority and the biblical assertion set aside or reinterpreted. The crucial step in the argument is to ask for the rationale for this traditional practice: what is the reason why conclusively proved physical truths are (traditionally and uncontroversially) given precedence over conflicting biblical assertions? Baronio's principle gives the answer and provides the rationale. That is, Baronio's principle explains why Augustine's principle is correct, and this explanation in turn justifies the former's plausibility.

Next, once one accepts Baronio's principle, one can apply it to give an answer to another question, yielding another corollary. What should one do when biblical

---

[42] Finocchiaro (1989, 96), or Galilei (2008, 119); cf. Favaro 5: 319.
[43] Finocchiaro (1989, 291), or Galilei (2008, 292); cf. Favaro 19: 405.

assertions contradict physical claims that have not yet been conclusively proved, but are capable of such a proof? The answer is that the natural philosopher should be free to examine such claims and search for a proof. This corollary must be regarded to be as well grounded as the traditional Augustinian principle; the same reasons that justify the latter, will justify the former. Clearly, if Galileo were in possession of a conclusive proof of Copernicanism, then he would not have had to write this letter, or to answer biblical criticism; he could have simply produced his proof, and the application of the traditional Augustinian principle would have easily and quickly resolved the problem. Thus, the mere writing of the letter is an indication that he felt Copernicanism was capable of conclusive proof, though not yet so proved. From this point of view, the *Dialogue* may be described as aiming to establish that the earth's motion is susceptible of conclusive proof (as distinct from establishing that this phenomenon is indeed conclusively proved).

This analysis also fits very well with another series of remarks in the *Letter to Christina* which describe Galileo's attitude toward Copernicanism, and which we examined earlier (Section 3.6). His explicit talk of "confirmation" is a way of stressing degrees of support and capacity of demonstration, as contrasted to conclusive proof and accomplished demonstration.

Thus, the *Letter to Christina* is essentially a plea for freedom of thought and inquiry in natural philosophy or science; it is not a misconceived attempt to force a non-scientific authority – the Church – to accept a scientific theory. Its central argument bears a direct logical connection with the procedural principle for which Galileo was condemned in 1633, and which is being justified. The justification is that one must be allowed to "hold and defend as probable an opinion … contrary to Holy Scripture"[44] because such probable reasoning is a necessary prerequisite for arriving at conclusively demonstrated physical truths, which provide direct understanding of the work of God and indirect help in interpreting His word. Hence, there is also a direct methodological connection with Galileo's effort in the *Dialogue*, whose procedure with respect to this issue is justified by the letter's central thesis. And from a more general point of view, the letter has also the following logical relation to the *Dialogue*: it provides the refutation of an anti-Copernican argument – the biblical objection – which is obviously not discussed in the book; it is in fact an instance of the criticism with which that work is filled. It is not an absurd attempt to justify a scientific theory – Copernicanism – by means of religious authorities and biblical texts.

9.4 Conclusion

This chapter began by formulating the scientific-methodological problem of Galileo's trial by distinguishing it from the theological and the legal problems. A major theological problem concerns the nature and validity of the concept of

---

[44] Finocchiaro (1989, 291), or Galilei (2008, 292); cf. Favaro 19: 405.

9.4 Conclusion

heresy under which he was found "vehemently suspected of heresy." The paramount legal issue involves the identity, relevance, and admissibility of the religious precept to which Galileo was bound by the developments of 1616. The scientific-methodological problem is the question whether Galileo's condemnation, though perhaps grounded on questionable theological and legal reasons, nevertheless embodies a deeper scientific wisdom. For example, though the substantive scientific belief for which Galileo was condemned – the earth's motion – is indeed factually true, perhaps his arguments were illogical and his procedure was not in conformity with the proper methodological principles.

The first thesis I established was that the trial involved explicitly at least one methodological disagreement, a dispute over whether it is proper to pursue a physical theory that is contrary to the Bible. I argued not only that this methodological issue is inherent in the relevant documentary and historical evidence, but also that to admit its existence allows one to avoid two opposite extremes: one is the anticlerical myth that Galileo was condemned for having seen and proved the truth, the other is the anti-Galilean suggestion that he deserved condemnation for his hasty, premature, and zealous commitment to the Copernican cause. Thus the Galileo affair has a doubly methodological component. One issue is implicit and involves the question whether Galileo was scientifically correct de jure as well as de facto in believing that the earth moves; to examine this problem required an analysis of Galileo's *Dialogue*. The other issue is explicit and turns on the question of exactly what relevant methodological principles Galileo felt he could be held to be accountable for, as regards the connection between science and the Bible; this required an analysis of the *Letter to Christina*.

My analysis of the *Dialogue* showed that Galileo cannot be charged with the methodological transgression of being biased, or of engaging in fallacious reasoning, or of not perceiving that his justification of Copernicanism is not a conclusive demonstration, or of having a hasty or excessive commitment to the earth's motion. My analysis of the *Letter to Christina* showed that it is in part a plausible plea for freedom of scientific inquiry because its central conclusion is that physical propositions capable of conclusive demonstration should not be condemned even if they conflict with the Bible; and in part it is a cogent threefold criticism of the biblical objection against Copernicanism because it shows that this argument from authority is based on a premise which does not support the conclusion, which in general cannot be known independently of the conclusion, and which is false per se.

Thus, Galileo was not only essentially right that the earth is in motion, but his reasoning justifying this astronomical proposition was also essentially correct. Similarly, he was basically correct in holding that Scripture is not an astronomical authority, and his arguments to justify this methodological principle were basically right. Although being essentially or basically right does not mean being perfectly or completely right, it does mean being more nearly right than wrong. And so it is not true that Galileo was right for the wrong reasons.

Chapter 10
Galileo as a Bad Theologian?

In the last chapter we saw that although Galileo was factually right in believing the thesis of the earth's motion, there are legitimate questions that can be raised about whether his reasons for this belief were logically and methodologically valid. However, although such questions are legitimate, that is not to say that the corresponding anti-Galilean criticisms are themselves correct, either factually, logically, or methodologically. Similarly, although Galileo was substantively right in believing that Scripture is not a scientific authority, his supporting arguments can and should be questioned. But again, these questions have not yielded tenable criticisms of his hermeneutical arguments.

On the other hand, the controversy over Galileo's hermeneutical views and arguments is more complex than the one over his cosmological views and arguments. For one thing, criticism of his hermeneutics has a longer history than the criticism of his astronomy. For another, it is harder to determine what his hermeneutical views were as compared with his astronomical views. Thus the hermeneutical controversy can more easily generate another genre of anti-Galilean criticism based on attributing to him views he did not hold; here the objection would be that he held theses that are substantively incorrect. This provides us with one reason why it will be instructive to examine the history of the substantive criticism of Galileo's hermeneutics. But there is another reason.

I have already mentioned, and it is worth repeating and stressing, that the Galileo affair involves not only the original controversy about the earth's motion, climaxing with the Inquisition's condemnation of Galileo in 1633, but also a subsequent controversy that began then and continues to our own day. This later affair is about the facts, causes, consequences, issues, and lessons of the original one; it reflects the issues of the original but has acquired a life and fascination of its own. Like the original controversy, the subsequent one is the subject of a considerable secondary literature, but since it covers four centuries, almost all studies focus on just particular sub-episodes.[1] Moreover, although there is now one comprehensive study of the subsequent controversy, that study is primarily an introductory survey of sources, facts, and issues.[2] That same study suggests that some important next steps would be to try to detect general trends or patterns, to formulate and justify deeper historical interpretations, and to elaborate better philosophical evaluations. This chapter is an example of that type of next step.

10.1 Relative Calm (1633–1784)

For about a century and one half after the 1633 condemnation, discussions of Galileo's trial were either private, subdued, merely implicit, brief, cursory, or vague. This is not to deny that the condemnation had important after-effects, but these do not exhibit the character of an intense on-going controversy. For example, as we have seen, immediately after the condemnation, the Church undertook an unprecedented attempt to publicize and disseminate Galileo's sentence and abjuration;[3] but obviously this was a one-sided phenomenon devoid of reasoned or critical discussion. Immediate was also the reaction of Catholic thinkers, who were forced to adjust their thinking, behavior, or rhetoric to the new reality created by Galileo's condemnation; Descartes' feeling of devastation combined with a creative re-orientation of his natural philosophy is only the most famous and spectacular case of such an adjustment.[4] We have also seen that the period of Galileo's life between his trial and his death was full of developments, such as the impassioned letters by Nicolas Claude Fabri de Peiresc to Cardinal Francesco Barberini, containing reasoned and eloquent pleas for Galileo's liberation;[5] but they were both unsuccessful and unknown to the world then. After Galileo's death, a dispute started regarding the legitimacy of his Christian burial, its location, and the decoration of his tomb; this dispute was not resolved until 1737, when his body was moved to an honorific mausoleum in the church of Santa Croce in Florence;[6] and this dispute is important and interesting as an episode of the historical aftermath of the trial; but it was not

---

[1] See, for example, Artigas and Sánchez de Toca (2008), Baldini (2000), Beltrán Marí (1998), Benítez (1999), Beretta (1999a), Borgato (1996), Brandmüller (1992, 147–198), Brandmüller and Greipl (1992), Fantoli (2003b, 345–373), Favaro (1887; 1891), Feldhay (1995, 13–25; 2000), Galluzzi (1993a,b; 1998; 2000), Garzend (1912), Hall (1980), Heilbron (1999), Howell (1996), Langford (1971, 159–188), Lerner (1998b), Maffei (1975; 1987); Mayaud (1997), Mercati (1926–1927), Monchamp (1892; 1893), Motta (1993; 1997b; 2000), Olivieri (1840; 1841a,b), Pepe (1996a,b), Pesce (1987; 1991a; 2000), Poupard (1992), Redondi (1994), Roberts (1870; 1885), Santillana (1955, 322–330), Segre (1989; 1997; 1998), Simoncelli (1992), Tabarroni (1983), and Wallace (1996, 392–396).
[2] See Finocchiaro (2005b) and Chapter 8.
[3] Section 8.2 above; cf. Finocchiaro (2005b, 26–33).
[4] Gaukroger (1995), Heilbron (1999), Finocchiaro (2005b, 43–51), and Section 8.3.
[5] Favaro (16: 169–171, 202), Finocchiaro (2005b, 53–55), and Section 8.3.
[6] Galluzzi (1993b; 1998), Finocchiaro (2005b, 81–82, 112–13), and Section 8.4.

a controversy about the facts, causes, and issues of the trial. Similar remarks apply to the various developments that occurred during the papacy of Benedict XIV, such as the republication (with the Church's approval) of the *Dialogue* as part of Galileo's works in 1744; the creation of a separate special file of proceedings of Galileo's trial out of a regular volume of Inquisition archives (in 1755); and the partial repeal of the anti-Copernican Decree of 1616 embodied in the 1758 edition of the *Index*.[7]

Such after-effects of the trial are interesting and important. But they do not constitute an explicit controversy. Nor does this character belong to the many comments and brief descriptions which began finding their way into well known works. For example, in 1644, John Milton published a classic argument for freedom of thought in an essay entitled *Areopagitica*, in which at one point he recalled his visit to Florence with these words: "There it was that I found the famous Galileo, grown old a prisoner to the Inquisition, for thinking in astronomy otherwise than the Franciscan and Dominican licensers thought."[8] In 1657, in the eighteenth of his *Provincial Letters*, in the context of the Jansenist controversy and more specifically while making a distinction between questions of fact, of reason, and of faith, Blaise Pascal addressed the Jesuits in these terms: "It was in vain that you obtained from Rome the decree against Galileo, which condemned his opinion regarding the earth's movement. It will take more than that to prove that it keeps still."[9] In 1704 in the *New Essays on Human Understanding*, in a section discussing the error of confusing hypotheses and facts Leibniz gives the following illustration:

> The Copernicans have learned from their experience of their adversaries that hypotheses which are recognized as such are still upheld with ardent zeal ... Our zeal in defense of our hypotheses seems to be merely a result of our passionate desire for self respect. It is true that those who condemned Galileo believed that the earth's state of rest was more than a hypothesis, for they held it to be in conformity with Scripture and with reason. But since then people have become aware that reason, at least, no longer supports it ... Yet they still go on suppressing the Copernican doctrine in Italy and Spain, and even in the hereditary domains of the emperor. This is greatly to the discredit of those nations; if only they had a reasonable amount of freedom in philosophizing, their minds could be raised to the most splendid discoveries.[10]

In 1734 Voltaire, as we have seen, published his *Philosophical Letters*, one of which is a critical comparison of Descartes and Newton; and he could not refrain from remarking that "the great Galileo, at the age of fourscore, groaned away in the dungeons of the Inquisition, because he had demonstrated by irrefragable proofs the motion of the earth."[11] Later (in 1753) Voltaire, in the *Essay on the Customs and Spirit of Nations*, gave a capsule (one-page) history of Galileo's trial and wrote the

---

[7] Mayaud (1997, 119–212), Beretta (1999a; 2005a), Baldini (2000, 281–347), Finocchiaro (2005b, 126–153), and Section 8.7.

[8] Milton (1644, 538); cf. Finocchiaro (2005b, 76–77).

[9] Pascal (1967, 295–296); cf. Finocchiaro (2005b, 108–110).

[10] Leibniz (1997, 515). Cf. Finocchiaro (2005b, 99–107) and Section 8.5.

[11] Voltaire (1901, 37: 167). Cf. Finocchiaro (2005b, 115–119) and Section 8.6.

memorable sentence that "the manner in which this great man was treated by the Inquisition towards the latter end of his days, would reflect perpetual disgrace on Italy, if this disgrace were not wiped off by the very glory of Galileo."[12] Finally, in 1751, in the "Preliminary Discourse" to the *Encyclopedia*, Jean D'Alembert mentioned Galileo's trial while criticizing those who claimed that religion should regulate not only faith and morals but also enlighten us about the world systems: "Theological despotism won out. A tribunal whose name still cannot be spoken without fear in France became powerful in the south of Europe, in the Indies, and the New World. It condemned a celebrated astronomer for having maintained that the earth moved and declared him a heretic."[13] Such comments are the seeds of the cause célèbre that the discussion of Galileo's trial was to become, but they do not constitute yet the real controversy as such.

10.2 Mallet du Pan's Thesis (1784)

Let us now examine what began happening about a century and one-half after the trial. In 1784, Jacques Mallet du Pan published in the *Mercure de France* a 10-page article entitled "Lies Printed on the Subject of the Persecution of Galileo." His explicit aim was to excuse or absolve the Inquisition for the alleged crime of having condemned Galileo. Mallet claimed that according to the popular view the condemnation was the result of barbarism, ineptitude, fear, and ignorance, but that this view was fiction. He argued that it was Galileo's meddling with the interpretation of the Bible, together with his petulance, that caused his misfortunes. In Mallet du Pan's own memorable words, "Galileo was persecuted not at all insofar as he was a good astronomer, but insofar as he was a bad theologian" (1784, 122). The wrong theological principle was that of supporting an astronomical thesis by means of biblical passages. The basis of Mallet's interpretation appeared to be various letters written by Galileo himself, by Piero Guicciardini, and by Francesco Niccolini, from which Mallet quoted.

For example, Mallet asserted that when Galileo was in Rome in 1615–1616, he wanted the pope to declare that the earth's motion is endorsed by the Bible; that Bellarmine's warning to Galileo was to stop trying to reconcile Scripture and the geokinetic idea, not to stop supporting this idea in other ways; that in 1616 Galileo "tried to make the question of the earth's rotation on its axis degenerate into a question of dogma" (Mallet du Pan 1784, 127); and that his defense in the 1633 proceedings consisted of biblical interpretations, not scientific arguments. Throughout his essay, Mallet stressed the kind and unusual treatment received by Galileo, such as the pope's appointment of the special commission to investigate the *Dialogue* before forwarding the case to the Inquisition; the advance notice of major developments

---

[12] Voltaire (1754–1757, part 4, chapter 1, 4–5).
[13] D'Alembert (1963, 73). Cf. Finocchiaro (2005b 120–125) and Section 8.6.

conveyed by the pope to Tuscan ambassador Niccolini; the fact that Galileo was never put in an actual prison; and the fact that when he was detained between the first and second depositions, he was allowed to lodge in the apartment of the Inquisition's prosecuting attorney.

The author of this article is as interesting as its content.[14] Mallet was a Swiss Protestant journalist, well known for his moderate views that displeased both radicals and reactionaries. Born in Geneva in 1749, he was educated in the republican and Calvinist environment of that city. In 1770, he published a pamphlet of radical political ideas, which was condemned and burned in Geneva. Voltaire recommended him for a professorship of history and *belles-lettres* to the landgrave Hesse-Cassel in Germany, but Mallet resigned his position within a few months and started working for S.N.H. Linguet's *Annales politiques, civiles et litteraires*. In 1783, he moved to Paris and became director of the *Mercure de France*, which was to become one of the most widely read magazines in France. When the revolution came, from the pages of this newspaper he was one of the very few who advanced reasonable views critical of both sides. In 1792, Louis XVI sent him on a diplomatic mission to the French émigrés, but when this failed Mallet was unable to return to France. He thus started an itinerant life, continuing his journalistic efforts, and ending up in London, where he died in 1800.

Mallet's article did not go unnoticed, and a critique of it appeared within 6 months. It is perhaps of some significance that the critical response was published in the same magazine. The author was Girolamo Ferri, an Italian writing from Rome, and the article was entitled "Apologia for Galileo."

Ferri began with the wise and insightful observation that "the exaggeration of truth is a lie" (1785, 54), and went on to argue that Mallet had indeed exaggerated the truth. Ferri admitted that several stories about Galileo were false, namely that he was tortured, that he had his eyes gouged for punishment, and that he ended his days in prison. But he refuted Mallet's claim that Galileo wanted to make a religious dogma out of Copernicanism by summarizing (correctly) the main point of the *Letter to Christina*. He advanced an interesting interpretation of the Galileo–Bellarmine meeting in February 1616: "the philosopher listened patiently to the Controversialist, and since he was not convinced, the Inquisition commissary in the presence of several witnesses forbade him to support orally or in writing the system of the earth's motion" (Ferri 1785, 57). Against Mallet's assertion that the publication of the *Dialogue* was "ridiculous disobedience" on Galileo's part, Ferri mentioned all the precautions Galileo had taken. Ferri also pointed out that the Inquisition's respectful treatment of Galileo was partly due to the fact that the grand duke was treating the matter as an affair of state. In regard to Mallet's claim that at the 1633 trial Galileo discussed not astronomical evidence but biblical interpretation, Ferri (with no knowledge of the proceedings) responded that Galileo had little choice but to do that. Mallet had based this particular claim on an alleged letter

---

[14] Cf. *Nouvelle Biographie Générale* (Paris, 1863); 33: 78–81; and *Enciclopedia italiana* (Rome, 1934), 22: 24–25.

written by Galileo to Vincenzo Renieri in December 1633, soon after returning to Arcetri. Ferri did not question the authenticity of this letter.

This letter had been circulating for a while, and the full text was published for the first time by Girolamo Tiraboschi in 1785, in vol. 8 of the Rome edition of his *History of Italian Literature* (Tiraboschi 1782–1797, 8: 147–149 n). The letter is sufficiently interesting and had such important repercussions that it deserves verbatim translation. Independently of Mallet's use of it, it represents an important document in the history of the subsequent Galileo affair. Moreover, my hunch is that the almost simultaneous publication of Mallet's article and of the Renieri letter is not an accident but is emblematic of some deeper development. The text of the letter is as follows.

10.3 Text of an Apocryphal Letter (1785)

You know very well, O most esteemed Father Vincenzo, that so far my life has been nothing but a series of accidents and episodes which only the patience of a philosopher can view with indifference, as necessary effects of the many strange revolutions to which is subject the world we inhabit. Despite the fact that we work hard at benefiting our fellow human beings, in one way or another they manage to pay us back with ingratitude, theft, accusations, and all that we find in the course of my life. Let this suffice to you, so that you will no longer have to ask me for news of the trial and of a crime which I do not know to have committed. In your last letter of June 17 of this year, you asked me to relate what happened to me in Rome and what was the attitude toward me of the father commissary Ippolito Lancio and of the assessor monsignor Alessandro Vitrici. These are the names of the officials I still remember, although I am now told that they have both changed and that monsignor Pietro Paolo Febei has been appointed assessor and father Vincenzo Macolani commissary. I find interesting a tribunal in which, because I was reasonable, I was regarded as little less than a heretic. Who knows? Perhaps people will turn me from a professional philosopher into a professional historian of the Inquisition! On the other hand, others want me so much to be an example of Italian ignorance and stupidity, that in the end I may have to pretend having these traits. Dear Father Vincenzo, I am not unwilling to put on paper my feelings on what you asked me, provided that in sending you this letter we take the precautions I took when responding to Signor Lottario Sarsi Sigensano; under this name lay hidden the Jesuit Father Orazio Grassi, author of the *Astronomical and Philosophical Balance*, who managed to wound me together with our common friend Signor Mario Guiducci. Letters were then not enough, and I had to publish *The Assayer* and place it under the wings of the bees of Urban VIII, so that they could sting Grassi and defend me. But this letter will suffice to you, for I do not feel like writing a book on my trial and the Inquisition, having little inclination to be a theologian and still less to be a criminologist.

Ever since I was a young man, I had thought of publishing a *Dialogue on the Two Systems, Ptolemaic and Copernican*; to this end, from the time I went to Padua as a professor I had constantly made observations and philosophized, guided principally by an idea that came to me, namely to account for the ebb and flow of the sea by means of some assumed motions of the earth. I said a few words about this when prince Gustavus of Sweden honored me by coming to my lectures in Padua; as a young man he was traveling incognito in Italy and stopped there with his entourage for many months; it so happened that I was able to be of service to him by means of my new speculations and the curious problems which I was formulating and solving every day; he also wanted me to teach him the Tuscan language. My ideas on the earth's motion became known to the public in Rome through a long essay addressed to the most excellent Lord Cardinal Orsini; thus I was accused of being a scandalous and temerarious writer. After the publication of my *Dialogues*, I was called to Rome by the Congregation of the Holy Office; having arrived on February 10, 1633, I was subjected to the supreme clemency of that tribunal and of the sovereign pontiff Urban VIII; he regarded me as worthy of his esteem even though I did not know how to compose epigrams and love sonnets. I was placed under house arrest at the delicious palace of Trinità de' Monti, residence of the Tuscan ambassador. The following day, the father commissary Lancio came to see me and took me with him for a coach ride; along the way he asked me various questions and appeared to be zealously concerned that I should remedy the scandal I had caused in all of Italy by supporting the opinion of the earth's motion; to all the solid and mathematical arguments I presented him with, he responded nothing but that "the Earth stays eternally still … the earth stays eternally still," as Scripture says. While carrying on such a dialogue, we arrived at the palace of the Holy Office. This is located to the east of the magnificent church of St. Peter. I was immediately introduced by the commissary to the assessor monsignor Vitrici, who had near him two Dominican friars. With civility, they ordered me to present my reasons to the full Congregation, and they told me that if I were found guilty there would be an opportunity for a defense.

The following Thursday, I appeared before the Congregation, but unfortunately, as I started to present my proofs, it decided they would not be heard; as much as I tried, I was never allowed to explain them. By means of zealous digressions, the Congregation tried to convince me of the scandal, and the text of Scripture was always advanced as the Achilles' heel of my crime. Having remembered a scriptural argument, I advanced it, but with little success. I said that there are in the Bible expressions which correspond to what the ancients believed in regard to astronomical science, and that the passage in question could be of this kind; for example, I added, in Job, chapter 37, verse 18, it is stated that the heavens are solid and polished like a mirror made of copper or bronze. Elihu is the one who says this. Thus we see that here he speaks in accordance with the system of Ptolemy, which has been demonstrated absurd by modern philosophy and by the most solid part of right reason. If, then, we focus on the stopping of the sun by Joshua to demonstrate that the sun moves, we must also consider this passage which says that heaven consists of many heavens that are like mirrors. This consequence seemed right to me, but it was overlooked; the only reply I got was a shrugging of shoulders, the typical resort of those who are prejudiced and have a pre-formed opinion. Finally, I was obliged to retract this opinion of mine like a true Catholic, and as a punishment my *Dialogue* was prohibited; having been in Rome five months (while the city of Florence was infected with the plague), for prison I was sent (out of generous mercy) to the residence of my dearest friend in Siena, Monsignor Archbishop Piccolomini. I enjoyed his most cordial conversation with such inner peace and satisfaction that I resumed my studies there and found and demonstrated most of my mechanical conclusions on the resistance of solids, together with other speculations. After about five months, when the plague in my motherland had ceased, at the beginning of December of this year 1633, His Holiness commuted my imprisonment from the limits of that house to the freedom of the countryside I enjoy so much; thus I returned to my villa at Bellosguardo and then to Arcetri where I am now living, breathing this salubrious air near my dear motherland Florence. Be well.[15]

---

[15] For my translation, I have used and compared the text found in Tiraboschi (1782–1797, 8: 147–149 n) and in Albèri (1842–1856, 7: 40–43).

10.4 Gaetani's Forgery

This letter deserves a detailed analysis, interpretation, and evaluation, to expose its many inaccuracies, to question the un-Galilean flavor of its language and style, to identify possible sources of various elements of it, and to trace the influence exercised by some of its specific claims. However, here the focus is as follows.

First and foremost, the letter is apocryphal.[16] An Italian man of letters named Pietro Giordani (1774–1848), who lived in Florence in 1824–1830, became suspicious of the authenticity of this letter primarily on account of the fact that the style seemed un-Galilean. When he expressed this doubt to the grand duke of Tuscany, the latter ordered an investigation. This centered around the library in Rome of the Gaetani[17] family, who held the title of dukes of Sermoneta. In fact, editors of the letter usually claimed that the original manuscript was kept there. When the original manuscript was located, it was discovered not only that the handwriting was not Galileo's, but also that there was a note at the end of the letter stating that it had been forged by one of the Gaetani dukes for the purpose of deliberately deceiving Girolamo Tiraboschi (who, as mentioned above, was the first to publish the letter in 1785).

This note does not specify which member of the Gaetani family perpetrated the forgery. But there were only two brothers in the relevant period, the older layman Francesco (1738–1810) and the younger clergyman Onorato (1742–1797), and Giovanni Battista Nelli implies that Onorato Gaetani was the responsible one. In fact, in his 1793 biography of Galileo, Nelli (1793, 1: 129 n. 1) quoted a brief passage of the letter to support his claim that Galileo taught the Italian language to Gustavus Adolphus (1594–1632), king of Sweden from 1611; Nelli noted that he had received a copy of the letter in March 1770, and that the original was in Rome in possession of monsignor Onorato Gaetani, duke of Sermoneta. Nelli was apparently unaware that the letter was apocryphal, for with this quotation he thought he was confirming Vincenzio Viviani's claim (1654, 629) that Galileo had taught the royal prince Gustavus. However, here it must be mentioned that this claim about king Gustavus Adophus is one of the letter's many inaccuracies; for later scholars have determined that king Gustavus was never in Italy, and that the person involved was probably the Gustavus who was the son of Erik XIV of Sweden and who lived all his life in exile.[18]

---

[16] My account relies on Albèri (1842–1856, 7: 40 n. 1), Nelli (1793, 1: 129 n. 1), Martin (1868, 160, 193, 212, 393, 400), Favaro (1905, 144–145), Fiorani (1969, pp. 77–78 n. 158, p. 157; 1973a,b), Mayaud (1997, 224 n. 35), Monsagrati (2000).

[17] I retain the spelling of this name used by Albèri, Martin, and Nelli, even though the *Dizionario biografico degli italiani* (Fiorani 1973a,b) and Fiorani (1969) prefer an initial 'C' (Caetani) rather than 'G'.

[18] Venturi (1818–1821, 1: 19–20), Casini (1985, 40 n. 23).

10.4 Gaetani's Forgery

In 1848, Eugenio Albèri for purely historiographical reasons included the letter in the eighth volume of his edition of Galileo's collected works; but he noted in the same context that the letter was a fabrication by a certain duke Gaetani, whose explicit purpose was to deceive Tiraboschi; Albèri also credited Pietro Giordani with the feat of exposing the forgery. Antonio Favaro, the editor of the standard National Edition of Galileo's collected works, agreed that the letter was apocryphal, and so it does not appear in that edition; and it is interesting to note that Favaro apparently did not find any letter by Renieri to Galileo, dated 17 June 1633, mentioned in the apocryphal text.

Second, the letter does advance an account of Galileo's condemnation as theologically based. It portrays the inquisitors at the 1633 trial as unwilling to even listen to any scientific arguments and concerned exclusively with biblical interpretation. It depicts Galileo at the trial as engaged in biblical exegesis. And the letter's account is vivid enough that anyone acquainted with it could easily arrive at the same conclusion as Mallet, even without having read his article; that is, the conclusion that Galileo was condemned not for being a good astronomer, but for being a bad theologian. Thus, although Mallet's article and Gaetani's forgery each contain several other points involving other important aspects of the Galileo affair, they agree, correspond, and overlap in regard to the following apologia of the Inquisition: Galileo was a bad theologian insofar as he advocated the erroneous theological principle that scriptural passages should be used to support astronomical conclusions; that the Inquisition condemned him for this reason; and that therefore the Inquisition was right to condemn him.

Thirdly, I should make it clear that I regard such an account as untenable, indeed false, and even perverse insofar as it does not merely fall short of the truth, but inverts and subverts the truth. I would want to argue that the Inquisition was wrong to condemn Galileo since he preached and practiced the principle that scriptural passages should *not* be used in astronomical investigation, but only when dealing with questions of faith and morals; since the Inquisition found *this* principle intolerable and abominably erroneous and held and wanted to uphold the *opposite* principle that Scripture *is* a scientific authority, as well as a moral and religious one; and since on this question of theological and epistemological principle Galileo was right and the Inquisition wrong. Such conclusions can be substantiated by a careful study of the original events of Galileo's trial and a sound analysis of the relevant texts.[19] In this regard, John Paul II's pronouncements are emblematic and significant. While his rehabilitation of Galileo turned out to be a disappointment in some respects,[20] the pope was clear that Galileo advanced sound principles regarding the relationship between biblical interpretation and scientific investigation, that Galileo was a good theologian, so to speak. In these particular statements John Paul was partly echoing previous papal declarations, such as the 1893 encyclical *Providentissimus Deus*.

---

[19] See Chapters 4 and 9 above. Cf. Fantoli (2003b), Finocchiaro (1989, 28–30), Howell (1996, 2002), Motta (2000), Pesce (1987; 1991a; 2000).

[20] See Finocchiaro (2005b, 338–357) and Section 8.17.

At any rate, the substantive and detailed re-refutation of such an account is not within the scope of this chapter. Here my aim is to suggest that the account is not merely false, but also a myth; to identify the myth, sketch its development, and glimpse at its pre-history; and to explore the possibility that the heated controversy we know today really began in the mid 1780's with Mallet's article and Gaetani's forged letter and did not really exist beforehand.

10.5 Diffusion and Development of a Myth (1790–1908)

1. One of the first places in which the Mallet–Gaetani myth took root was Antoine Bérault-Bercastel's history of the church. This was a monumental work of 24 volumes published in Paris in the period from 1778 to 1790. Volume 21 was published in 1790 and deals with the period from the beginning of Jansenism in 1630 to the Treaty of Westphalia in 1648. On pp. 140–146 of this volume we find a section entitled "The Affair of Galileo with the Inquisition." The section begins with these words: "At about the time when the Holy Office was endorsed by Urban VIII, this institution rendered (in the name of this pope) a judgment on which all the efforts of a crowd of historians or orators have propagated nothing but the densest shadows. After almost two centuries during which, regarding the subject of the celebrated Galileo, one has uttered shouts of barbarism and ignorance against the Inquisition, one has almost annihilated the memory of what really happened in the course of this affair. Thus it will not be useless to relate it. Here it is." (Bérault-Bercastel 1778–1790, 21: 140). Then the author gives an account which reads like an abridged reprint of Mallet's article. What saves Bérault-Bercastel from our charge of plagiarism is the fact that he explicitly gave a precise reference to Mallet's article and expressed his reliance on Mallet. For example, the church historian ended his account by saying: "There you have the truth of the story of Galileo and his judges, which had been so strangely deformed. We owe its discovery to the sound criticism and fair-mindedness of a citizen of Geneva, who in such a matter is a guarantor beyond suspicion" (Bérault-Bercastel 1778–1790, 21: 146).

Although Bérault-Bercastel's appropriation of Mallet's account is quite uncritical, because it is more succinct, less polemical, and more focused on the biblical issue, it gains somewhat in rhetorical effectiveness. That is, certain aspects of the account emerge more clearly and incisively. For example, whereas the following contrast between Copernicus and Galileo does not appear until the fifth paragraph of Mallet's article, in Bérault-Bercastel it appears in the initial paragraph, as the first substantive thesis of the account, immediately following the introductory remarks quoted in the previous paragraph. Here is Bérault-Bercastel's version: "Copernicus was the first who promulgated that the earth turned round the sun, but did so in a manner purely as an opinion in relation to physics, and had never been reproved by any tribunal. Galileo was not satisfied with adopting that theory and publicizing it extensively; he undertook to confirm it by the Holy Scriptures. He converted a subject of a speculative kind in natural philosophy into a dogmatic controversy, and dared even to attempt to oblige the Inquisition to make a decision in favor of his views."[21]

Similarly, Bérault-Bercastel's abridged account makes it easier to detect the (real or invented) documentary basis of the account. He mostly retains Mallet's references to and quotations from the Renieri apocryphal letter, which thus constitutes a proportionately larger part of the abridged account. Moreover, a different problem gets aired, involving the letter dated 4 March 1616, which the Tuscan ambassador Piero Guicciardini wrote to the Tuscan government at the time of the anti-Copernican decree of the Index. Mallet had referred to, paraphrased, and quoted extensively from this letter, which is highly critical of Galileo. By contrast Bérault-Bercastel (1778–1790, 21: 141) quotes only a brief sentence from Guicciardini, namely that while in Rome in 1616 Galileo "demanded that the pope and the Holy Office declare the system of Copernicus founded on the Bible." A comparison with Mallet's corresponding passage and with the text of Guicciardini's letter reveals that the just quoted sentence does not appear in the original letter; in fact, it is a portion of Mallet's own quotation which must be regarded as a fabrication.[22] The error or swindle is easy to overlook when one reads Mallet, but becomes more striking in Bérault-Bercastel. Thus, while even the anti-Galilean Guicciardini had not blamed Galileo for wanting Copernicanism endorsed by the Church on biblical grounds, in the hands of Mallet and Bérault-Bercastel, Galileo becomes guilty of this fault as well.

It is not difficult to understand and sympathize with Bérault-Bercastel's uncritical appropriation of Mallet's account. The Catholic historian could have hardly failed to be impressed by the fact that a Protestant, apparently motivated by the desire for historical truth, had come to the defense of the Inquisition. In fact, Bérault-Bercastel was not the only one to be impressed in this manner.

2. At about the same time, Mallet's article started being turned into encyclopedia entries. One of these involved Nicolas Bergier's *Dictionnaire de théologie* (1788–1790). In its entry on Galileo, it is stated that the usual view of Galileo's trial "is a calumny which we will refute … in the entry SCIENCE."[23] It formulates the usual view as the thesis that "this scientist was persecuted and imprisoned by the Inquisition for having taught with Copernicus that the earth turns around the sun."[24] To be more precise the relevant entry is entitled "Sciences Humaines,"[25] whose last third deals with Galileo's trial and is essentially a digest of Mallet's article, with an explicit acknowledgment to him. Two things are noteworthy in Bergier's account. He explicitly connects the interpretation of Galileo's trial with the question of the relationship between science and religion, when he begins by saying that one of the main things alleged to prove that Christianity is the enemy of science is Galileo's

---

[21] Bérault-Bercastel (1778–1790, 21: 140–146), here quoted from the English translation in Madden (1863, 144–145).

[22] Mallet du Pan (1784, 124), Favaro 12: 241–243.

[23] Here quoted from Bergier (1823, vol. 3, column 464).

[24] Here quoted from Bergier (1823, vol. 3, column 464).

[25] Reprinted in Bergier (1823, vol. 7, pp. 365–370).

trial. And he explicitly repeats Mallet's most characteristic slogan, when he goes on to say that this incompatibility thesis is untenable because Galileo was not persecuted for being a good astronomer but for being a bad theologian.[26]

Another similar encyclopedia entry with wide repercussions appeared in François Feller's *Dictionnaire historique* (1797).[27] Although he discussed Galileo's life in general and his scientific work in particular, and although he referred to many primary and secondary works, a large part of the entry dealt with the trial and relied on Mallet's article and the Renieri apocryphal letter. Feller was even acquainted with and mentioned Ferri's criticism of Mallet, but dismissed it. The only noteworthy item to add here is that Feller included several other myths besides that of Galileo as a bad theologian. That is, he spoke of Galileo's illegitimate birth, even though it was about 50 years since this claim had been conclusively refuted, and various scholars had explicitly retracted and corrected the claim;[28] and Feller also presented, without qualification as if it were true, the legendary claim that immediately after the abjuration at the trial, Galileo uttered the words *e pur si muove*.[29]

3. The Mallet–Gaetani myth received a significant amount of publicity in 1797 when the Bérault-Bercastel version was reprinted by Tiraboschi (1782–1797, 10: 362–364 n. 1) in vol. 10 of the Rome edition of his *History of Italian Literature*. In fact, Bérault-Bercastel's account appeared quoted in full in a long supplementary footnote which Tiraboschi added to the printed text of one of his lectures. Such publicity may be regarded as a second indirect contribution by Tiraboschi to the diffusion of the bad-theologian myth, besides his complicity a decade earlier when he had published the Renieri apocryphal letter. However, there is another important development that began to unfold; that is, Tiraboschi apparently was encouraged to elaborate and publish his own apologia of the Church, which differed from the bad-theologian thesis. Tiraboschi's account was summarized earlier (Section 8.8), when we saw that it is important in its own right, for it constitutes an apologia that is elegant, well-argued, and original, and that became influential. Here I only want to stress its divergence from and debt to Mallet du Pan.

Tiraboschi advanced his account in two lectures which he delivered in 1792 and in 1793, and which were later (1797) published in an Appendix to the last volume (no. 10) of the Rome edition of his *History of Italian Literature*.[30] In the first lecture, entitled "On the First Proponents of the Copernican System," Tiraboschi argued that

---

[26] Here quoted from Bergier (1823, vol. 7, p. 369).

[27] Although I have been able to consult only the eighth edition of this encyclopedia published in 1832, the article in question may be dated at 1797, the date of the second edition, because this edition mentions Mallet and the first one appeared before Mallet's article; Feller died in 1802. Cf. Feller (1782; 1797; 1832).

[28] The claim had been advanced in Erythraeus (1643); Brucker (1742–1744) repeated it, but later retracted it (Brucker 1766–1767, vol. 4, part 2, p. 634); cf. Nelli (1793, 1: 25–26).

[29] For more on the myth of *e pur si muove*, see Section 8.6.

[30] Tiraboschi (1782–1797, 10: 362–383); cf. Finocchiaro (2005b, 165–172).

before Galileo the Church was supportive of the new theory. In the second lecture, entitled "On the Condemnation of Galileo and of the Copernican System," he articulated and defended the thesis that Galileo was condemned primarily because he was too aggressive, zealous, and rash in advocating Copernicanism, and that if he had not defended the earth's motion in that manner, he would not have gotten in trouble with the Church. In regard to the biblical issue, he implicitly rejected Mallet's thesis when he said that the main point of Galileo's letters to Castelli and to Christina is that the literal interpretation of the Bible is binding only for questions of faith and morals and not for physical questions. On this principle Tiraboschi commented that "this proposition may in a sense be acknowledged to be true; nevertheless, it was deemed to be and was in fact dangerous, especially at that time, when the memory was still fresh of the painful losses which the Roman Church had suffered in the North, and which originated in large measure from the innovators' freedom to interpret Sacred Scripture as they wished and to give it the sense that suited them best."[31]

Despite his rejection of the bad-theologian thesis, Tiraboschi was probably influenced by Mallet's article in the sense of being encouraged to develop an anti-Galilean account. In support of such influence, I would point out that Tiraboschi's own earlier (1780) account of Galileo's life and work in the first edition of his *History of Italian Literature*, although it included the story of the trial, was relatively matter-of-fact and stayed away from controversial and sensitive issues.[32]

4. The Tiraboschi type of apologia later began to compete with the Mallet type. For example, in 1820 in the context of the Settele controversy, the Inquisition's commissary who took a relatively progressive stand, Maurizio Olivieri, stated that Bérault-Bercastel's account was incorrect and indicated his inclination toward that of Tiraboschi (cf. Brandmüller and Greipl 1992, 200–201). Olivieri later (in 1840) elaborated his own alternative view of Galileo as a bad scientist.[33] And in 1835, Scottish physicist David Brewster developed his own version of Tiraboschi's anti-Galilean rashness thesis; working with no apparent knowledge of Tiraboschi and in the context of elaborating an account that is both anti-clerical and anti-Galilean, as we have seen, Brewster portrayed Galileo as reckless and too bold in 1613–1616, insofar as he failed to appreciate the understandable mental inertia of his opponents.[34]

5. An important repercussion of the Mallet–Gaetani myth occurred in 1838 when it showed signs of having spread beyond continental Europe to the British Isles. This happened with the publication in the *Dublin Review* of an article by Peter Cooper entitled "Galileo – The Roman Inquisition." Of course, we would not expect a simple repetition or mere English translation of the account, but rather some kind of adaptation in the light of the different conditions in which the author found himself and the about half a century of intellectual and political developments

---

[31] In Finocchiaro (2005b, 168); cf. Tiraboschi (1782–1797, 10: 378).

[32] Tiraboschi (1772–1782, 8: 123–144; 1782–1797, 8: 143–172).

[33] See Olivieri (1840; 1841a,b), Bonora (1872).

[34] Brewster (1835; 1841), Section 8.11.

of European history. Moreover, Cooper's article also deserves attention in its own right insofar as it may be the first lengthy apologetic account of Galileo's trial in the English language; and it is also interesting as an expression of the attitude of English or Irish Catholics.

Cooper's article was occasioned in part by the publication of John Elliot Drinkwater Bethune's book *Life of Galileo* (in 1829–1832) and of the brief account of the trial included in William Whewell's *History of the Inductive Sciences* (in 1837).[35] In fact these works are listed at the beginning of the article in the style in which this happens when a review article is published. However, the initial paragraph makes it obvious that Cooper is also animated by the same motive that had inspired Mallet, namely to dispel the many falsehoods spread in regard to the trial of Galileo, for example that he was held in prison for 5 years, that his eyes were gouged out as punishment, and that it is indicative of religion's jealousy and mistrust toward science. Moreover, the broad scope of Cooper's article is suggested by the fact that he shows direct acquaintance with some of the works available to Mallet, such as Angelo Fabroni's collection of letters (of 1773–1775) and Giovanni Targioni Tozzetti's (1780) book on Tuscan science in the seventeenth century, but also some acquaintance with works published after Mallet's article, such as those of Giovanni Battista Nelli (1793), Jean Biot (1816), Giambattista Venturi (1818–1821), and Jean Delambre (1821).

Cooper elaborates and defends the following theses: Copernicanism was never properly and officially declared a heresy (this of course is absolutely correct). Galileo was the one to begin interfering with biblical interpretation (this is a mythological distortion à la Mallet). In 1615–1616, Galileo acted imprudently, and that's why the Church silenced him (this is an interpretation similar to Tiraboschi's apologia). Galileo wanted the Church to endorse Copernicanism (again, this is an element of Mallet's myth). Galileo was condemned insofar as he was a bad theologian, not insofar as he was a good astronomer (this is the canonical formulation of the Mallet thesis). He believed wrongly that Copernicanism had been conclusively proved. He believed wrongly that the tidal argument provided a conclusive proof. In 1633, he was condemned for disobedience, not for heresy. The Church favored science by directly supporting Copernicanism before Galileo and by indirectly supporting first the Lincean Academy and later the Accademia Fisicomatematica.

In regard to Whewell, besides other objections, Cooper focused on Whewell's contrast between the poor condition of science in Italy and the good situation in England and Germany. Cooper's criticism consists primarily of a long, bitter, and sarcastic recitation of various problems in Great Britain such as the regressive manner of dealing with geological questions in early nineteenth century; British persecution of Catholics; Britain's two-centuries' delay of the Gregorian reform of the calendar; the High Ecclesiastic Commission Council for the suppression of heresy, or Star Chamber, abolished by Charles I; and the persecution and torture of clergyman Edmund Peacham.

---

[35] Drinkwater Bethune (1830; 1832; 1833), Whewell (1837). Cooper's article led Whewell to revise his original account; see Whewell (1847; 1857a,b).

10.5 Diffusion and Development of a Myth (1790–1908)

In the context of such a wide-ranging discussion, and in addition to an average amount of inaccuracies, however, we find uncritical references to and uncritical acceptance of the Renieri apocryphal letter, Bergier's and Bérault-Bercastel's accounts, and the fabricated sentence from Guicciardini's letter of 4 March 1616 (cf. Cooper 1838, 88, 95 n, 108 n). Moreover, Cooper exploits the key point of Mallet's thesis to formulate a novel and original apology. His own words are worth quoting:

> Of the evidence, then, which we have adduced – and in stating it we have held back no one circumstance of the slightest importance – the following appears to us to be the legitimate summary: that the distinguished individual with whose story we have been all this while occupied, was never condemned – never indeed so much as arraigned – but once; and then not for his science, or his religion, or any other mere matter of opinion whatsoever, but for the *moral* fault of having in a most flagrant manner transgressed a solemn injunction placed on him by the highest tribunal in the land; a tribunal to which he had himself appealed – whose decision he loudly and pertinaciously demanded, and at last succeeded in extorting. For the transgression of an injunction like this, aggravated, too, by circumstances of insult and contumely against the authority that awarded it, was he condemned for the first and last time, towards the close of his life, [in] 1633; in one word for a grievous contempt of court. [Cooper 1838, 104]

Despite some incoherence between this account and Mallet's, the contradiction is more apparent than real. The apparent tension emerges when, besides denying the traditional view that Galileo was condemned for his science, Cooper also denied that Galileo was condemned for his theology. The inconsistency evaporates if we note that Cooper was relatively clear about the chronology of the situation, and that this explanation was meant to apply to the one and only time that Galileo was formally tried, namely to the condemnation of 1633. Cooper was saying that in 1633 Galileo was not condemned for his Copernican astronomy, nor for his biblical hermeneutics, but for his immoral disobedience of the special injunction of 1616. Now, let us ask Cooper why the alleged disobedience is immoral. I believe his answer would be that Galileo's disobedience was immoral primarily because he himself believed that the Inquisition had the right to pass judgment on such matters; and the evidence that Galileo believed this is that in 1615–1616 he supposedly himself tried to have the Inquisition rule favorably on the earth's motion, and do so on biblical grounds. But this last claim is essentially Mallet's thesis.

Mallet himself may have been somewhat unclear as to whether his interpretation applied to the first or the second phases of the trial, but once one distinguishes the two phases, then Cooper can go on holding the bad-theologian thesis for the first phase and the immoral-disobedience claim for the second phase. That is how, underlying the apparent inconsistency between Cooper and Mallet, there is a deeper agreement; Cooper is using Mallet's thesis to elaborate and justify his own novel criticism of Galileo. Since the disobedience thesis went on to gain wide acceptance and continues to be highly popular in our own day,[36] this discussion of Cooper's

---

[36] See for example, Blackwell (1998a, 355), Brandmüller (1992, 144–146), Gingerich (1995, 342), Mayaud (1997, 313). Cf. Section 12.6 below.

apparently original formulation, and of how it was grafted onto Mallet's older account, is of some importance.

6. Soon after Cooper's account was first published in 1838, it (and with it the Mallet–Gaetani myth) spread to America.[37] During the U.S. Congress debates, in 1838–1846, over the founding of the Smithsonian Institution, congressman and ex-president John Quincy Adams delivered several speeches in various cities in favor of astronomy in general and an astronomical observatory in particular. One of these lectures was given at the Cincinnati Astronomical Society on 10 November 1843. Some of Adams's assertions offended and upset Catholics, and within several months a reply was published in the form of an anonymous book consisting of Cooper's 1838 article from the *Dublin Review*, preceded by a long introduction explicitly and severely critical of Adams's lecture. Although the book did not indicate any name of editor or author of the introduction, the title page was revealing enough: *Galileo – The Roman Inquisition: A Defence of the Catholic Church from the Charge of Having Persecuted Galileo for His Philosophical Opinions; from the Dublin Review, with an Introduction by an American Catholic*, Cincinnati: Published for the Catholic Book Society by Monfort and Conahans (1844). This unnamed "American Catholic" may be identified as John B. Purcell, archbishop of Cincinnati.[38]

There is little question that Adams's account of Galileo is relatively superficial, contains factual inaccuracies, and expresses interpretations unflattering to the Catholic Church. Purcell tries to capitalize on the factual inaccuracies to discredit Adams in general, but also tries to refute his interpretations by using Cooper's account (and also Brewster's), including the claim that Galileo was not condemned for his scientific ideas but for theologically unsound hermeneutics. One irony of this exchange is that Purcell failed to see that Adams happened to share a key part of this claim, and so their disagreement involved deeper questions, rather than merely factual issues.

In fact, Adams had explained the 1633 condemnation in these terms: "At 70 years of age, Galileo was compelled by the sentence of these inquisitor cardinals to crave pardon for having maintained the truth, and abjured it as absurdity, error and heresy, upon his knees, with his hands upon the gospel" (quoted in Purcell 1844, 10). And then Adams had derived the following more general lesson: "In the lives of Copernicus, of Tycho Brahe, of Kepler, and of Galileo, we see the destiny of almost all the great benefactors of mankind. We see, too, the irrepressible energies of the human mind, in the pursuit of knowledge and of truth, in conflict with the prejudices, the envy, the jealousy, the hatred, and the lawless power of their contemporaries upon the earth" (quoted in Purcell 1844, 11). To this Purcell replied:

> Galileo strove not for truth, but for victory! For the vindication of the Church from the odious charge of persecuting science in the person of Galileo, we do not choose to rest

---

[37] See Adams (1843), Bemis (1956, 518–20), Cooper (1844), Portolano (2000).

[38] I owe this information to private correspondence from Dr. Philip Shoemaker, who refers to the *Daily Cincinnati Enquirer*, 13 November 1843, p. 2, columns 1 and 2.

content with the palliative statement of Hallam that "for eighty years the theory of the Earth's motion had been maintained without censure; and it could only be the greater boldness of Galileo which drew down upon him the notice of the Church."[39] Nor with the admission of Sir David Brewster that "the Church party were not disposed to interfere with the prosecution of Science, however much they may have dreaded its influence."[40] Nor yet with the very candid and eloquent exposition of the Edinburgh Review (October 1837) which, more than anything we have seen from Protestant authority, presents this point in its true light ... The facts connected with Galileo's first attempt to force from Rome the concession that the Copernican doctrine was consistent with Scripture, set this important matter at rest. We will not here anticipate the admirable exposition of this point and its accompanying circumstances made by the writer in the Dublin Review. [Purcell 1844, 19–20]

The penultimate sentence in this quotation embodies the Mallet–Gaetani myth. However, this was a thesis with which Adams seemed to agree. In fact, when describing the initial phase (1613–1616) of Galileo's troubles, Adams had asserted:

He was denounced, before the tribunal of the inquisition, and in his own defence, wrote memoir upon memoir, to prevail upon the Pope, and the inquisitors, to declare the Copernican system, to be in strict conformity with the Holy Scriptures. As the Pope, and seven cardinals,[41] appointed by him to solve this knotty question, pronounced, that the doctrine of the earth's motion, was an absurdity in physics, and a damnable heresy in religion, Galileo was expressly forbidden, ever again to maintain, by word of mouth, or in writing, that the rotary motion of the earth was countenanced by the holy scriptures. [Quoted in Purcell 1844, 10]

The difference between Adams and Purcell seems to be that for Adams the Inquisition had no right to silence Galileo for his opinions, be it the physical idea that the earth moves, or the theological claim that the earth's motion is compatible with Scripture; hence he interprets the condemnation as an abuse of power. On the other hand, for Purcell the Inquisition did have the right to limit Galileo's freedom of thought, especially regarding the question of the compatibility of the earth's motion with Scripture; and that's why portraying Galileo as playing the theologian (and a bad one at that) becomes crucial for justifying the Inquisition's condemnation.

7. Although the Mallet–Gaetani myth became increasingly untenable as more and more documentation and reflection on the trial were published, it proved relatively impervious to rational and critical refutation, at least in some of its forms. Its next important appearance occurred in Marino Marini's infamous apologia of 1850. We have already come across this work in connection with the torture question (Section 8.11), whereas now I want to explain how it advanced the bad-theologian thesis.

---

[39] Quoted by Purcell from Hallam (1837–1839, 4: 16).
[40] Purcell gives no bibliographical reference for this quotation; his article refers to two different editions of Brewster's account (1835; 1841), which have almost identical text.
[41] Adams is here confusing the 1633 proceedings with the 1616 ones. It was the 1633 proceedings which concluded with an Inquisition sentence signed by 7 (out of 10) cardinal-inquisitors. On the other hand, it was in February 1616 that the physical absurdity and the religious erroneousness of the earth's motion were declared (unanimously) in a committee report of 11 Inquisition consultants, but the cardinal-inquisitors did not endorse that report and so did not themselves issue a formal decree. See Finocchiaro (1989, 146, 291).

Marini's book reached new heights of extremism, as may be seen from the fact that he ends his account of the trial by claiming to have shown that "to render due praise to the justice, wisdom, and moderation of the Inquisition, we must affirm that perhaps there has never been a judicial action as just and as wise as this one" (Marini 1850, 141), meaning Galileo's trial. What inflamed the situation to an unprecedented level was the fact that Marini's book was the Church's first offer to the public after the file of original Inquisition proceedings was given back to her in 1843, following Napoleon's transfer (in 1810) of the Vatican Archives to France and his plan to have the trial proceedings published and translated into French; we have seen that this plan did not come to fruition, and that after the fall of Napoleon the file was lost and believed to have been destroyed.[42] When it resurfaced in 1843 and was returned to Rome, the general expectation was that the Church herself would publish it. As prefect of the Vatican Secret Archives, Marini instead published his book. The work does contain an unusual amount of quotations from the proceedings, which were not accessible to anyone else. It was obvious, however, that these quotations were highly selective and taken out of context.

Marini gave a general justification of the Inquisition based on its historical role in saving Europe from heresy, and on a comparison between its practices and the treatment of heretics by Protestant churches and of ordinary criminals by lay state courts. He argued at length that, although Galileo was threatened with torture, he was not actually tortured, criticizing the contrary claims of Guglielmo Libri. He charged Galileo with all kinds of inconsistency, insincerity, and imprudence. He argued that the Inquisition acted justly and wisely in opposing, trying, and condemning Galileo, because (1) he supported a physical theory by means of biblical passages and expected the Church to do the same and to officially endorse it; (2) he violated Bellarmine's injunction, by publishing the *Dialogue*; and (3) he ridiculed pope Urban VIII through the character of Simplicio.

Marini's discussion of the last allegation (3) is a good example of excessive indulgence in portraying the situation in the worst light against Galileo; it soon gave rise to a new apologetic line. In discussing the disobedience thesis (2), Marini failed to mention that the special injunction document lacks Galileo's signature and that in the first deposition Galileo denied knowledge of the crucial clause in the special injunction; and Marini acted as if the injunction did not contradict Bellarmine's certificate. Marini's discussion of the first thesis (1) just mentioned brings us to the main thread of my present analysis.

Marini was probably aware of the apocryphal character of the Renieri letter, because he did not quote or even mention it at all. On the other hand, he relied uncritically on Mallet's 1784 article, and on his repeaters, Feller, Bergier, and Cooper.[43] Marini explicitly referred to them with approval several times, and he was

---

[42] For the details of this fascinating story, see Barbier (1811; 1812; 1814a,b), Delambre (1821, 1: xix–xxxii), Favaro (1887), Finocchiaro (2005b, 175–192), Gebler (1879, 319–329), Marini (1850, 143–153), Mercati (1926–1927), Pagano (1984, 10–26), and Section 8.9.

[43] For Mallet, see Marini (1850, 39 n. 2, 141); for Bergier, see Marini (1850, 39 n. 2, 94); for Feller, see Marini (1850, 54); for Cooper, see Marini (1850, 6).

especially effusive about Mallet. In fact, Marini ended his book's central chapter dealing with Galileo's trial as follows. After asserting, as quoted above, that the Inquisition's condemnation of Galileo was the wisest and most fair judicial act in recorded history, he added: "To remove any suspicion of partiality [on my part] toward the Inquisition, I refer the reader to the dissertation by the Genevan Mallet du Pan, included in Geneva's literary review of 1784, aimed against Voltaire and the French encyclopedists; in it are rebutted all the insults which some bad writers are accustomed to vomit against the Inquisition when the discussion turns to Galileo; in his affair, that Protestant affirms, he was wrong on all counts" (Marini 1850, 141). Marini also quoted the adulterated passage from Guicciardini's letter, containing the fabricated sentence that Galileo "wanted that the pope and the Holy Office declare the system of Copernicus founded on the Bible."[44]

However, Marini also did something unique and admittedly ingenious and clever, exploiting the documentation available only to him. This involves the Inquisition minutes of 16 June 1633, at which meeting the pope decided how to bring the trial to a conclusion; the minutes contain the ambiguous and much discussed order to interrogate Galileo one more time under torture or threat of torture. A less widely discussed part of the minutes contains a sentence stating that Galileo "is to be enjoined that in the future he must no longer treat in any way (in writing or orally) of the earth's motion or sun's stability, *nor of the opposite*, on pain of relapse."[45] The crucial phrase is "nor of the opposite" and refers to the opposite of the Copernican hypothesis, namely to the geostatic thesis. Why was such a phrase included? The reason is, Marini answered, that Galileo was to be told that it is wrong for him to argue in favor of the earth's rest, as well as in favor of the earth's motion. Galileo's preferred manner of arguing was allegedly to use biblical texts in support of a physical claim. The Church thus was telling Galileo to stop this theologically unsound practice. Of course, this is the Mallet–Gaetani myth. Marini's own formulation is worth quoting:

> The Inquisition's indignation against Galileo was not elicited by his teaching a heretical doctrine, such as the Copernican one, regarding which the Church had been silent for so many years, but by his manner of elaborating it. For neither the Inquisition nor the Congregation of the Index prohibited the supposition that the sun stands still and the earth moves. Instead, what moved the Inquisition to indignation was to want to guarantee this supposition by means of scriptural texts, so as to show it to be in accordance with the meaning of the sacred word. Thus, Galileo's abuse of Scripture had brought against him the two Congregations, the Inquisition and the Index. [Marini 1850, 53]

As if the abuse of wanting to support an astronomical hypothesis with biblical texts were not enough, Marini went on to claim that "Galileo wanted to have it proclaimed a dogma of the faith ... but the Inquisition did not want to transform a scientific opinion into a dogma, and on account of this it rejected Galileo's absurd pretensions" (1850, 5).

---

[44] Marini (1850, 94). Cf. Favaro (12: 241–243), Mallet du Pan (1784, 124).
[45] Here quoted from Finocchiaro (2005b, 247), italics added. Cf. Favaro 19: 283, Pagano (1984, 154). The crucial phrase is "nor of the opposite," *et e contra* in the original Latin.

In conclusion, although Marini may be said to have advanced a new shred of evidence in favor of the Galileo-bad-theologian thesis, it is questionable whether this ought to be counted to his credit; for such an effort is accompanied by a stunning blindness to the mountains of evidence against it. One of these mountains (though not the only one) is the whole *Dialogue*, where it is clear that Galileo stays away from biblical-based arguments.

8. The German-speaking world was not immune to the diffusion of the myth. This came about through the work of Alfred von Reumont, a German who spent much of his life in Italy, in part as an official in the Prussian legation in Florence. He authored a book-length essay entitled "Galileo and Rome," first published in 1849 in a Berlin periodical and then in a revised version as part of his multi-volume work entitled *Contributions to Italian History* (see Reumont 1849; 1853). The essay is a work of documentation as well as interpretation, containing German translations of relevant correspondence. It discusses many topics and issues, and strives with some success to be impartial.

Here its main interest is that it represents the clear dissociation of the interpretive and evaluative claims from much of its original (pseudo) documentary base. In fact, on the one hand Reumont is aware of the apocryphal character of the Renieri letter. On the other hand, regarding one of the key issues he says that "Galileo's great mistake was, that he insisted on bringing into conformity with the Scriptures the doctrine of the earth's motion … So that in the interpretation of certain passages in the Bible an arbitrary discretion was assumed, which the Church, according to her invariable principles, could not concede to an astronomical doctrine as yet unproved."[46] Such a view could be supported by a partial (and ultimately incorrect) reading of Galileo's letters to Castelli and to Christina.

9. The myth acquired another boost in 1863, this time in the English-speaking world, with the publication of what is perhaps the first English-language book entirely devoted to a discussion of the Galileo affair. An Irishman named Richard Madden was the author, and his purpose was also multi-faceted. For example, a major theme of the work is the origin, content, and interpretation of the Inquisition archives that had found their way into the library of Trinity College, Dublin. Another useful aspect of the book is that it has several chapters containing a critical discussion of and lengthy quotations from various works on Galileo's trial.

In regard to the Galileo-bad-theologian thesis, one can get an idea of the author's view from the fact that he thought very highly of Marini's book. In the appendix where Madden quoted and discussed it, he mentioned that Marini discussed two important issues: whether Galileo was tortured and whether the root cause of his condemnation was his good astronomy or his bad theology. Madden remarks that "these two matters are treated in a manner by Mons. Marini that leaves nothing to be desired, and, I may add, nothing to be said henceforward, by any man who makes a study of this subject for truth's sake, against the conclusions that Mons. Marini has arrived at" (Madden 1863, 209). Equally revealing and emblematic is the fact that

---

[46] Reumont (1853), as quoted and translated in Madden (1863, 129).

10.5 Diffusion and Development of a Myth (1790–1908)

Madden gives a complete English translation of Bérault-Bercastel's account, which he praises in the following terms: "the Abbé Bercastel in his '*Histoire de l'Eglise*' (Tome xi, pp. 80 to 83), a work of authority and good repute among Roman Catholics, treats this subject extensively, and evidently with a good knowledge of the main facts of it" (Madden 1863, 144). And since, as we have seen, that account ends with a reference to the article of a "citizen of Geneva," not otherwise identified, Madden identifies him for the reader with these words: "the trustworthy witness referred to in the last passage was a Protestant gentleman of celebrity for his literary talents, Mallet du Pan. See article in '*Mercure de France*,' 17 July, 1784 … The article of Mallet du Pan ought to have been published *in extenso* by Bercastel" (Madden 1863, 147–48).

Moreover, it ought to come as no surprise that in the chapter where Madden formulated his own synthetic account "On the Justifiability of the Proceedings against Galileo," he had explicit references to and quotations from Feller, Cooper, and Reumont, besides Marini and Bérault-Bercastel. Madden's own summary conclusion does not lack a certain eloquence, however convoluted and perverse:

> Might it not, I say, have sufficed for every legitimate purpose to have said – it was a hard thing to compel that old man to retract an opinion on a scientific subject which was not at variance with any religious dogma or doctrine, and which he was firmly persuaded was true. But supposing the condemnation, in point of fact, was not of the opinion, but virtually – though not nominally and in distinct terms, as it ought to have been – of those adventitious circumstances surrounding it – such as Galileo's alleged attempts, of which there can be no reasonable doubt, to produce Scriptural authority in support of an astronomical theory, and to get those attempts sanctioned by the highest ecclesiastical authorities – it would be impossible on any just grounds for Christian men of any church, who had a just sense of the dignity and divine origin of their religion, to find fault with such a condemnation.[47]

10. After having crossed national and continental boundaries, linguistic cultures, and problem orientations, and after having withstood historical and political upheavals, the Mallet–Gaetani myth crossed social boundaries. What I mean is that it spread from the relatively high culture of scholarship and erudition to the level of popular culture, where one judges that the experts have established some proposition and feels entitled to propagate it without much argument or evidence.

On 21 April 1887, a marble column commemorating Galileo was inaugurated in Rome, near the Villa Medici Palace. It reads: "The palace next to this spot,/which belonged formerly to the Medici,/was a prison for Galileo Galilei,/guilty of having seen/the earth turn around the sun./SPQR / MDCCCLXXXVII." The historical vicissitudes and conceptual content of this monument will be discussed later (Section 12.7)[48] as an example of the conflict between science and religion in the subsequent Galileo affair. Here I want merely to document the fact that an ecclesiastic reaction to the monument indicates that the bad-theologian thesis was alive and well at that time.

---

[47] Madden (1863, 156); for a more detailed critical analysis of Madden's account, see Finocchiaro (2005b, 237–240).

[48] For more details on this development, see Berggren and Sjöstedt (1996, 145–147).

On 23 April 1887, in an unsigned article entitled "Epigrafi ed offese" (i.e., "Epigraphs and Insults"), *L'Osservatore Romano* criticized the Villa Medici inscription for various reasons. Primarily it objected that Galileo was not condemned for having "seen the earth turn around the sun" because this proposition was never formally declared a heresy by any pope. "What, then, was Galilei's fault ...? If you do not believe us, you should believe Sarpi himself, the impartial Balbo, and Guicciardini, who recognize that the great Tuscan philosopher had the fault and the recklessness of wanting to change the physical and astronomical question into a theological one. You should believe Galileo himself, who, in the last years of his life, regretted having engaged in arbitrary interpretations of the Bible based on private judgment, which were especially dangerous at that time when this was the practice of the heretics in many parts of Europe."[49] While the last reference is unclear, it probably refers to the Renieri letter; the reference to Guicciardini is probably a reference to his letter of 4 March 1616. Since I have previously mentioned that, although critical of Galileo, this letter does not attribute to him an attempt to base Copernicanism on the Bible, the article is probably relying on those accounts which misquote the letter by fabricating a crucial sentence. And indeed in the previous paragraph, the unsigned article mentions several authors, among whom are Marini and Mallet.

This account in a popular and nonscholarly medium such as *L'Osservatore Romano* would perhaps be dismissed by some as inconsequential. On the other hand, the authoritativeness of this newspaper has no peers. Moreover, it should be obvious by now that the Galileo affair is not and should not be treated merely as a controversy in the history of science, of historiography, of scholarship, and of erudition, for it became a cause célèbre in the history of culture. Thus, this evidence from the Vatican newspaper is highly significant.

11. At any rate, there is evidence that the myth also continued to flourish even in the context of high culture and first-rate scholarship and erudition. I am referring to the work of Duhem. His account has already been previewed in earlier chapters, and it will be presented and evaluated in detail later as a classic example of the genre of criticism of Galileo as a bad epistemologist (Chapter 11). Here it is useful to establish its connection with the Mallet myth.

One of Duhem's key theses is that Galileo's condemnation was the result not of principled disagreement between people holding opposite standards, but rather of substantive disagreement between epistemological realists, that is between the "impenitent realism of Galileo" and "the intransigent realism of the Peripatetics of the Holy Office."[50] Now, in a terminologically arbitrary move, under the label of realism Duhem includes the principle, among others, that acceptable astronomical hypotheses must be reconcilable with the Bible.[51] Thus, when Galileo adopted the Copernican system, besides using other requirements (such as compatibility with

---

[49] "Epigrafi ed offese," *L'Osservatore Romano*, 23 April 1887.

[50] Duhem (1908, 135); cf. Duhem (1969, 112).

[51] Duhem (1908, 126–127); cf. Duhem (1969, 105).

physics), "he was induced to reconcile his assertions with the texts of the Bible. In due course, he turned theologian."[52] That is, he supposedly adopted biblical compatibility as the second of "the two signs which both Copernicans and Ptolemaics by common agreement required of all acceptable astronomical hypotheses."[53] Duhem mentions Galileo's *Letter to Christina* as the basis for this attribution.

In the history of the Mallet–Gaetani myth, Duhem's originality lies in the fact that he attributes the bad theological principle first stressed by Mallet not only to Galileo but also to the Inquisition. They were both bad theologians, unlike Cardinal Bellarmine and Pope Urban, who according to Duhem rejected realism and advocated an epistemological instrumentalism that could have avoided the tragedy.

10.6 Metamorphosis of the Myth (1909–1959)

The myth of Galileo as a bad theologian seems to have disappeared in the course of the twentieth century. I know of no clear or significant statements of it after Duhem's version in 1908. Even such compendia of anti-Galilean criticism as Adolf Müller's books of 1909 and 1911 and Arthur Koestler's *Sleepwalkers* of 1959 do not include it. However, they do speak as if Galileo interfered improperly in theological matters and regard this interference as a factor contributing to his condemnation. This thesis is a far cry from the original Mallet myth but may be taken to be its metamorphosis.

For Müller, it is as if Galileo's fault now becomes his being a good theologian (so to speak), despite his lack of theological training and position:

> although he had never undertaken theological studies and indeed had frequently declared his incompetence in this subject, one must agree with Grisar that (except for a few inaccuracies) he advances principles which all theologians would accept today, and which even then (at least in theory) were shared by the more open-minded theologians. But ... for Galileo to present himself with theological weapons was extremely presumptuous given the situation at the time. There has been much discussion about who brought the question into the theological arena ... In 1879 Reusch said that "this was not done by Galilei but by his enemies. In his published writings Galilei did not touch at all on the theological aspect of the question, and if he did it in the letters to Castelli and to Christina of Lorraine, he did it only because he was forced by his enemies."[54]

And then Müller goes on to criticize Reusch's thesis by arguing that Galileo's enemies must be subdivided into at least two groups (philosophers and clergymen), and that although philosophers Sizzi and Colombe preceded Galileo, he preceded clergymen Lorini and Caccini, and in any case he outdid everyone else in exacerbating the problem.

---

[52] Duhem (1908, 127); cf. Duhem (1969, 105).
[53] Duhem (1908, 128); cf. Duhem (1969, 106).
[54] Müller (1911, 139–140); cf. Reusch (1879, 55).

Müller's portrayal of Galileo as a theological intruder was refined by Koestler in an ingenious, more substantive, but quite sophistical way. Koestler tried to show, as we have seen (Section 8.15), that the chief aim of the *Letter to Christina* is to illegitimately shift the burden of proof in the Copernican controversy: "it is no longer Galileo's task to prove the Copernican system, but the theologians' task to disprove it. If they don't, their case will go by default, and Scripture must be reinterpreted" (Koestler 1959, 437).

10.7 Demise of the Myth (1979–1992)

Finally, as suggested earlier (Section 8.17), some of pope John Paul II's recent pronouncements on the Galileo affair perhaps mark the death knell of Mallet's myth. My point here should not be misunderstood. There is little question that these recent Vatican actions turned out to be disappointing in the end, for the papal re-evaluation was informal, incomplete, and mixed. However, from the point of view of what principles Galileo held regarding the relationship among science, religion, and the Bible, John Paul spoke with unprecedented clarity and remarkable accuracy. His words are worth repeating.

For example, in the 1979 Einstein centennial speech, the pope said: "He who is rightly called the founder of modern physics declared explicitly that the two truths, of faith and of science, can never contradict each other … The Second Vatican Council does not express itself otherwise" (section 7, paragraph 2).[55] Moreover, "Galileo formulated important norms of an epistemological character, which are indispensable to reconcile Holy Scripture and science. In his letter to the grand-duchess mother of Tuscany, Christine of Lorraine, he … introduces the principle of an interpretation of the sacred books which goes beyond the literal meaning but is in conformity with the intention and the type of exposition characteristic of them … The Ecclesiastical Magisterium admits the plurality of the rules for the interpretation of Holy Scripture. It teaches expressly in fact, with Pius XII's encyclical *Divino afflante spiritu*, the presence of different literary styles in the sacred books" (section 8, paragraphs 1–2).

And in 1992, at the conclusion of his re-assessment, the pope had not changed his mind in this regard but reaffirmed the point with these words: "paradoxically, Galileo, a sincere believer, showed himself to be more perceptive in this regard than the theologians who opposed him" (section 5, paragraph 4); whereas "the majority of theologians did not recognize the formal distinction between Sacred Scripture and its interpretation, and this led them unduly to transpose into the realm of the doctrine of the faith a question that in fact pertained to scientific investigation"

---

[55] As stipulated in Section 8.17, references to the pope's 1979 and 1992 speeches will be given by just mentioning section numbers (printed in all editions) and paragraph numbers (easily supplied by the reader).

(section 9, paragraph 1). Moreover, from the Galileo affair "another lesson we can draw is that the different branches of knowledge call for different methods ... The error of the theologians of the time when they maintained the centrality of the earth was to think that our understanding of the physical world's structure was in some way imposed by the literal sense of Sacred Scripture" (section 12, paragraph 1). So much for the Mallet–Gaetani myth. R.I.P.

10.8 Conclusion

This chapter has described the contents and basis, as well as the pre-history, birth, diffusion, development, metamorphosis, and apparent demise of an account of Galileo's trial that flourished from the end of the eighteenth century to the beginning of the twentieth. The account claims that Galileo was not condemned for his astronomical conclusion that the earth moves, but for his theologically unsound practice of supporting an astronomical view with biblical passages. This account was held and elaborated by such authors as Mallet du Pan (1784), Bergier (1788–1790a,b), Bérault-Bercastel (1790), Feller (1797), Cooper (1838), Purcell (1844), Marini (1850), Reumont (1853), Madden (1863), the unknown writer of the 1887 article in *L'Osservatore Romano*, and Duhem (1908).

This account is untenable insofar as it is typically based on such things as Gaetani's forged letter to Renieri, a fabricated sentence in Guicciardini's letter of 4 March 1616, and misinterpretations and exaggerations of genuine documents (Galileo's letters to Christina and Castelli). The account is also false; for example, it conflicts with the evidence from the 1633 sentence, the *Dialogue*, and the *main* theme of these letters. The account is perverse insofar as it inverts and subverts the truth; that is, it claims that Galileo supported astronomical conclusions with Biblical passages, whereas he preached and practiced the *opposite* principle that Scripture should *not* be used to support physical propositions. But the account was relatively widespread and long-lasting. For these reasons, it is appropriate to label it a myth, with some of the pejorative connotations that this label carries, but also with some of the dignity attached to it by professional scholars of mythology; they stress that myths perform valuable and important functions in defining and preserving the cultural cohesiveness of social groups.[56]

Besides sketching the development of this myth, I have also alluded to another even more important development. That is, after this myth was first formulated and publicized with the appearance of Mallet's article and the Renieri apocryphal letter, other pro-clerical apologetic accounts began to appear explicitly and to develop. For example, as discussed above, Tiraboschi embellished the Galileo-reckless the-

---

[56] See Carvalho-Neto (1965) for a general account of myths, and Lessl (1999) and Finocchiaro (2009a) for an application to the case of Galileo.

sis in 1793, and it was followed by other similar versions, such as Brewster's account (1835/1841); in 1838 Cooper elaborated the account of Galileo as immorally disobedient, which has achieved great popularity; and in 1850 Marini articulated an explanation in psychological terms involving human feelings and frailties, namely that Galileo's *Dialogue* insulted or offended pope Urban VIII by caricaturing him in portraying the character Simplicio,[57] soon adopted and elaborated by Biot (1858).

It is also important to note another significant development that followed the appearance of Mallet's article and Gaetani's forgery in 1784–1785. That is, works began to be published elaborating accounts that were explicitly anti-clerical and/or pro-Galilean. I have already mentioned Girolamo Ferri's article in 1785 replying to Mallet and published on the pages of the same magazine. A more spectacular example of much wider scope is Nelli's two-volume biography of Galileo, published in 1793. The next significant example of this genre came about half a century later (in 1841), with the works of Guglielmo Libri, which received wide dissemination in French, Italian, and German; among other questions, Libri explicitly elaborated the torture thesis, which led to intense and heated discussions.

Thus, the evidence and analysis presented in this chapter suggest that the subsequent Galileo affair has experienced at least three different periods: one of relatively subdued discussions from 1633 to about 1784; a period from about 1784 to about 1908 during which the debate became relatively heated and polarized and accounts were elaborated that were explicitly pro- and anti-Galilean and pro- and anti-clerical; and the period since then during which the heated and polarized discussions continue but the less plausible and less tenable accounts (on both sides) are discarded. The Mallet–Gaetani myth was one of those accounts that were devised during the second period but did not survive. However, it seems to have acted as a catalyst for the subsequent Galileo affair to become the cause célèbre it is today.

---

[57] An interpretation along these lines had been mentioned in 1661 by Thomas Salusbury, but made no impact; whereas after the point was advanced by Marini, it was soon adopted and elaborated by Biot (1858, 3: 1–59), who regarded it as a great lesson, so much so that it led him to reject and revise his own earlier (1816) interpretation. For more on Salusbury, see Wilding (2008) and Section 12.7.

Chapter 11
Galileo as a Bad Epistemologist?

One of the most famous, emblematic, and influential critiques of Galileo is that of Pierre Duhem. Historically, it is one of the best documented accounts, and philosophically one of the most sophisticated analyses. We have already had the occasion to mention, preview, or briefly summarize it on several occasions. Now, it is time to examine it fully and systematically.

What makes Duhem's critique especially distinctive and relevant is the fact that it stresses Galileo's alleged epistemological errors, and it does so in the context of Galileo's defense of Copernicus and condemnation by the Inquisition. In fact, the analysis of Galileo's scientific work from an epistemological point of view has a much older history than Duhem's account and a much more general significance than the issues inherent in the trial.[1] However, such a general topic is beyond the scope of this book. Here the focus is on Galileo's defense of Copernicus and the context of the trial, in order to understand the original content and structure of that defense, its contemporary and subsequent reception, and its critical and evaluative validity. Thus, here Duhem's account provides just the right focus.

11.1 Introduction: Duhem on Saving the Phenomena

There is a growing consensus that Duhem's general epistemological position about the aim and structure of physical theory represents a judicious balance between the twin extremes of instrumentalism and essentialism; correspondingly, the older interpretation of him as simply an instrumentalist[2] is becoming increasingly untenable, and indeed may be seen as being itself an oversimplification.

---

[1] See, for example, the references to Galileo's method in Hume's *History of England*, 1754–1762, vol. iv, chapter xlix, appendix (cf. Hume 1851–1860, 4: 521–527) and in Kant's *Critique of Pure Reason*, preface to the second edition, 1787, pp. Bxii–Bxvi (cf. Kant 1965, 20–22). See also Koyré (1943), Popper (1963, 97–119), Feyerabend (1975, 69–108; 1988, 55–151), Finocchiaro (1980, 103–166; 1997a, 1–7, 335–356).
[2] Agassi (1957), Popper (1963, 104).

The most substantial and convincing case for the new, moderate interpretation has been made by Maiocchi, though we can find intuitions of and gropings toward the same interpretation in other authors.[3] I regard such an anti-instrumentalist interpretation as established in its essential point, and shall therefore take it for granted in what follows.

There is another thesis which will constitute part of what may be called the background knowledge for my analysis below. This thesis is a criticism by various scholars[4] of Duhem's history of the controversy between realists and instrumentalists, that is, of his account of the views and practices of the proponents and the opponents of "saving the phenomena." As early as 1947, Morpurgo-Tagliabue objected to Duhem's attempt to trace the idea to the ancients, though this critic was willing to attribute it to later commentators such as Proclus and Simplicius; he also argued specifically that Ptolemy intended his theory as neither apodictically real, nor merely instrumentalist, but as "likely." Later, Popper briefly expressed similar reservations. Then, in a more elaborate and more erudite study, Lloyd reinforced this criticism about classical antiquity and extended it to the cases of Proclus and Simplicius as well; he argued that they were all realist oriented, and that Duhem is partial, one-sided, and often demonstrably incorrect in his selection and interpretation of the texts. More recently, Jardine has extended the criticism of Duhem's history to the sixteenth century; he rejects Duhem's classification of views into realist or instrumentalist as an oversimplification, arguing that "Duhem's tidy partition involves ... both an underestimation of the diversity of positions adopted and a failure to appreciate the extent to which they represent responses to specific problems in the astronomy of the period rather than the promotion of general metaphysical or epistemological theses" (Jardine 1984, 225). Finally, Goddu has undermined Duhem's interpretation of Copernicus as a metaphysical essentialist who held that astronomical hypotheses must be true and demonstrated to be true; this interpretation is shown to contradict the best recent scholarship, which holds that Copernicus viewed his hypothesis as probable, fallible, and revisable.

Now, in the light of this negative, critical consensus about Duhem's history of "saving the phenomena," it would be surprising if he should have managed to get Galileo's methodology or epistemology and the Galileo affair right and avoid pitfalls similar to those that affect his account of other figures, periods, and episodes. And in fact, we find, for example, Westfall arguing that Duhem does not do justice to the position of Cardinal Bellarmine, a central figure in the Galileo affair; although Duhem does pay attention to certain parts of Bellarmine's letter to Foscarini that do express the cardinal's instrumentalism vis-à-vis the Copernican theory, Duhem ignores other parts of the letter where Bellarmine clearly exhibits a realist attitude toward the geostatic "theory."[5] The adequacy of Duhem's account of

---

[3] Maiocchi (1985; 1990). Cf. Darling (2003), Jaki (1969, xxv), McMullin (1990a).
[4] Morpurgo-Tagliabue (1947–1948; 1981, 41–52), Popper (1963, 99 n. 6), Lloyd (1978), Goddu (1990).
[5] Westfall (1989, 16–17); see also Stoffel (2001).

11.1 Introduction: Duhem on Saving the Phenomena

Galileo will be seen presently. Before I undertake that task, a third introductory remark is in order.

To the positive and appreciative consensus about Duhem's epistemological analysis of the aim and structure of physical theory, and to the negative and critical consensus about his history of "saving the phenomena," there does not correspond a consensus in regard to the aim of his work bearing the title *To Save the Phenomena*. Brenner (1990, 173–174) claims that this work is basically a history of methodology aimed to justify Duhem's own instrumentalist epistemology by identifying historical precedents, and to justify his continuism by criticizing the occurrence of any methodological revolution in the sixteenth and seventeenth centuries. Maiocchi (1985, 268–277) finds it to be primarily an exercise in religious apologetics, and then he is faced with the task of explaining it away, or at least finding some redeeming features; these he finds in the connections between that work and another one which Duhem simultaneously produced, *Le mouvement absolu et le mouvement relatif* of 1909. Martin (1987; cf. 1991) advances an interpretation which is both anti-instrumentalist and anti-religious, when he argues that Duhem's account of the Galileo affair was meant to be an illustration and a confirmation of the epistemological principle that pure logic is not the only guide for our scientific judgments; Martin's evidence is partly direct and textual, but also indirect and circumstantial, stressing the curious fact that the journal where the book first appeared in the form of articles was placed on the *Index* soon thereafter, thus suggesting that Duhem was opposed to the Church's anti-modernism. To these three scholarly views, we should add Duhem's own self-conception found in a note added to the second edition of *The Aim and Structure of Physical Theory*; there he regards his book on saving the phenomena as a "development" (Duhem 1954, 39 n. 9) of the discussion of the opinions of physicists on the nature of physical theory, but he formulates his own epistemological thesis by saying "that physical theories are by no means explanations, and that their hypotheses [are] not judgments about the nature of things, only premises intended to provide consequences conforming to experimental laws" (Duhem 1954, 39).

Such dissensus cannot, of course, be put to use in the same way as the above mentioned instances of consensus; but that is not to say that it is irrelevant to the present inquiry. In fact, the present inquiry aims to determine what is the interpretation of Galileo's epistemology advanced by Duhem in his book on saving the phenomena, what are his supporting arguments and evidence, and whether they are adequate. This determination may help us decide among the alternative interpretations of the whole book.

In any case, the critical examination of Duhem's interpretation of Galileo is important in order to help us arrive at a correct view of this classic case. Moreover, it has been enormously influential, and many accounts of the Galileo affair in particular[6] or of Galilean methodology in general are largely variations on Duhemian themes.[7]

---

[6] E.g., Langford (1971), Brandmüller (1982; 1987; 1992), Brandmüller and Greipl (1992, 15–130), Rowland (2003).
[7] E.g., Feher (1982), McMullin (1978; 1990b), Wallace (1981b; 1984a).

11.2 Unificationism

It is difficult not to be struck by Duhem's memorable thesis "that logic was on the side of Osiander, Bellarmine, and Urban VIII, and not on the side of Kepler and Galileo; that the former had understood the exact import of the experimental method; and that, in this regard, the latter were mistaken" (136/113).[8] The sense in which the former were right and the latter wrong is that the former understood and appreciated "that the hypotheses of physics are merely mathematical contrivances intended to save the phenomena" (140/117), and not descriptions of reality capable of being true and of being false. If this were the end of the story, then critics of Galileo would indeed have in Duhem's account the anti-Galilean goldmine it has often been taken to be and continues to be; but serious students of Duhem would then be faced with the difficulty that his account would appear to be a reaffirmation of instrumentalism, which he is supposed to have transcended by the time he published *The Aim and Structure of Physical Theory* a few years before writing the present work *To Save the Phenomena*.

But in fact, this is not the end of the story. The above mentioned thesis is presented at the beginning of the concluding chapter of *To Save the Phenomena* in order to motivate the main conclusion. This conclusion is a constructive and positive lesson that can be derived from the scientific practice of Kepler and Galileo, and which they may be thus said to have intuited. The practice in question is their sustained effort to unify terrestrial and celestial physics, their intuition that "the same dynamics must represent the motion of stars, the oscillations of the ocean, and the fall of heavy bodies by means of the same set of mathematical formulas" (140/116–17). The principle in question is that physical hypotheses must be required "to save *all* the phenomena of the inanimate universe *in the same way*" (140/117). And this principle is an expression of the same anti-instrumentalist, anti-skeptical, anti-model-oriented attitude which Duhem upheld, for example, against the British mechanical models of electromagnetic phenomena.

In short, Duhem's account of Galileo is not one-sidedly critical and negative, but contains an appreciative and positive element as well. It is critical of what Duhem likes to label Galileo's realism, and which he thinks he can find in Galileo's reflective pronouncements about the status of Copernicanism; but he is appreciative of Galileo's practical scientific success, whose import however Galileo presumably did not properly understand but only vaguely felt, while its proper definition was left to Duhem himself to define and formulate. Similarly, Duhem is trying to be even-handed vis-à-vis Galileo and his opponents; they were both partly right and partly wrong. Presumably, Galileo was substantively correct in the specific

---

[8] Duhem (1908, 136), my translation; cf. Duhem (1969, 113). Subsequent references to Duhem (1908) will be given in parenthesis in the text, without citing Duhem's name or the year of publication; the translation will be my own, unless otherwise noted; but for the convenience of the reader, references will also be given to the English translation of this book, Duhem (1969), by writing the latter's page number after a slash.

cosmological thesis, as well as methodologically correct in regard to the unification principle; but he was wrong in his view of the aim of physical theory and in his methodology of demonstration (as we shall examine presently). And among his opponents, Bellarmine and Urban VIII were of course wrong about the particular scientific hypothesis they favored, but they were right about the logic of demonstration and the aim and nature of physical theory.

Duhem's attribution to Galileo of the unification principle is plausible and essentially correct. One might wish to question whether the degree of awareness on the part of Galileo was as slight as Duhem claims, but that is relatively unimportant.[9] Similarly, one could question the exact formulation of the unification principle, since it seems that if carried too far the principle corresponds neither to current scientific practice nor to the most defensible epistemological judgment. For example, the birth of modern chemistry in the eighteenth century necessitated a subdivision of the inanimate into physical and chemical domains, and consequently a limitation of the range and scope of phenomena to be saved by means of the same set of hypotheses; and in the twentieth century, physics has found it necessary to make analogous subdivisions between the domain of intermediate-size bodies and, on the one hand, the domain of macroscopically large or cosmological phenomena, and on the other hand, the domain of microscopically small or subatomic bodies.

11.3 The Condemnation of Galileo

Duhem's interpretation of Galileo's trial is that it was not the result of methodological disagreements between people holding opposite principles, but rather of a substantive disagreement between realists, between the "impenitent realism of Galileo" and "the intransigent realism of the Peripatetics of the Holy Office" (135/112). Moreover, he sees little difference between the condemnation of 1616 and that of 1633: the former allegedly amounted to a prohibition to use the key Copernican hypotheses in any way whatever, "even for the sole purpose of 'saving the phenomena'" (128/106), whereas "the condemnation of 1633 was a confirmation of the sentence of 1616" (135/112). The rationale underlying these condemnations was a pair of two realist principles which, according to Duhem, "both Copernicans and Ptolemaics by common agreement required of all acceptable astronomical hypotheses" (127/106), namely that they be compatible with sound physics and with Scripture. The difference between the two sides was in the application of these two principles, which yielded a negative answer for the Inquisition and a positive one for Galileo.

By explaining the condemnation in terms of realism, Duhem is in a sense blaming that tragic episode on realist epistemology, and thus adding further discredit to realism. Moreover, and here we come to a second major Duhemian thesis, the condemnation need not have occurred, and indeed would not have occurred, if the

---

[9] Some of this counterevidence can be found in Finocchiaro (1980), and in Chapters 3 and 9 above.

realists on both sides had listened to the voices of reason, moderation, and prudence represented by people like Cardinal Bellarmine and Pope Urban VIII. Their criticism of Galileo's realism was based on precepts stemming from an ancient tradition, for "these precepts had been formulated by Posidonius, Ptolemy, Proclus, and Simplicius, and an uninterrupted tradition had brought them to Osiander, Reinhold, and Melanchthon" (128/106).

I have already mentioned the widely shared misgivings about the existence of such a tradition, as well as some reservations about whether such precepts can be attributed even to Bellarmine and Urban. For the moment let us note simply that Duhem's counterfactual proposition is obviously meant to strengthen the antirealist tradition of which he himself feels to be a part.

This Duhemian interpretation of Galileo's trial has considerable simplicity, elegance, and originality. Its simplicity lies primarily in the fact that it explains both phases of the episode (in 1616 and in 1633) by means of the same epistemological interpretation in terms of realism. Its elegance lies in the fact that it makes sense of the key provision in the special injunction stating that the Inquisition had "ordered and enjoined the said Galileo ... to abandon completely the above mentioned opinion that the sun stands still at the center of the world and the earth moves, and henceforth not to hold, teach, or defend it in any way whatever, either orally or in writing."[10] Duhem is suggesting the phrase "in any way whatever" was referring to the discussion of the earth's motion as an hypothesis meant merely to save the appearances. The sentence of 1633 then becomes justified as a literal application of this injunction to the *Dialogue*. Finally, it is a measure of the originality of Duhem's interpretation, but also of its extremist and radical character, that hardly anyone has followed his theses on the trials and condemnation, although his account of the epistemology of Galileo, Bellarmine, and Urban has acquired a large following.

This I find strange because I believe that the main weakness in Duhem's account of the affair is precisely in the methodological and epistemological doctrines he attributes to the participants, primarily Galileo, Bellarmine, and Urban. If his epistemological attributions were accurate, his interpretation of the affair would be worth pursuing by refining and updating various historical and documentary details. In any case, the epistemology of the situation is intrinsically important; so let us turn to it.

11.4 Metaphysics

A preliminary difficulty with Duhem's epistemological account is the assimilation of Galileo and Kepler. I am not referring to their unificationist attitude toward terrestrial and celestial physics, which I mentioned above and is part of Duhem's appreciation. Rather I am referring to the character of their realism. What is at stake

---

[10] Finocchiaro (1989, 147), or Galilei (2008, 176). Cf. Favaro 19: 322, Pagano (1984, 101).

11.4 Metaphysics

is not the existence of an overlap in their position, but the extent of that overlap. There is no question that they were both opposed to regarding hypotheses as mere instruments of calculation and prediction, and that they both favored the evaluation of astronomical hypothesis in terms of physics. The issue concerns the distinction between physics and metaphysics, or to be more exact, the use of metaphysical considerations in astronomical and physical investigations.

Duhem is certainly correct in claiming that "Kepler takes every possible occasion to support his hypotheses with arguments drawn from physics and metaphysics ... He wanted, as we know, the science of the celestial motions to rest on foundations guaranteed by physics and metaphysics" (1969, 104). Although this thesis would need to be elaborated along the lines supplied by Jardine (1984) and by Hatfield (1990), the essential point would remain that Kepler made significant use of metaphysical considerations.

The same cannot be said of Galileo. It is probably correct, as Burtt (1932) argued, that metaphysical principles are presupposed or implicitly used in his mathematical and physical argumentation; for example, it is probably true, as Duhem argues in another work (1909, 5–7), that the meaning and truth of the propositions "the earth is motionless" and "the earth moves" are questions that ultimately belong to metaphysics insofar as they presuppose metaphysical theories of place, space, and motion. However, this does not imply any significant metaphysical involvement, unless one ignores the important difference between explicit and implicit metaphysics and unless one uses an anachronistic conception of metaphysics. And it is clearly true that Galileo's writings are full of concrete epistemological reflections, but this cannot be equated with significant metaphysics either, on pain of conflating epistemology and metaphysics and on pain of ignoring that Galileo's epistemological considerations were always practical and concrete oriented and not abstract.[11]

A particular piece of evidence supporting this anti-metaphysical interpretation is the content and structure of Galileo's *Dialogue*. This work is a strengthening of Copernicanism by a criticism of the anti-Copernican arguments and by the elaboration of several pro-Copernican arguments. Now, none of the favorable arguments are (explicitly) metaphysical; very few of the unfavorable arguments criticized are; and in his answer to those that are, Galileo stays away from abstract metaphysical considerations as much as possible. Such is the case, for example, of the two teleological arguments, that the heavens are unchangeable because heavenly changes would be useless, and that the Copernican system is impossible because in a Copernican universe the space between the outermost planet and the nearest fixed star would be useless.[12]

Galileo's a-metaphysical attitude is important not only vis-à-vis Kepler but also vis-à-vis the "intransigent realism" of the Inquisition. In fact, when Duhem reports

---
[11] For an elaboration of the a-metaphysical interpretation of Galileo sketched here, see Finocchiaro (1980, 103–166; 1997a, 335–356), Hatfield (1990, 117–143).
[12] Favaro (7: 83–87, 385–99), Galilei (1967, 58–62, 358–372), Galilei (1997, 247–264).

that the consultants of the Holy Office judged the two main Copernican hypotheses to be "foolish and absurd in philosophy" (127/106), he clarifies this only in terms of physics, and explains that they reached this judgment because Copernicanism conflicted with the only physics they accepted, Aristotelian physics. In so doing, Duhem is restricting the term "philosophy" to just natural philosophy and excluding metaphysics or first philosophy, and there is no textual or historical justification for such a prejudicial interpretation. They certainly had in mind, among others, the teleological arguments as well as the objection that a simple body like the earth can have one and only one natural motion.

To summarize the present point, Galileo was indeed a realist, but not in the sense that he thought it proper to confirm or disconfirm astronomical hypotheses on the basis of whether or not they conform to metaphysical principles. By identifying his realism with that of Kepler and of the Inquisition in this regard, Duhem fails to appreciate this important point.

This point is all the more important in the light of the fact that Duhem's entire epistemology is based on a failure to properly distinguish scientific explanation from metaphysical explanation. From the very first chapter of *The Aim and Structure of Physical Theory*, it is obvious that he is unwilling or unable to see that not all explanation is metaphysical. To be sure, his motivation is anti-metaphysical in the sense that he properly wants to make physics autonomous and independent of metaphysics, but this aim is not served by his failure to distinguish metaphysical from nonmetaphysical explanations and metaphysical thinkers like Kepler from nonmetaphysical ones like Galileo.

11.5 Biblical Authority

One of the crucial issues in Galileo's trial was, as we have seen, that of the role of the Bible in physical inquiry. Duhem attributes to the realists the principle that acceptable astronomical hypotheses must be reconcilable with the Bible. To be sure, he does regard this principle as somewhat derivative, in the sense that it is derivable from the principle that acceptable hypotheses must be compatible with physical principles; for the latter principle would imply that hypotheses aim to be true descriptions of reality, and since the Bible is also true and two truths cannot contradict each other, it follows that astronomical hypotheses cannot contradict biblical assertions (126–27/105). Despite the connection, Duhem does list the biblical requirement separately, as deserving special attention.

Now, Duhem is somewhat ambivalent in including biblical reconciliation in Galileo's realism. In fact, he first examines Galileo's *Treatise on the Sphere, or Cosmography*, the introductory textbook in which Galileo favors a geostatic position; and Duhem concludes that in this work Galileo, besides requiring hypotheses to save the appearances, requires them to conform to physics, but does not in any way appeal to the Bible. Not without a touch of subdued praise and pride, Duhem notes that this is one respect, indeed the only one, in which this work by a Catholic

11.5 Biblical Authority

is different from the similar one by Protestant George Horst, the *Exposition of the Spherical Doctrine* (126/105).

However, Duhem goes on to claim that when Galileo adopted the Copernican system, besides retaining the physicalist requirement, "he was induced to reconcile his assertions with the texts of the Bible. In due course, he turned theologian" (127/105). That is, he supposedly adopted the biblical principle as the second of "the two signs which both Copernicans and Ptolemaics by common agreement required of all acceptable astronomical hypotheses" (128/106). Duhem mentions Galileo's *Letter to Christina* as the basis for his attribution.

Now, it is true, as we have seen, that in the years 1613–1615, in a number of letters, Galileo attempts to reconcile the main Copernican claims with a number of biblical assertions. For example, in the *Letter to Christina* he argues that the Joshua miracle is quite compatible with Copernicanism, or at least much more compatible with it than with the Ptolemaic system.

However, as I have argued before, this is not an attempt to support Copernicanism on the basis of biblical assertions, nor is it an explicit or implicit usage of the principle in question. Rather Galileo is criticizing the biblical objection to the earth's motion, which claimed that the earth's motion is unacceptable because it contradicts the Bible. He first objects to the inferential soundness of the reasoning: that is, even if the premise were true, the conclusion would not follow because the Bible is not a scientific authority, and hence its assertions about the physical world do not constitute proper evidence that the world corresponds to them. The principle that Galileo does accept is formulated with the memorable words (attributed to one Cardinal Baronio) that "the intention of the Holy Spirit is to teach us how one goes to heaven and not how heaven goes."[13] Galileo then makes a second criticism of the argument: its premise that the earth's motion contradicts the Bible (for example, in Joshua 10:12–13) is incorrect. In other words, there are at least two flaws in the argument: it is a nonsequitur and it contains a false premise.

This second criticism would indeed license the following additional comment: even if one thinks that the Bible is a scientific authority (that the objection is inferentially sound), the anti-Copernican conclusion would still be unjustified because the premise is unjustified.

Now, all of this Galileo can validly say, and does say, without committing himself in the least to the principle that in physical inquiry it is proper either to confirm or to disconfirm hypotheses by means of biblical texts. It follows that in regard to this principle, there was a methodological disagreement between Galileo and the Inquisition officials. Their dispute was not merely a substantive one about how a commonly shared principle was to be applied to the case at hand, but rather primarily about whether or not the Bible had any role in physical inquiry. Galileo's realism did not include the biblical principle.

My interpretation (see Chapters 4 and 9 above) is further strengthened by the fact that the final sentence in the trial of 1633 condemned Galileo for having held and

---

[13] Finocchiaro (1989, 96), or Galilei (2008, 119); cf. Favaro 5: 319.

believed not only the Copernican hypotheses, but also the principle "that one may hold and defend as probable an opinion after it has been declared and defined contrary to Holy Scripture."[14] This is merely a different way of formulating Baronio's aphorism, and it is precisely the opposite of what Duhem seems to be attributing to Galileo.

11.6 Certainty

Another important element of Galileo's realism is, according to Duhem, that "Galileo thought that the reality of the earth's motion is not only demonstrable, but already demonstrated" (130/108). Literally, this is of course, just a particular judgment about a particular hypothesis, but it seems clear that Duhem means to attribute to Galileo a general epistemological position. However, it would not make sense to generalize the judgment into the principle that physical hypotheses or theories are not only demonstrable but also already demonstrated, since the truth of this generalization would depend at best on the hypothesis in question. So, for a proper formulation of the general principle, we must drop any reference to the "already demonstrated." Moreover, the "demonstrable" needs to be qualified, otherwise the principle would be insufficiently specific. In any case, when Duhem goes on to criticize this idea, he objects that hypotheses cannot be proved to be "certainly true" (132/110). So it seems that Duhem is attributing to Galileo the principle that physical hypotheses can be demonstrated to be true with certainty. I suppose the linguistic connection with realism is that this also means that the reality of physical hypotheses can be demonstrated with certainty.

Having clarified this much, Duhem's refutation of Galileo's realism is easy to guess. It is the following, which Duhem regards as the lesson derivable from Hipparchus by way of Thomas Aquinas, Osiander, and Pope Urban VIII. He points out that "the confirmations of experience, regardless of how numerous and precise we suppose them to be, can never transform an hypothesis into a certainty because one must also demonstrate another proposition; that is, that the same facts of experience flatly contradict any other hypothesis imaginable" (134/111). And he suggests that the latter demonstration can never be given; that, for example, at the time of Galileo the observations did not contradict the theory of Tycho Brahe (133/110).

I accept the cogency of Duhem's refutation of the ideal of certainty. What is less clear, however, is his evidence that Galileo does in fact subscribe to that ideal. Duhem quotes two paragraphs from Galileo's "Considerations on the Copernican Opinion," drafted in 1615 in an attempt to reply to Bellarmine's letter to Foscarini. These paragraphs must be quoted in full in order to appreciate the extent of Duhem's distortion:

> 6. Not to believe that there is a demonstration of the earth's mobility until it is shown is very prudent, nor do we ask that anyone believe such a thing without a demonstration. On the contrary, we only seek that, for the advantage of the Holy Church, one examine with

---

[14] Finocchiaro (1989, 291), or Galilei (2008, 292); cf. Favaro 19: 405.

the utmost severity what the followers of this doctrine know and can advance, and that nothing be granted them unless the strength of their arguments greatly exceeds that of the reasons for the opposite side. Now if they are not more than ninety percent right, they may be dismissed; but if all that is produced by philosophers and astronomers on the opposite side is shown to be mostly false and wholly inconsequential, then the other side should not be disparaged nor deemed paradoxical, so as to think that it could never be clearly proved. It is proper to make such a generous offer since it is clear that those who hold the false side cannot have in their favor any valid reason or experiment, whereas it is necessary that all things agree and correspond with the true side.

7. It is true that it is not the same to show that one can save the appearances with the earth's motion and the sun's stability, and to demonstrate that these hypotheses are really true in nature. But it is equally true, or even more so, that one cannot account for such appearances with the other commonly accepted system. The latter system is undoubtedly false, while it is clear that the former hypotheses, which can account for the appearances, may be true. Nor can one or should one seek any greater truth in a position than that it corresponds with all particular appearances.[15]

Duhem's interpretation involves nothing less than taking "*may* be true" to mean "*must* be true" or "is certainly true." Galileo clearly states that, comparing the geokinetic and the geostatic systems, the latter is undoubtedly false while the former *may be* true, but Duhem "interprets" this to mean that "the destruction of one of the two opposing systems assures the certainty of the other one" (131/109). Moreover, Galileo's talk of percentages makes it crystal clear that he is talking of degrees of plausibility or probability or likelihood, which represent a movement away from certainty and mathematical demonstration; whereas Duhem claims Galileo is talking of a situation "just as in mathematics" (131/109). Finally, the last sentence makes it clear that Galileo is suggesting that one should not expect to be able to show any more than that a given hypothesis *may be* true; Duhem admits this fact, but claims that what Galileo has in mind is not what he says, but rather the opposite of what he says because "in the preceding lines his thinking emerges clearly" (131/109). It is true that Galileo's thinking emerges clearly in the preceding lines of the passage, but it is equally the case that what emerges is a consistent thought, which is precisely an alternative to the certainty ideal.

Unfortunately, with few exceptions,[16] Duhem's inverted and distorted reading has been uncritically accepted by most scholars. Therefore, it is worth stressing that it is based on an inadmissible and intolerable interpretation of this passage.

11.7 Proof Strategies

Duhem also attributes to Galileo a methodology of proof, a theory of how hypotheses are established. This attribution may be regarded as distinct from (though related to) the previous one because the previous point dealt with the aim of physical

---

[15] Galilei (2008, 165–166); cf. Finocchiaro (1989, 85) and Favaro 5: 368–369.
[16] E.g., Morpurgo-Tagliabue (1947–1948; 1981; 1985).

theory (whether the aim is certain and indubitable truth), whereas the present point concerns the means for accomplishing an aim that may be taken as uncontroversial (that is, the aim of "establishing" a theory, without being any more specific about the character of such establishment). The connection between the two is that an aim may be evaluated on the basis of whether or not there are means of bringing it about, as was just done for the ideal of certainty; on the other hand, a strategy may be evaluated on the basis of whether or not it has the desired effect.

Duhem's own strategy for demonstrating his own (historical) thesis is to elaborate the strategy followed by Galileo in supporting the earth's motion: first, he claims that "Galileo thought that the reality of the earth's motion is not only demonstrable, but already demonstrated" (130/108); then Duhem tries to "learn how he understood the demonstration would proceed" (130/108), and finds two strategies. One is a strategy described both as the empirical analogue of mathematical *reductio ad absurdum* and as the Baconian methodology of crucial experiments: "he conceives the proof of an hypothesis on the model of *reductio ad absurdum* in mathematics: by convicting a system of error, experience confers certainty to the opposite system; positive science progresses by a series of dilemmas each of which is resolved by means of a *crucial experiment*" (132/109). The other strategy is the strategy of saving the appearances, illustrated in "the following argument of Galileo: since heavenly phenomena are all in accordance with Copernicus's hypotheses, whereas they cannot be saved by Ptolemy's system, Copernicus's hypotheses are undoubtedly true" (133/110).

Finally, Duhem points out the flaws in these strategies. The *reductio ad absurdum* strategy fails because competing theories, such as the Copernican and the Ptolemaic systems, are normally not contradictories but merely contraries; they do not exhaust all the alternatives and could both be false. And the strategy of saving the appearances fails for the reason mentioned earlier, namely that perhaps the phenomena can be saved by means of some other hypothesis.

This entire construction hinges on the accuracy of the first interpretive thesis in this series, the claim that Galileo believed to have shown, not only that Copernicanism was demonstrable, but also that it had already been demonstrated. Unfortunately the documentation provided by Duhem is inadequate. He refers to the beginning of Galileo's "Considerations on the Copernican Opinion," where Galileo states that he intends to expose two errors: the view that Copernicus regarded the earth's motion merely as an instrument of calculation and prediction and not as a description of reality, and the view that the earth's motion "is such an immense paradox and obvious foolishness that no one can doubt in any way that it cannot be demonstrated now or ever, or indeed that it can never find a place in the mind of sensible persons."[17] Unfortunately for Duhem, this second view only means the earth's motion is not demonstrable; that is, Galileo wants to refute the idea that the earth's motion is indemonstrable, which is to say he believes that the earth's motion is demonstrable, but this is not to say that it has already been demonstrated.

---

[17] Finocchiaro (1989, 70), or Galilei (2008, 148); cf. Favaro 5: 351.

The same weaker interpretation is suggested by the fact that in the same passage Galileo also formulates the position he wants to criticize in terms of the demonstrative status of the geostatic view. He wants to refute the idea that "the earth's stability and sun's motion are so well demonstrated in philosophy that we can be sure and indubitably certain about them."[18] To say that the earth's rest has not been demonstrated with certainty is equivalent to saying that the earth's motion has not been disproved with certainty, namely is *capable* of being proved.

Given this lack of direct evidence for the Galilean belief that Copernicanism had already been demonstrated, and given that the Galilean arguments and strategies defined by Duhem do not validly support that belief, we lose any justification for attributing the belief to Galileo. If those arguments had been valid they might have been used as indirect evidence that Galileo held the belief, but as things stand Duhem wants to have it both ways. He wants those strategies to prove the existence of Galileo's erroneous reasoning as well as the existence of his erroneous belief. But to prove one he has to presuppose the other.

Another difficulty with Duhem's account is a curious incoherence. The strategy of proof by saving the appearances which he attributes to Galileo is assuming some version of anti-realism or instrumentalism to the effect that an hypothesis is true or acceptable if it saves the phenomena. Only this assumption would license the inference that because Copernicanism saves the phenomena, therefore it is true. But then Duhem is attributing an anti-realist principle to Galileo, which contradicts the realist portrayal he also wants to advance.

11.8 Conclusion

In summary, Duhem's account of Galileo's epistemology does have the judicious character one would expect from the best recent Duhem scholarship; this account attributes to Galileo a sound principle of unification, as well as a realism which is deemed unsound. The appreciative element of Duhem's account is essentially correct, and extremely important if we want ourselves to avoid one-sidedness in our interpretation of Duhem. Moreover, the realist methodology attributed by him to Galileo enables him to advance an interpretation of the Inquisition's condemnation which has considerable simplicity, elegance, and originality. Finally, the realism attributed by Duhem to Galileo consists in part of the idea that physical theories and hypotheses are not merely instruments of calculation and prediction, but also descriptions of physical reality; and this idea is indeed an important part of Galilean methodology.

Duhem also attributes to Galileo other epistemological principles involving metaphysics, the Bible, the ideal of certainty, and the "methodology of crucial experiments." However, Galileo did not hold such principles, and, less importantly, it is somewhat arbitrary to subsume them under the label of "realism."

---

[18] Finocchiaro (1989, 70), or Galilei (2008, 148); cf. Favaro 5: 351.

Galileo did not believe that astronomical (or physical) theories should be confirmed or disconfirmed by means of metaphysical considerations. Duhem seems to suggest the contrary by his assimilation of Galileo with Kepler and with the methodology of the Inquisition's consultants. But this interpretation is refuted by Galilean practice and fails to properly distinguish natural philosophy and first philosophy.

Galileo did not think that astronomical (or physical) theories should be confirmed or disconfirmed by means of biblical texts. Duhem is somewhat ambivalent in attributing this principle to Galileo, but the attribution prevails for the more central and mature parts of Galileo's career. This interpretation is the result of a misunderstanding of the logic of Galileo's critique of the biblical objection to Copernicanism, and involves a flawed reading of the *Letter to Christina*.

Galileo did not always hold that physical theories in general, and Copernicanism in particular, are demonstrable with certainty. Duhem makes his contrary attribution largely on the basis of a passage from Galileo's "Considerations on the Copernican Opinion." However, this passage happens to be an explicit and unambiguous statement that all we can determine about the epistemological status of Copernicanism or of physical theory is that it "may be true."

Finally, Galileo did not think Copernicanism had been adequately established, nor that he could establish it (or a physical theory in general) simply by refuting one inadequate alternative and saving a few phenomena in Copernican terms. Duhem supports his contrary interpretation by assuming that Galileo regarded Copernicanism as established and reconstructing his supporting argument. Unfortunately, there is a vicious circularity between this assumption and this reconstruction; the assumption is unwarranted; and the reconstruction is not only logically inconclusive, but also internally incoherent insofar as it attributes to Galileo the very same methodology of saving the phenomena whose neglect was allegedly his central error. At any rate, a proper examination of the issue would have to examine Galileo's mature case in favor of Copernicanism (in the *Dialogue* of 1632), and this would reveal both Galileo's reservations about the epistemological status of Copernicanism, as well as the complexity and multifaceted nature of the case in its favor. He did regard Copernicanism as capable of being demonstrated, and as being on its way toward being demonstrated, but not as capable of being demonstrated with certainty and not as already demonstrated with adequacy.

Chapter 12
Galileo as a Symbol of Science Versus Religion?

Ever since the condemnation of Galileo in 1633, his trial has commonly been viewed as epitomizing the conflict between science and religion. In the twentieth century, some Catholic officials have gone to the other extreme of claiming that instead the trial really exemplifies the *harmony* between science and religion. Each view yields both pro-Galilean and anti-Galilean conclusions, when combined with other appropriate assumptions. For example, from the conflict thesis one can easily derive the conclusion that Galileo is one of the supremely instructive examples in the struggle for individual freedom and civil liberties against religious oppression, as well as the conclusion that Galileo is one of the major figures to be blamed for the disastrous separation in modern Western culture between scientific knowledge and cultural values, or more generally, between facts and values. Similarly, the harmony thesis could be used to infer that Galileo is first and foremost a religious hero, who understood theoretically and practiced in his actual life the proper harmonious relationship between science and religion; and also to infer that Galileo's religiosity and piety constitute just one more example of his many scientific, philosophical, and practical errors. Obviously such theses and conclusions presuppose different conceptions of science, religion, and their interaction, and they are based on different limited portions of the available historical evidence and documentation. This chapter aims primarily to bring some order into this tangle of issues, and secondarily to lay the foundations for an approach that would give Galileo neither undue blame nor undeserved praise.

12.1 The "Interaction" Between "Science" and "Religion"

Many scholars have recently questioned the fruitfulness of traditional accounts of the history of the relationship between science and religion.[1] I believe a general consensus has now emerged to the effect that the two main traditional approaches

---

[1] The views stated in this paragraph are gleaned from these authors: Brooke (1991; 1996; 1998), Brooke and Cantor (1998), Lindberg and Numbers (1986; 1987), Livingstone (1997), Moore (1992), Numbers (2009), Osler (1998), Rudwick (1981), Wilson (1996; 2000), Wykstra (1996a,b).

to the topic are both oversimplifications: that is, the approach that interprets the relationship as one of conflict, and the approach that construes the connection as one of harmony. A key flaw of both accounts is that they are really hasty generalizations: there is indeed conflict in some historical episodes, but not in others; and the same is true for harmony. Another important flaw is that traditional approaches tend to presuppose definitions of science and religion which are essentialist, anachronistic, or unhistorical. There also seems to be a general consensus that the variety of relationships is much richer than what is conveyed by just the notions of conflict and harmony. For example, one cannot neglect such other possibilities as separation, dialogue, integration, and subordination.[2] Even within the concept of conflict, we should add the notion of (peaceful) competition to that of warfare; and within the notion of harmony, it is extremely important to distinguish the direction of influence, whether from religion to science or from science to religion. And in any case, there are several different kinds of influences: presupposition, endorsement, motive, prescription, and substantive source. And the relata are no less complex than the relationship: science can refer to the contexts of discovery, justification, popularization, and so on; and religion can refer to theology, metaphysics, world view, myth, ritual, and ecclesiastic institutions, for example.

I am not sure I would go so far as to agree with the proposal to abandon the terms "science" and "religion" altogether, advanced by some scholars.[3] And I am not sure the situation for this topic is any more problematic than other situations, such as those involving the questions of the relationship between science and society, science and politics, science and philosophy, science and rhetoric, and science and art.[4] Nevertheless, the collective weight of these historiographical discussions on science and religion is such that one can hardly conduct business as usual when one is engaged in studying some historical episode relevant to both science and religion. It is important, therefore, to analyze the Galileo affair in the light of the above mentioned literature, while remaining open to the possibility of making additional historiographical, conceptual, and methodological distinctions as needed.[5]

---

[2] Besides the authors mentioned in the previous note, this variety of relationships has also been discussed by Ruse (1997).
[3] Osler (1998), Wilson (1996; 1999).
[4] For a flavor of such problems, see for example Finocchiaro (1988; 1990).
[5] Cantor (1995), Brooke and Cantor (1998, 106–138), Lindberg (2003), and Shea and Artigas (2006) have also carried out such an exercise. My account can thus be read in conjunction with theirs. I believe, however, that despite our overlapping topic, aim, and approach, there are some differences. By and large, their main concern seems to be to give a statement and criticism of the conflict thesis; whereas here my main concern is evaluation (both constructive and critical), and my targets are the harmony thesis as well as the conflict thesis, and anti-clerical as well as apologetic accounts.

12.2 Conflictual Accounts

Let us begin by examining the conflict thesis. Because Galileo's trial involved a conflict between one of the founders of modern science and one of the world's great religions, it has traditionally been seen as an example of the conflict between science and religion, or at least science and Christianity, or science and Catholicism.[6] This interpretation is initially plausible, but I am not sure it is ultimately correct. For the relevant documents show that many churchmen were on his side and many scientists were critical of him. For example, in 1615–1616 he received the support of Monsignor Piero Dini, Carmelite Father Paolo Antonio Foscarini, and Dominican friar Tommaso Campanella; and after the condemnation of 1633 his tragedy was significantly alleviated by Ascanio Piccolomini, archbishop of Siena, and Fra Fulgenzio Micanzio, theologian to the Republic of Venice. Moreover, the attitude of some of the key ecclesiastic players was nuanced and complex: Cardinal Robert Bellarmine helped Galileo in some ways and hindered him in others in 1613–1616, and Pope Urban VIII was his friend and patron until 1632 and turned against him only thereafter. On the other hand, the opposition to Galileo from Jesuits Christopher Scheiner and Orazio Grassi involved primarily scientific issues such as the discovery and interpretation of sunspots and the interpretation of comets; so it may be regarded largely as the opposition of scientific peers who happened to disagree with him. Thus, we may say that there was a split within both science and religion, and the real conflict was between two things which I shall call a conservative and a progressive attitude.

These terms are not actor's categories, and their employment runs the risk of anachronism. However, they are useful notions and refer to a phenomenon which is an essential aspect of the way in which human history develops (cf., e.g., Kuhn 1977). This is the tension between the old and the new, between tradition and innovation, between preserving what already exists and changing it in some way.

Thus it is not surprising that similar comments would apply to attempts, such as that of Stillman Drake,[7] to show that the root cause of the affair was a conflict between science and philosophy. By philosophy here Drake means academic professional philosophy, namely professors of philosophy. But in regard to "science," it is unclear whether Drake means professors of mathematics or natural philosophers who did not hold a university position. Thus, it is unclear whether the alleged contrast is between professors of philosophy and professors of mathematics, or between

---

[6] See, for example, Draper (1875), White (1896, 130–152; 1965); because of their explicitness and militancy, Draper and White can be regarded as important sources of the conflict thesis, and they have been the main targets of the above mentioned recent criticism. However, the conflict thesis may be gleaned from other more significant authors, for example, Einstein (1953), Milton (1644, 35), Popper (1963, 97–119). Moreover, Draper and White have explicit precursors, such as Voltaire's 1728 letter on Descartes and Newton and D'Alembert's 1751 preliminary discourse to the *Encyclopedia*; cf. Finocchiaro (2005b, 115–125), and Section 8.6 above.

[7] See Drake (1976; 1978; 1980). Drake's account is only a recent version of a type of interpretation that goes back at least to L'Epinois (1867, 143–145; 1877; 1878).

academic and nonacademic natural philosophers. To delimit the discussion, let us focus on the alleged conflict between Galileo and his "philosophical" opponents, that is on his relationship with persons who held academic positions in philosophy or had some other kind of philosophical pretension or claim.

When we do this we find that, on the one hand some philosophers were critical of Galileo and displayed various degrees of opposition to him, from the militant anti-Galileanism of Ludovico delle Colombe in 1611 to the philosophical criticism of René Descartes in the 1630s, with intermediate cases such as Cosimo Boscaglia and Scipione Chiaramonti. On the other hand, as we have seen (Chapter 4), Campanella (despite his own troubles) was constantly supportive of Galileo, publishing an *Apologia pro Galileo* in 1622; and Campanella was first and foremost a philosopher. Moreover, the attitude of some leading establishment philosophers was, again, nuanced and complex, if one examines all aspects of their interaction with Galileo; this would be the case for two professors of philosophy at the University of Padua, Cesare Cremonini and Fortunio Liceti. Finally, of course, one cannot ignore Galileo's own philosophical pretensions, for example his "claim to have spent more years studying philosophy than months studying pure mathematics" (Favaro 10: 353), as well as his insistence and success in obtaining the title of Philosopher to the Grand Duke of Tuscany, besides Chief Mathematician. Thus, we cannot replace the alleged warfare between science and religion by the alleged conflict between science and philosophy; rather we may want to resort, once again, to the conflict between a conservative and a progressive attitude.

The importance of the dialectic of conservation and innovation is also shown by the fact that this notion manages to re-assert itself in the context of recent studies which break new ground by turning away from the tradition of conflicts between science and other disciplines. For example, while criticizing the tradition of portraying the original affair as a clash between reason and unreason, Rivka Feldhay argues that the Church was not a monolithic institution, but that the Dominicans represented its conservative wing, and the Jesuits its progressive wing.[8] Here I would want to stress that the Dominicans and the Jesuits were not monolithic entities either, a point admitted by Feldhay but not sufficiently exploited by her.

For example, there seems to have been some disagreement between two of Galileo's Jesuit enemies: Christopher Scheiner, with whom Galileo was embroiled in a controversy about sunspots, and who is sometimes seen as the instigator of the 1633 trial; and Melchior Inchofer, who in April 1633 wrote one of three consultant reports on Galileo's *Dialogue*, used by the Inquisition as documentation that Galileo had defended and come close to holding the thesis of the earth's motion. A recent study calls attention to a document recently discovered in the Archives of the Society of Jesus in Rome.[9] It is Scheiner's evaluation of a book manuscript by Inchofer.

---

[8] Feldhay (1995). The progressiveness of the Jesuits, as well as their influence on Galileo, has been documented by Wallace (1984a). The non-monolithic character of the Catholic Church has been explicitly stressed in various ways by other authors, such as Segre (1991b, 30).

[9] Gormam (1996); see also Blackwell (2006, 65–92). On Inchofer, see Blackwell (2006, 45–64, 105–206), Cerbu (2001), Shea (1984).

12.2 Conflictual Accounts

The evaluation was written on 9 August 1633, and the book was published the same year under the title of *Tractatus syllepticus*. Scheiner's evaluation was generally positive, and he recommended publication. However, he expressed two reservations: that Inchofer went too far and should moderate his claims that (1) questions of the location and behavior of the earth and sun are matters of faith, and that (2) biblical authority "is greater than the capacity of any human mind" (Gorman 1996, 316). Aside from suggesting that Scheiner was not the moving force behind Galileo's trial of 1633, this suggests that there was a split within the Jesuits and that Inchofer was part of the conservative wing. However, it also suggests that Scheiner was partly progressive and partly conservative. Indeed, I would go further and find a split between conservative and innovative tendencies within the minds of many key players, including Galileo, Cardinal Bellarmine, and Pope Urban VIII.[10]

Before concluding this discussion of the conflict thesis, it is important to note that there are sophisticated versions of it which are hard to fault and may very well be unavoidable. They have been advanced by scholars who may be considered to be fully cognizant of the complexity of the historical relationship between science and religion and of the untenability of extreme, oversimplified accounts. They support the claim, not that conflict is necessary or unavoidable, but that the *potential* for conflict between science and religion is *always* present, and that new cases similar to the Galileo affair *may* arise in the future.

For example, Richard Blackwell has argued plausibly that, partly because of the trial of Galileo, "the authority behind scripturally based religion, at least in the Catholic tradition, became highly centralized, monolithic, esoteric, non-fallible, resistant to change, and self-protective ... Meanwhile in the centuries since Galileo's day, modern science has also become thoroughly institutionalized, but in a quite different fashion. By contrast, modern science as an institution is pluralistic, democratic, public, fallibilistic, innovative, and self-corrective."[11] In other words, if we focus not on particular beliefs, but on the mindset fostered by science and religion, respectively, as they have in fact developed in the West since the time of Galileo's trial, we discover a difference and a potential conflict. I believe Blackwell is even willing to admit that science and religion might have developed differently from the way they have actually developed, but that given their actual development in the last four centuries, the possibility of trouble can never be dismissed.

A complementary account has been given by Marcello Pera (1998). Approaching the topic from a very different angle, Pera argues that the trial of Galileo involved a conflict between two principles: (1) that science can investigate any factual question and end up rejecting any factual claim; and (2) that some factual questions are essential to religious faith and cannot be rejected by a believer, on pain of abandoning religion. Examples of the latter are the questions whether the physical universe is infinite in time (that is, whether it is eternal or had a beginning), and whether the

---

[10] It is interesting that, for the case of Darwinism, a similar phenomenon has been found by Moore (1979, 102–103); this point has been stressed by Lindberg and Numbers (1987, 147).

[11] Blackwell (2006, 103). See also Blackwell (1991; 1998a; 1998b).

soul survives after death.[12] He attributes the first principle to Galileo, and the second one to Bellarmine. Although these attributions have some foundation, they are questionable. However, the important point here is that the Galileo affair has implications that are problematic for the science-religion relationship. For, if we accept Galileo's claim that the Bible is not an authority on astronomical questions, then presumably we would want to extend this claim to include physical questions; then it is tempting to embark on the slippery slope of generalizing to questions of biology, psychology, and history. At some point the conflict with some religious belief is unavoidable.

12.3 John Paul II's Harmony Thesis

Let us now see how the harmony thesis fares, for, strange as it may seem, there have been those who have attempted to reverse the traditional conflictual interpretation by claiming that the Galileo affair illustrates the *harmony* between science and religion. This thesis was first advanced in a clear and explicit form by Agostino Gemelli in 1942, during what I have called (Section 8.14) the tricentennial rehabilitation of Galileo. Such harmony was also a major theme in Pope John Paul II's rehabilitation of Galileo in 1979–1992, whose general highlights I summarized earlier (Section 8.17). That episode now needs to be re-focused in terms of this theme.

In his 1979 speech to the Pontifical Academy of Sciences for the commemoration of the Einstein centennial, John Paul II expressed his regret for Galileo's suffering "at the hands of men and organisms of the Church,"[13] and he quoted the Second Vatican Council's general condemnation of such interferences with freedom of speech and of thought. He went on to state his full support for new and deeper studies of the affair, conducted in a spirit which he described as "loyal recognition of wrongs from whatever side they come."[14] Then he focused on three interrelated points: (1) that Galileo not only believed that religious and scientific truths cannot contradict each other, but the reason he gave for this belief was essentially identical to the reason given by the Second Vatican Council; (2) that he conducted his scientific research in the same spirit of piety and divine worship which the same Council recommended as exemplary; and (3) that he formulated important epistemological norms about the relationship between science and the Bible, which the Church later recognized as correct. The pope summarized his own interpretation

---

[12] Although Pera does not mention Giordano Bruno here, it is interesting to note that such claims played a key role in Bruno's eventual condemnation and execution, and so the situation discussed by Pera has more than mere hypothetical interest; cf. Fiorani (1993), and Finocchiaro (2002). Beretta (1998, 103; 2005b, 239) has stressed the historical connection between Galileo's trial and the condemnation of the thesis of the mortality of the soul in 1513 at the Fifth Lateran Council.

[13] John Paul II (1979), section 6, paragraph 1.

[14] John Paul II (1979), section 6, paragraph 2.

12.3 John Paul II's Harmony Thesis

of the episode with these words: "in this affair the agreements between religion and science are more numerous and above all more important than the incomprehensions which led to the bitter and painful conflict that continued in the course of the following centuries."[15] In the eloquent words of a churchman who later contributed to elaborating this interpretation, "Galileo did not, for his part, have a personal 'Galileo affair'" (Coyne 1985, 178).

Note that this is not merely a denial of the traditional view claiming that the affair exemplifies the warfare between science and religion; the pope was also *reversing* the traditional view by claiming that the same episode really proves their harmony. His supporting argument seems to be that Galileo believed that science and religion are essentially in harmony rather than in conflict, and advanced plausible and eloquent reasons for such harmony.

John Paul's argument is important and cannot be just dismissed. However, I do not think it succeeds in establishing the harmony thesis. For the conflict reappears when we contrast Galileo's view to that of his opponents and critics (including churchmen), who did believe that there *is* a conflict. Admittedly, the alleged contradiction is not between scientific and religious *truths*, which in such terms is a conceptual impossibility, given the standard meaning of the notions of *truth* and *contradiction*; but the conflict is between biblical and scientific *propositions*. In other words, contextually the conflict remains, and it is between those who affirmed and those who denied that there is a conflict between physical inquiry and biblical statements.

This re-emergence of the conflict is a good example of the kind of complication with which the history of the science-religion relationship abounds, and which has been well documented and eloquently discussed by such scholars as John Brooke, David Lindberg, and Ron Numbers.[16]

Another "complication" along the same lines is that, on the other hand, there is something importantly right in the pope's interpretation of Galileo's view of the science-religion relationship. Now, this is worth stressing because there are scholars who persist in attributing to Galileo incoherences and absurdities on the topic which he did not fall into.[17] There is no need to elaborate once again the Galilean position (from the letters to Castelli and to Christina), which has already been done on two earlier occasions (Sections 4.5 and 9.3); but it is useful to summarize and highlight it here.

Galileo's key claim is the memorable principle he attributes to Cardinal Baronio: that "the intention of the Holy Spirit is to teach us how one goes to heaven and not how heaven goes."[18] This principle explicitly and immediately invalidates the inferential

---

[15] John Paul II (1979), section 7, paragraph 1.

[16] Brooke (1991), Brooke and Cantor (1998), Lindberg and Numbers (2003), Numbers (2009).

[17] The interpretation I am criticizing can be gleaned from such works as Carroll (1997; 1999; 2001), Brooke (1991, 77–80), Koestler (1959), Langford (1971), McMullin (1967c; 1995; 1998; 2005c), Moss (1983; 1986; 1993, 181–211). Of course, the views of all these authors do not coincide in every respect, but they do contain a common strand. For more details on the alternative interpretation I advance here, see Chapters 4 and 9 above; cf. Fantoli (2003b), Howell (1996), Pesce (1987; 1991a; 2005).

[18] In Finocchiaro (1989, 96), or Galilei (2008, 119); cf. Favaro 5: 319.

soundness of the scriptural argument against Copernicanism; in other words, it undermines the logical relevance of the biblical premise to the astronomical conclusion. Galileo also criticizes the truth of the argument's premise, by showing that the Joshua passage is inconsistent with the geostatic view but compatible with the geokinetic view. This exegetical exercise, far from conflicting with Baronio's principle, complements it elegantly, since one can always criticize a reason offered to support a conclusion by questioning either the logical relevance or the factual truth of the reason.

However, Galileo also gives a justification of Baronio's principle. One argument is that it provides the explanation of the correctness of the uncontroversial Augustinian principle asserting the priority of demonstration. A second reason is that God revealed Himself in two ways, through the Book of Nature as well as through Holy Scripture. Then Galileo uses Baronio's principle to derive a corollary about demonstrability, as distinct from demonstration: when a physical proposition is capable of conclusive proof (even if not yet proved), natural philosophers should be free to study the relevant natural phenomena and search for a proof. Finally, one may also ask what to do in regard to physical claims which, besides lacking a conclusive proof, are not even capable of being conclusively proved; for this class of propositions, Galileo sees no difficulty in conceding to accept the Bible's word.

Thus, John Paul's view of the Galileo affair is partly right, namely in regard to the fact that Galileo firmly believed, and had good reasons why, scientific inquiry and biblical interpretation can be seen to be harmonious. However, the conflict emerges at another level, namely in regard to the fact that many of Galileo's contemporaries, especially churchmen, disagreed with his claim and/or reasons. Now, this conflict existed not only in the historical context of four centuries ago, but also it does not seem to have disappeared in the present situation. In fact, however ambitious John Paul's original intentions may have been in 1979, and however pro-Galilean his own personal conclusions may have been at the end of the episode in 1992, the whole process was mixed because, as we saw earlier (Section 8.17), there were other ecclesiastic officials who acted and spoke in a more conservative and apologetic manner, thus tending to oppose the pope's innovative inclinations.[19]

12.4 Morpurgo-Tagliabue's Version of Harmony

A less well known, but more promising attempt at a harmony thesis is an account that can be gleaned from the work of an Italian scholar named Guido Morpurgo-Tagliabue.[20] To best appreciate this thesis, let us begin by calling attention to a

---

[19] For more details on this account, see, Finocchiaro (2005b, 338–357; 2008a), and Section 8.17 above. For accounts that are more negative, see Reston (1994, 139–144, 283–286), DiCanzio (1996, 321–330), Segre (1997), Beltrán Marí (1998), Finocchiaro (1986; 1999a), Benitez (1999, 85–110), Fantoli (2001), Coyne (2005), McMullin (2005a, 7). For the best documentation and a relatively well-balanced account, see Artigas and Sánchez de Toca (2008).

[20] Morpurgo-Tagliabue (1947–1948; 1981, 60–61, 64, 68–69, 70, 85–87, 88, 167, 189–90).

12.4 Morpurgo-Tagliabue's Version of Harmony

change or development in Galileo's epistemological and methodological view of the nature of science. He certainly began with a version of the Aristotelian ideal of science as demonstration. This is relatively well known and uncontroversial, although recently new evidence and interesting nuances have been elaborated, largely due to the efforts of William Wallace.[21] I also believe that by the end of his life, and certainly in the mature science of the *Dialogue*, Galileo held a fallibilist, probabilist, and hypothetical epistemology.[22] This later epistemology is still realist, so that in that regard I would say there was no change. Moreover, the later Galilean epistemology is non-exclusivist, in the sense that it does not claim that all scientific knowledge must be fallibilist; it is rather somewhat eclectic in allowing necessity and demonstration if and when they are attainable, as is the case for the new science of motion sketched in the *Two New Sciences*. But the crucial point is that revisable and merely probable hypotheses are not automatically denied scientific status.

In my opinion, this change in Galileo's career is not only a historically real development that can be documented, but it is also a progressive development, a change for the better, as it were. Now, let us ask how and why Galileo underwent this development. The answer lies, I believe, in the pressure from the Catholic Church. That is, the various conservative and reactionary elements which opposed, criticized, and even persecuted Galileo did in one sense happen to perform a valuable service, by making him see the light, as it were, in epistemological matters. It is not that these ecclesiastical elements themselves held a fallibilist epistemology and convinced Galileo of it; rather they subscribed to the Aristotelian demonstrative ideal, but their criticism of Galileo's arguments helped him understand that the case in favor of Copernicanism was still not conclusive; this perception suggested that the search for a conclusive demonstration is an important stage of scientific investigation. This account may be regarded as a version of the harmony thesis because it elaborates an example of a beneficial influence by religion on science.

The main difficulty with this thesis is that it is a hypothesis that has not yet been fully documented. However, I see no difficulties with its documentation. In any case, the present context is not one where such documentation is the main point of the discussion. I mention the thesis primarily as a promising version of the harmony thesis in regard to the Galileo affair, suggesting primarily the lesson that the history of the relationship between science and religion is a very complicated business.[23]

---

[21] Wallace (1981b; 1984a; 1992a,b).

[22] Finocchiaro (1980; 1997a), Gingerich (1995; 2000), Pitt (1992).

[23] An analogous version of such a thesis, but at a more general level, may be gleaned from Heilbron (1999, 202–207); although Heilbron's main purpose lies elsewhere, and although his intention is not an apologia of the Catholic Church, he does suggest that the lip service to the hypothetical status of Copernicanism required by the anti-Copernican decree of 1616 and by Galileo's condemnation of 1633 fostered an attitude of instrumentalism that was sound.

12.5 Feyerabend's Version of Conflict

In fact, I now go on to discuss another example of complication pointing toward the same suggestion and lesson, but involving instead an updated and revised conflictual thesis. It was advanced by Paul Feyerabend, in an essay with the revealing title of "Galileo and the Tyranny of Truth."[24] The essay was a contribution to a 1985 conference on "The Galileo Affair: A Meeting of Faith and Science," sponsored by the Cracow Pontifical Academy of Theology and the Vatican Astronomical Observatory (cf. Coyne et al. 1985). The conference thus appears to have had an apologetic or pro-clerical aim, in the sense that it was meant to substantiate and elaborate Pope John Paul's harmony thesis. Feyerabend did contribute a thesis which is in one sense apologetic and pro-clerical, but which remains conflictual, and so is critical and anti-clerical in another sense. This of course is the kind of irony and iconoclasm at which Feyerabend was a master. And we shall soon see the broader significance of this particular irony.

As we saw earlier (in the penultimate section of the Introduction), Feyerabend interprets Galileo's trial as involving a conflict between two philosophical attitudes toward, and historical traditions about, the role of experts. Supposedly, Galileo advocated the uncritical acceptance by society of the views of experts, whereas the Church advocated the evaluation by society of the views of experts in the light of human and social values. Feyerabend attributes the latter principle to Cardinal Bellarmine, based on the evidence of his letter to Foscarini. Feyerabend (1985, 164) concludes that "the Church would do well to revive the balance and graceful wisdom of Bellarmine, just as scientists constantly gain strength from the opinions of … their own pushy patron saint Galileo."

Note now that Feyerabend is advocating a conflictual account, and thereby rejecting the harmony thesis. Rather than reversing the traditional type of *interpretation*, he reverses what may be called the traditional *evaluation*. In fact, he is (in the historical context) siding with the Church and against Galileo insofar as he thinks that the rule advocated by the Church was sounder than the one advanced by Galileo. At the same time, since the Church in the meantime has herself switched sides, the result is that Feyerabend is upholding the past Church against the present-day Church. The content and nature of Feyerabend's evaluation became more obvious later, in the 1988 edition of *Against Method*, where he explicitly criticized the rehabilitation efforts of Pope John Paul II with the (in)famous words I quoted in the Introduction: "the Church at the time of Galileo not only kept closer to reason as defined then and, in part, even now; it also considered the ethical and social consequences of Galileo's views. Its indictment of Galileo was rational and only opportunism and a lack of perspective can demand a revision."[25] And this substantiates

---

[24] Feyerabend (1985; 1987, 247–264).
[25] Feyerabend (1988, 129). Cf. Ratzinger (1994, 98), which has a slightly different wording since he is translating from a German edition of Feyerabend's work.

my earlier remark that Feyerabend's account is apologetic and pro-clerical in one sense, but critical and anti-clerical in another sense. However, a clarification about these terms is in order.

Here and throughout this book, the labels *pro-clerical, anti-clerical, apologetic, pro-Galilean*, and *anti-Galilean* are intended to have a descriptive, informative, and piecemeal connotation, rather than a loaded, inflammatory, holistic, or name-calling meaning. Thus, note that I apply these terms primarily to theses and not to persons, and that in my account authors often advance views that are a mixture of such orientations; moreover, pro-clerical and pro-Galilean are not meant to be opposite. For example, to limit myself to authors explicitly discussed in this chapter, note that here I am describing Feyerabend's account as pro-clerical in one sense, but anti-clerical in another. Note also that whereas above (Section 12.2) I discussed a thesis by Richard Blackwell which might be labeled anti-clerical, below (Section 12.6) he also appears to be subscribing to another thesis which is pro-clerical. And note that below (Section 12.7) I point out how Viviani's interpretation is both pro-clerical and pro-Galilean. The non-invidious and nonloaded character of these terms may also be seen from the fact that I would have little difficulty describing as pro-clerical certain parts of this chapter (for example, my justification in Section 12.3 of Pope John Paul II's interpretation of Galileo's views on science and Scripture) and pro-Galilean certain other parts (for example, my account of the heresy versus disobedience issue in Section 12.6).

Recall that earlier (in the penultimate section of the Introduction), I criticized Feyerabend's account of the Galileo affair as textually untenable (by reading the principle about the role of experts into Bellarmine's letter to Foscarini) and as logically fallacious (by equivocating on two different meanings of that principle). That criticism still applies, but my main point here is that Feyerabend's account provides a good illustration of how an intelligent scholar can formulate an interesting thesis, which is an updated and sophisticated version of the traditional, otherwise discredited, conflictual account.

12.6 Heresy or Disobedience?

A common way of attempting to defuse the conflict between science and religion in Galileo's trial is to claim that he was condemned not for heresy but for disobedience.[26] For short, this may be called the disobedience thesis, but it should be noted that it contains two parts: a denial of heresy, and an assertion of disobedience. Presumably, he was not condemned for heresy because Copernicanism was never declared a formal heresy. But he was condemned for disobedience because in 1616 he had promised to obey the Church's orders on the subject; because with his

---

[26] See, for example, Blackwell (1998a, 355), Brandmüller (1992, 144–146), Gingerich (1995, 342), Mayaud (1997, 313).

Dialogue he broke that promise; and because the 1633 sentence held him accountable for such a violation. Such a thesis may be taken to lessen the seriousness of the censure on Galileo and so the depth of the conflict between him and the Church; it is thus an attempt to undermine or tone down the conflict thesis. It is also a type of apologetic or pro-clerical statement in the sense that, for example, the Church can be spared the error of having declared, or at least of having implicitly considered, heretical a physical truth;[27] and this not only saves its own doctrine of infallibility, which of course is of little concern to non-Catholics, but it also upholds to some degree its reputation in the eyes of non-Catholics.

My first reservation about the disobedience thesis is that it seems to presuppose that the two crimes (heresy and disobedience) are mutually exclusive, so that Galileo could have been condemned for only one of the two alternatives, but not for both. But I believe that this assumption is questionable. For the disobedience in question involves intellectual matters of what to hold or defend, and so it is not merely a disciplinary matter but becomes a doctrinal one. Moreover, heresy ultimately and essentially is simply to believe what the Church commands not to believe, or not to believe what it commands us to believe, in short disobedience in matters of belief.[28] So there is no necessary contradiction between the two alleged crimes, but they are rather different ways of looking at the situation. In short, at best the disobedience thesis is correct in what it affirms, and incorrect in what it denies.

Secondly, "heresy" is not a single univocal crime, but a family of crimes. So while it is indeed true that Galileo was not condemned for heresy in the sense of "formal heresy," it is equally true that he was condemned for heresy in the sense of "vehement suspicion of heresy." Let me elaborate.

To begin with, Copernicanism was never officially declared to be a formal heresy. In 1616 the Inquisition consultants' report did indeed state that the heliocentric and heliostatic thesis was formally heretical (though it did not attribute the same degree of censure to the geokinetic thesis).[29] However, the Congregation of the Holy Office did not endorse this recommendation and never went through with such a formal declaration. Instead, it was the Congregation of the Index that took some action. It issued a decree that condemned and permanently banned Foscarini's *Letter* and temporarily banned Copernicus's book until corrected (and these corrections were published in 1620). These were indeed disciplinary measures. However, the Index's decree contained an unprecedented doctrinal pronouncement: it declared the earth's motion false philosophically (i.e., physically) and also contrary to Scripture. Now, although strictly and technically speaking

---

[27] The attribution of some such error is a recurring theme in the controversy about Galileo's trial. See, for example, Arnauld (1775–1783, 9: 307–314), Mivart and Jackson (1885), Beretta (1998, 278; 1999a, 481–485; 2001b, 316–17; 2003a, 180–183), and Camerota (2004, 517–518).

[28] For this conception of heresy, see Garzend (1912), Finocchiaro (2005b, 272–274), and Section 8.13 above.

[29] Favaro 19: 320–321, Pagano (1984, 99–100), Finocchiaro (1989, 146–147).

12.6 Heresy or Disobedience?

"contrary to Scripture" was not equivalent to "heretical," an equally valid technical point is that contrariety to Scripture was a standard ground to justify the attribution of heresy.[30] Moreover, it is a historical fact that many people, including educated clergymen and Church officials, did not make the fine distinction between heresy and contrariety to Scripture.[31]

It is also true that in 1633, at the conclusion of the trial, Galileo was not condemned for formal heresy. But it is equally true that he was condemned for vehement suspicion of heresy, as the sentence explicitly states[32] and as we have had occasion to mention many times before. What was meant by vehement suspicion of heresy? I believe the sentence was not merely or primarily saying that he was suspected of being a formal heretic, and that the trial proceedings were unable to confirm or deny the suspicion. Rather the sentence was convicting him of a specific category of religious crime, admittedly not the most serious type (called "formal heresy"), but an intermediate type, more serious for example than what was called "slight suspicion of heresy."[33]

Moreover, the sentence goes on to explain what the heresy was and claims it to have been twofold: the first part is a physical claim, the key Copernican thesis of heliocentrism and geokineticism; the second part is the methodological principle that "one may hold and defend as probable an opinion after it has been declared and defined contrary to Holy Scripture."[34] This prescription is basically a way of stating Baronio's principle, for in the context of astronomy and natural philosophy the two principles basically imply each other. For if we start by agreeing with Baronio, then the purpose of the Bible is not to teach astronomical propositions, and so the Church has no business passing judgment on such propositions, and in astronomical inquiry it becomes permissible to disregard such Biblical statements and ecclesiastic judgments. Conversely, if we start with the above quoted principle, it should be noted at the outset that the opinion in question is an astronomical proposition; the principle is saying that it is permissible to hold and defend an astronomical opinion even if the Church has declared it contrary to the Bible; this can only be if such ecclesiastic declarations are irrelevant; and this in turn can only happen if the purpose of the Bible is not to teach astronomical knowledge. This talk of heresy in the sentence would be completely unfounded and arbitrary if we take it to mean formal heresy, but makes some sense if we think in terms of suspected heresy.

My third objection to the disobedience thesis questions the exact nature, content, and reality of Galileo's disobedience. It is certainly true that in 1616 the Church gave him an order which he promised to obey. Furthermore, it is true that as a result of the plea bargaining after the first deposition in 1633, he pleaded guilty of having

---

[30] This point has been stressed by Beretta in many places, e.g., Beretta (1998, 97–108; 1999a, 473–474).

[31] Finocchiaro (2005b, 26–42), and Section 8.2 above.

[32] Favaro 19: 405, Finocchiaro (1989, 291), Galilei (2008, 292).

[33] Masini (1621, 16–18), Garzend (1912), Giacchi (1942), Neveu and Mayaud (2002, 288), Finocchiaro (1989, 38; 2005b, 12–13), and Sections 7.2 and 8.1 above.

[34] Finocchiaro (1989, 291), or Galilei (2008, 292); cf. Favaro 19: 405.

disobeyed the order not to hold or defend the geokinetic thesis. And the sentence certainly includes talk about some transgression as a result of the publication of the *Dialogue*. But several clarifications are in order here.

One is that the disobedience for which he was condemned did not pertain to the mere discussion of the topic. The inquisitors did not press that charge after Galileo produced Bellarmine's certificate and in the light of the fact that the special-injunction document lacks Galileo's signature. Nor did Galileo admit having been ordered not to discuss the topic.

Another clarification regards Galileo's holding and defending Copernicanism. Even under the verbal threat of torture, he denied having done so intentionally or deliberately; and his denial seems to have been accepted by the tribunal.

A final clarification regards the fact that, just because the Inquisition found him guilty of having held and defended the earth's motion, and just because he admitted having done so, these assertions do not make it so. I am not sure the prosecution ever proved that the *Dialogue* holds and defends Copernicanism. I would argue as follows.

To begin with, we must clearly distinguish the notion of *holding* a view from the notion of *defending* it. This is relatively uncontroversial, and the Inquisition consultants who wrote reports on the *Dialogue* made the distinction.[35] But what this means is that the evidence proving that Galileo *defended* Copernicanism cannot be the same as the evidence proving that he *held* it.

Now, in my opinion, in order to prove the case about Galileo having *held* the earth's motion, one would have to prove that he regarded his arguments as conclusive. For as distinct from defending a view, to hold it suggests belief and commitment, and a sufficient degree of belief and commitment that one is not going to abandon it lightly. I realize, of course, that this is both a long story and a controversial one. Let me mention just one point,[36] because it is usually neglected, namely the fact that it is clear that Galileo gives several arguments in favor of the earth's motion; this multiplicity is, I believe, an indication that he did not regard any one of them as conclusive, not even the argument from tides. Thus, it is not obvious that in the *Dialogue* Galileo *holds* the opinion of the earth's motion.

To prove that he was guilty of *defending* the earth's motion, one would have to overcome the following problem. The *Dialogue* discusses all the scientific and philosophical arguments on both sides of the controversy. His discussion takes the form of not only presenting and analyzing the arguments, but also of evaluating them. Now, his evaluation is indeed basically unfavorable and negative for the anti-Copernican arguments, and favorable and positive to various degrees for the pro-Copernican arguments. If his evaluations are correct, is the discussion really a defense? Is not what he is doing merely to articulate the thesis that the pro-Copernican arguments are stronger than the anti-Copernican ones? Is it his fault if the arguments on one side are stronger than those on the other side?

---

[35] Favaro 19: 342–376, Pagano (1984, 139–153), Finocchiaro (1989, 262–276).

[36] For more details about my view and a discussion of alternative interpretations, see Finocchiaro (1997a, 52–58).

In other words, one might say that there are three ways of defending an opinion: one is to give reasons or evidence supporting its truth; another is to refute objections or counter-arguments against its truth; a third way is to do both, giving supporting evidence and refuting objections. Galileo's *Dialogue* was clearly a defense of the earth's motion in the third sense, and so (in that sense) he disobeyed the letter of Bellarmine's certificate. But perhaps the spirit of Bellarmine's warning was to prohibit only one-sided defenses of the first two kinds; in this case, Galileo committed no real violation or disobedience.

Moreover, even if we focus on the literal disobedience, as long as we also focus on the (third) comprehensive kind of defense, we cannot avoid asking the question whether Galileo's defense was fair, impartial, reasonable, and plausible. If it was, that means that the earth's motion is more likely or more probable than the earth's rest. In this case, the presupposition of Bellarmine's warning no longer subsists. That presupposition, explicitly stated in the Index's decree, was that the earth's motion is false (philosophically or physically). Thus the warning becomes unjustified, and its literal violation implies no crime or guilt.

Such distinctions and issues not only have general logical and cultural importance (cf. Finocchiaro 2005a, 292–326), but also were discussed in the Inquisition consultants' reports submitted in April 1633.[37] However, their case for guilt on this score is far from conclusive or convincing. The case is certainly not proved beyond a reasonable doubt, and I do not think it is proved even by a preponderance of the evidence, to use present-day legal jargon. At any rate, these issues were never aired at the trial. My conclusion is that even Galileo's alleged disobedience is questionable, and to that extent the disobedience thesis does not succeed in defusing the conflict between him and the Church.

12.7 Science Versus Religion in the Subsequent Affair?

What is the upshot of these reflections? Does anything more interesting or substantial follow than that the trial of Galileo, like the relationship between science and religion, or indeed like any episode in human history or phenomenon in human society, is a very complicated business? Although complexity is a fact of life that cannot be ignored or brushed aside, the search for and drive toward simplification is also a fact of (mental) life. As long as simplification is not confused with, and does not lapse into, over-simplification, simplifying ideas are quite legitimate.

My first simplifying hypothesis is this. Ultimately it may turn out that, as suggested above, underlying the apparent conflict between science and religion in the trial of Galileo, this episode exhibits the deep structure of nothing less, and nothing more, than the universal and eternal conflict between conservation and innovation.

However, independently of this simplification, I believe we need to study the subsequent controversy *about* Galileo's trial more seriously than simply to find

---

[37] Favaro 19: 348–360, Pagano (1984, 139–153), Finocchiaro (1989, 262–276).

instances of conflict or harmony, and of conservations or innovation, in the *original* episode. In fact, the subsequent controversy (1633–1992), although it obviously began with the condemnation of Galileo and reflects the issues of the original controversy (1613–1633), acquired a life of its own and possesses a fascination rivaling that of the original. In other words, the historiography of Galileo's trial is itself a complex phenomenon and development of Western cultural history of the last four centuries. Now, it may be that, underlying the diversity and complexity of opinion on the trial of Galileo, it is the subsequent controversy that possesses the characteristics of a conflict between science and religion. But even if such an elegant possibility is not the case, the subsequent controversy or historiography of the trial is likely to prove profoundly instructive in other ways as well.[38] However, before such a deep structure is demonstrated, the variety, complexity, and longevity of the trial's aftermath should be ascertained and exhibited. A chronological overview and synoptic sketch was given earlier (Chapter 8), and that was a mere summary of a longer work already accomplished.

Here the following analysis is offered as an example of how that mass of data could be mined in the search for the possible existence of a (subsequent) conflict between science and religion, or of some other deep structure. And to make the discussion more focused, I begin by stating my simplifying idea: as regards this *subsequent controversy*, I claim that the science versus religion conflict is indeed an essential feature of it; that this conflict is more of an integral part of it than of the original trial; that, again, underlying such surface structure there is a *cultural deep-structure*; but that this deep structure is the phenomenon of the birth and evolution of *cultural myths* and their interaction with documented facts.[39]

My first example is provided by Vincenzio Viviani.[40] In 1654 he made the first serious attempt to write a biography of Galileo, and in the process he formulated an account of the trial. His account is important for several reasons: for example, he had lived with Galileo during the last few years of his life; he had unparalleled access to the documents; and although it was not published until 1717, his was the first biography of Galileo.

Here I want to focus on the following features of Viviani's account of the trial: it exhibits a conflict between Galileo and the Church, as well as the overcoming of this conflict; it attempts to be both pro-Galilean and pro-clerical, thus illustrating Viviani's own effort at compromise;[41] but it is so outrageous as to suggest a historiographical category that I would label "mythological." This, in fact, is Viviani's entire account:

---

[38] Brooke and Cantor (1998, 106, 130, and 132) seem to advance such a suggestion in their own analysis of "the contemporary relevance of the Galileo affair."

[39] See also Lessl (1999), Benítez (1999, 85–110), Finocchiaro (2009a).

[40] Brooke and Cantor (1998, 123–126) do discuss Viviani, but they focus on his general interpretation of Galileo's science and methodology, and do not mention at all Viviani's interpretation of the affair. Although my focus here is different, their critique is correct, as may also be seen from Biagioli (1993, 87–88) and Segre (1989; 1991a).

[41] Cf. Section 8.5 above, which discusses other aspects of Viviani's attempt at compromise, and other attempts by Auzout and Leibniz.

12.7 Science Versus Religion in the Subsequent Affair?

> For his other admirable speculations Mr. Galileo had been raised to heaven with immortal fame, and for his many discoveries he had been regarded by men as a god; thus, the Eternal Providence allowed him to prove himself human by letting him commit an error when, in discussing the two systems, he showed himself more inclined to believe the Copernican hypothesis, which had been condemned by the Church as incompatible with Divine Scripture. Because of this, after the publication of his *Dialogue* Mr. Galileo was called to Rome by the Congregation of the Holy Office. Having arrived there around 10 February 1633, through the great generosity of that Tribunal and of the Sovereign Pontiff Urban VIII (who already knew him as highly meritorious in the republic of letters), he was kept under arrest at the residence of the Tuscan ambassador in the delightful palace of Trinità dei Monti. Having been shown his error, he quickly retracted this opinion like a true Catholic. As a punishment his *Dialogue* was banned. After this five-month detention in Rome (while the city of Florence was infected with the plague), he was generously assigned for house arrest the residence of Monsignor Archbishop Piccolomini, who was the dearest and most esteemed friend he had in the city of Siena. He enjoyed the latter's highly cordial conversation with so much ease and emotional satisfaction that he resumed his studies and discovered and demonstrated most of his mechanical conclusions on the resistance of solids, among other speculations. After about five months, when the plague in his homeland had completely ceased, at the beginning of December 1633 His Holiness commuted his house arrest from the restriction of that residence to the freedom of country living, which he so much enjoyed. So he returned to his villa in Arcetri, where he had been living already most of the time on account of the healthy air and the great accessibility to the city of Florence, and where consequently he could easily receive visits by friends and relatives which always brought him great comfort and consolation. [Favaro 19: 617]

Another instructive account is the one advanced by Thomas Salusbury in 1661 in the foreword to his *Mathematical Collections and Translations*. The first part of the first volume of this work contains the first published English translation of Galileo's *Dialogue*, as well as the *Letter to Christina*, Foscarini's *Letter*, Kepler's introduction to the *Astronomia nova*, and Zúñiga's commentary on Job 9: 6. The two most relevant paragraphs read as follows:

> The first Book which offers it self to your view in this Tome is that singular and unimitable Piece of Reason and Demonstration the Systeme of Galileo. The subject of it is a new and Noble part of Astronomy, to wit the Doctrine and Hypothesis of the Mobility of the Earth and the Stability of the Sun; the History whereof I shall hereafter give you at large in the Life of that famous Man. Only this by the by; that the Reader may not wonder why these Dialogues found so various entertainment in Italy (for he cannot but have heard that though they have been with all the veneration valued, read & applauded by the Iudicious yet they were with much detestation persecuted, suppressed & exploded by the Superstitious) I am to tell him that our Author having assigned his intimate Friends Salviati and Sagredo the more successfull Parts of the Challenger, and Moderater, he made the famous Commentator Simplicius to parsonate the Peripatetick. The Book coming out, and Pope Urban the VIII taking his Honour to be concerned as having in his private Capacity bin very positive in declaiming against the Samian Philosophy, and now (as he supposed) being ill delt with by Galileo who had summed up all his Arguments, and put them into the mouth of Simplicius; his Holiness thereupon conceived an implacable Displeasure against our Author, and thinking no other revenge sufficient, he employed his Apostolical Authority, and deals with the Consistory to condemn him and proscribe his Book as Heretical; prostituting the Censure of the Church to his private revenge. This was Galileo's fortune in Italy: but had I not reason to hope that the English will be more hospitable, on the account of the Principle which induceth them to be civil

to (I say not to dote on) Strangers, I should fear to be charged with imprudence for appearing an Interpreter to that great Philosopher. And in this confidence I shall forbear to make any large Exordium concerning him or his Book: & the rather in regard that such kind of Gauderies become not the Gravity of the Subject; as also knowing how much (coming from me) they must fall short of the Merits of it, or him: but principally because I court only persons of Judgment & Candor, that can distinguish between a Native Beauty, and spurious Vernish. This only let me premise, though more to excuse my weakness in the menaging, than to insinuate my ability in accomplishing this so arduous a Task, that these profound Dialogues have bin found so uneasy to Translate, that neither affectation of Novelty could induce the French, nor the Translating humour perswade the Germans to undertake them. This difficulty, as I conceived, was charged either upon the Intricacy of this manner of Writing, or upon the singular Elegance in the stile of Galileo, or else upon the miscarriage of the unfortunate Mathias Berneggerus who first attempted to turn them into Latine for the benefit of the Learned World.

I shall not presume to Censure the Censure which the Church of Rome past upon this Doctrine and its Assertors. But, on the contrary, my Author having bin indefinite in his discourse, I shall forbear to exasperate, and attempt to reconcile such persons to this Hypothesis as devout esteem for Holy Scripture, and dutifull Respect to Canonical Injunctions hath made to stand off from this Opinion: and therefore for their sakes I have at the end of the Dialogues by way of supplement added an Epistle of Galileo to Her Most Serene Highness Christina Lotharinga the Grand Dutchesse Mother of Tuscany; as also certain Abstracts of John Kepler, Mathematician to two Emperours, and Didacus a Stunica a famous Divine of Salamanca, with an Epistle of Paulo Antonio Foscarini a learned Carmelite of Naples, that shew the Authority of Sacred Scripture in determining of Philosophical and Natural Controversies: hoping that the ingenious & impartial Reader will meet with full satisfaction in the same. And least what I have spoken of the prohibiting of these Pieces by the Inquisition may deterre any scrupulous person from reading of them, I have purposely inserted the Imprimatur by which that Office licenced them. And for a large account of the Book or Author, I refer to the Relation of his Life, which shall bring up the Reare in the Second Tome.[42]

Salusbury did publish the biography promised here,[43] as he also did publish a part one of the second volume that included Galileo's *Two New Sciences*. Aside from the sincerity of Salusbury's confession of the difficulty of the translation, four points deserve stress. The irony of his descriptions, especially regarding his reasons for including the exegetical appendices, exhibits an obvious anti-clerical animosity on his part. Second, he found the *Dialogue* to be a "singular and unimitable Piece of Reason and Demonstration," and this judgment had constantly gained currency in the 30 years since its original publication. Less universal, but widely shared, was Salusbury's judgment that the work was objective, balanced, and unbiased ("with all the veneration valued, read & applauded by the Iudicious" and "indefinite in his discourse").

---

[42] Salusbury (1661–1665, vol. 1). The foreword bears no page numbers and precedes immediately the text of the *Dialogue*. I have retained the spelling, punctuation, and style of Salusbury's archaic but comprehensible and pleasant English; the only exception is that he printed the foreword in italics, with the names of persons in roman type, and I have dropped both of those conventions.

[43] See the recent discovery, and fascinating speculations, discussed in Wilding (2008).

12.7 Science Versus Religion in the Subsequent Affair?

Finally, Salusbury was the first to formulate publicly the thesis that the root cause of the condemnation was Pope Urban's personal offence for being caricatured by the character of Simplicio. Now, admittedly this allegation is psychologically plausible; and in some pre-trial documents[44] there is mention of its factual basis, namely of the fact that at the end of the book Urban's favorite argument is uttered by Simplicio. However, there is no documentary evidence that it played a role in the trial; rather Urban's feeling offended for this reason is mentioned for the first time in a document 2 years after the trial.[45] Moreover, it is unlikely that Urban attached too much importance to that feeling because he had much more important things to worry about: for example, the charge that the book was a clear violation of the special injunction of 1616 represented a serious administrative problem; and the politics of the 30 Years War had the pope facing the threat of impeachment proceedings against him by the college of cardinals.

However, the inherent plausibility of the allegation is simply too great, and so this thesis became and continues to be one of the most common and widespread explanations. In my view, the status of this thesis about the later proceedings of 1632–1633 is similar to the status of another famous allegation, about the earlier phase of the trial in 1613–1616; that is, that on 21 December 1614, at the Church of Santa Maria Novella in Florence, Dominican Friar Tommaso Caccini preached a sermon on the suggestive biblical verse, "Ye men of Galilee, why stand ye gazing up into heaven?" (Acts 1:11).[46] In both cases, even if there is no chronologically proper documentary evidence, the possibilities mentioned could have happened and are so emotionally appealing that a common response is that, not only they *could* have happened, but they *should* or *must* have happened. Again, we are in the presence of the phenomenon of myth formation.

My next example consists of William Whewell's writings on the Galileo affair. Given Whewell's general importance in the history of the history and philosophy of science and his keen interest in the relationship between science and religion, the potential rewards of such a study are very great.

A clue to the complexity of Whewell's case is provided by the location and chronology of his writings on the subject. There is first of all an essay entitled "The Copernican System Opposed on Theological Grounds," which exists in three different versions, corresponding to the three editions of his *History of the Inductive Sciences* (see, e.g., Whewell 1837, 1857a). Then there is a piece entitled "Case of Galileo" in the *Philosophy of the Inductive Sciences*, which is a response to Peter Cooper's criticism of the first-edition account.[47] Finally, there is a discussion entitled

---

[44] Magalotti to Guiducci, 7 August 1632, in Favaro 14: 368–371; Magalotti to Guiducci, 4 September 1632, in Favaro 14: 379–382; "Special Commission's Report on the *Dialogue* (September 1632)," in Favaro 19: 324–327, in Pagano (1984, 105–108), in Finocchiaro (1989, 218–222), and in Galilei (2008, 272–276).

[45] Castelli to Galileo, 22 December 1635, in Favaro 16: 363–364; cf. Pieralisi (1875, 365–366).

[46] Fabroni (1773–1775, 1: 47 n. 1); cf. Finocchiaro (2005b, 115).

[47] Whewell (1847). The criticism had appeared in Cooper (1838); cf. Section 10.5 above.

"Were the Papal Edicts against the Copernican System Repealed?" which was added to the third edition of the *History* (Whewell 1857b).

Whewell's sketch of the events of the affair reads like an uncritical summary of the Inquisition sentence, which had always been well publicized and contained a version of the events from 1613 to 1633. Moreover, in the first edition of the *History* he explains Galileo's troubles as stemming mostly from an important characteristic of the Italian Catholic Church: "in Italy the Church entertained the persuasion that her authority could not be upheld at all, without maintaining it to be supreme on all points" (Whewell 1837, 399). Using present-day concepts, we might equate this characteristic with holism or totalitarianism, as distinct from authoritarianism; and historically speaking, there can be no doubt that Whewell is adopting this point from Riccioli's *Almagestum novum*.[48] But later Whewell stresses an explanation in terms of what he calls "decorum," advanced in the context of interpreting the legendary *e pur si muove*: "this is sometimes represented as the heroic soliloquy of a mind cherishing its conviction of the truth, in spite of persecution; I think we may more naturally conceive it uttered as a playful epigram in the ear of a cardinal's secretary, with a full knowledge that it would be immediately repeated to his master" (Whewell 1847, 699).

Here, Whewell's nonchalant portrayal of ecclesiastic totalitarianism is a portrayal of an essential and insurmountable conflict between Galilean science and Catholic religion. Whewell's revision of his earlier conflictual and anti-clerical account in response to the criticism of a Catholic scholar is an expression of science-religion conflict in the history of the subsequent controversy. And Whewell's "decorous" twist and embellishment of the *e pur si muove* may be viewed as myth-making in action.

One of the most instructive examples of the conflict between science and religion in the subsequent controversy involves Villa Medici in Rome. This palace is one of the most impressive buildings in the city, and for a long time it was the property of the Grand Duchy of Tuscany, ruled by the House of Medici. The Medici also owned another palace, Palazzo Firenze, nearer the center of the city, which served as the residence and office of the Tuscan ambassador to Rome.[49] After 1610, when Galileo received the title of Philosopher and Chief Mathematician to the Most Serene Grand Duke of Tuscany, during his visits to Rome he frequently resided at Villa Medici and occasionally at Palazzo Firenze. During most of the 1633 trial (from his arrival in February to the sentencing on June 22), he stayed at Palazzo Firenze. But on June 23, the imprisonment that was part of the sentence was commuted to house arrest at Villa Medici, and so he stayed there for about another week. Then his place of detention was commuted to the archbishop's residence in Siena, where he arrived on July 9.

---

[48] Cf. Riccioli (1651, 2: 290) and Section 8.4 above.

[49] For little known factual information and important clarifications about the difference between Villa Medici and Palazzo Firenze, see Shea and Artigas (2003, 30, 74, 106–107, 134–135, 179–180, 195).

12.7 Science Versus Religion in the Subsequent Affair?

Next to Villa Medici, at the edge of the street, stands today a commemorative column, erected near the end of the nineteenth century, which reads as follows: "The palace next to this spot,/which belonged formerly to the Medici,/was a prison for Galileo Galilei,/guilty of having seen/the earth turn around the sun./SPQR/ MDCCCLXXXVII."

The origin of this monument goes back to 1872,[50] when the municipal government of Rome decided to affix an inscription on the facade of the palace in memory of Galileo. However, the building then belonged to the French government, which opposed the project, feeling that the monument would offend the pope. Thus it took another 15 years before the monument came into being, and it had to take the form of a free-standing commemorative column on public property on the adjacent street.

The monument was applauded by the anti-clerical press but sharply criticized by the Vatican newspaper *L'Osservatore Romano*, in an unsigned article entitled "Epigraphs and insults" dated 23 April 1887. It objected that Galileo was not "imprisoned" in Villa Medici; that he was not found guilty of having "seen the earth turn around the sun"; but that his guilt was to have attempted "to change the physical and astronomical question into a theological one."[51]

Here it is obvious that there is a clash between science and religion, or at least between some people's perceptions of science and some institutional practices of religion. But I believe that there is another aspect of the episode which is even more basic, and which I shall label its mythological dimension.

On the one hand, the Villa Medici column echoes several common myths about Galileo's trial. One involves the unqualified talk of imprisonment, by contrast to house arrest. A second one pertains to the one-sided focus on the astronomical issue that disregards the methodological (or hermeneutical, or theological, or philosophical) question. More importantly, the Villa Medici inscription is a good formulation of the empiricist myth, namely the claim that Galileo saw the earth move around the sun; this is a phenomenon which is impossible to observe directly even today, let alone in Galileo's time. Such empiricism is a myth in the sense that it reflects the principle that observation is paramount in science. This principle is strictly speaking false; for example, to begin observing one has to have some idea of where, when, and how to look; and conflicts between observation and theory are not necessarily resolved by rejecting or revising the theory, but sometimes by repeating or refining the observation. And yet many working scientists pay lip service to the empiricist principle; so believing it seems to perform a good function in science.

On the other hand, in opposing this myth, the religious side succumbed to its own anti-Galilean myth. This is the claim that Galileo was condemned not for being a good astronomer, but for being a bad theologian. Presumably, he was a good astronomer insofar as he believed that the earth moves, which is indeed true. But he was allegedly a bad theologian insofar as he supported the earth's motion on the basis of scriptural passages.

---

[50] For more details, see Berggren and Sjöstedt (1996, 19–20, 145–147).
[51] "Epigrafi ed Offese," *L'Osservatore Romano*, 23 April 1887.

As we have seen (Chapter 10), the inventors and perpetrators of this myth were led to this conclusion in part by the fact than on various occasions Galileo engaged in biblical exegesis, arguing that the biblical texts adduced by his critics against the earth's motion are more in accordance with the Copernican than with the Ptolemaic system. However, the difficulty here is that such Galilean exercises at biblical exegesis are taken out of context and are misunderstood by these myth-makers, as we have seen on numerous occasions. In truth, Galileo held the opposite principle, as we saw earlier.

However, such a misinterpretation of Galileo's view perhaps helps many Catholics make sense or justify the Inquisition condemnation, and so it enhances their religiosity and piety. This is precisely the function of cultural myths.

What I am suggesting is that, whereas in Galileo's original trial the science versus religion conflict was a small (but irreducible) aspect of the process, in the subsequent controversy this conflict is pervasive, overwhelming, and undeniable. Advocates of the harmony thesis (e.g., Agostino Gemelli and Pope John Paul II) may bemoan this fact, but they themselves do not deny the historical existence and reality of the subsequent conflict. I am not denying it either, but I am trying to explain it on the basis of something that I regard more fundamental. In my cultural and deep-structural approach, I take both science and religion as important elements of culture and try to identify the myths that are operative in them. As is the case for other cultural institutions, myths play a significant role.

The identification of the relevant myths is not the end of the story, of the cultural story of the subsequent Galileo affair, but merely the beginning. Equally important is the examination of the interaction between such myths and facts, namely documented facts, as established by historians and scholars. Thus, I would want to note that the two particular myths just mentioned seem to have been eventually discarded. Nowadays, no serious spokesman for science would claim that fundamental facts like the earth's motion can be observed directly. For example, to refer to two cultural icons, Albert Einstein and Karl Popper, they have both emphasized the conceptual aspects of the scientific method.[52] Similarly, no churchman would today attribute to Galileo the methodologically unsound principle that Scripture *is* a scientific authority. For example, recall that (Section 8.17) in John Paul II's rehabilitation of Galileo, the pope was clear and explicit that Galileo not only held the correct principles about the relationship between science and Scripture, but gave insightful reasons in support of such principles; in short, that Galileo was a good theologian.

On the other hand, this is not the end of the story either. In fact, in regard to the theory versus observation issue, now the pendulum seems to have swung to the other extreme; so a common slogan nowadays is that all observation is theory-laden, which makes theorizing paramount in scientific research (cf. Brown 1979). Thus, a new portrayal of Galileo has been developed, as someone who not only had no observations to prove his views and refute his opponents, but did not even have

---

[52] Einstein (1934; 1953), Popper (1963, 33–65, 97–119).

good arguments, and instead was an epistemological anarchist for whom anything goes or a sophistical rhetorician who had mastered the art of making the worse argument appear stronger.[53] In short, a new myth about Galileo's trial has arisen or been revived. Interestingly, however, to complete the picture, we must note that on the religious side something analogous has happened. That is, John Paul's re-examination of the Galileo affair has given rise to the widespread belief that the Church has officially, fully, and judicially rehabilitated him. This did not in fact happen, and as some acute scholars[54] have written, this belief is really the very latest myth in an un-ended and un-ending story.

12.8 Conclusion

The histories of Galileo's original trial (1613–1633) and of the subsequent Galileo affair (1633–1992) suggest several lessons. One of the most obvious and general of these can be formulated once we realize that the subsequent affair is largely co-extensive with the *historiography* of the original trial: that is, in studying the trial scholars bring to their subject general claims about the relationship between science and religion. As we have seen, it was not hard to discern this pattern when we examined the views of such authors as Draper, White, Drake, Feyerabend, Morpurgo-Tagliabue, Salusbury, Viviani, and Whewell.

A more specific and controversial lesson regards the claim that science and religion are incompatible. Some historians do try to interpret Galileo's trial in the light of this conflict, but we have seen that this interpretation is untenable. However, my discussion suggests that the root difficulty of such an interpretation is that it is formulated in an oversimplified manner, for other conflictual interpretations are similarly untenable when simplistically formulated; and we have seen that this is the case for the alleged conflict between science and philosophy, and even for that between Jesuits and Dominicans. Moreover, the oversimplified version of the harmony interpretation is also untenable. On the other hand, sophisticated versions of both harmonious and conflictual interpretations involving science and religion remain viable. Thus the lesson here appears to be the undesirability of oversimplification.

This plea for sophistication and against oversimplification, if it is to stand, must identify relevant criteria. To begin with, such criteria should not be expected to be mechanical rules the following of which guarantees correctness and whose violation automatically leads one astray; instead such "criteria" ought to be conceived as general guidelines which should always be kept in mind but whose end result can come about only through concrete historical investigation. Having said that, the

---

[53] See Feyerabend (1975; 1988; 1993), and the criticism in Finocchiaro (1980). Cf. the penultimate section of the Introduction and Section 12.5 above.
[54] Benítez (1999, 85–110), Segre (1998).

guidelines I would want to stress here are the points that were mentioned in the beginning of this chapter, which are worth reiterating. Thus, secondly, one should avoid hasty generalizations, or concluding that the conflict (or the harmony) that has been found to be true in some cases is true of all. Thirdly, one must be mindful of the fact that what is meant by science or religion in different historical periods and cases may differ. Fourthly, there are other kinds of relationships between science and religion besides conflict and harmony, for example, separation, dialogue, integration, and subordination. Fifthly, each of these kinds of relationships is itself multi-faceted; for example, conflict can take the form of either open warfare or peaceful competition; and harmony can involve influence by science on religion and also by religion on science. Sixthly, influence can manifest itself in several ways: presupposition, endorsement, motivation, prescription, or substantive source. Finally, "science" can refer to the context of discovery, justification, popularization, and so on; and "religion" can refer to theology, metaphysics, world view, myth, ritual, or ecclesiastic institutions.

Another less obvious, but more important, lesson is that the subsequent affair may be seen as relevant to the question of the interaction of science and religion in another way. To see this, we need to consider that the subsequent controversy has been a recurring theme in the Western cultural history of the last 400 years; and we need to distinguish the two controversies, the original one that climaxed with Galileo's condemnation in 1633 and the subsequent one that began then and continues to our day. Now, although it is not the case that the original trial exemplifies the conflict between science and religion,[55] the subsequent controversy may very well do that since the latter is defined primarily by the way the trial was *perceived* and the traditional and most common perception (whether correct or incorrect) has been one of conflict between science and religion. To undermine such a conflict (in the subsequent affair) it is not enough to criticize the factual correctness of the corresponding accounts. I believe the most promising way is not to deny or explain away the conflict, but to regard it as the surface manifestation of something deeper. Nor do I think that that deeper structure is the same one underlying the original controversy, namely the dialectic between conservation and innovation. That deeper structure lies, I would suggest, in the phenomenon of the origin, diffusion, and development of cultural myths, and their interaction with documented facts.

---

[55] In the sense that the science versus religion conflict is the most important aspect of the trial. But such a conflict might have to be allowed as a first approximation, as a simplification so to speak, in a context in which one distinguishes simplification from oversimplification – an eighth guideline to be added to those mentioned in the previous paragraph.

Selected Bibliography

Accarisius, J. 1637. *Terrae quies, solisque motus demonstratus primum theologicis, tum plurimis philosophicis rationibus.* Rome.
Accattoli, L. 1990. Tutti d'accordo su Galileo. *Corriere della sera*, 30 March, 15.
Acloque, P. 1982. L'histoire des expériences pour la mise en évidence du mouvement de la Terre. *Cahiers d'Histoire et de Philosophie des Sciences*, new series, no. 4, 1–141.
Adams, J.Q. 1843. *An oration, delivered before the Cincinnati Astronomical Society.* Cincinnati.
Agassi, J. 1957. Duhem versus Galileo. *British Journal for the Philosophy of Science* 8: 237–248.
Agassi, J. 1971. On explaining the trial of Galileo. *Organon* 8: 137–166.
Albèri, E. (ed). 1842–1856. *Le opere di Galileo Galilei*, 15 tomes in 16 vols. Florence.
Andres, G. 1776. *Saggio della filosofia del Galileo.* Mantua.
Anfossi, F. 1820. *Le fisiche rivoluzioni della natura o la palingenesi filosofica di Carlo Bonnet convinta di errore.* 2nd edn. Rome.
Anfossi, F. 1822. *Se possa difendersi, ed insegnare, non come semplice ipotesi, ma come verissima e come tesi, la mobilità della terra, e la stabilità del sole da chi ha fatta la professione di fede di Pio IV.* Rome.
Angeli, S. degli. 1667. *Considerationi sopra la forza di alcune ragioni fisico mattematiche.* Venice.
Angeli, S. degli. 1668a. *Seconde considerationi sopra la forza dell'argomento fisico mattematico del M. Rev. P. Gio. Battista Riccioli.* Padua.
Angeli, S. degli. 1668b. *Terze considerationi sopra una lettera del molto illustre et eccellentissimo Signor Gio. Alfonso Borelli.* Venice.
Angeli, S. degli. 1669. *Quarte considerationi sopra la confermatione d'una sentenza del Sig. Gio. Alfonso Borelli.* Padua.
Anonymous. 1615. Judicium de epistola F. Pauli Foscarini de mobilitate Terrae. Now in Biblioteca Corsiniana, Rome.
Anonymous. 1882. Judicium de epistola F. Pauli Foscarini de mobilitate Terrae. In Berti 1882, 72–73.
Anonymous. 1991. An Unidentified Theologian's Censure of Foscarini's Letter. In Blackwell 1991, 253–254.
Argoli, A. 1644. *Pandosium sphaericum, in quo singula in elementaribus regionibus, aetque aetherea, mathematice pertractantur.* Padua.
Ariew, R. 1987. The phases of Venus before 1610. *Studies in History and Philosophy of Science* 18: 81–92.
Ariew, R., and D. Garber, trans and eds. 1989. *Philosophical essays [of Leibniz].* Indianapolis: Hackett.
Arnauld, A. 1775–1783. *Oeuvres de Messire Antoine Arnauld*, 49 vols. Paris.
Artigas, M., and M. Sánchez de Toca. 2008. *Galileo y el Vaticano.* Madrid: Biblioteca de Autores Cristianos.
Atti del convegno di studio su Pio Paschini nel centenario della nascita, 1878–1978. 1979. Udine: Pubblicazioni della Deputazione di Storia Patria per il Friuli.

Auzout, A. 1665. *Lettre a Monsieur l'Abbé Charles*. Paris. Rpt. in *Memoires de l'Academie Royale des Sciences, depuis 1666 jusqu'à 1699*, vol. 7, pt. 1, 1–68, Paris, 1729–1733.
Baldini, U. 1996a. La formazione scientifica di G.B. Riccioli. In Pepe 1996a, 123–182.
Baldini, U. 1996b. Sul contesto storico e scientifico del caso Settele. In *Giornata galileiana* 1996, 21–58.
Baldini, U. 2000. *Saggi sulla cultura della Compagnia di Gesù (secoli XVI–XVIII)*. Padua: CLEUP.
Baldini, U., and G.V. Coyne (eds). 1984. *The Louvain lectures (Lectiones Lovanienses) of Bellarmine and the autograph copy of his 1616 Declaration to Galileo*. Vatican City: Specola Vaticana.
Baldini, U. and L. Spruit. 2001. Nuovi documenti galileiani degli archivi del Sant'Ufficio e dell'Indice. *Rivista di Storia della filosofia* 56: 661–699.
Baliani, G.B. 1638. *De motu naturali gravium solidorum*. Genoa.
Baliani, G.B. 1646. *De motu naturali gravium solidorum et liquidorum*. Genoa.
Barbier, A.A. 1811. Report to Napoleon Bonaparte, 12 March. In Favaro 1887, 198.
Barbier, A.A. 1812. Letter to the Minister of Religions, 16 October. In Favaro 1887, 200.
Barbier, A.A. 1814a. Letter to Count de Blacas, 5 December. In Favaro 1887, 204.
Barbier, A.A. 1814b. Letter to Count de Blacas, 16 December. In Favaro 1887, 208.
Baretti, G. 1757. *The Italian Library*. London.
Barker, P. and B.R. Goldstein. 1998. Realism and instrumentalism in sixteenth-century astronomy. *Perspectives on Science* 6: 232–258.
Barni, J. 1862. *Les martyrs de la libre pensée*. Geneva.
Basile, B. 1983. Galileo e il teologo 'copernicano' Paolo Antonio Foscarini. *Rivista di letteratura italiana* 1: 63–96.
Basile, B. 1987. Galilei e il teologo Foscarini. In idem, *L'invenzione del vero*. Roma: Salerno.
Beltrán Marí, A., trans and ed. 1994. *Diálogo sobre los dos máximos sistemas del mundo, ptolemaico y copernicano*. Madrid: Alianza Editorial.
Beltrán Marí, A. 1998. 'Una reflexión serena y objectiva'. *Arbor* 160(629): 69–108.
Beltrán Marí, A. 2001a. *Galileo, ciencia y religión*. Barcelona: Paidós.
Beltrán Marí, A. 2001b. Tratos extrajudiciales, determinismo procesal y poder. In Montesinos and Solís 2001, 463–490.
Beltrán Marí, A. 2006. *Talento y poder: Historia de las relaciones entre Galileo y la Iglesia católica*. Pamplona: Laetoli.
Bemis, S.F. 1956. *John Quincy Adams and the Union*. New York: Knopf.
Benítez, H.H. 1999. *Ensayos sobre ciencia y religión*. Santiago, Chile: Bravo y Allende.
Bentley, E. (ed). 1966a. *Bertolt Brecht, Galileo*. New York: Grove.
Bentley, E. 1966b. Introduction. In Bentley 1966a, 7–42.
Bérault-Bercastel, A.H. de. 1778–1790. *Histoire de l'Eglise*, 24 vols. Paris.
Beretta, F. 1998. *Galilée devant le tribunal de l'Inquisition*. Doctoral Dissertation, Faculty of Theology, University of Fribourg, Switzerland.
Beretta, F. 1999a. Le procès de Galilée et les archives du Saint-Office. *Revue des sciences philosophiques et théologiques* 83: 441–490.
Beretta, F. 1999b. La Siège Apostolique et l'affaire Galilée. *Roma moderna e contemporanea* 7: 421–461.
Beretta, F. 1999c. De l'inerrance absolue à la vérité salvifique de l'Écriture. *Freiburger Zeitschrift für Philosophie und Theologie* 46: 461–501.
Beretta, F. 2000. L'archivio della Congregazione del Sant'Ufficio. In Del Col and Paolin 2000, 119–144; rpt. in Beretta 2001e.
Beretta, F. 2001a. Urbain VIII Barberini protagoniste de la condamnation de Galilée. In Montesinos and Solís 2001, 549–574.
Beretta, F. 2001b. 'Omnibus christianae, catholicae philosophiae amantibus, D.D.' *Freiburger Zeitschrift für Philosophie und Theologie* 48: 301–325.
Beretta, F. 2001c. Un nuovo documento sul processo di Galileo Galilei. *Nuncius* 16: 629–641.

Beretta, F. 2001d. Giordano Bruno e l'Inquisizione romana. *Bruniana & Campanelliana* 7: 15–49.
Beretta, F. 2001e. L'archivio della Congregazione del Sant'Ufficio. *Rivista di storia e letteratura religiosa* 37: 29–58.
Beretta, F. 2001f. Galileo Galilei und die römische Inquisition (1616–1633). In Wolf 2001, 141–158.
Beretta, F. 2003a. L'affaire Galilée et l'impasse apologétique. *Gregorianum* 84: 169–192.
Beretta, F. 2003b. Une deuxième abjuration de Galilée, ou l'inaltérable hiérarchie des disciplines. *Bruniana & Campanelliana* 9: 9–43.
Beretta, F. 2004. Rilettura di un documento celebre. *Galilaeana* 1: 91–115.
Beretta, F. 2005a. The documents of Galileo's trial. In McMullin 2005a, 191–212.
Beretta, F. 2005b. Galileo, Urban VIII, and the prosecution of natural philosophers. In McMullin 2005a, 234–261.
Beretta, F. (ed). 2005c. *Galilée en procès, Galilée réhabilité?* Saint-Maurice: Éditions Saint-Augustin.
Beretta, F., and M.-P. Lerner. 2006. Un *édit* inédit: Autour du placard de mise à l'Index de Copernic par le Maître du Sacre Palais Giacinto Petroni. *Galilaeana* 3: 199–216.
Berggren, L., and L. Sjöstedt. 1996. *L'ombra dei grandi*. Rome: Artemide Edizioni.
Bergier, N.S. 1788–1790a. Galilée. Rpt. in Bergier 1823, vol. 3, column 465.
Bergier, N.S. 1788–1790b. Sciences humaines. Rpt. in Bergier 1823, 7: 365–370.
Bergier, N.S. 1823. *Dictionnaire de Théologie*, 8 vols. Toulouse.
Bernini [or Bernino], D. 1709. *Historia di tutte l'heresie*, vol. 4. Rome.
Berti, D. 1876a. *Copernico e le vicende del sistema copernicano in Italia*. Rome.
Berti, D. 1876b. *Il processo originale di Galileo Galilei pubblicato per la prima volta*. Rome.
Berti, D. 1877. La critica moderna e il processo di Galileo Galilei. *Nuova antologia di scienze, letter ed arti*, year 12, 2nd series, vol. 4, 5–34.
Berti, D. 1878. *Il processo originale di Galileo Galilei: Nuova edizione accresciuta, corretta e preceduta da un'avvertenza*. Rome.
Berti, D. 1882. Antecedenti al processo galileiano e alla condanna della dottrina copernicana. *Atti della R. Accademia dei Lincei: Memorie della classe di scienze morali, storiche e filosofiche*, 1881–1882, year 279, 3rd series, 10: 49–96.
Bertolla, P. 1979. Le vicende del 'Galileo' di Paschini (dall'epistolario Paschini-Vale). In *Atti del convegno di studio su Pio Paschini*, 173–208.
Besomi, O., and M. Helbing (eds). 1998. *Dialogo sopra i due massimi sistemi del mondo, tolemaico e copernicano*, 2 vols. Critical edition and commentary. Padua: Antenore.
Biagioli, M. 1993. *Galileo Courtier*. Chicago: University of Chicago Press.
Biagioli, M. 1996. Playing with the evidence. *Early Science and Medicine* 1: 70–105.
Biagioli, M. 2000. Replication or monopoly? *Science in Context* 13: 547–590.
Biagioli, M. 2003. Stress in the book of nature. *MLN: Modern Language Notes* 118: 557–585.
Biagioli, M. 2006a. *Galileo's Instruments of Credit*. Chicago: University of Chicago Press.
Biagioli, M. 2006b. From print to patents. *History of Science* 44: 139–186.
Bianchi, L. 2000. Interventi divini, miracoli e ipotesi soprannaturali nel 'Dialogo' di Galileo. In Canziani et al. 2000, 239–251.
Bianchi, L. 2001. Agostino Oreggi, qualificatore del Dialogo, e i limiti della conoscenza scientifica. In Montesinos and Solís 2001, 575–586.
Biot, J.B. 1816. Galilée. In *Biographie universelle*, 52 vols (Paris, 1811–1828), 16: 318–337.
Biot, J.B. 1858. *Mélanges scientifiques et littéraires*, 3 vols. Paris.
Blackwell, R.J. 1991. *Galileo, Bellarmine, and the Bible*. Notre Dame: University of Notre Dame Press.
Blackwell, R.J., trans and ed. 1994. *A Defense of Galileo, the Mathematician from Florence by Thomas Campanella*. Notre Dame: University of Notre Dame Press.
Blackwell, R.J. 1998a. Could There Be Another Galileo Case? In Machamer 1998a, 348–366.
Blackwell, R.J. 1998b. *Science, Religion and Authority*. Milwaukee: Marquette University Press.

Blackwell, R.J. 2006. *Behind the Scenes at Galileo's Trial*. Notre Dame: University of Notre Dame Press.
Boaga, E. 1990. Annotazioni e documenti sulla vita e sulle opere di Paolo Antonio Foscarini teologo 'copernicano' (1562c.–1616). *Carmelus* 37: 173–216.
Bocchini Camaiani, B., and A. Scattigno (eds). 1998. *Anima e paura*. Macerata: Quodlibet.
Bonora, T. (ed). 1872. *Di Copernico e di Galileo*. Bologna.
Borelli, G.A. 1668a. *Confermazione d'una sentenza del Signor Gio. Alfonso Borelli*. Naples.
Borelli, G.A. 1668b. *Risposta ... alle considerazioni fatte sopra alcuni luoghi del suo libro della Forza della Percossa dal R. P. F. Stefano de gl'Angeli*. Messina.
Borgato, M.T. 1996. La prova fisica della rotazione della Terra e l'esperimento di Guglielmini. In Pepe 1996a, 201–261.
Borgato, M.T. (ed). 2002. *Giambattista Riccioli e il merito scientifico dei gesuiti nell'età barocca*. Florence: Olschki.
Boyer, C.B. 1967. Galileo's place in the history of mathematics. In McMullin 1967b, 232–255.
Bradley, J. 1729. A letter giving an account of a new-discovered motion of the fix'd stars. *Philosophical Transactions of the Royal Society of London* 35(406): 637–661.
Brandmüller, W. 1982. *Galilei und die Kirche, oder das Recht auf Irrtum*. Regensburg: F. Pustet.
Brandmüller, W. 1987. *Galileo y la Iglesia*. Madrid: RIALP.
Brandmüller, W. 1992. *Galilei e la Chiesa, ossia il diritto di errare*. Vatican City: Libreria Editrice Vaticana.
Brandmüller, W., and E.J. Greipl (eds). 1992. *Copernico e la Chiesa: Fine della controversia (1820). Gli atti del Sant'Uffizio*. Florence: Olschki.
Brecht, B. 1994. *Life of Galileo* (trans: Willett, J., ed.: Willett, J. and Manheim, R.). New York: Arcade.
Brenner, A.A. 1990. *Duhem*. Paris: Vrin.
Brewster, D. 1835. Galileo. In *Eminent and Literary Scientific Men of Italy, Spain, and Portugal*, 3 vols., 2: 1–62. In *The Cabinet Cyclopedia*, ed. D. Lardner. London.
Brewster, D. 1841. *The martyrs of science, or the lives of Galileo, Tycho Brahe, and Kepler*. London.
Brice, C., and A. Romano (eds). 1999. *Sciences et religions de Copernic à Galilée (1540–1610)*. Rome: Ecole Française de Rome.
Brooke, J.H. 1991. *Science and religion*. Cambridge: Cambridge University Press.
Brooke, J.H. 1996. Religious belief and natural science. In van der Meer 1996, 1: 1–26.
Brooke, J.H. 1998. The historiography of religion and science interaction. Paper presented at the conference "Science in Theistic Contexts," Pascal Centre for Advanced Studies in Faith and Science, Redeemer University College, Ancaster, Ontario, Canada, 21–25 July.
Brooke, J. and G. Cantor. 1998. *Reconstructing nature*. Edinburgh: T & T Clark.
Brooke, J., and E. Ihsanoglu (eds). 2005. *Religious values and the rise of science in Europe*. Istanbul: Center for Islamic History.
Brown, H.I. 1979. *Perception, theory and commitment*. Chicago: University of Chicago Press.
Brucker, J.J. 1742–1744. *Historia critica philosophiae*, 4 vols. Leipzig.
Brucker, J.J. 1766–1767. *Historia critica philosophiae*. 2nd edn., 6 vols. Leipzig.
Bucciantini, M. 1994a. Dopo il *Sidereus Nuncius*. *Nuncius* 9: 15–35.
Bucciantini, M. 1994b. Galileo e la Chiesa. *Memorie domenicane*, new series, 25: 471–476.
Bucciantini, M. 1995. *Contro Galileo*. Florence: Olschki.
Bucciantini, M. 1997. Scienza e filologia. *Giornale critico della filosofia italiana* 76: 424–445.
Bucciantini, M. 1998. Celebration and conservation. In Hunter 1998, 21–34.
Bucciantini, M. 1999. Teologia e nuova filosofia. In Brice and Romano 1999, 411–452.
Bucciantini, M. 2001. Novità celesti e teologia. In Montesinos and Solís 2001, 795–808.
Bucciantini, M. 2003. *Galileo e Keplero*. Turin: Einaudi.
Bucciantini, M. 2004. Reazioni alla condanna di Copernico. *Galilaeana* 1: 3–19.
Bucciantini, M. 2005. Copernicani in Europa. *Galilaeana* 2: 307–313.
Bucciantini, M. 2009. Istruzioni per denigrare la scienza. *Il Sole 24 Ore*, 10 May, 38.

Bucciantini, M., and M. Torrini (eds). 1997. *La diffusione del copernicanesimo in Italia, 1543–1610*. Florence: Olschki.
Burtt, E.A. 1932. *The metaphysical foundations of modern physical science*, 2nd edn. London: Routledge.
Butts, R.E., and J.C. Pitt (eds). 1978. *New perspectives on Galileo*. Dordrecht: Reidel.
Cabeo, N. 1646. *In quatuor libros meteorologicorum aristotelis commentaria*. Rome.
Cabibbo, N. 2009. Dialogo nel nome di Stensen. *Il Sole 24 Ore*, 10 May, 38.
Calandrelli, G. 1806a. *Osservazioni e riflessioni sulla parallasse annua dell'Alfa della Lira*. In Calandrelli and Conti 1803–1824.
Calandrelli, G. 1806b. *Risultato di varie osservazioni sopra la parallasse annua di Wega, o Alfa della Lira*. Rome.
Calandrelli, G., and A. Conti. 1803–1824. *Opuscoli astronomici*, 8 vols. Rome.
Calmet, A. 1720. Dissertation sur le système du monde des anciens hébreux. In idem, *Dissertations qui peuvent servir de prolégomènes de l'Écriture Sainte*, 1: 438–459. Paris.
Calmet, A. 1734. *Prolegomena et dissertationes in omnes et singulos Scripturae libros*, 2 vols. Venice.
Calmet, A. 1744. Dissertazione sovra il sistema del mondo degli antichi ebrei. In Galilei 1744, 4: 1–20.
Camerota, M. 2004. *Galileo Galilei e la cultura scientifica nell'età della Controriforma*. Rome: Salerno.
Campanella, T. 1622. *Apologia pro Galilaeo*. Frankfurt.
Campanella, T. 1628. Campanella to Pope Urban VIII, 10 June. In Campanella 1927, 218–228.
Campanella, T. 1637. *Disputationum in quatuor partes suae philosophiae realis libri quatuor*. Paris.
Campanella, T. 1821. Pezzi migliori dell'*Apologia* del Campanella a favore del Galileo. In Venturi 1818–1821, 1: 1–6.
Campanella, T. 1853. *Apologia pro Galileo*. In Galilei 1842–1856, vol. 5, pt. 2, 495–558.
Campanella, T. 1911. *Apologia di Galileo* (trans: Ciampoli, D.). Lanciano: Carabba.
Campanella, T. 1927. *Lettere*, ed. V. Spampanato. Bari: Laterza.
Campanella, T. 1937. *The Defense of Galileo* (trans: McColley, G.). In *Smith College Studies in History*, vol. 22, nos. 3–4.
Campanella, T. 1968. *Apologia di Galileo*, ed. Luigi Firpo. Turin: UTET.
Campanella, T. 1971. *Apologia per Galileo* (trans and ed.: Femiano, S.). Milan: Marzorati.
Campanella, T. 1992. *Apologia di Galileo* (trans: Lotto, A., ed.: Ditadi, G.). Este: Isonomia.
Campanella, T. 1994. *A Defense of Galileo* (trans and ed.: Blackwell, R.J.). Notre Dame: University of Notre Dame Press.
Campanella, T. 1997. *Apologia pro Galileo*, ed. P. Ponzio. Milan: Rusconi.
Campanella, T. 1998. *Le poesie*, ed. F. Giancotti. Turin: Einaudi.
Campanella, T. 1999. *Apologia di Galileo* (trans and ed.: Ernst, G.). In Ernst 1999, 357–407.
Campanella, T. 2001a. *Apologia pro Galileo* (trans and ed.: Lerner, M.-P.). Paris: Les Belles Lettres.
Campanella, T. 2001b. *Apologia per Galileo* (trans and ed.: Ponzio, P.). Milan: Bompiani.
Campanella, T. 2006. *Apologia pro Galileo* (trans: Ernst, G., ed.: Lerner, M.-P.). Pisa: Scuola Normale Superiore.
Campanella, T. N.d. *Modo di filosofare*. In Campanella 1998, 43–45.
Cantor, G. 1995. Science, religion and history. *The University of Leeds Review* 38(1995–1996): 1–19.
Cantor, M. 1864. Galileo Galilei. *Zeitschrift für Mathematik und Physik* 9(3): 172–197.
Canziani, G., M.A. Granada, and Y.C. Zarka (eds). 2000. *Potentia Dei: L'onnipotenza divina nel pensiero dei secoli XVI e XVII*. Milan: FrancoAngeli.
Caramuel Lobkowitz, J. 1644. *Sublimium ingeniorum crux*. Louvain.
Carlen, C. (ed). 1981. *The Papal Encyclicals, 1740–1981*, 5 vols. Wilmington: McGrath.
Carli, A., and A. Favaro (eds). 1896. *Bibliografia galileiana (1568–1895)*. Rome.
Caroti, S. 1987. Un sostenitore napoletano della mobilità della terra. In Lomonaco and Torrini 1987, 81–121.

Carroll, W.E. 1997. Galileo, science, and the Bible. *Acta Philosophica* 6: 5–37.
Carroll, W.E. 1999. Galileo and the interpretation of the Bible. *Science and Education* 8: 151–187.
Carroll, W.E. 2001. Galileo and biblical exegesis. In Montesinos and Solís 2001, 677–692.
Carvalho-Neto, P. de. 1965. *The concept of folklore* (trans: Wilson, J.M.P.). Coral Gables, FL: University of Miami Press.
Casini, P. (ed). 1985. *Paolo Frisi, Elogi*. Rome: Theoria.
Casini, P. 1987. Frisi tra illuminismo e rivoluzione scientifica. In *Ideologia e scienza nell'opera di Frisi*, Barbarisi, G., 2 vols, 1: 15–33. Milan: FrancoAngeli.
Cavalieri, B. 1642. *Sphaera, seu doctrinae sphaericae tractatus*. Ms. Bologna University Library, ms. lat. 1858, no. 9.
Cerbu, T. 2001. Melchior Inchofer, 'un homme fin et rusé'. In Montesinos and Solís 2001, 587–614.
Chalmers, A., and R. Nicholas. 1983. Galileo and the dissipative effects of a rotating earth. *Studies in History and Philosophy of Science* 14: 315–340.
Chasles, P. 1862. *Galileo Galilei*. Paris.
Chasles, P. 1867. Etudes nouvelles sur Galilée. *Revue des cours littéraires de la France et de l'etranger*, 13 April, 4(20): 307–310.
Chiaramonti, S. 1633. *Difesa al suo Antiticone*. Florence.
Christianson, J.R., A. Hadravova, P. Hadrava, and M. Sole (eds). 2002. *Tycho Brahe and Prague*. Frankfurt: Verlag Harri Deutsch.
Clagett, M. (ed). 1959. *Critical problems in the history of science*. Madison: University of Wisconsin Press.
Clavelin, M. 1974. *The natural philosophy of Galileo* (trans: Pomerans, A.J.). Cambridge, MA: MIT Press.
Clavelin, M. 1996. *La philosophie naturelle de Galilèe*. 2nd edn. Paris: Albin Michel.
Cohen, I.B. 1960. *The birth of a new physics*. Garden City: Doubleday.
Colombe, L. delle. 1610–1611. Contro il moto della Terra. In Favaro 3: 251–290.
Cooper, P. 1838. Galileo – the Roman Inquisition. *Dublin Review* 5(9): 72–116.
Cooper, P. 1844. *Galileo – The Roman Inquisition*. Cincinnati.
Copernicus, N. 1992. *On the revolutions* (trans and ed.: Rosen, E.). Baltimore: Johns Hopkins University Press.
Couturat, L. (ed). 1903. *Opuscules et fragments inédits de Leibniz*. Paris.
Coyne, G.V. 1985. Conclusion. In Coyne, Heller, and Zycínski 1985, 177–179.
Coyne, G.V. 2005. The Church's most recent attempt to dispel the Galileo myth. In McMullin 2005a, 340–359.
Coyne, G.V., M. Heller, and J. Zycínski (eds). 1985. *The Galileo affair*. Vatican City: Specola Vaticana.
Crombie, A.C. 1967. The mechanistic hypothesis and the scientific study of vision. In *Historical aspects of microscopy*, eds. S. Bradbury and G. L'E Turner, 3–112. Cambridge, England: Heffer.
Czermak, J. (ed). 1993. *Philosophy of mathematics*. Vienna: Holder-Pichler-Tempsky.
D'Addio, M. 1983. Considerazioni sui processi a Galileo (parte I). *Rivista di storia della chiesa in Italia* 37: 1–52.
D'Addio, M. 1984. Considerazioni sui processi a Galileo (parte II). *Rivista di storia della chiesa in Italia* 38: 47–114.
D'Addio, M. 1985. *Considerazioni sui processi a Galileo*. Rome: Herder.
D'Addio, M. 1993. *Il caso Galilei*. Rome: Studium.
D'Alembert, J. 1751. Discours preliminaire. In Diderot and D'Alembert 1751–1780, 1: i–xlv.
D'Alembert, J. 1963. *Preliminary discourse to the Encyclopedia of Diderot* (trans and ed.: Schwab, R.N.). Indianapolis: Bobbs-Merrill.
Dahlstrom, D.O. (ed). 1991. *Nature and scientific method*. Washington: Catholic University of America Press.
Darling, K.M. 2003. Motivational realism: the natural classification for Pierre Duhem. *Philosophy of Science* 70: 1125–1136.

Selected Bibliography

De Caro, M. 1993. Galileo's mathematical platonism. In Czermak 1993, 13–22.
Delambre, J.B. 1821. *Histoire de l'astronomie moderne*, 2 vols. Paris.
Del Col, A., and G. Paolin (eds). 2000. *L'Inquisizione romana: metodologia delle fonti e storia istituzionale*. Trieste: Edizioni Università di Trieste.
Denzinger, H., and A. Schönmetzer (eds). 1967. *Enchiridion symbolorum, definitionum, et declarationum de rebus fidei et morum*, 34th ed. Herder: Freiburg im Breisgau.
Descartes, R. 1637. *Discours de la méthode*. Leiden.
Descartes, R. 1644. *Principia philosophiae*. Amsterdam.
Descartes, R. 1897–1913. *Oeuvres*, 13 vols., ed. C. Adam and P. Tannery. Paris: Cerf.
DiCanzio, A. 1996. *Galileo: His science and his significance for the future of man*. Portsmouth: Adasi Publishing.
Diderot, D., and J. D'Alembert (eds). 1751–1780. *Encyclopédie, ou Dictionnaire raisonné des sciences, des arts et des métiers*, 35 vols. Paris.
Dinis, A. de Oliveira. 1989. The cosmology of Giovanni Battista Riccioli (1598–1671). Unpublished Ph. D. Dissertation, University of Cambridge.
Dinis, A. de Oliveira. 2002. Was Riccioli a secret Copernican? In Borgato 2002.
Dinis, A. de Oliveira. 2003. Giovanni Battista Riccioli and the science of his time. In Feingold 2003.
Ditadi, G. (ed). 1992. *Tommaso Campanella, Apologia di Galileo*. Este: Isonomia.
Donovan, A., L. Laudan, and R. Laudan (eds). 1988. *Scrutinizing science*. Dordrecht: Kluwer.
Dooley, B. 1996. Processo a Galileo. *Belfagor* 51: 1–21.
Dooley, B. 2002. *Morandi's last prophecy and the end of Renaissance politics*. Princeton: Princeton University Press.
Dorn, M. 2000. *Das Problem der Autonomie der Naturwissenschaften bei Galilei*. Stuttgart: Franz Steiner.
Drake, S. 1957. *Discoveries and opinions of Galileo*. Garden City: Doubleday.
Drake, S. 1970. *Galileo studies*. Ann Arbor: University of Michigan Press.
Drake, S. 1976. *Galileo against the philosophers*. Los Angeles: Zeitlin & Ver Brugge.
Drake, S. 1978. *Galileo at work*. Chicago: University of Chicago Press.
Drake, S. 1980. *Galileo*. New York: Hill and Wang.
Drake, S. 1983. *Telescopes, tides & tactics*. Chicago: University of Chicago Press.
Drake, S. 1986a. Galileo and the projection argument. *Annals of Science* 43: 77–79.
Drake, S. 1986b. Reexamining Galileo's Dialogue. In Wallace 1986, 155–175.
Drake, S. 1987. Galileo's steps to full Copernicanism, and back. *Studies in History and Philosophy of Science* 18: 93–105.
Drake, S. 1999. *Essays on Galileo and the history and philosophy of science*, 3 vols., eds. N.M. Swerdlow and T.H. Levere. Toronto: University of Toronto Press.
Drake, S., and C.D. O'Malley, trans and eds. 1960. *The controversy on the comets of 1618*. Philadelphia: University of Pennsylvania Press.
Draper, J.W. 1875. *History of the conflict between religion and science*. New York.
Drinkwater Bethune, J.E. 1830. *Life of Kepler, with the conclusion of Galileo*. London.
Drinkwater Bethune, J.E. 1832. *Life of Galileo Galilei*. Boston.
Drinkwater Bethune, J.E. 1833. *Life of Galileo Galilei*. In *Lives of eminent persons*, 1–106. London.
Dubarle, D. 1964. Le dossier Galilée. *Signes du temps* 14: 21–26.
Duhem, P. 1908. *SOZEIN TA PHAINOMENA: Essai sur la notion de theorie physique de Platon à Galilée*. Paris: Hermann.
Duhem, P. 1909. *Le mouvement absolu et le mouvement relatif*. Montligeon: Imprimerie-Librairie de Montligeon.
Duhem, P. 1954. *The aim and structure of physical theory* (trans: Wiener, P.). Princeton: Princeton University Press.
Duhem, P. 1969. *To Save the Phenomena* (trans: Doland, E. and Maschler, C.). Chicago: University of Chicago Press.
Einstein, A. 1934. On the method of theoretical physics. In idem, *The World as I See It*. New York: Covici Friede Publishers.

Einstein, A. 1953. Foreword. In Galilei 1953, vi–xx.
Ennis, R.H. 1996. Critical thinking dispositions. *Informal Logic* 18: 165–182.
Epigrafi ed offese. 1887. L'Osservatore Romano, 23 April.
Ernst, G. (ed). 1999. *Tommaso Campanella*. Rome: Poligrafico e Zecca dello Stato.
Erythraeus, J.N. [G.V. de' Rossi]. 1643. *Pinacotheca imaginum illustrium, doctrinae vel ingenii laude, virorum*. Coloniae Agrippinae.
Estève, P. 1755. *Histoire générale et particulière de l'astronomie*, 3 vols. Paris.
Fabris, R. 1986. *Galileo Galilei e gli orientamenti esegetici del suo tempo*. Vatican City: Pontifical Academy of Sciences.
Fabroni, A. (ed). 1773–1775. *Lettere inedite di uomini illustri*, 2 vols. Florence.
Fahie, J.J. 1903. *Galileo*. London: John Murray.
Fahie, J.J. 1929. *Memorials of Galileo Galilei, 1564–1642*. London: The Courier Press.
Fantoli, A. 1994. *Galileo: for Copernicanism and for the Church* (trans: Coyne, G.V.). Vatican City: Vatican Observatory Publications.
Fantoli, A. 1996. *Galileo: for Copernicanism and for the Church*. 2nd edn. (trans: Coyne, G.V.). Vatican City: Vatican Observatory Publications.
Fantoli, A. 2001. Galileo e la Chiesa cattolica. In Montesinos and Solís 2001, 733–752.
Fantoli, A. 2003a. *Il caso Galileo*. Milan: Rizzoli.
Fantoli, A. 2003b. *Galileo: for Copernicanism and for the Church*. 3rd edn. (trans: Coyne, G.V.). Vatican City: Vatican Observatory Publications.
Fantoli, A. 2005. The disputed injunction and its role in Galileo's trial. In McMullin 2005a, 117–149.
Favaro, A. 1887. *Miscellanea galileiana inedita*. Venice.
Favaro, A. (ed). 1890–1909. *Le opere di Galileo Galilei*, 20 vols. Florence: Barbèra.
Favaro, A. 1891. *Nuovi studi galileiani*. Venice.
Favaro, A. 1902. Amici e corrispondenti di Galileo Galilei: IV. Alessandra Bocchineri – V. Francesco Rasi – VI. Giovanfrancesco Buonamici. *Atti del Reale Istituto Veneto di scienze, lettere ed arti*, academic year 1901–1902, vol. 61, pt. 2, 665–701.
Favaro, A. 1905. Amici e corrispondenti di Galileo Galilei: XII. Vincenzo Renieri. *Atti del Reale Istituto Veneto di scienze, lettere ed arti*, academic year 1904–1905, vol. 64, pt. 2, 111–195.
Favaro, A. 1908. Galileo Galilei e la determinazione del peso dell'aria. *Rivista di fisica, matematica e scienze naturali* 9(108): 577–588.
Favaro, A. 1911a. Alla ricerca delle origini del motto 'E pur si muove'. *Atti del Reale Istituto Veneto di scienze, lettere ed arti*, vol. 70, pt. 2, 1219–1232.
Favaro, A. 1911b. E pur si muove. *Il Giornale d'Italia*, 12 July, 4.
Feher, M. 1982. Galileo and the demonstrative ideal of science. *Studies in History and Philosophy of Science* 13: 87–110.
Feingold, M. (ed). 2003. *Jesuit science and the republic of letters*. Cambridge, MA: MIT Press.
Feldhay, R. 1995. *Galileo and the Church*. Cambridge: Cambridge University Press.
Feldhay, R. 1998. The use and abuse of mathematical entities. In Machamer 1998a, 80–145.
Feldhay, R. 2000. Recent narratives on Galileo and the Church. *Science in Context* 13: 489–507.
Feller, F.X. de. 1782. Article on Galileo. In idem, *Dictionnaire Historique*. Augsburg.
Feller, F.X. de. 1797. Article on Galileo. In idem, *Dictionnaire Historique*. 2nd edn., 4: 251–253. Liège.
Feller, F.X. de. 1832. Galilée-Galilei. In idem, *Dictionnaire Historique*. 8th edn., 6: 26–28. Lille.
Femiano, S., trans and ed. 1971. *Apologia per Galileo*. Milan: Marzorati.
Ferngren, G.B., E.J. Larson, D.W. Amundsen, and A.-M. E. Nakhla (eds). 2000. *The history of science and religion in the western tradition*. New York: Garland.
Ferri, G. 1785. Apologie de Galilée. *Mercure de France*, 8 January, 54–63.
Feyerabend, P.K. 1975. *Against method*. London: NLB.
Feyerabend, P.K. 1981. *Rationalism, realism and scientific method*. Cambridge: Cambridge University Press.
Feyerabend, P.K. 1985. Galileo and the tyranny of truth. In Coyne, Heller, and Zycinski 1995, 155–166.

Feyerabend, P.K. 1987. *Farewell to reason.* London: Verso.
Feyerabend, P.K. 1988. *Against method.* Revised edn. London: Verso.
Feyerabend, P.K. 1993. *Against method.* 3rd edn. London: Verso.
Field, J.V. 1988. *Kepler's geometrical cosmology.* Chicago: University of Chicago Press.
Findlen, Paula. 2007. The sun at the center of the world. In Martin 2007, 655–677.
Finocchiaro, M.A. 1973a. *History of science as explanation.* Detroit: Wayne State University Press.
Finocchiaro, M.A. 1973b. Essay-review of Lakatos's *Criticism and the growth of knowledge. Studies in History and Philosophy of Science* 3: 357–372.
Finocchiaro, M.A. 1980. *Galileo and the art of reasoning: rhetorical foundations of logic and scientific method.* Dordrecht: Reidel.
Finocchiaro, M.A. 1985. Wisan on Galileo and the art of reasoning. *Annals of Science* 42: 613–616.
Finocchiaro, M.A. 1986. Toward a philosophical interpretation of the Galileo affair. *Nuncius* 1: 189–202.
Finocchiaro, M.A. 1988. Science and society in Newton and in Marx. *Inquiry* (Oslo) 31: 103–121.
Finocchiaro, M.A., trans and ed. 1989. *The Galileo affair: a documentary history.* Berkeley: University of California Press.
Finocchiaro, M.A. 1990. Varieties of rhetoric in science. *History of the Human Sciences* 3: 177–193.
Finocchiaro, M.A. 1995. Review of A. Fantoli's *Galileo: for Copernicanism and for the Church,* and J. Reston, Jr's. *Galileo: a life. Isis* 86: 486–488.
Finocchiaro, M.A., trans and ed. 1997a. *Galileo on the world systems: a new abridged translation and guide.* Berkeley: University of California Press.
Finocchiaro, M.A. 1997b. Review of M. Bucciantini's *Contro Galileo. Isis* 88: 141–142.
Finocchiaro, M.A. 1999a. The Galileo affair from John Milton to John Paul II: problems and prospects. *Science and Education* 8: 189–209.
Finocchiaro, M.A. 1999b. Review of R. Feldhay's *Galileo and the Church. Isis* 90: 596–597.
Finocchiaro, M.A. 1999c. Review of P.-N. Mayaud's *La condamnation des livres coperniciens et sa revocation. Isis* 90: 363–364.
Finocchiaro, M.A. 2001. Review of O. Besomi and M. Helbing's critical edition of Galileo's *Dialogue on the Two Chief World Systems,* and of P. Machamer's *Cambridge companion to Galileo. Philosophy of Science* 68: 578–580.
Finocchiaro, M.A. 2002. Philosophy versus religion and science versus religion: the trials of Bruno and Galileo. In Gatti 2002, 51–96.
Finocchiaro, M.A. 2003. Review of J. Renn's *Galileo in context. Isis* 94: 523–524.
Finocchiaro, M.A. 2004. Review of W.R. Shea and M. Artigas's *Galileo in Rome. Isis* 95: 494–495.
Finocchiaro, M.A. 2005a. *Arguments about arguments: systematic, critical, and historical essays in logical theory.* New York: Cambridge University Press.
Finocchiaro, M.A. 2005b. *Retrying Galileo, 1633–1992.* Berkeley: University of California Press.
Finocchiaro, M.A. 2005c. Review essay of Michele Camerota's new biography of Galileo. *Early Science and Medicine* 10: 545–557.
Finocchiaro, M.A. 2006a. Review of M. Biagioli's *Galileo's instruments of credit. Historical Studies in the Physical and Biological Sciences* 37: 174.
Finocchiaro, M.A. 2006b. Review of P.-N. Mayaud's *Le conflit entre l'astronomie nouvelle et l'Écriture sainte aux XVIe et XVIIe siècles. Isis* 97: 750–752.
Finocchiaro, M.A. 2006c. Review of E. McMullin's *The Church and Galileo. Isis* 97: 353–355.
Finocchiaro, M.A. 2007a. Mill on liberty of argument. In *Reason Reclaimed,* eds. H.V. Hansen and R.C. Pinto, 121–134. Newport News: Vale Press.
Finocchiaro, M.A. 2007b. Review of A. Beltrán Marí's *Talento y poder: historia de las relaciones entre Galileo y la Iglesia católica. Isis* 98: 838–839.
Finocchiaro, M.A. 2007c. Review of W.R. Shea and M. Artigas's *Galileo observed. Renaissance Quarterly* 60: 1413–1414.

Finocchiaro, M.A. 2008a. The Church and Galileo. *The Catholic Historical Review* 94: 260–282.
Finocchiaro, M.A., trans and ed. 2008b. *The Essential Galileo*. Indianapolis and Cambridge (MA): Hackett Publishing Co.
Finocchiaro, M.A. 2008c. Review of R.J. Blackwell's *Behind the scenes at Galileo's trial*. *The Journal of Modern History* 80: 687–688.
Finocchiaro, M.A. 2008d. Review of M. Bucciantini's *Galileo e Keplero*. *Isis* 99: 833–834.
Finocchiaro, M.A. 2009a. Myth 8: that Galileo was imprisoned and tortured for advocating Copernicanism. In Numbers 2009, 68–78, 249–252.
Finocchiaro, M.A. 2009b. The Galileo affair: Maurice Finocchiaro discusses the lessons and the cultural repercussions of Galileo's telescopic discoveries. *Physics World*, March, 22(3): 54–57.
Finocchiaro, M.A. 2009c. Review of J. Speller's *Galileo's Inquisition trial revisited*. *Early Science and Medicine* 14: 576–578.
Finocchiaro, M.A. 2009d. Review of M. Artigas and M. Sánchez de Toca's *Galileo y el Vaticano: historia de la comisión pontificia de estudio del caso Galileo (1981–1992)*. *Catholic Historical Review* 95:
Fiorani, L. 1969. *Onorato Caetani*. Rome: Istituto di Studi Romani.
Fiorani, L. 1973a. Caetani, Francesco. *Dizionario biografico degli italiani* 16: 168–170.
Fiorani, L. 1973b. Caetani, Onorato. *Dizionario biografico degli italiani* 16: 209–212.
Firpo, L. 1993. *Il processo di Giordano Bruno*, ed. D. Quaglioni. Rome: Salerno.
Firpo, L., trans and ed. 1968. *Apologia di Galileo di Tommaso Campanella*. Turin: UTET.
Fisher, A. 1991. Testing fairmindedness. *Informal Logic* 13: 31–36.
Fisher, A., and M. Scriven. 1997. *Critical thinking*. Point Reyes, CA: Edgepress; Norwich, UK: Centre for Research in Critical Thinking.
Fleck, L. 1979. *Genesis and development of a scientific fact*. Chicago: University of Chicago Press.
Flora, F. (ed). 1953. *Galileo Galilei: Opere*. Milan: Ricciardi.
Foscarini, P.A. 1615a. *Lettera sopra l'opinione de' pittagorici e del Copernico della mobilità della Terra e stabilità del Sole e del nuovo pittagorico sistema del mondo*. Naples.
Foscarini, P.A. 1615b. Ad Illustrissimum et Reverendissimum Dominum Cardinalem Bellarminium pro defensione epistolae fratis Pauli Antonii Foscareni. In Boaga 1990, 205–214.
Foscarini, P.A. 1635. *Epistola circa pythagoricum et Copernici opinionem de mobilitate Terrae* (trans: Diodati, E.). In Galilei 1635.
Foscarini, P.A. 1641. *Epistola circa pythagoricum et Copernici opinionem de mobilitate Terrae* (trans: Diodati, E.). In Galilei 1641.
Foscarini, P.A. 1661. *An Epistle concerning the Pythagorian [sic] and Copernican opinion of the mobility of the Earth*. In Salusbury 1661 1665, vol. 1, pt. 1, 471–503.
Foscarini, P.A. 1663. *Epistola circa pythagoricum et Copernici opinionem de mobilitate Terrae* (trans: Diodati, E.). In Galilei 1663.
Foscarini, P.A. 1710. Lettera sopra l'opinione de' pittagorici. In Galilei 1710, 36–68.
Foscarini, P.A. 1811. Lettera sopra l'opinione de' pittagorici. In Galilei 1808–1811, 13: 73–135.
Foscarini, P.A. 1853. Lettera sopra l'opinione de' pittagorici. In Galilei 1842–1856, vol. 5, pt. 2, 455–494.
Foscarini, P.A. 1882. Defensio Epistulae Pauli Antonii Foscareni. In Berti 1882, 73–78.
Foscarini, P.A. 1911–1913. Defensio Epistulae Pauli Antonii Foscareni. In Franco 1911–1913, 496–504.
Foscarini, P.A. 1991a. Defense (trans: Blackwell, R.J.). In Blackwell 1991, 255–263.
Foscarini, P.A. 1991b. A letter concerning the opinion of the Pythagoreans and Copernicans about the mobility of the earth (trans: Blackwell, R.J.). In Blackwell 1991, 217–251.
Foscarini, P.A. 1992. *Lettera sopra l'opinione de' Pittagorici*, ed. L. Romeo. Montalto Uffugo: Grafiche Aloise.
Foscarini, P.A. 1997. Lettera. In Ponzio 1997, 199–237.
Foscarini, P.A. 2001. Lettera. In Ponzio 2001, 199–237.

Franco, A. 1911–1913. Excerpta historiae ordinis. *Analecta Ordinis Carmelitarum* 2: 461–468, 493–504, 524–527.
Franklin, A. 1998. Mechanics, Aristotelian. In *Routledge Encyclopedia of Philosophy*, ed. E. Craig, vol. 6. London: Routledge.
Frisi, P. 1756. *Dissertation sur le mouvement diurne de la terre*. Berlin.
Frisi, P. 1766. Saggio sul Galileo. *Il Caffè*, 2(3): 17–27.
Frisi, P. 1775. *Elogio del Galileo*. Milan and Leghorn. Rpt. in Casini 1985, 31–92.
Frisi, P. 1777. Galilée, Philosophie de. In *Supplément à l'Encyclopédie* 3: 172–176. Amsterdam.
Frova, A., and M. Marenzana (eds). 2006. *Thus spoke Galileo*. Oxford: Oxford University Press.
Galilei, G. 1635. *Systema cosmicum*. Strasbourg.
Galilei, G. 1636. *Nov-antiqua sanctissimorum patrum, & probatorum theologorum doctrina de Sacrae Scripturae testimoniis* (trans: Diodati, E., ed.: Bernegger, M.). Strasbourg.
Galilei, G. 1641. *Systema cosmicum*. Lyons.
Galilei, G. 1663. *Systema cosmicum*. London.
Galilei, G. 1710. *Dialogo sopra i due massimi sistemi del mondo*. Florence [Naples].
Galilei, G. 1744. *Opere*, 4 vols., ed. G. Toaldo. Padua.
Galilei, G. 1808–1811. *Opere*, 13 vols. Milan.
Galilei, G. 1842–1856. *Opere*, 15 vols. with 1-vol. supplement, ed. E. Albèri. Florence.
Galilei, G. 1890–1909. *Le opere di Galileo Galilei*, 20 vols., ed. A. Favaro. Rpt. 1929–1939, 1968. Florence: Barbèra.
Galilei, G. 1953. *Dialogue concerning the two chief world systems* (trans and ed.: Drake, S.). Berkeley: University of California Press.
Galilei, G. 1960. *On motion and on mechanics* (trans and eds.: Drabkin, I.E. and Drake, S.). Madison: University of Wisconsin Press.
Galilei, G. 1967. *Dialogue concerning the two chief world systems* (trans and ed.: Drake, S.). 2nd revised edn. Berkeley: University of California Press.
Galilei, G. 1970. *Dialogo sopra i due massimi sistemi del mondo, tolemaico e copernicano*, ed. L. Sosio. Turin: Einaudi.
Galilei, G. 1974. *Two new sciences* (trans and ed.: Drake, S.). Madison: University of Wisconsin Press.
Galilei, G. 1989. *Sidereus Nuncius or the Sidereal Messenger* (trans and ed.: van Helden, A.). Chicago: University of Chicago Press.
Galilei, G. 1997. *Galileo on the world systems: a new abridged translation and guide* (trans and ed.: Finocchiaro, M.A.). Berkeley: University of California Press.
Galilei, G. 2008. *The essential Galileo* (trans and ed.: Finocchiaro, M.A.). Indianapolis and Cambridge, MA: Hackett Publishing Co.
Galileo a Padova, 1592–1610: Celebrazioni del IV centenario. 1995, 5 vols. Trieste: Lint.
Galluzzi, P. 1973. Il platonismo del tardo cinquecento e la filosofia di Galileo. In *Ricerche sulla cultura dell'Italia moderna*, ed. P. Zambelli, 39–79. Bari: Laterza.
Galluzzi, P. 1977. Galileo contro Copernico. *Annali dell'Istituto e Museo di storia della scienza* 2: 87–148.
Galluzzi, P. (ed). 1984. *Novità celesti e crisi del sapere*. Florence: Giunti Barbera.
Galluzzi, P. 1993a. Gassendi e *l'affaire Galilée* delle leggi del moto. *Giornale critico della filosofia italiana* 72/74: 86–119.
Galluzzi, P. 1993b. I sepolcri di Galileo. In *Il pantheon di Santa Croce a Firenze*, ed. L. Berti, 145–182. Florence: Cassa di Risparmio di Firenze.
Galluzzi, P. 1998. The sepulchers of Galileo. In Machamer 1998a, 417–447.
Galluzzi, P. 2000. Gassendi and l'*Affaire Galilée* of the laws of motion. *Science in Context* 13: 509–545.
Galluzzi, P. 2009. Galileo, il caso è riaperto. *Il Sole 24 Ore*, 10 May, 38.
Garcia, S. 2000. L'Edition strasbourgeoise du *Systema cosmicum* (1635–1636). *Bulletin de la Société de l'histoire du protestantisme francais* 146: 307–334.
Garcia, S. 2001. Elie-Diodati-Galilée. In Montesinos and Solís 2001, 883–894.

Garcia, S. 2004. *Galiléi-Elie Diodati*. Geneva: Georg.
Garcia, S. 2005. Galileo's relapse. In McMullin 2005a, 265–278.
Garin, E. 1975. *Rinascite e rivoluzioni*. Rome: Laterza.
Garzend, L. 1912. *L'Inquisition et l'hérésie*. Paris: Desclée de Brouwer.
Gassendi, P. 1642. *De motu impresso a motore translato*. Paris.
Gassendi, P. 1646. *De proportione qua gravia decidentia accelerantur epistolae tres*. Paris.
Gassendi, P. 1649. *Apologia in Io. Bap. Morini librum*. Lyons.
Gatti, H. 1997. Giordano Bruno's *Ash Wednesday supper* and Galileo's *Dialogue of the two major systems*. *Bruniana & Campanelliana* 3: 283–300.
Gatti, H. 1999. *Giordano Bruno and Renaissance science*. Ithaca: Cornell University Press.
Gatti, H. (ed). 2002. *Giordano Bruno: philosopher of the Renaissance*. Aldershot: Ashgate.
Gatti, H. 2008. Copernico (sez. Giordano Bruno). *Bruniana & Campanelliana* 14: 511–520.
Gaukroger, S. 1978. *Explanatory structures*. Atlantic Highlands: Humanities.
Gaukroger, S. 1995. *Descartes*. Oxford: Clarendon.
Gebler, K. von. 1876. *Galileo Galilei und die römische Curie nach den autentischen Quellen*. Stuttgart. (Gebler 1876–1877, vol. 1).
Gebler, K. von. 1876–1877. *Galileo Galilei und die römische Curie: nach den autentischen Quellen*, 2 vols. Stuttgart.
Gebler, K. von. 1877. *Die Acten des Galilei'schen Processes, nach der Vaticanischen Handschrift*. Stuttgart. (Gebler 1876–1877, vol. 2).
Gebler, K. von. 1878. Ist Galilei gefoltert worden? *Die Gegenwart*, vol. 13, nos. 18, 19, 24, and 25.
Gebler, K. von. 1879a. *Galileo Galilei and the Roman Curia* (trans: Mrs. George Sturge). London.
Gebler, K. von. 1879b. *Galileo Galilei e la curia romana*, 2 vols (trans: Prato, G.). Florence.
Gemelli, A. 1942. Scienza e fede nell'uomo Galilei. In *Nel terzo centenario della morte di Galileo Galilei*, 1–27.
Genovesi, E. 1966. *Processi contro Galileo*. Milano: Casa Editrice Ceschina.
Gherardi, S. 1870. *Il processo di Galileo riveduto sopra documenti di nuova fonte*. Florence.
Giacchi, O. 1942. Considerazioni giuridiche sui due processi contro Galileo. In *Nel terzo centenario della morte di Galileo Galilei*, 383–406.
Giard, L. (ed). 1995. *Les jésuites à la Renaissance*. Paris: PUF.
Gingerich, O. (ed). 1975. *The nature of scientific discovery*. Washington: The Smithsonian Institution.
Gingerich, O. 1981. The Censorship of Copernicus' *De revolutionibus*. *Annali dell'Istituto e Museo di Storia della Scienza di Firenze* 6: 45–61.
Gingerich, O. 1982. The Galileo affair. *Scientific American*, August, 132–143.
Gingerich, O. 1986. Galileo's astronomy. In Wallace 1986, 111–126.
Gingerich, O. 1995. Hypothesis, proof, and censorship. In *Galileo a Padova*, 4: 325–344.
Gingerich, O. 2000. Copernican revolution. In Ferngren et al. 2000.
Gingerich, O. 2002. *An annotated census of Copernicus'* De revolutionibus. Boston: Brill.
Gingerich, O. 2003a. From *occhiale* to printed page. *Journal for the History of Astronomy* 34: 251–267.
Gingerich, O. 2003b. The Galileo sunspot controversy. *Journal for the History of Astronomy* 34: 77–78.
Giornata galileiana, 16 giugno 1994. 1996. Pontificia Academia Scientiarum, Commentarii, vol. 3, no. 34. Vatican City: Pontifical Academy of Sciences.
Goddu, A. 1990. The realism that Duhem rejected in Copernicus. *Synthese* 83: 301–316.
Goddu, A. 1996. The logic of Copernicus's arguments and his education in logic at Cracow. *Early Science and Medicine* 1: 28–68.
Goddu, A. 2006. Reflections on the origins of Copernicus's cosmology. *Journal for the History of Astronomy* 37: 37–53.
Godman, P. 2000. *The saint as censor*. Leiden: Brill.

Gorman, M.J. 1996. A matter of faith? *Perspectives on Science* 4: 283–320.
Govi, G. 1872. Il Sant'Uffizio, Copernico e Galileo. *Atti della reale accademia delle scienze di Torino* 7: 565–590, 808–838.
Granada, M.A. 1996. Il problema astronomico-cosmologico e le sacre scritture dopo Copernico. *Rivista di storia della filosofia* 51: 789–828.
Granada, M.A. 1997. Giovanni Maria Tolosani e la prima reazione romana di fronte al 'De revolutionibus'. In Bucciantini and Torrini 1997, 11–35.
Granada, M.A. (ed). 2001. *Cosmologia, teologia y religion en la obra y en el proceso de Giordano Bruno*. Barcelona: Universidad de Barcelona.
Granada, M.A. 2002. *Sfere solide e cielo fluido*. Milan: Guerrini.
Granada, M.A. 2006. Did Tycho eliminate the celestial spheres before 1586? *Journal for the History of Astronomy* 37: 125–145.
Grant, E. 1984. In defense of the earth's centrality and immobility. *Transactions of the American Philosophical Society* 74(4): 1–67.
Grisar, H. 1882. *Galileistudien*. Regensburg.
Grua, G. (ed.) 1948. *Textes inédits d'après les manuscrits de la bibliotèque provinciale de Hanovre*. 2 vols. Paris: Presses Universitaires de France.
Guasco, M., E. Guerriero, and F. Traniello (eds). 1991. *Storia della Chiesa*, vol. XXIII. Cinisello Balsamo: Edizioni Paoline.
Guglielmini, G. 1789. *Riflessioni sopra un nuovo esperimento in prova del diurno moto della terra*. Rome.
Guglielmini, G. 1792. *De diurno Terrae motu experimentis physico-mathematicis confirmato opusculum*. Bologna.
Haeckel, E.H. 1878–1879. *Gesammelte populäre Vorträge aus dem Gebiete der Entwicklungslehre*. Bonn.
Hagen, J.H. 1911. *La rotation de la Terre*. Rome: Specola Astronomica Vaticana.
Hall, A.R. 1980. Galileo in the eighteenth century. In *Transactions of the Fifth International Congress on the Enlightenment*, ed. H. Mason, 1: 81–99. Oxford: The Voltaire Foundation at the Taylor Institution.
Hall, E.H. 1903. Do falling bodies move south? *Physical Review* 17: 179–190.
Hall, E.H. 1904. Experiments on the deviations of falling bodies. *Proceedings of the American Academy of Arts and Sciences* 39: 341–349.
Hall, E.H. 1910. Air resistance to falling inch spheres. *Proceedings of the American Academy of Arts and Sciences* 45: 379–384.
Hallam, H. 1837–1839. *Introduction to the literature of Europe in the fifteenth, sixteenth and seventeenth centuries*. London.
Harris, W.H., and J.S. Levey (eds). 1975. *The new Columbia encyclopedia*. New York: Columbia University Press.
Hatfield, G. 1990. Metaphysics and the new science. In Lindberg and Westman 1990, 93–166.
Headley, J.M. 1997. *Tommaso Campanella and the transformation of the world*. Princeton: Princeton University Press.
Heilbron, J.L. 1999. *The Sun in the church*. Cambridge, MA: Harvard University Press.
Heilbron, J.L. 2005. Censorship of astronomy in Italy after Galileo. In McMullin 2005a, 279–322.
Hevelius, J. 1647. *Selenographia sive Lunae descriptio*. Dantzig.
Hilgers, J. 1904. *Der Index der verbotenen Büche in seiner neuen Fassung dargelegt und rechtlich-historisch gewürdigt*. Freiburg im Breisgau: Herder.
Hill, D.K. 1984. The projection argument in Galileo and Copernicus. *Annals of Science* 41: 109–133.
Hobbes, T. 1642. [*De Mundo*]. Unpublished ms. First published in Hobbes 1973.
Hobbes, T. 1973. *Critique du "De Mundo" de Thomas White*, ed. J. Jacquot and H. W. Jones. Paris: Vrin/CNRS.
Howell, K.J. 1996. Galileo and the history of hermeneutics. In van der Meer 1996, 4: 245–260.

Howell, K.J. 2002. *God's two books*. Notre Dame: University of Notre Dame Press.
Howell, K.J. 2005. Natural knowledge and textual meaning in Augustine's interpretation of Genesis. Paper presented at the workshop "Interpreting Nature and Scripture," Pascal Centre for Advanced Studies in Faith and Science, Redeemer University College, Ancaster, Ontario, 18–23 July.
Hujer, K. 1967. Galileo's trial in the epistemology of Einsteinian physics. In *Symposium internazionale di storia, metodologia, logica e filosofia della scienza*, 289–295.
Hull, D., M. Forbes, and R.M. Burian (eds). 1995. *PSA 1994*. 2 vols. East Lansing: Philosophy of Science Association.
Hume, D. 1851–1860. *The history of England from the invasion of Julius Caesar to the abdication of James the Second, 1688*, 6 vols. New York: Harper.
Hunter, M. (ed). 1998. *Archives of the scientific revolution*. Woodbridge: The Boydell Press.
Huygens, C. 1673. *Horologium oscillatorium*. Paris.
Inchofer, M. 1633. *Tractatus syllepticus*. Rome
Inchofer, M. 1635. *Vidiciarum S. Sedi Apostolicae, Sacrorum Tribunalium et Authoritatum*. Rome: Biblioteca Casanatense, MS 182.
Ingoli, F. 1616. *De situ et quiete terrae contra Copernici systema disputatio*. In Favaro 5: 403–412.
Ingoli, F. 1618a. Replicationes de situ et motu Terrae contra Copernicum ad Joannis Kepleri caesarei mathematici impugnationes. In Bucciantini 1995, 177–205.
Ingoli, F. 1618b. De emendatione sex librorum Nicolai Copernici De Revolutionibus. In Vatican Library, Barb. Lat. 3151, ff 58r–61v; Hilgers 1904, 540–542; English trans. Gingerich 1981, 51–56; Bucciantini 1995, 207–209; Mayaud 1997, 71–72.
Jaki, S.L. 1969. Introduction. In Duhem 1969, ix–xxvi.
Jardine, N. 1984. *The birth of history and philosophy of science*. Cambridge: Cambridge University Press.
John Paul II. 1979. Deep harmony which unites the truths of science with the truths of faith. *L'Osservatore Romano*, weekly edition in English, 26 November, 9–10.
John Paul II. 1992. Faith can never conflict with reason. *L'Osservatore Romano*, weekly edition in English, 4 November, 1–2.
Kant, I. 1965. *Critique of pure reason* (trans: Kemp Smith, N.). New York: St. Martin's Press.
Kelter, I.A. 1995. The refusal to accommodate. *Sixteenth Century Journal* 26: 273–283.
Kelter, I.A. 1997. A Catholic theologian responds to Copernicanism. *Renaissance and Reformation* 21(2): 59–70.
Kelter, I.A. 2005. The refusal to accommodate. In McMullin 2005a, 38–53.
Kepler, J. 1618. Responsio ad Ingoli disputationem. In Favaro 1891, 173–184; Mayaud 2005, 3: 259–264.
Kepler, J. 1619. Admonitio ad bibliopolas exteros, praesertim italos. In Albèri 1842–1856, vol. 5, pt. 2, 633–634; Mayaud 2005, 3: 265–266.
King, H.C. 1955. *The history of the telescope*. London: Griffin.
Koestler, A. 1959. *The sleepwalkers*. New York: Macmillan.
Koyré, A. 1943. Galileo and Plato. *Journal of the History of Ideas* 5: 400–428.
Koyré, A. 1955. A documentary history of the problem of fall from Kepler to Newton. In *Transactions of the American Philosophical Society*, new series, vol. 45, pt. 4, 329–395. Philadelphia: American Philosophical Society.
Koyré, A. 1966. *Etudes galiléennes*. Paris: Hermann.
Koyré, A. 1978. *Galileo studies* (trans: Mepham, J.). Hassocks: Harvester.
Kozhamthadam, J. 1994. *The discovery of Kepler's laws*. Notre Dame: University of Notre Dame Press.
Kuhn, T.S. 1957. *The Copernican revolution*. Cambridge, MA: Harvard University Press.
Kuhn, T.S. 1962. *The structure of scientific revolutions*. Chicago: University of Chicago Press.
Kuhn, T.S. 1970. *The structure of scientific revolutions*. 2nd edn. Chicago: University of Chicago Press.
Kuhn, T.S. 1977. *The essential tension*. Chicago: University of Chicago Press.

Lakatos, I. 1978. *The methodology of scientific research programmes*. Cambridge: Cambridge University Press.
Lakatos, I., and E. Zahar. 1975. Why did Copernicus' research program supersede Ptolemy's? In Westman 1975a, 354–383.
Lamalle, E. 1964. Nota introduttiva all'opera. In Paschini 1964a, 1: vii–xv.
Langford, J.J. 1971. *Galileo, science and the Church*, revised edn. Ann Arbor: University of Michigan Press.
Lansbergen, J. van. 1633. *Apologia pro commentationibus Philippi Lansbergii in motum Terrae diurnum et annum*. Middleburg.
Lattis, J. 1994. *Between Copernicus and Galileo*. Chicago: Univesity of Chicago Press.
Laudan, L. 1977. *Progress and its problems*. Berkeley: University of California Press.
Laudan, L. 1984. *Science and values*. Berkeley: University of California Press.
Laudan, L., et al. 1986. Scientific change. *Synthese* 69: 141–223.
Le Cazre, P. 1645a. *Physica demonstratio qua ratio, mensura, modus ac potentia accelerationis motus in naturali descensu gravium determinantur*. Paris.
Le Cazre, P. 1645b. *Vindiciae demonstrationis physicae de proportione qua gravia decidentia accelerantur*. Paris.
Le Tenneur, L.A. 1646. *Disputatio physico-mathematica de aequabili motus acceleratione et motu telluris*. National Library, Paris, Ms. Fonds Lat. 6740.
Le Tenneur, L.A. 1649. *De motu naturaliter accelerato*. Paris.
Leibniz, G.W. 1923ff. *Sämtliche Schriften und Briefe*. Darmstadt, Leipsig, and Berlin.
Leibniz, G.W. 1997. *New essays on human understanding* (trans and eds.: Remnant, P. and Bennet, J.). Cambridge: Cambridge University Press.
Leo XIII. 1893. *Providentissimus Deus*. In Carlen 1981, 2: 325–339.
L'Epinois, H. de. 1867. Galilée: son procès, sa condamnation d'après des documents inédits. *Revue des questions historiques*, year 2, vol. 3, 68–171.
L'Epinois, H. de. 1877. *Les pièces du procès de Galilée précédées d'un avant-propos*. Paris.
L'Epinois, H. de. 1878. *La question de Galilée*. Paris.
Lerner, M.-P. 1980. L'Achille des coperniciens. *Bibliothèque d'humanisme et renaissance* 42: 313–327.
Lerner, M.-P. 1992. *Tre saggi sulla cosmologia alla fine del Cinquecento*. Naples: Bibliopolis.
Lerner, M.-P. 1995a. L'entrée de Tycho Brahe chez les jésuites ou le chant du cygne de Clavius. In Giard 1992, 145–185.
Lerner, M.-P. 1995b. La Science galiléenne selon Tommaso Campanella. *Bruniana & Campanelliana* 1: 121–156.
Lerner, M.-P. 1996–1997. *Le monde des spheres*, 2 vols. Paris: Les Belles Lettres.
Lerner, M.-P. 1998a. Essay review: 'Copernicus is not susceptible to compromise'. *Studies in the History and Philosophy of Science* 29: 636–672.
Lerner, M.-P. 1998b. Pour une edition critique de la sentence et de l'abjuration de Galilée. *Revue des sciences philosophiques et théologiques* 82: 607–629.
Lerner, M.-P. 1999. L' 'hérésie' héliocentrique. In Brice and Roman 1999, 69–91.
Lerner, M.-P. 2001a. La réception de la condamnation de Galilée en France au XVIIe siècle. In Montesinos and Solís 2001, 513–548.
Lerner, M.-P., trans and ed. 2001b. *Tommaso Campanella, Apologia pro Galileo*. Paris: Les Belles Lettres.
Lerner, M.-P. 2001c. Introduction. In Lerner 2001b, ix–clxv.
Lerner, M.-P. 2001d. Le moine, le cardinal et le savant. *Les cahiers de l'humanisme* 2: 71–94.
Lerner, M.-P. 2001e. Verité des philosophes et verité des théologiens selon Tommaso Campanella. *Freiburger Zeitschrift für Philosophie und Theologie* 48: 281–300.
Lerner, M.-P. 2002a. Aux origines de la polémique anticopernicienne (I). *Revue des sciences philosophiques et théologiques* 86: 681–721.
Lerner, M.-P. 2002b. Tycho Brahe censured. In Christianson et al. 2002, 95–101.
Lerner, M.-P. 2004. Copernic suspendu et corrigé. *Galilaeana* 1: 21–89.

Lerner, M.-P. 2005. The heliocentric 'heresy'. In McMullin 2005a, 11–37.
Lerner, M.-P. (ed). 2006. *Tommaso Campanella, Apologia pro Galileo.* Pisa: Scuola Normale Superiore.
Lessl, T.S., 1999. The Galileo legend as scientific folklore. *Quarterly Journal of Speech* 85: 146–168.
Levere, T.H., and W.R. Shea (eds). 1990. *Nature, experiment, and the sciences.* Dordrecht: Kluwer.
Libri, G. 1838–1841. *Histoire des sciences mathématiques en Italie,* 4 vols. Paris.
Libri, G. 1840–1841. Review of D. Brewster's *Life of Galileo. Journal des savants,* 1840, 556–569, 589–602; 1841, 157–171, 203–223.
Libri, G. 1841a. *Essai sur la vie et les travaux de Galilée.* Paris.
Libri, G. 1841b. *Galileo, sua vita e sue opere.* Milan.
Libri, G. 1841c. *Histoire des sciences mathématiques en Italie,* vol. 4. Paris.
Libri, G. 1842. *Galileo Galilei zu seinem Gedächtniss im zweiten Säcularjahr seines Todes, sein Leben und seine Werke* (trans and ed.: Carove, F.W.). Siegen und Wiesbaden.
Libri, G. 1865. *Histoire des sciences mathématiques en Italie,* vol. 4. 2nd edn. Halle.
Limborch, P. van. 1692. *Historia Inquisitionis.* Amsterdam.
Limborch, P. van. 1731. *The history of the Inquisition,* 2 vols. (trans: Chandler, S.). London.
Lindberg, D.C. 1992. *The beginnings of Western science.* Chicago: University of Chicago Press.
Lindberg, D.C. 2003. Galileo, the church, and the cosmos. In Lindberg and Numbers 2003, 33–60.
Lindberg, D.C., and R.L. Numbers (eds). 1986. *God and nature.* Berkeley: University of California Press.
Lindberg, D.C., and R.L. Numbers. 1987. Beyond war and peace. *Perspectives on Science and Christian Faith* 39: 140–149.
Lindberg, D.C., and R.L. Numbers (eds). 2003. *When science & Christianity meet.* Chicago: University of Chicago Press.
Lindberg, D.C., and R.S. Westman (eds). 1990. *Reappraisals of the scientific revolution.* Cambridge: Cambridge University Press.
Livingstone, D.N. 1997. Science and religion. *Christian Scholar's Review* 26: 270–292.
Lloyd, G.E.R. 1978. Saving appearances. *Classical Quarterly* 28: 202–222.
Lomonaco, F., and M. Torrini (eds). 1987. *Galileo e Napoli.* Naples: Guida.
Lorinus, J. 1605. *Acta apostolorum commentaria.* Lyons.
Lorinus, J. 1606. *Commentaria in Ecclesiasten.* Lyons.
Maccagni, C. (ed). 1972. *Saggi su Galileo Galilei,* vol. 3, tome 2. Florence: Barbèra.
Maccarrone, M. 1979a. Mons. Paschini e la Roma ecclesiastica. In *Atti del convegno di studio su Pio Paschini,* 49–93.
Maccarrone, M. 1979b. Mons. Paschini e la Roma ecclesiastica. *Lateranum* 45: 154–218.
Machamer, P. (ed). 1998a. *The Cambridge companion to Galileo.* Cambridge: Cambridge University Press.
Machamer, P. 1998b. Galileo's machines, his mathematics, and his experiments. In Machamer 1998a, 53–79.
Machamer, P. 2005. Review of Finocchiaro's *Retrying Galileo. Science* 309: 58.
MacLachlan, J.H. 1977. Mersenne's solution for Galileo's problem of the rotating earth. *Historia Mathematica* 4: 173–182.
MacLachlan, J.H. 1990. Drake against the philosophers. In Levere and Shea 1990, 123–144.
Madden, R.R. 1863. *Galileo and the Inquisition.* London and Dublin.
Maffei, P. 1975. Il sistema copernicano dopo Galileo e l'ultimo conflitto per la sua affermazione. *Giornale di astronomia* 1: 5–12.
Maffei, P. (ed). 1987. *Giuseppe Settele, il suo Diario e la questione galileiana.* Foligno: Edizioni dell'Arquata.
Maiocchi, R. 1985. *Chimica e filosofia.* Florence: La Nuova Italia.
Maiocchi, R. 1990. Pierre Duhem's *The Aim and Structure of Physical Theory. Synthese* 83: 385–400.
Mallet du Pan, J. 1784. Mensonges imprimées au sujet de la persécution de Galilée. *Mercure de France,* 17 July, 121–130.

Margolis, H. 1987. *Patterns, thinking, and cognition*. Chicago: University of Chicago Press.
Margolis, H. 1991. Tycho's System & Galileo's *Dialogue*. *Studies in History and Philosophy of Science* 22: 259–275.
Margolis, H. 1993. *Paradigms and barriers*. Chicago: University of Chicago Press.
Margolis, H. 2002. *It started with Copernicus*. New York: McGraw-Hill.
Marini, M. 1850. *Galileo e l'Inquisizione*. Rome.
Martin, T.H. 1868. *Galilée, les droits de la science et la méthode des sciences physiques*. Paris.
Martin, R.N.D. 1987. Saving Duhem and Galileo. *History of Science* 25: 301–319.
Martin, R.N.D. 1991. *Pierre Duhem*. La Salle: Open Court.
Martin, J.J. (ed). 2007. *The Renaissance world*. London: Routledge.
Martini, C.M. 1972. Galileo e la teologia. In Maccagni 1972, 441–451.
Martini, C.M., G. Ghiberti, and M. Pesce. 1995. *Cento anni di cammino biblico*. Milan: Vita e Pensiero.
Masini, E. 1621. *Sacro arsenale overo prattica dell'officio della Santa Inquisizione*. Genoa.
Mayaud, P.-N. 1997. *La condamnation des livres coperniciens et sa révocation à la lumière de documents inédits des Congrégations de l'Index et de l'Inquisition*. Rome: Editrice Pontificia Università Gregoriana.
Mayaud, P.-N. (ed). 2005. *Le conflit entre l'astronomie nouvelle et l'Écriture Sainte aux XVIe et XVIIe siècles*. 6 vols. Paris: Honoré Champion.
McColley, G., trans and ed. 1937. *The defense of Galileo*. Smith College Studies in History, vol. 22, nos. 3–4.
McMullin, E. 1967a. Bibliografia Galileiana 1940–1964. In McMullin 1967b, i-xciii.
McMullin, E. (ed). 1967b. *Galileo: man of science*. New York: Basic Books.
McMullin, E. 1967c. Introduction. In McMullin 1967b, 3–51.
McMullin, E. 1978. The conception of science in Galileo's Work. In Butts and Pitt 1978, 209–258.
McMullin, E. 1980. Galileo's slim chance to win a belated acquittal, Letter to the Editor, *New York Times*, 10 November.
McMullin, E. 1981. How should cosmology relate to theology? In Peacocke 1981, 17–57.
McMullin, E. 1987. Bruno and Copernicus. *Isis* 78: 55–74.
McMullin, E. 1990a. Comment: Duhem's middle way. *Synthese* 83: 421–430.
McMullin, E. 1990b. Conceptions of science in the scientific revolution. In Lindberg and Westman 1990, 27–92.
McMullin, E. 1995. Scientific classics and their fates. In Hull, Forbes, and Burian 1995, 2: 266–274.
McMullin, E. 1998. Galileo on science and scripture. In Machamer 1998a, 271–347.
McMullin, E. 1999. From Augustine to Galileo. *Modern Schoolman* 76: 169–194.
McMullin, E. (ed). 2005a. *The Church and Galileo*. Notre Dame: University of Notre Dame Press.
McMullin, E. 2005b. The Church's ban on Copernicanism, 1616. In McMullin 2005a, 150–190.
McMullin, E. 2005c. Galileo's theological venture. In McMullin 2005a, 88–116.
Mercati, A. 1926–1927. Come e quando ritornò a Roma il codice del processo di Galileo. *Atti della pontificia accademia delle scienze dei nuovi Lincei* 80: 58–63.
Mercati, A. (ed). 1942. *Il sommario del processo di Giordano Bruno*. Vatican City: Biblioteca Apostolica Vaticana.
Mereu, I. 1979. *Storia dell'intolleranza in Europa*. Milan: Mondadori.
Mersenne, M. 1634. *Les questions théologiques, physiques, morales, et mathématiques*. Paris.
Mersenne, M. 1647. *Novarum observationum physico-mathematicarum tomus tertius*. Paris.
Mill, J.S. 1997. *The spirit of the age, on liberty, the subjection of women*, ed. A. Ryan. New York: Norton.
Miller, D.M. 2008. The Thirty Years War and the Galileo affair. *History of Science* 46: 49–74.
Millman, A.B. 1976. The plausibility of research programs. In *PSA 1976*, ed. F. Suppe and P.D. Asquith, 1: 40–148. East Lansing: Philosophy of Science Association.
Milton, J. 1644. *Areopagitica*. In Milton 1953–1982, 2: 485–570.
Milton, J. 1953–1982. *Complete prose works*, 8 vols. New Haven: Yale University Press.

Miscellanea galileiana, 3 vols. 1964. Pontificiae Academiae Scientiarum Scripta Varia, no. 27. Vatican City: Ex Aedibus Adademicis in Civitate Vaticana.
Mivart, St. George Jackson. 1885. Modern Catholics and scientific freedom. *Nineteenth Century* 18: 30–47.
Monchamp, G. 1892. *Galilée et la Belgique*. Saint-Trond, Brussels, and Paris.
Monchamp, G. 1893. *Notification de la condamnation de Galilée datée de Liège, 20 septembre 1633*. Cologne and Saint-Trond.
Monsagrati, G. 2000. Giordani, Pietro. *Dizionario biografico degli italiani* 55: 219–226.
Montesinos, J., and C. Solís (eds). 2001. *Largo campo di filosofare*. La Orotava: Fundación Canaria Orotava de Historia de la Ciencia.
Moore, J.R. 1979. *The post-Darwinian controversies*. Cambridge: Cambridge University Press.
Moore, J.R. 1992. Speaking of 'Science' and 'Religion' – then and now. *History of Science* 30: 311–323.
Morin, J.B. 1631. *Famosi et antiqui problematis de Telluris motu vel quiete, hactenus optata solutio*. Paris.
Morin, J.B. 1640. *Astronomia iam a fundamentis integre et exacte restituta*. Paris.
Morin, J.B. 1643. *Alae Telluris fractae*. Paris.
Morin, J.B. 1650. *Résponse à une longue lettre de Monsieur Gassend*. Paris.
Morpurgo-Tagliabue, G. 1947–1948. I processi di Galileo e l'epistemologia. *Rivista di storia della filosofia*, vol. 2, nos. 2–3; vol. 3, no. 1.
Morpurgo-Tagliabue, G. 1981. *I processi di Galileo e l'epistemologia*. Rome: Armando.
Morpurgo-Tagliabue, G. 1985. Sussiste ancora una questione galileiana?. *La nuova civiltà delle macchine* 3(1–2): 91–99.
Mosley, A. 2007. *Bearing the heavens: Tycho Brahe and the astronomical community of the late sixteenth century*. Cambridge: Cambridge University Press.
Moss, J.D. 1983. Galileo's letter to Christina. *Renaissance Quarterly* 36: 547–576.
Moss, J.D. 1986. The rhetoric of proof in Galileo's writings on the Copernican system. In Wallace 1986, 179–204.
Moss, J.D. 1993. *Novelties in the heavens*. Chicago: University of Chicago Press.
Motta, Franco. 1993. La ricezione della condanna di G. Galilei nel XVII secolo. Thesis in History of Christianity, Faculty of Political Science, University of Bologna, academic year 1992–1993.
Motta, Franco. 1996. Review of two 1992 books by W. Brandmüller. *Rivista di storia e letteratura religiosa* 32: 673–683.
Motta, Franco. 1997a. Bellarminiana. *Rivista di storia e letteratura religiosa* 33: 131–160.
Motta, Franco. 1997b. Copernico, i gesuiti, le sorgenti del Nilo. In Venturi Barbolini 1977, 109–170.
Motta, Franco. 1997c. 'Geographia Sacra'. *Annali di storia dell'esegesi* 14: 477–506.
Motta, Franco (ed). 2000. *Galileo Galilei, Lettera a Cristina di Lorena*. Genoa: Marietti.
Motta, Franco. 2001. I criptocopernicani. In Montesinos and Solís 2001, 693–718.
Motta, Uberto. 1993. Querenghi e Galileo: L'ipotesi copernicana nelle imagini di un umanista. *Aevum* 67: 595–616.
Mousnier, P. 1646. *Tractatus physicus de motu locali*. Lyons.
Mousnier, P., and H. Fabri. 1648. *Metaphysica demonstrativa*. Lyons.
Müller, A. 1909a. *Der Galilei-Prozess (1632–1633) nach Ursprung, Verlauf und Folgen*. Freiburg im Breisgau: Herdersche Verlagshandlung.
Müller, A. 1909b. *Galileo Galilei und das Kopernikanische Weltsystem*. Freiburg im Breisgau: Herder.
Müller, A. 1911. *Galileo Galilei* (trans: Perciballi, P.). Rome: Max Bretschneider.
Naylor, R. 2003. Galileo, Copernicanism and the origins of the new science of motion. *British Journal for the History of Science* 36: 151–181.
Naylor, R. 2007. Galileo's tidal theory. *Isis* 98: 1–22.
Nel terzo centenario della morte di Galileo Galilei. 1942, ed. Università Cattolica del Sacro Cuore. Milan: Società Editrice Vita e Pensiero.
Nelli, G.B.C. 1793. *Vita e commercio letterario di Galileo Galilei*, 2 vols. Lausanne [Florence].

Neveu, B., and P.-N. Mayaud. 2002. L'affaire Galilée et la tentation inflationniste. *Gregorianum* 82: 287–311.
Nonis, P. 1979. L'ultima opera di Paschini. In *Atti del convegno di studio su Pio Paschini*, 158–172.
Nonnoi, G. 2000. *Saggi galileiani*. Cagliari: AM&D Edizioni.
Numbers, R.L. (ed). 2009. *Galileo goes to jail and other myths about science and religion*. Cambridge, MA: Harvard University Press.
Olivieri, M.B. 1840. *Di Copernico e di Galileo: Scritto postumo*. In Bonora 1872, 1–133.
Olivieri, M.B. 1841a. Galilée et l'inquisition romaine. *L'Université catholique*, 1st series, 11: 219–227.
Olivieri, M.B. 1841b. Der heilige Stuhl gegen Galilei und das astronomische System der Copernicus. *Historisch-politische Blätter für das katolische Deutschland*, vol. 7, 385ff., 449ff., 513ff., and 577ff.
Oreggi, A. 1629. *De Deo uno tractatus primus*. Rome.
Osler, M.J. 1998. Mixing metaphors. *History of Science* 36: 91–113.
Osler, M.J. (ed). 2000. *Rethinking the scientific revolution*. Cambridge: Cambridge University Press.
Pagano, S.M. (ed). 1984. *I documenti del processo di Galileo Galilei*. Vatican City: Pontificia Academia Scientiarum.
Pagnini, P. (ed). 1964. *Opere [di Galileo Galilei]*, 5 vols. Florence: Salani.
Palmerino, C.R. 1999. Infinite degrees of speed. *Early Science and Medicine* 4: 269–328.
Palmieri, P. 1998. Re-examining Galileo's theory of tides. *Archive for the History of Exact Sciences* 53: 223–375.
Palmieri, P. 2008a. Galileus deceptus, non minime decepit: A re-appraisal of a counter-argument in *Dialogo* to the extrusion effect of a rotating Earth. *Journal for the History of Astronomy* 39: 425–452.
Palmieri, P. 2008b. *Reenacting Galileo's experiments*. Lewiston: Edwin Mellen.
Pantin, I. 1999. New philosophy and old prejudices. *Studies in History and Philosophy of Science* 30: 237–262.
Parasin, M.M. 1648. *Systema mundi*. Stockholm.
Pascal, B. 1967. *The provincial letters* (trans and ed.: Krailsheimer, A.J.). Hammondsworth: Penguin.
Paschini, P. 1943. L'insegnamento di Galileo. *Studium*, April, 39: 94–97.
Paschini, P. 1964a. *Vita e opere di Galileo Galilei*, 2 vols., ed. E. Lamalle. In *Miscellanea galileiana*, vols. 1 and 2.
Paschini, P. 1964b. *Vita e opere di Galileo Galilei*, 2 vols. Vatican City: Pontificia Accademia delle Scienze.
Paschini, P. 1965. *Vita e opere di Galileo Galilei*, ed. M. Maccarrone. Rome: Herder.
Pastor, L. von. 1891–1953. *History of the popes from the close of the middle ages*, 40 vols. Vols. 1–2, London, 1891; vols. 3–40, London: Kegan Paul, 1894–1953. Also vols. 1–40, St. Louis: Herder, 1898–1953.
Paul, R.W. 1990. *Critical thinking*, ed. A.J.A. Binker. Rohnert Park: Center for Critical Thinking and Moral Critique, Sonoma State University.
Peacocke, A.R. (ed). 1981. *The sciences and theology in the twentieth century*. Notre Dame: University of Notre Dame Press.
Pepe, L. (ed). 1996a. *Copernico e la questione Copernicana in Italia dal XVI and XIX secolo*. Florence: Olschki.
Pepe, L. 1996b. Ferrara e le celebrazioni copernicane, 1871–1973. In Pepe 1996a, 281–291.
Pera, M. 1998. The God of the theologians and the God of the astronomers. In Machamer 1998a, 367–388.
Pesce, M. 1987. L'interpretazione della Bibbia nella Lettera di Galileo a Cristina di Lorena e la sua ricezione. *Annali di storia dell'esegesi* 4: 239–284.
Pesce, M. 1989. Esegesi storica ed esegesi spirituale nell'ermeneutica biblica cattolica dal pontificato di Leone XIII a quello di Pio XII. *Annali di storia dell'esegesi* 6: 261–291.

Pesce, M. 1991a. Momenti della ricezione dell'ermeneutica biblica galileiana e della *Lettera a Cristina* nel XVII secolo. *Annali di storia dell'esegesi* 8: 55–104.
Pesce, M. 1991b. Una nuova versione della lettera di G. Galilei a B. Castelli. *Nouvelles de la république des lettres*, no. 2, 89–122.
Pesce, M. 1991c. Il rinnovamento biblico. In Guasco, Guerriero, and Traniello 1991, 575–610.
Pesce, M. 1992a. Il *Consensus veritatis* di Christoph Wittich e la distinzione tra verità scientifica e verità biblica. *Annali di storia dell'esegesi* 9: 53–76.
Pesce, M. 1992b. Le redazioni originali della Lettera 'copernicana' di G. Galilei a B. Castelli. *Filologia e critica* 17: 394–417.
Pesce, M. 1995a. Dalla enciclica biblica di Leone XIII 'Providentissimus Deus' (1893) a quella di Pio XII 'Divino afflante spiritu' (1943). In Martini, Ghiberti, and Pesce 1995, 39–100.
Pesce, M. 1995b. L'indisciplinabilità del metodo e la necessità politica della simulazione e della dissimulazione in Galilei dal 1609 al 1642. In Prodi 1995, 161–184.
Pesce, M. 1996. Una rinnovata difesa dell'esegesi storica ed esigenza di un'interpretazione teologica. *Studia patavina* 43: 25–42.
Pesce, M. 1998. Il primo Galileo e l'ermeneutica biblica. In Bocchini Camaiani and Scattigno 1998, 331–345.
Pesce, M. 2000. Introduzione. In Motta 2000, 7–66.
Pesce, M. 2001. Gli ingegni senza limiti e il pericolo per la fede. In Montesinos and Solís 2001, 637–660.
Pesce, M. 2005. *L'ermeneutica biblica di Galileo e le due strade della teologia cristiana*. Rome: Edizioni di Storia e Letteratura.
Pessel, A. (ed). 1985. *Questions inouyes, questions harmoniques, questions theologiques, les mecaniques de Galilée, les préleudes de l'armonie universelle*. Paris: Fayard.
Pieralisi, S. 1875. *Urbano VIII e Galileo Galilei*. Rome.
Pieralisi, S. 1876. *Correzioni al libro* Urbano VIII e Galileo Galilei. Rome.
Pieralisi, S. 1879. *Sopra la nuova edizione del processo originale di Galileo Galilei fatta da Domenico Berti*. Rome.
Pitt, J. 1992. *Galileo, human knowledge, and the book of nature*. Dordrecht: Kluwer.
Poincaré, H. 1904. La Terre tourne-t-elle? *Bulletin de la Societé astronomique de France* 18: 216.
Poincaré, H. 1952. *Science and hypothesis*. New York: Dover.
Ponzio, P. (ed). 1997. *Tommaso Campanella, Apologia pro Galileo*. Milan: Rusconi.
Ponzio, P. 1998. *Copernicanesimo e teologia*. Bari: Levante.
Ponzio, P., trans and ed. 2001. *Apologia per Galileo*. Milan: Bompiani.
Popper, K.R. 1963. *Conjectures and refutations*. New York: Harper.
Portolano, M. 2000. John Quincy Adams's rhetorical crusade for astronomy. *Isis* 91: 480–503.
Poupard, P. (ed). 1983. *Galileo Galilei*. Tournai: Desclée International.
Poupard, P. (ed). 1987. *Galileo Galilei* (trans: Campbell, I.). Pittsburgh: Duquesne University Press.
Poupard, P. 1992. 'Galileo case' is resolved. *L'Osservatore Romano*, weekly edition in English, 4 November, 8.
Price, D.J. de S. 1959. Contra-Copernicus. In Clagett 1959, 197–218.
Prodi, P. (ed). 1995. *Disciplina dell'anima, disciplina del corpo e disciplina della società tra medioevo ed età moderna*. Bologna: Il Mulino.
Purcell, John B. (ed). 1844. *Galileo – The Roman inquisition*. Cincinnati.
Ranke, L. 1841. *The ecclesiastical and political history of the popes of Rome*, 2 vols. (trans: Austin, S.). Philadelphia.
Ratzinger, J. 1994. *A turning point for Europe?* (trans: McNeil, B.). San Francisco: Ignatius Press.
Redondi, P. 1983. *Galileo eretico*. Turin: Einaudi.
Redondi, P. 1987. *Galileo heretic* (trans: Rosenthal, R.). Princeton: Princeton University Press.
Redondi, P. 1994. Dietro l'immagine. *Nuncius* 9: 65–116.
Redondi, P. 2004a. *Galileo eretico*. 2nd edn. Turin: Einaudi.
Redondi, P. 2004b. Fede lincea e teologia tridentina. *Galilaeana* 1: 117–141.
Reeves, E. 1991. Augustine and Galileo on reading the heavens. *Journal of the History of Ideas* 52: 563–579.

Reeves, E. 1997. *Painting the heavens*. Princeton: Princeton University Press.
Reeves, E., and A. van Helden. Forthcoming. *Galileo and Scheiner on sunspots*. Chicago: University of Chicago Press.
Renieri, V. 1647. *Tabulae motuum caelestium universales*. Florence.
Renn, J. (ed). 2001. *Galileo in context*. Cambridge: Cambridge University Press.
Reston Jr., J. 1994. *Galileo: a life*. New York: HarperCollins.
Reumont, A. von. 1849. Galilei und Rom. *Berliner Calender für 1849*, 139–240.
Reumont, A. von. 1853. Galilei und Rom. In idem, *Beiträge zur italianischen Geschichte*, 6 vols, 1: 303–424. Berlin, 1853–1857.
Reusch, F.H. 1879. *Der Process Galilei's und die Jesuiten*. Bonn.
Riccioli, G.B. 1651. *Almagestum novum*, 2 vols. Bologna.
Riccioli, G.B. 1668. *Argomento fisicomattematico del padre Gio. Battista Riccioli della Compagnia di Gesù contro il moto diurno della Terra*. Bologna.
Riccioli, G.B. 1669. *Apologia pro argumento physico-mathematico contra systema copernicanum*. Venice.
Roberts, W.W. 1870. *The pontifical decrees against the motion of the Earth considered in their bearing on advanced ultramontanism*. London. Reprinted in Roberts 1885, 55–116.
Roberts, W.W. 1885. *The pontifical decrees against the doctrine of the Earth's movement, and the ultramontane defence of them*. Oxford and London.
Robinet, A. 1988. *G.W. Leibniz*. Florence: Olschki.
Rommel, C. von, (ed). 1847. *Leibniz und Landgraf Ernst von Hessen-Rheinfels*. 2 vols. Frankfurt am Main.
Ronchi, V. 1958. *Il cannocchiale di Galileo e la scienza del seicento*. Turin: Boringhieri.
Rosen, E. 1947. *The naming of the telescope*. New York: Schuman.
Rosen, E. (ed). 1959. *Three Copernican treatises*, 2nd edn. New York: Dover.
Rosen, E. 1975. Was Copernicus's *Revolutions* approved by the Pope? *Journal of the History of Ideas* 36: 531–542.
Rosen, E., trans and ed. 1992. *Nicholas Copernicus: on the revolutions*. Baltimore: Johns Hopkins University Press.
Ross, S. 1962. Scientist: the story of a word. *Annals of Science* 18: 65–85.
Rowland, W. 2003. *Galileo's mistake*, revised edn. New York: Arcade.
Rudwick, M. 1981. Senses of the natural world and senses of God. In Peacocke 1981, 241–261.
Ruse, M. 1997. Introduction. In Russell 1997.
Russell, B. 1997. *Religion and science*. New York: Oxford University Press.
Russell, J.L. 1989. Catholic astronomers and the Copernican system after the condemnation of Galileo. *Annals of Science* 46: 365–386.
Saka, P. 2006. Letter to the Editor, 27 February. *American Scientist on Line*, http://www.americanscientist.org/template/BookshelfLetterTypeDetail/assetid/50287. Consulted on 6 May.
Salusbury, T., trans and ed. 1661–1665. *Mathematical collections and translations*, 2 vols. London.
Salvini, S. 1717. *Fasti consolari dell'accademia fiorentina*. Florence.
Santillana, G. de. 1955. *The crime of Galileo*. Chicago: University of Chicago Press.
Santillana, G. de. 1960. Galileo e i moderni. *Tempo presente* 5: 322–328.
Scartazzini, J.A. 1877–1878. Il processo di Galileo Galilei e la moderna critica tedesca. *Rivista europea*, year 8, 4: 829–861; year 9, 5: 1–15 and 221–249, 6: 401–423, and 10: 417–453.
Schroeder, H.J. 1978. *Canons and decrees of the Council of Trent*. Rockford, IL: Tan Books and Publishers.
Scriven, M. 1976. *Reasoning*. New York: McGraw-Hill.
Segre, M. 1989. Viviani's life of Galileo. *Isis* 80: 207–232.
Segre, M. 1990. Redondi's theory and new perspectives in Galilean studies. *Archives internationales d'histoire des sciences* 40: 3–10.
Segre, M. 1991a. *In the wake of Galileo*. New Brunswick: Rutgers University Press.
Segre, M. 1991b. Science at the Tuscan Court, 1642–1667. In *Physics, Cosmology and Astronomy, 1300–1700*, ed. S. Unguru, 295–308. Dordrecht: Kluwer.
Segre, M. 1997. Light on the Galileo case? *Isis* 88: 484–504.

Segre, M. 1998. The never ending Galileo story. In Machamer 1998a, 388–416.
Segre, M. 1999. Galileo: a 'rehabilitation' that has never taken place. *Endeavour* 23(1): 20–23.
Segre, M. 2003. Zwischen Trient und Vatikanum II. *Berichte zur Wissenschaftsgeschichte* 26: 129–136.
Segre, M. 2005. Il caso Galileo. *Nuncius* 19: 733–747.
Settele, G. 1819 [actual publication date 1821]. *Elementi di ottica e di astronomia*, vol. 2, *Astronomia*. Rome
Settele, G. 1820–1833. *Diario*. In Maffei 1987, 283–421.
Shank, M.H. 2005. Setting the stage. In McMullin 2005a, 57–87.
Sharratt, M. 1994. *Galileo, decisive innovator*. Cambridge, MA: Blackwell.
Sharratt, M. 2005. Galileo's 'rehabilitation'. In McMullin 2005a, 323–339.
Shea, W.R. 1972. *Galileo's intellectual revolution*. New York: Science History Publications.
Shea, W.R. 1984. Melchior Inchofer's 'Tractatus Syllepticus'. In Galluzzi 1984, 283–292.
Shea, W.R. 2005. Review of Finocchiaro's *Retrying Galileo*. *Isis* 96: 644.
Shea, W.R., and M. Artigas. 2003. *Galileo in Rome*. Oxford: Oxford University Press.
Shea, W.R., and M. Artigas. 2006. *Galileo observed*. Sagamore Beach: Science History Publications.
Simoncelli, P. 1992. *Storia di una censura*. Milan: FrancoAngeli.
Sinke Guimarães, A. 2005. The swan's song of Galileo's myth. At www.traditioninaction.org/History/A_003_Galileo.html. Consulted on 15 July.
Snow, C.P. 1964. *The two cultures and a second look*. Cambridge: Cambridge University Press.
Sobel, D. 1999. *Galileo's daughter*. New York: Walker & Company.
Sobel, D., trans and ed. 2001. *Letters to father*. New York: Walker & Company.
Socci, A. 1993. Scienza e fede: Accademia o politica? *30 giorni* 11(1): 60–63.
Soccorsi, F. 1946. Il processo di Galileo. *La Civiltà Cattolica* 97: 175–184, 429–438.
Soccorsi, F. 1947. *Il processo di Galileo*. Rome: Edizioni La Civiltà Cattolica.
Soccorsi, F. 1963. *Il processo di Galileo*. Rome: Edizioni La Civiltà Cattolica.
Soccorsi, F. 1964. Il processo di Galileo. In *Miscellanea galileiana*, 3: 849–929.
Sosio, L. (ed). 1970. *Dialogo sopra i due massimi sistemi del mondo tolemaico e copernicano*. Turin: Einaudi.
Speller, J. 2008. *Galileo's inquisition trial revisited*. Frankfurt: Peter Lang.
Stimson, D. 1917. *The gradual acceptance of the Copernican theory of the universe*. New York: Baker and Taylor.
Stoffel, J.-F. 2001. Pierre Duhem interprète de l' 'Affaire Galilée'. In Montesinos and Solís 2001, 765–782.
Sylla, E. 1991. Galileo and probable reasons. In Dahlstrom 1991, 211–234.
Symposium internazionale di storia, metodologia, logica e filosofia della scienza. 1967. Florence: Gruppo Italiano di Storia delle Scienze.
Tabarroni, G. 1983. Giovanni Battista Guglielmini e la prima verifica sperimentale della rotazione terrestre (1790). *Angelicum* 60: 462–486.
Tamburini, F. 1990. La riforma della Penitenzieria nella prima metà del secolo XVI e i cardinali Pucci in recenti saggi. *Rivista di storia della Chiesa in Italia* 44: 110–140.
Targioni Tozzetti, G. 1780. *Notizie degli aggrandimenti delle scienze fisiche accaduti in Toscana nel corso di anni LX del secolo XVII*, 3 tomes in 4 vols. Florence.
Thomason, N. 1992. Could Lakatos, even with Zahar's criterion for novel fact, evaluate the Copernican research programme? *British Journal for the Philosophy of Science* 43: 161–200.
Tiraboschi, G. 1772–1782. *Storia della letteratura italiana*, 9 tomes in 13 vols. Modena.
Tiraboschi, G. 1782–1797. *Storia della letteratura italiana*, 10 tomes in 13 vols. Rome.
Torricelli, E. 1644. De motu gravium naturaliter descendentium, et proiectorum. In idem, *Opera geometrica*, 95–243 (irregular pagination). Florence.
Torrini, M. 1993. Galileo copernicano. *Giornale critico della filosofia italiana* 72/74: 26–42.
Toulmin, S. and J. Goodfield. 1961. *The fabric of the heavens*. New York: Harper.
Van der Meer, J.M. (ed). 1996. *Facets of faith and science*, 4 vols. Lanham: University Press of America.

Van Helden, A. 1984. Galileo and the telescope. In Galluzzi 1984, 149–158.
Van Helden, A. 1994. Telescopes and authority from Galileo to Cassini. *Osiris*, second series, 9: 7–29.
Varenius, B. 1650. *Geographia generalis in qua affectiones generales Telluris explicantur*. Amsterdam.
Venturi, G. (ed). 1818–1821. *Memorie e lettere inedite finora o disperse di Galileo Galilei*, 2 vols. Modena.
Venturi Barbolini, A.R. (ed). 1997. *Girolamo Tiraboschi*. Modena: Biblioteca Estense Universitaria.
Viganò, M. 1969. *Il mancato dialogo fra Galileo e i teologi*. Rome: Edizioni La Civiltà Cattolica.
Viviani, V. 1654. *Racconto istorico della vita di Galileo*. In Favaro 19: 599–632.
Viviani, V. 1701. *De locis solidis*. Florence.
Viviani, V. 1717. *Racconto istorico della vita di Galileo*. In Salvini 1717, 397–431.
Viviani, V. 1992. *Vita di Galileo*, ed. L. Borsetto. Bergamo.
Voelkel, J.R. 2001a. *The composition of Kepler's* Astronomia nova. Princeton: Princeton University Press.
Voelkel, J.R. 2001b. Giovanni Antonio Magini's 'Keplerian' tables of 1614 and their implications for the reception of Keplerian astronomy in the seventeenth century. *Journal for the History of Astronomy* 32: 237–262.
Voltaire 1754–1757. *The general history and state of Europe*, 6 parts in 3 vols. London.
Voltaire 1824. *A philosophical dictionary*. 2nd edn., 6 vols. London.
Voltaire 1877–1883. *Oeuvres complètes*, 52 vols., ed. L. Moland. Paris.
Voltaire 1901. *The works of Voltaire*, 42 vols. (trans: Fleming, W.F., ed.: Smollett, T. et al.). Paris: Du Mont.
Wallace, W.A. 1981a. Does Galileo's trial beg for reopening? *Los Angeles Times*, 11 April, Part I-B, 4.
Wallace, W.A. 1981b. *Prelude to Galileo*. Dordrecht: Reidel.
Wallace, W.A. 1983a. Galileo and Aristotle in the *Dialogo*. *Angelicum* 60: 311–332.
Wallace, W.A. 1983b. Galileo's science and the trial of 1633. *The Wilson Quarterly* 7: 154–164.
Wallace, W.A. 1984a. *Galileo and his sources*. Princeton: Princeton University Press.
Wallace, W.A. 1984b. Galileo's early arguments for geocentrism and his later rejection of them. In Galluzzi 1984, 31–40.
Wallace, W.A. (ed). 1986. *Reinterpreting Galileo*. Washington: Catholic University of America Press.
Wallace, W.A. 1987. Galileo and the professors of the Collegio Romano at the end of the sixteenth century. In Poupard 1987, 44–60.
Wallace, W.A. 1992a. *Galileo's logical treatises*. Dordrecht: Kluwer.
Wallace, W.A. 1992b. *Galileo's logic of discovery and proof*. Dordrecht: Kluwer.
Wallace, W.A. 1996. *The modeling of nature*. Washington: Catholic University of America Press.
Wallace, W.A. 1999. Galilei, Galileo. In *Encyclopedia of the Renaissance*, ed. P.F. Grendler, 6 vols, 3: 2–9. New York: Scribner's.
Ward, S. 1635. *In Ismaelis Bullialdi astronomiae philolaicae fundamenta, inquisitio brevis*. Oxford.
Wendelen, G. 1644. *Eclipses lunares ab anno 1573 ad 1643 obervatae*. Antwerp.
Wendelen, G. 1647. *De causis naturalibus pluviae purpureae bruxellensis*. 2nd edn. Brussels.
Westfall, R.S. 1989. *Essays on the trial of Galileo*. Vatican City: Vatican Observatory Publications.
Westman, R.S. 1972. Kepler's theory of hypothesis and the 'realist dilemma'. *Studies in History and Philosophy of Science* 3: 233–264.
Westman, R.S. (ed). 1975a. *The Copernican achievement*. Berkeley: University of California Press.
Westman, R.S. 1975b. The Melanchthon circle, Rheticus, and the Wittenberg interpretation of the Copernican theory. *Isis* 66: 165–193.
Westman, R.S. 1975c. Three responses to the Copernican theory. In Westman 1975a, 285–345.
Westman, R.S. 1975d. The Wittenberg interpretation of the Copernican theory. In Gingerich 1975.
Westman, R.S. 1984. The reception of Galileo's 'Dialogue'. In Galluzzi 1984, 329–371.
Westman, R.S. 1986. The Copernicans and the Churches. In Lindberg and Numbers 1986, 76–113.

Westman, R.S. 1987. La Préface de Copernic au pape: Esthétique humaniste et réforme de l'Eglise. *History and Technology* 4: 359–378.
Westman, R.S. 1990. Proof, poetics, and patronage. In Lindberg and Westman 1990, 167–205.
Westman, R.S. 1994. Two cultures or one? *Isis* 85: 79–115.
Westman, R.S. Forthcoming. *The Copernican question.* Berkeley: University of California Press.
Whewell, W. 1837. The Copernican system opposed on theological grounds. In idem, *History of the inductive sciences*, 1st edn., 3 vols., 1: 397–404. London.
Whewell, W. 1847. Case of Galileo. In idem, *Philosophy of the inductive sciences founded upon their history*, 2nd edn., 2 vols., 1: 696–700. London.
Whewell, W. 1857a. The Copernican system opposed on theological grounds. In idem, *History of the inductive sciences*, 3rd edn., 3 vols., 1: 303–312. London.
Whewell, W. 1857b. Were the papal edicts against the Copernican system repealed? In idem, *History of the inductive sciences*, 3rd edn., "Additions to the Third Edition," 3 vols., 1: 393–394. London.
White, A.D. 1869. First of the course of scientific lectures. *New York Daily Tribune*, 18 December, 4.
White, A.D. 1876. *The warfare of science.* New York.
White, A.D. 1896. *A history of the warfare of science with theology in Christendom*, 2 vols. New York.
White, A.D. 1915. *La condanna di Galileo Galilei e le responsabilità della Chiesa.* Mendrisio: Casa Editrice Cultura Moderna.
White, A.D. 1965. *A history of the warfare of science with theology in Christendom.* Abridged edn. by B. Mazlish New York: Free Press.
White, T. 1642. *De mundo dialogi tres.* Paris.
Wilding, N. 2008. The return of Thomas Salusbury's *Life of Galileo* (1664). *British Journal for the History of Science* 41: 241–265.
Wilson, D.B. 1996. On the importance of eliminating *science* and *religion* from the history of science and religion. In van der Meer 1996, 1: 27–48.
Wilson, D.B. 1999. Galileo's religion *versus* the church's science? *Physics in Perspective* 1: 65–84.
Wilson, D.B. 2000. The historiography of science and religion. In Ferngren et al. 2000.
Wisan, W.L. 1978. Galileo's scientific method. In Butts and Pitt 1978, 1–57.
Wisan, W.L. 1984. On the art of reasoning. *Annals of Science* 41: 483–487.
Wohlwill, E. 1870. *Der Inquisitionsprocess des Galileo Galilei.* Berlin.
Wohlwill, E. 1877. *Ist Galilei gefoltert worden?* Leipzig.
Wohlwill, E. 1909. *Galilei und sein Kampf für die Kopernickanischen Lehre*, vol. 1. Leipzig: Voss.
Wohlwill, E. 1926. *Galilei und sein Kampf für die Kopernickanischen Lehre*, vol. 2. Leipzig: Voss.
Wolf, H. (ed). 2001. *Inquisition, Index, Zensur.* Paderborn: Ferdinand Schöningh.
Wykstra, S.J. 1996a. Have worldviews shaped science? In van der Meer 1996, 1: 91–114.
Wykstra, S.J. 1996b. Should worldviews shape science? In van der Meer 1996, 2: 123–171.
Zerilli, D. 1668. *Confermazione d'una sentenza.* Naples.

Index

A

aberration of starlight, xxi-xxii, 238
abjuration of Galileo, xviii, 151, 156, 159, 160, 161, 162, 163, 176, 180, 181, 182, 188, 190, 196, 209, 210, 211, 212, 226, 252, 262
abstract entities, 115–20
accommodation, principle of, 67, 77, 82, 182, 205
Adams, John Quincy, 266–67
ad hoc, 23, 41, 58, 124, 241
ad hominem argument, 81
aesthetics, methodological criterion of, 35, 38–64
aether, 7, 8, 9, 11, 26, 31, 130
affirming the consequent, xiv, 240
Agassi, Joseph, 233, 277, 315
Agnani, G.D., 181
air, xxiii, 4–5, 6, 7, 8, 9, 10, 11, 12, 26, 29, 31
Ajalon, 8
Albergotti, Ulisse, 70
Albèri, Eugenio, 197, 259, 316
Alcinous, 168
Amorites, 32, 78, 81
Anfossi, Filippo, 191–95
Angeli, Stefano degli, xxi
annual motion, 15, 21, 22, 35, 81, 129, 241
anthropocentrism, 96, 213
anti-Copernican decree of 1616, xvii, xxxviii, 45, 59, 61, 69, 72, 74–75, 79, 89, 94–95, 141–42, 143, 144, 145, 157, 162, 168, 172, 173, 176, 183, 187, 199, 231, 232, 237, 243, 253, 261, 299, 302, 305
antipodes, 178
Apelles, 56
apocryphal letter to Renieri, xxv, 185, 186, 256–60, 261, 262, 265, 268, 270, 275

Apologia pro Galileo (Campanella), 90, 94, 294
approach, Galileo's, xxxvi-xxxvii, xxxix, xliii, 90
apriorism, 86, 133
Aquarius, 17
Aquinas, Saint Thomas, 75, 86, 92, 93, 94, 286
Arcadian Journal, 193
Arcetri, xxxv, 164, 166, 212, 256, 257, 307
Archimedes, 124
argument, definition of, ix, xli, 132
Aries, 17
Ariew, Roger, 174, 315
Aristarchus, xiii, 22, 133, 178
Aristotle, xli, 3, 4, 19, 46, 47, 48, 50, 60, 132
Armellini, Giuseppe, 216
ascension, 57
Assayer (Galileo), xvii, 45, 116, 143, 144, 256
Astronomical and Philosophical Balance (Grassi), 256
atomic bomb, xix, xxxi, 211
Attavanti, Giannozzo, 71, 139, 140
Augustine, Saint, xxiv, 67, 82, 84, 85, 86, 87, 92, 93, 94, 182, 205, 227, 245, 247
authority: astronomical, 153, 204, 249; biblical or scriptural, 69, 77, 86, 87, 91, 92, 93, 94, 182, 197, 271, 284–86, 295; Church or ecclesiastic, 63, 174; institutional, x, xx, xxxvii, xxxix; philosophical, 80, 85, 89, 182, 183, 213; scientific, xxiii, xxiv, xxv, xxx, xxxvii, xxxviii, 67, 77, 86, 87, 89, 94, 95, 131, 174, 186, 204, 305, 207, 210, 234, 245, 248, 251, 259, 285, 312; theological, 83
Auzout, Adrien, 171, 172–73, 194, 334, 306
axis: defined 12; of ecliptic, 16–17

339

B

bad-theologian thesis, 186, 251–76. *See also* theological issues in trial of Galileo
Baldigiani, Antonio, 66, 171–72
Baldini, Ugo, xxii, 143, 150, 170, 183, 184, 222, 224, 252, 253
Baliani, Giovanni, xx, xxiii, 44, 57, 60
Barberini, Antonio, Sr., 161, 162
Barberini, Francesco, xxvi, 152, 162, 165, 201, 252
Barberini, Maffeo, 143. *See also* Urban VIII
Baretti, Giuseppe, 175–76
Barker, Peter, xiv, 35, 316
Baronio, Cesare, 67, 247, 285, 297, 298, 303
begging the question, 80, 94, 127, 129, 139, 247
Bellarmine, Robert, 71, 76, 79, 96, 143, 255, 293, 295, 296; biblical literalism, 222; certificate for Galileo, 142, 143, 149, 150, 158, 160, 199, 201, 202, 222, 231–32, 268, 304, 305; as epistemologist, xxviii, 223, 224, 227, 273, 278–79, 280, 281, 282; Fabri's principle, 172, 173; Feyerabend's interpretation, xxxii, 300; Galileo's reply, 140, 189; letter to Foscarini, xxviii, xxxii, 33, 75, 79, 172, 173, 278, 286, 301; Louvain lectures, 222, 224; warning to Galileo, xvii, xxxvii-xxxviii, 61, 141, 143, 146, 149, 157, 160, 163, 190, 199–201, 202, 215, 232, 254, 305
Beltrán Marí, Antonio, xvii, xlii, 34, 49, 70, 72, 81, 138, 140, 142, 143, 144, 148, 151, 201, 298
Benedict XIV, xviii-xix, 179, 183, 253
Benedict XVI, xxxii, xxxiii, xxxiv. *See also* Ratzinger, Joseph
Benítez, Hermes, 252, 298, 306, 313, 316
Bentley, Eric, 213
Bérault-Bercastel, Antoine, 260–61, 262, 263, 265, 271, 275
Beretta, Francesco, xxvii, xlii, 68, 138, 141, 142, 150, 151, 159, 160, 183, 201, 252, 253, 396, 302, 303
Bergier, Nicolas, 261, 262, 265, 268, 275
Bernegger, Matthias, 166, 168, 308
Bernini, Domenico, 175
Berti, Domenico, 198, 199, 201, 202, 203
Bertolla, Pietro, 216, 219
Besomi, Ottavio, xlii, 34, 144, 181, 317, 323
Bessel, Friedrich, xxii, 28
Biagioli, Mario, xvii, xxiv, xlii, 48, 49, 81, 116, 143, 306

Biot, Jean, 197–98, 264, 276
birds' flight, 50, 194
Blacas d'Aulps, Luis Casimir de, 189
Blackwell, Richard, xlii, 69, 71, 73, 76, 79, 89, 90, 92, 140, 141, 151, 216, 218, 265, 294, 295, 301
Bologna, xii, xxii, 162, 179
book of nature, 92, 194, 115, 116, 119, 298
Borgia, Gaspare, 152
Boscaglia, Cosimo, 70, 294
Bradley, James, xxi, 223
Brandmüller, Walter, 81, 222, 223–24, 225, 227
Brecht, Bertolt, xix, xxxi, 178, 211–14
Brewster, David, 197–98, 263, 266, 267, 276
Brooke, John, xii, 291, 292, 297, 306
Brown, Harold, 312, 318
Brucker, Johann, 169, 262
Bruno, Giordano, xiv-xv, 96, 296
Bucciantini, Massimo, xvii, xxxvi, xlii, 35, 36, 48, 49, 51, 72, 73, 318–19, 323, 324
Buonamici, Giovanfrancesco, 162, 163
Burtt, Edwin, 283

C

Cabeo, Niccolò, xx
Caccini, Tommaso, 71, 139, 140, 176, 215, 273, 309
Caetani, Bonifazio, 89
Caetani, Onorato. *See* Gaetani, Onorato
Calabria, 76
Calandrelli, Giuseppe, xxii
Calmet, Augustin, 180, 181–82
Calvinism, 255
Camerota, Michele, xii, xvii, xlii, 34, 48, 49, 64, 66, 69, 70, 71, 115, 124, 138, 143, 144, 302, 319
Campanella, Tommaso, xvii, xxvi, xxx, 65, 66, 89–93, 94, 96, 230, 293, 284
Cancer, 17
Canis Major, 14
Cantor, Geoffrey, 291, 292, 297, 306, 318
capable of being proved, 83, 84, 85, 244, 248, 249, 289, 290, 298. *See also* demonstrable; provable
Capellari, Bartolomeo, 193
Capricorn, 17
Carafa, Petrus, 162, 163
Carroll, William, xxiv, 79, 244, 297, 320
Casini, Paolo, 178, 258, 320
Castelli, Benedetto, 70, 71, 138; Galileo's letter to, 44, 71, 79, 91, 139, 140, 153, 204, 233, 263, 270, 273, 275, 297
Catholic Counter-Reformation, 96, 141

Catholic University of Milan, xxxi, 207
celestial sphere, 5, 10, 12, 13, 17, 23, 27, 28.
  See also stellar sphere
centrifugal force, 30, 31, 99, 105
centripetal motion, 98, 103, 107, 113
Cesi, Federico, 44, 54, 66
Chalmers, Alan 101, 103, 104, 320
charity, principle of, 233
Chasles, Philarète, 197, 198
Chiaramonti, Scipione, 30, 294
Christina of Lorraine, 70, 138; Galileo's letter
  to, xxiv, 45, 60, 67, 68, 79–89, 91,
  133, 140, 166, 167, 168, 180, 182,
  204, 205, 233, 234, 235, 236, 243–48,
  255, 263, 270, 273, 274, 275, 285,
  290, 297, 307, 308
Church Fathers, anti-Copernican argument,
  xiv, 32–33, 34, 73–74, 80, 84, 141
circular motion, 7, 8, 9, 10, 14, 31, 46,
  70, 98, 101
circumstantialism, xix, 96, 197–98, 214
Clavelin, Maurice, 117, 320
Clavius, Christopher, 65
Colombe, Ludovico delle, 69, 273, 294
Columbus, Christopher, 4
column commemorating Galileo in Rome,
  271, 311
comets, xvii, 116, 143, 144, 170, 293
compass, Galileo's proportional, 166, 168
composition of motion, 15, 23, 37, 97, 128.
  See also superposition of motions
Comte, Auguste, xix
confession by Galileo, 150, 151, 152, 158,
  160, 161, 189–90, 201, 202, 209, 210,
  212, 215
conservation and innovation, dialectic of, xxx,
  xxxix, 227, 233, 294, 305–306, 314
conservation of motion, xxiii, xxxvii, 29, 30,
  32, 39, 97, 128
Considerations on the Copernican Opinion
  (Galileo), 33, 44, 76, 87, 89, 140, 286,
  288, 290
consistency, methodological criterion of,
  38–64
constellation, xxiii, 17; defined 14
Conti, Carlo, 70
Cooper, Peter, 263–64, 268, 271, 275, 276, 309
Cosimo II de' Medici, 70, 185, 142
Council of Trent, 73, 75
cowardice, Galileo accused of, 197–98
Coyne, George, 143, 222, 224, 297, 298, 300
Cracow conference on Galileo, 300
creationism, 243
Cristaldi, Belisario, 193

criteriology, 68
critical approach, Galileo's, x, xv, xvii-xviii,
  xxxvi-xxxvii, xxxviii, xxxix, xl, 49, 64,
  85, 121–23, 124, 132–34, 137, 149,
  252, 280, 292
critical reasoning, ix, x, xi, xli, xlii, 21, 121,
  124, 132–34
critical thinking, xxiii, xxxvii
Crollis, D. de, 193

D
D'Alembert, Jean, xix, xxix, 175, 178–79,
  185, 254, 293
Darwinism, 295
daughter, Galileo's elder, 164, 165, 212–13
death of Galileo, xxi
deception of the senses, anti-Copernican
  argument, 25–26, 131
declination, 15; defined, 13
deferent, 19; defined, 18
Delambre, Jean, 189–90, 196, 264
demonstrable, vs. demonstrated, 86, 93, 286,
  288. See also capable of being proved;
  provable
demonstrative ideal, xxviii, 88, 299
demonstrative sciences, xxviii, 88
denying the consequent, 125, 127
depositions in trial of Galileo, 45, 71, 139,
  140, 150, 151, 189–90, 196, 199, 201,
  202, 203, 210, 215, 255, 168, 303
Descartes, René, xix, 144, 162, 163–64, 166,
  176–77, 183, 252, 253, 293, 294
DiCanzio, Albert, xi, xxvii, 101, 107,
  298, 321
Diderot, Denis, xix, 178
Dini, Piero, 44, 293
Diodati, Elia, 166, 167
Discourse on the Tides (Galileo), xxviii, 45,
  59, 87, 140, 145. See also tides
*Discourses and Mathematical Demonstrations
  on Two New Sciences* (Galileo). See
  Two New Sciences
disobedience thesis, xxv, xxvii, 201, 205, 297,
  255, 263–66, 269, 301–305
diurnal motion, 5, 12, 14, 15, 22, 35, 81, 82,
  215, 241
divine omnipotence, anti-Copernican
  argument, xiv, xliii, 33–34, 144–45,
  148, 149, 150
Dog Star, 14
downward, in geostatic world view, 7, 9, 29,
  31, 32, 70, 99, 100, 102, 103, 106, 108,
  112, 113, 128

Drake, Stillman, xxvii, 49, 51, 55, 56, 64, 70, 96, 101, 124, 129, 145, 178, 232, 233, 236, 237, 240, 245, 246, 293
Drinkwater Bethune, John Eliot, 264
Dubarle, Dominique, 218
Dublin Review, 263, 266, 267
Duhem, Pierre, xxvii-xxviii, xxxix, 205, 222, 223, 272-73, 275, 277-90

E

E pur si muove, 175-76, 262, 310
earth, shape of, 4
earth-heaven dichotomy, anti-Copernican argument, 26, 52, 55, 130, 132,
east-wet gunshots, anti-Copernican argument, 29, 194
eccentric, 44, 55; defined, 19
Ecclesiastes, 32, 70, 75, 77, 78, 91-92, 94
eclipse, 55, 56, 170
ecliptic, 13, 16-17, 27, 28, 241
Einstein, Albert, 220, 222, 223, 224, 233, 274, 293, 296, 313
element, in geostatic world view, 4-5, 6, 7, 8, 26, 31, 170
ellipse, 35, 215
Elzevier publishing company, 168
empirical accuracy, methodological criterion of, 38-64
empiricism, 42, 86, 91, 311
Encyclopedia (Diderot and D'Alembert), xix, xxix, 175, 178, 254, 293
Enlightenment, 175-79, 182, 183, 198, 227
Ennis, Robert, xxxvii, 322
epicycle, 19, 23, 27, 44, 54, 123, 215; defined, 18
epistemology, x, xi, xiv, xv, xvi, xxiv, xxvii-xxviii, xxxvii, xxxix, xl, xlii, 24, 25, 33, 34, 35, 50, 67-68, 71, 82, 81, 85, 86, 87, 94, 96, 98, 105, 116, 122, 131, 137, 140, 141, 156, 181, 194, 205, 220, 221, 222, 223, 226, 227, 236, 237, 259, 272, 274, 277, 290, 296, 299, 313
equator, xxi, 5, 14, 15, 16, 31, 99, 100, 101, 104, 108, 111, 112, 125; defined 12-13
equinox, 15-16
equivocation, xxxii, xxxiv, 222, 239
Erik XIV of Sweden, 258
Erythraeus, Janus Nicius, 169-70
escape velocity, 107
Estève, Pierre, 176
Euclidean geometry, 106, 114
evolution, theory of, 94

ex cathedra, papal pronouncements, 80, 142, 206, 209, 221, 230
exegesis, 65, 79, 181, 185, 259, 312; vs. hermeneutics, 67-68
experiments, xv, xix, xxii, xxviii, xlii, 30, 31, 99, 117, 119, 125, 219, 280, 288, 289
explanatory coherence, xvi, 30, 41, 42, 51, 58, 62, 63, 123-24; defined, 23
exsecant, 100-113
external force, definition of, 10-11
extrinsic denomination, principle of, 78
extruding power of whirling, anti-Copernican argument, xvi, 30-31, 50, 97-120, 133

F

Fabri, Honoré, xx, 172, 173, 204
Fabroni, Angelo, 176, 185, 186, 264
fair-mindedness, Galileo's, xxxvii, xxxix, xl, xli, 132, 133, 134, 137
fallibilism, xxviii, 78, 88, 121
fallibility, xxxvii, 134
Fantoli, Annibale, xxix, xlii, 33, 48, 49, 70, 138, 142, 151, 181, 216, 252, 297, 298, 322
Favaro, Antonio, 217, 259
Febei, Pietro Paolo, 256
Feldhay, Rivka, xxx, xlii, 101, 117, 252, 294, 322
Feller, François, 262, 268, 271, 275
Ferri, Girolamo, 255-56, 262, 276
Feyerabend, Paul, xix, xxxii-xxxix, 37, 38, 124, 240, 300-301, 313
Findlen, Paula, xii, 323
fire, as an element, 4-5, 6-7, 8, 9, 10, 31
firmament, 73; defined, 5
Fisher, Alec, xxxvii, 324
fixed stars, 5, 6, 12, 13, 14, 15, 17, 18, 21, 23, 27, 28, 52, 130, 132, 170, 240, 241, 283
Florence, xii, xvi, xix, xxxv, 71, 95, 138, 139, 142, 145, 146, 148, 161, 166, 169, 171, 176, 252, 253, 257, 258, 270, 307, 309
Fontenelle, Bernard Le Bovier de, 177
forced motion, 8-12, 29, 30, 32, 97. *See also* violent motion
forgery: Gaetani's 185, 186, 258-60, 276; special injunction, 199-200, 202-203
formal heresy, xviii, 74, 95, 141, 148, 151, 152, 157, 159, 161, 205, 297, 230, 301, 302, 303, 346

Index 343

Foscarini, Paolo, xvii, xxviii, xxx, xxxii, 45, 65, 67, 69, 75, 76–79, 87, 90, 93, 94, 96, 140, 142, 168, 172, 183, 184, 193, 231, 278, 286, 293, 300, 301, 302, 307, 308
Foucault, Léon, xxii, xxvii, 238
Frattini, Candido, 193
Frisi, Paolo, 178
Froidmont, Libert, 162
Frova, Andrea, xii, 325
fundamentalism, xxiii, 96, 166

G

Gaetani, Onorato, xxv, 258–60, 262, 266, 267, 269, 271, 273, 275, 276
Galluzzi, Paolo, xii, xx, xxi, xxxvi, 115, 172, 175, 252, 325
Garber, Daniel, 174, 315
Garcia, Stéphane, 85, 168, 325–26
Garzend, Léon, 205–207
Gassendi, Pierre, xx, 169
Gasser, Achilles, 69
Gatti, Hilary, xi, xii, xiv, 326
Gaukroger, Stephen, 101, 104, 144, 164, 252, 326
Gebler, Karl von, xxiii, 198–99, 203–204, 222
Gemelli, Agostino, xxix, 207–208, 209, 216, 223, 296, 312
Gemini, 17
geocentric, definition of, 3
geokinetic, definition of, 21
geostatic, definition of, 3
Gherardi, Silvestro, 200–201, 203
Ghilini, Girolamo, 166
Gibeon, 32, 131
Gingerich, Owen, 53, 72, 232, 240, 242, 265, 299, 301, 326
Giordani, Pietro, 258, 259
Giovasco, Luigi Maria, 180, 181
Goddu, André, xiv, 34, 35, 278, 326
Goldstein, Bernard, xiv, 35, 316
Granada, Miguel, 69, 319, 327
Grandi, Antonio Maria, 191, 192
Grassi, Orazio, 215, 256, 293
gravitation, xxii, 32, 107, 129, 238
gravity, in geostatic world view, 10, 50, 99, 103, 107
great circle, 12, 13, 16, 17
Gregorian reform of calendar, 264
Gregory XVI, 193
Greipl, Egon Johannes, 222, 263
Grisar, Hartmann, 273
Guglielmini, Giambattista, xxii

Guicciardini, Piero, 185–86, 254, 261, 265, 269, 272, 275
Guiducci, Mario, 161, 163, 256
Gustavus Adolphus, 147, 256, 258

H

Hagen, J.H., xxii
Hallam, H., 267
Hatfield, Gary, 283, 327
Hautt, David, 168
heaven, definition of, 5
heavenly bodies, definition of, 6
heavenly spheres, definition of, 6
heavy bodies, definition of, 9
Heilbron, John, 252, 299, 327
Helbing, Mario, xlii, 34, 144, 181, 317
heliocentric, definition of, 22
heresy. *See* formal heresy; vehement suspicion of heresy
hermeneutics: vs. exegesis, 67–68; history of, 86; issues in trial of Galileo, xxv, 79, 85, 86, 87, 137, 140, 152, 187, 204, 236, 243–48, 251–76, 284–86, 311; meta-hermeneutics, xxiv, 67–68, 137; vindication of Galileo, xix, xxiv, xxv, xxxviii, xl, 203–205, 219, 220, 226, 227
Hill, David, 101, 113, 327
History and Demonstration Concerning Sunspots (Galileo), xvi, 44, 51, 54, 55, 66, 71, 91, 139, 140, 157, 199
Holy Office, Congregation of the, 70, 142, 150, 175, 216, 217, 231, 232, 234, 242, 143, 257, 260, 261, 269, 272, 281, 284
Holy Roman Empire, 147, 167, 168
horizontal motion, 30, 128
Horky, Martin, 69
Hoskin, Michael, xi
house arrest of Galileo, xviii, 152, 159, 164, 165, 166, 196, 212, 257, 307, 310, 311
Howell, Kenneth, 69, 72, 73, 77, 78, 82, 84, 86, 252, 259, 297, 327–28
Huygens, Christiaan, xxi, 103, 104
hypothesis, concept of, 146–47
hypothesis, treating as a, 145, 146, 150, 164, 180, 191, 194
hypothetical discussion, 144, 146, 157

I

ignoratio elenchi, 102
illegitimate birth, Galileo's alleged, 262

imprimatur, 158, 167, 181, 182, 187, 191, 192, 193, 225, 308
imprisonment of Galileo, xviii, 149, 159, 165, 175, 176, 177, 187, 188, 253, 255, 257, 261, 264, 271, 310, 311. *See also* house arrest
imprudence attributed to Galileo, xxv, 96, 186, 187, 188, 207, 264, 268, 308
Inchofer, Melchior, 141, 294, 295
Index decrees: of 1616 against Copernicanism, 45, 59, 61, 72, 79, 94, 131, 141, 142, 143, 144, 145, 157, 162, 168, 199, 215, 231, 232, 261, 302, 305; of 1620 revising Copernicus's book, 72, 162, 180, 181; of 1634 banning Galileo's *Dialogue*, 159
Index of Prohibited Books, 152, 167, 179, 183, 185, 191, 193, 225, 253, 279
Index, Congregation of, xvii, xxix, xxxvii, 72, 89, 142, 171, 183, 184, 224, 269, 302
inductive argument, 59, 60, 239, 240, 264, 309
inertia, xv, xxiii, 32, 97, 107, 124, 128, 197
infallibility of pope or Church, 188, 206, 221, 302
inferior planets, definition of, 23
infinity, 3, 11, 34, 96, 100, 107, 295
Ingoli, Francesco, 45, 72–76, 93, 94, 145, 230
injunction to Galileo, xxv, xxvi, xxxvii–xxxviii, 146, 148, 149, 150, 156, 157, 158, 160, 163, 187, 188, 190, 199, 201, 202, 203, 208, 209, 222, 231, 232, 234, 265, 268, 282, 304, 309
innovation versus conservation, dialectic of, xxx, xxxix, 227, 233, 294, 305–306, 314
instantaneous motion, 11
instrumentalism, epistemological, xiv, xxvii, xxviii, xxxix, 34–35, 36, 66, 69, 95–96, 146–47, 180, 215, 222, 235, 273, 277–90, 299
intelligent design, 243
internal force, definition of, 10–11
Israel, 32, 33, 78, 81, 131

J

Jansenism, 260
Jardine, Nicholas, xii, 35, 278, 283, 328
Jerome, Saint, 86
Jesuits, xxii, xxxv; conspiracy theory, 169, 172; and Dominicans, xxx, 294, 313; and Galileo, 166, 215, 219, 253, 293; internal split, 295; and regular clergy, 217
Job, 70, 168, 181, 257, 307

John Paul II, xix, xxiv, xxix, xxx, xxxi, xxxvi, 76, 205, 207, 220–27, 259, 274–75, 296–99, 300, 301, 312
John XXIII, 217
Joshua miracle, xxiv, 32, 69, 71, 73, 74, 75, 78, 81, 82, 93, 94, 131, 139, 204, 247, 257, 285, 298
judgment. *See* judiciousness
judiciousness, 37, 43, 64, 116, 123, 129–31, 203, 227, 234, 2243, 277, 289; defined, 134
Jupiter, 5, 14, 17, 22, 23, 173, 238; moons of, xv, 52, 56, 66, 70, 130

K

Kant, Immanuel, x, 97, 277
Kepler, Johannes, xv, xxviii, 35, 36, 44, 48, 49, 51, 53, 54, 72, 168, 183, 184, 193, 197, 214, 215, 223, 237, 238, 266, 280, 282, 283, 284, 290, 307, 308
kinesthetic sense, 26
kinetic energy, 107
Koestler, Arthur, xix, xxiv, xxxi, 214–15, 240, 246, 273, 274
Koyré, Alexandre, xxi, 96, 115, 240, 277
Kuhn, Thomas, xxx, 4, 12, 22, 37, 38, 121, 123, 293

L

Ladislaus IV (king of Poland), 167
Lagalla, Giulio, 70
Lamalle, Edmond, 217, 219, 220
Lambertini, Prospero, 117
Langford, Jerome, 245–46
last will and testament of Galileo, 149, 169
Lateran Council, Fifth, 68, 92, 141, 296
latitude, xxi, 32, 56, 99; defined, 13
Laudan, Larry, 37, 38, 39, 40, 41, 321, 329
Laughton, Charles, 211
Lazzari, Pietro, 183–85, 194, 195
Le Cazre, Pierre, xx
L'Epinois, Henri de, 198, 199, 203
legal issues, xxv, xxxviii, 151–52, 167, 169, 230–31, 249; legal-impropriety thesis, xxvi, xxvii, 146, 199–200, 201, 202, 203, 222
Leibniz, Gottfried, xix, 171, 173–74, 184, 253, 306
Leiden, xv, 166
Lenoble, Robert, 219
Leo X, 92

Index

Leo XIII, xix, xxiv, 67, 86, 182, 204, 205, 234
Leo, constellation of, 17
Lerner, Michel, xxiv, 69, 72, 81, 89, 90, 91, 92, 252, 329–30
Le Tenneur, L.A., xx
levity, definition of, 9
liberalism, 166
Libra, 17
Libri, Guglielmo, 195–96, 268, 276
Liceti, Fortunio, 45, 294
Liège, 162
light bodies, definition of, 9
limited scriptural authority, principle of, 77, 87, 91, 92
Lincean Academy, xxxv, 44, 144, 264
Lindberg, David, 12, 291, 292, 295, 297
literalism, biblical, 82, 96, 170, 172, 183, 204, 222
Locher, Johannes, 71
longitude, 167; defined, 13
Lorini, Niccolò, 70, 71, 79, 139, 140, 215, 273
Louis XVI, 255
Louvain, 222, 224
luminaries, 14, 77
lunar orbit argument, 52, 54
Luther, Martin, 69

M

Maccarrone, Michele, 217, 218
Machamer, Peter, xxxiii, 330
MacLachlan, James, 100, 101, 103, 113, 145
Maculano da Firenzuola, Vincenzo, 201
Madden, Richard, 166, 270–71, 275
magnitude, 54; defined, 14
Mallet du Pan, Jacques, xxv, 185, 186, 188, 204, 254–76
manner of reasoning, Galileo's, 90–91
Margolis, Howard, 4, 331
Marini, Marino, 189, 196, 197, 198, 267–70, 271, 272, 275, 276
Mars, 5, 15, 17, 22, 238, 240; anti-Copernican argument, 26–27, 54, 130, 132
Master of the Sacred Palace, 190, 191, 192, 193
mathematics, xv, xxi, 3, 97–98, 106, 120, 138, 161, 164, 171, 225, 293; definitions of, 98, 114–15; Galileo's philosophy of, 115–20, 287–89, 294; mathematical hypothesis, 181; mathematical impossibility of extrusion, 1000, 106, 112, 113, 117; physical-mathematical reasoning, 97–120; vs. physics, 117–18
mausoleum for Galileo, xix, 95, 169, 175, 252

Mayaud, Pierre-Noël, xlii, 33, 70, 72, 75, 81, 142, 173, 179–84, 193, 252, 253, 258, 265, 301, 303, 367, 331
Mazzetti, Giuseppe, 193
McMullin, Ernan, xxiv, xxviii, xlii, 67, 70, 77, 79, 81, 82, 86, 87, 88, 95, 141, 210, 234, 244, 245, 278, 279, 297, 298, 331
mechanics, xiv, xv, xvi, xvii, xxi, xxiii, xxxvii, 24, 34, 49, 130, 141, 194, 195, 211, 237, 239, 241, 242, 257, 280, 307; anti-Copernican arguments, 28–32; extruding-power argument, 97–120; vertical-fall argument, 124–29
Medeiros, Joao, xi
Medicean planets or stars, 52, 57
Medici, 138; Giuliano de', 44, 53; Leopold de', 171, 172, 176; Villa, 149, 152, 233, 271, 272, 310, 311
medium, motion through, 11
Melanchthon, Philip, 282
Mercure de France, xxv, 186, 254, 255, 271
Mercure François, 162
Mercury, 5, 14, 17, 22, 53, 54, 238
meridian, 35; defined 13
Mersenne, Marin, xx, 103, 162, 164, 166
meta-hermeneutics, xxiv, 67–68, 137
metaphysics, x, 68, 144, 164, 173, 232, 282–84, 289, 292, 314
Micanzio, Fulgenzio, 167, 293
Michelangelo, 175
Milky Way, 15, 66
Mill, John Stuart, xxxvii
Miller, David, 148, 331
Milton, John, xix, 168, 253, 293
miracles, 34, 71, 78, 81, 82m 93, 131, 139, 285
mixed motion, 7
momentum, 29, 32, 128
moon, 22, 32, 54, 70, 77, 170; in geostatic world view, 5, 6, 8, 12, 14, 15, 17, 18, 22; lunar orbit argument, 52, 54; phases, xvi, 26, 53, 240; similarity to earth, 52; surface of, xv, 66, 130; tides, xxii, 129
Morin, Jean Baptiste, xx, 34, 162
Morpurgo-Tagliabue, Guido, xxviii, xlii, 34, 35, 144, 147, 164, 278, 287, 298–99, 313
Motta, Franco, xli, 85, 169, 182, 226, 252, 260, 332
Mousnier, Pierre, xx
Müller, Adolf, 273–74
Murillo, B.E., 176
mysticism, 214

N

Napoleon Bonaparte, xix, 188–90, 196, 197, 200, 268
natural circular motion, 9, 31
naturally accelerated motion, xxi, 32, 103, 116, 117, 118, 119, 128, 239
natural motion, anti-Copernican arguments, 31–32, 127–28, 236, 284. *See also* forced motion; violent motion
natural philosophy, xxi, 65–66, 68, 70, 87, 89, 91, 93, 115, 164, 169, 176, 178, 184, 235, 238, 248, 249, 252, 260, 284, 290, 303
natural place, 5, 6–7, 8, 9, 10, 50, 72
natural state, 8, 10, 97, 127
Naylor, Ron, xi, 129, 240, 332
nebulas, xv, 66
Nelli, G.B.C. de', 258, 264, 276
Neptune, 14
nested spheres, 6, 10, 17, 18, 19, 35
neutral motion, 39, 50, 58, 128
Newman, John Henry, xix
Newton, Isaac, xxi, xxii, 103, 121, 155, 176, 177, 223, 238, 253
Niccolini, Francesco, xxvi, 95, 254, 255
nonsequitur, 82, 88, 94, 139, 247, 285
North Star, 14
Nov-antiqua (Galileo), 85, 86, 325. *See also* Christina of Lorraine, Galileo's letter to
Numbers, Ronald, 291, 295, 297, 330, 333

O

odd numbers, law of, 116
Olivieri, Maurizio, xxiii, 185, 191–95, 230, 263
On the Revolutions of the Heavenly Spheres (Copernicus), xii, xvii, 3, 21, 35, 46, 72, 94, 121, 155, 162, 171, 181, 231
open-mindedness, Galileo's, x, xxxvii, xxxix, xl, xli, 132, 133, 134, 137
Oppenheimer, Robert, 212
opposites, in geostatic world view, 7, 8
orb, 33, 60. *See also* orbit
orbit, xiii, xv, xvi, xxii, 4, 14, 15, 16, 17, 18, 19, 21, 22, 23, 24, 26, 27, 28, 31, 32, 35, 36, 52, 54, 55, 56, 57, 60. *See also* orb
orbital velocity, 107
Orsini, Alessandro, 45, 257
Osiander, Andreas, xxviii, 35, 280, 282, 286
Osler, Margaret, 65, 291, 292, 333
Ostini, Pietro, 193
out-of-court settlement, 150, 151, 201
overarching thesis, x, xii, xxxvii–xliii, 64, 134
oversimplification, ix, 138, 177, 222, 277, 278, 292, 313, 314

P

Padua, xii, xxi, 165, 166, 179, 180, 181, 256, 294
Palazzo Firenze, 149, 152, 310
Palmieri, Paolo, xii, xvii, 101, 129, 240, 333
parabola, xv, 125
parallax of fixed stars, xvi, xxii, 51, 72, 239; anti-Copernican argument, 27–28, 57, 130, 237
parallel (on a spherical surface), 16; defined 13
pardon for Galileo, 165, 177, 266
Parma, 70
Pascal, Blaise, xxiii, 253
Paschini, Pio, 207, 222, 224; affair, 216–20; defense of Galileo, xxvii, 208–209
pastoral aspects of Galileo affair, xxv, 213, 226–27
Paul V, 142, 143
Paul VI, 217
Peiresc, Nicholas Claude Fabri de, xxvi, 164, 165, 166, 177, 180, 252
pendulum, Foucault's, xxii, xxvii, 238
penitential psalms, xviii, 159, 164
Peripatetic, 4
Pesce, Mauro, xxiii, xxiv, 70, 79, 81, 169, 170, 252, 259, 297, 333–34
Philosopher and Chief Mathematician, 138, 148, 310
philosophical play, xxvii, 165, 180
physics: Aristotelian, xiv, xv, xxxvii, 10–12, 32, 97, 124, 127, 128, 129, 195, 284; mathematical, 102–120; vs. mathematics, 117–18; modern classical, 32, 103, 104, 107, 128, 274; new Galilean, xv, xx, 32, 63, 97, 124–29
Piccolomini, Ascanio, 257, 293, 307
Pieralisi, Sante, 201, 203
Pieroni, Giovanni, 167
Pisa, 46, 47, 70, 138, 142
Pisces, 17
Pitt, Joseph, 145, 299, 319, 334
plague, 148, 149, 257, 307
Plato, 3, 47
plea bargaining, 150, 151, 303–304
pluralism, methodological, 53, 225–26, 227
Polaris, 14
Pontifical Academy of Sciences, xxxi, xxxvi, 207, 208, 209, 216, 217, 218, 220, 222, 224, 234, 296

Pontifical Lateran University, xxxi,
    xxxv, 207
Popper, Karl, 278, 312
Posidonius, 282
potential energy, 107
Poupard, Paul, 224, 225, 226, 227
practical application, methodological criterion
    of, 38–64
predictive novelty, methodological criterion of,
    40–64
primum mobile, 81, 82, 241
priority of demonstration, principle of, 67, 87,
    88, 94, 298
priority of Scripture, 87, 88
probabilism, xxviii, 87, 88, 152, 180, 215,
    237, 299
problem-solving success, methodological
    criterion of, 40–64
Proclus, 278, 282
progressiveness, methodological criterion of,
    37–64
projectile motion, xv, 12, 29, 97
Protestantism, xxv, 32, 147, 166, 167, 172,
    173, 186, 187, 255, 261, 267, 268, 269,
    271, 285
provable, vs. proved, xxiii, xxxvii, 89.
    See also capable of being proved;
    demonstrable
Providentissimus Deus (Leo XIII),
    xix, xxiv, 67, 76, 86, 182, 204,
    205, 234, 259
Psalm, xxxv, 32, 75, 245
Ptolemy, Claudius, xii, xli, 3, 4, 48, 50, 57, 60,
    82, 102, 182, 257, 278, 282, 288;
    Ptolemaics, 10, 35, 273, 381, 285;
    Ptolemaic system, xvi, xxi, 3, 18, 19,
    21, 22, 23, 27, 44, 49, 50, 51, 58, 60,
    66, 76, 94, 145, 182, 208, 209, 241,
    242, 247, 256, 285, 312
Purcell, John, 266, 267, 275
Pythagoreanism, xiii, 22, 48, 53, 70, 76,
    77, 78, 109

Q

quantitative invalidity, 97
quantitative precision, 103
quintessence, 7

R

rationality, xxxiii, 76, 121–34, 233, 234
rational-mindedness, Galileo's, xli, 121,
    132, 133, 134, 137

Ratzinger, Joseph, xxxiii, xxxiv, 300, 334.
    See also Benedict XVI
realism: Copernican, 35, 66, 142, 299; in
    Dialogue, 146, 235–37; Duhem's
    critique, xxvii–xxviii, xxxix, 272–73,
    277–90; mathematical, 115–20
reburial of Galileo, xix, 175, 252
rectilinear motion, 9, 32, 50, 107. *See also*
    straight motion
Redondi, Pietro, 148, 160, 175, 176,
    252, 334
reductio ad absurdum, 188
Reeves, Eileen, xvi, 54, 55, 56, 334–35
rehabilitations of Galileo, xix, xxiv, xxvii,
    xxix, xxxi, xxxii, xxxv, xxxvi, xxxix,
    76, 207–11, 217–18, 220–27, 259, 296,
    300, 312, 313
Reinhold, Erasmus, 282
relativity of motion, 164
Renaudot, Théophraste, 162, 163
Renieri, Vincenzo, 185, 186, 255–56,
    256–57, 259, 261, 262, 263, 265,
    269, 270, 272, 275
Renn, Jürgen, xi, xlii, 335
Reply to Ingoli: Galileo's, 45, 145;
    Kepler's 72
retrograde motion, 18–19, 23–24
Reumont, A. von, 270, 271, 275
Reusch, Franz, 273
Rheticus, Georg, 69
Riccioli, Giovanni Battista, xxi, xxiii, 162,
    170–71, 181, 182, 204, 219, 310
Richelieu, Armand Jean de, 167
rigorous examination undergone by Galileo,
    158–59, 161, 196, 202. *See also* torture
    of Galileo
Rinuccini, Francesco, 45
Roman College, xxii, 171, 183, 212
Rosen, Edward, 6, 69, 131
Rossi, Arcangelo, xii
Rossi, Giovanni Vittorio de', 169
Rowland, Wade, xxviii, 279

S

Sagittarius, 17
Sagredo, 116, 146, 307
Saint-Amant, Gerard, 166
Salusbury, Thomas, 168, 276, 307–309, 313
Salviati, 116, 146, 307
Santillana, Giorgio de, 49, 70, 160, 185, 232,
    233, 252
Sarpi, Paolo, 167, 272
Sarsi, Lotario, 256

Saturn, 5, 14, 17, 22, 54–55, 238
saving the appearances (or phenomena), xlii, 277–79, 281, 288, 289, 290
Scartazzini, J.A., 202, 203
Scheiner, Christopher, 56, 71, 215, 293, 294, 295
Scholasticism, 68, 73, 164
scientific revolution, 63–64, 214
scientist (the word), 65–66, 117
Scorpio, 17
Scriven, Michael, xxxvii, 324, 335
secant, 99–112
secant-tangent theorem, 108–112
Segre, Michael, xi, xii, xxx, 252, 294, 298, 306, 313, 335–36
Seneca, Lucius, 168
sentence against Galileo, xviii, xxxviii, 164, 172–73, 175, 180, 181, 187, 199, 243, 247, 266, 275, 310; and disobedience, 301–302, 304; dissemination of, 161–63, 252; and heresy, 205, 229–31, 303; and methodology, 232–35, 285–86; and proceedings of 1616, 282; revision of, 227; rigorous examination, 196; signatures of judges, 267; summary, 151–52, 156–60; Tiraboschi on, 188
Serarius, Nicolaus, 71
Settele affair, 185, 190–95, 204, 222, 225, 230, 263
Shea, William, xii, xxii, xlii, 48, 115, 145, 149, 152, 235, 239, 292, 294, 310, 310
ship analogy argument, 30, 50
Shoemaker, Philip, 266
Sidereal Messenger (Galileo), xv, xvii, 44, 51, 53, 57, 66, 69, 70, 91, 138
Siena, 152, 164, 165, 166, 257, 293, 307, 310
signatures on sentence, 160
signatures on special injunction, 146, 199, 268, 304
Simoncelli, Paolo, 216, 218, 252
Simplicio, 116, 117, 119, 145, 187, 215, 268, 276, 309
simplicity, 22–23, 38, 41, 42, 51, 54, 62, 63, 115, 123, 129, 174, 239, 282, 289
Simplicius, 278, 282, 307
Sirius, 14
Sizzi, Francesco, 70, 273
Sleepwalkers (Koestler), 214–215
Slezak, Peter, xi, xii
Snow, C.P., 214
Sobel, Dava, 164

Soccorsi, Filippo, 95–96, 207, 209–211, 224, 226
Socrates, 165, 177
solstice, xxxiv; defined, 15, 16
sophistry, 215, 234, 237, 240, 274, 313
Sorbonne, 167
special commission on Galileo's *Dialogue*, 148, 254
special injunction, xxv, xxvi, xxxviii, 146, 148, 149, 150, 156, 157, 158, 160, 163, 187, 188, 190, 199, 201, 202, 203, 207, 208–209, 222, 230, 231, 232, 234, 265, 268, 282, 304, 309
Speller, Jules, xvii, xxx, xlii, xliii, 33, 34, 138, 142, 143, 144, 147, 148, 151, 152, 159, 324, 336
Spruit, Leen, 150, 316
squares, law of, 116, 119
states' reactions to Galileo's condemnation, 167
stellar sphere, 5, 7, 8, 10, 16, 17, 21, 22, 35, 48. *See also* celestial sphere
straight motion, 7, 8, 9, 10, 31, 46, 70, 125, 127, 128. *See also* rectilinear motion
Strasbourg, 76, 166, 168
straw-man fallacy, 215, 240
Strehler, Giorgio, 297
strict demonstration, xviii, 61, 236, 237
sublunary, 8
summons issued to Galileo, 148, 158, 202
Sunspots Letters (Galileo). See *History and Demonstration Concerning Sunspots*
sunspots, 38, 54, 55, 57, 58, 60, 66, 70, 81, 129, 215, 241, 242, 293, 294
superlunary, 8
superposition of motions, xv, 30, 128. *See also* composition of motion
Systema cosmicum (Galileo), 166, 168

T
tangent, 99–114
tangent-secant theorem, 108–12
Targioni Tozzetti, Giovanni, 264
Taurus, 17
telescope, xvii, xxxv, xxxvii, 26, 44, 45, 103, 132, 133, 145, 167, 176, 240, 244; anti-Copernican arguments, 26–28; discoveries and implications, xv-xvi, 51–64, 66, 68, 69, 76, 97, 208; and judgment, 129–31

theological apologias of Galileo's condemnation, xxi, xxiii, xxv, 170–71, 205–207, 301–305
theological arguments against Copernicanism, xvi, xliii, 24, 32–34, 44, 65–96, 97, 131, 138, 141, 157, 236–37, 243–48
theological despotism, 178, 254
theological issues in trial of Galileo, xxiii, xxix, 152, 169, 178–79, 198, 229–35, 291–314
theological vindication of Galileo, xix, xxiv, xxvii, xxxi, xxxviii, 203–205, 207–11, 220–27
theology, x, xx, xxxix, 74, 76, 91, 96, 137, 161, 198, 205, 223, 266, 270, 292, 300, 314; as queen of the sciences, 83–84, 141, 246; vs. natural philosophy, 68, 178, 197
Thirty Years' War, 147, 152, 309
Thomason, Neil, xii, 23, 336
tides, xv, xxii, xxviii, 34, 38, 49, 58, 59–60, 129, 145, 237, 237, 239–40, 304. See also Discourse on the Tides
Tiraboschi, Girolamo, xxv, xxvi, 185, 186–88, 256, 258, 259, 262–63, 264, 275–76
Toaldo, Giuseppe, 181–82
Tolosani, Giovanni, 69
Torricelli, Evangelista, xxi, xxiii
torture question, 45, 151, 158–59, 195–97, 203, 212, 232, 255, 267, 268–69, 270, 276, 304
trade winds, 59, 140
Treatise on the Sphere, or Cosmography (Galileo), 4, 44, 50, 284
Trinity College, 270
Turiozzi, Fabrizio, 191–92
Two New Sciences, xv, xvii, xxiii, 62, 66, 116, 118, 119, 165, 166, 167, 211, 212, 213, 299, 308
two-cultures thesis, xix, xxxi, 214
Tycho Brahe, xiv, 35, 44, 49, 57, 65, 72, 116, 162, 174, 182, 197, 240, 266, 286
Tychonic system, xxi, 35, 58, 66, 96, 170, 215, 240–42

U
uniformly accelerated motion, 116, 119
upward, 7, 8, 9, 29, 128
Uranus, 14

Urban VIII, 61–62, 66, 95, 143–44, 147–48, 151, 161, 180, 213, 256, 257, 260, 293, 295, 307; divine-omnipotence argument, xlii, 33, 144, 236–37; Duhem's instrumentalist interpretation, xxxviii, 224, 273, 280–82, 286; and hypotheses, 144–45, 146–47, 149, 150; Simplicio caricature, 268, 276, 309

V
vacuum, 11
Vale, Giuseppe, 218
Van Helden, Albert, xi, xvi, 52, 54, 55, 56, 131, 335, 337
Vatican Astronomical Observatory, xxii, xxxv, 216, 222, 300
Vatican Commission on Galileo, 221–26
Vatican Council, Second, 209, 217, 218, 220, 274, 296
Vatican Radio, xxxi, 207
Vatican Secret Archives, 196, 200, 268
vehement suspicion of heresy, xviii, 95, 159, 169, 229–30, 302–303
Venice, xv, 70, 142, 162, 293
Venturi, Giambattista, 189–90, 196, 197, 264
Venus, 5, 14, 17, 22, 54, 55, 56, 238; anti-Copernican argument, 26, 132–33; phases, xvi, 39, 44, 53, 57, 58, 60, 66, 87, 130, 239; Scheiner's view, 56
vertical fall, anti-Copernican argument, xvi, 29, 124–29, 333
Villa Medici, 149, 152, 233, 271–72, 310–11
violent motion, 7–8, 10, 11, 31, 50. See also forced motion; natural motion;
Virgil, Bishop (8th century A.D.), 178
Virgo, 17
Viviani, Vincenzio, 181, 258, 313; account of trial, 180, 301, 306–307
Voltaire, François, xix, 175, 176–78, 179, 181, 185, 233, 253–54, 255, 269, 293

W
Wallace, William, xxii, xxviii, 50, 69, 225, 235, 252, 279, 294, 299
wandering star, 15, 52; defined, 13–14
water: as element in geostatic world view, 4–12, 31; and tides, 59, 239–40
Westfall, Richard, xlii, 222, 278

Westman, Robert, xiv, xxvi, 35, 121, 165
Whewell, William, 66, 264, 309, 310, 313
Wilding, Nick, 168, 276, 308, 338
wind: anti-Copernican argument, 59, 194; pro-Copernican argument, 59, 140
wisdom after the event, 215
Wohlwill, Emil, 199–200, 202

X
Ximenes, Ferdinando, 39, 140
Ximenes, Sebastiano, 71

Z
zodiac, 17
Zúñiga, Diego de, 70, 183, 184, 193